Teacher Editi

MATH 1

Fifth Edition

Part
2

MATH 1 Teacher Edition
Part 2
Fifth Edition

Writers
Charlene McCall
Lindsey Dickinson, MEd
Rita Lovely

Writer Consultant
L. Michelle Rosier

Biblical Worldview
Ben Adams, MEd
Tyler Trometer, MDiv

Academic Integrity
Jeff Heath, EdD

Instructional Design
Rachel Santopietro, MEd
Danny Wright, DMin

Educational Technology
Ken Matesevac, MS

Editors
Krystal Allweil
Elizabeth Kinney
Carol Myers

Design Coordinator
Michael Asire

Concept Designer
Kathryn Ratje

Production Designers
Sarah Centers
Lydia Thompson
Clarissa Wisener

Illustration, Design & Production
QBS Learning

Permissions
Ruth Bartholomew
Stacy Stone

Project Coordinators
Stephanie Evans
Kyla J. Smith

Postproduction Liaison
Peggy Hargis

Photo Credits
Cover FatCamera/E+ via Getty Images
Worktext photo credits appear at the end of this book.

The text for this book is set in Acumin Variable Concept by Robert Slimbach, Adobe Minion Pro, Adobe Myriad Pro, Calibri by Monotype, Chalkboard, Circe by Paratype, Free 3 of 9 by Matthew Welch, Gill Sans, Grenadine MVB by Akemi Aoki, Helvetica, Helvetica Neue, Minion Math by Typoma GmbH, PreCursive, Report by Ray Larabie, Symbol, and Times New Roman PSMT.

ISBN 978-1-64626-398-1 (two parts)

15 14 13 12 11 10 9 8 7 6 5 4

21 Addition with Three-Digit Numbers

CHAPTER 12: ADDITION WITH TWO-DIGIT NUMBERS

PAGES	OBJECTIVES	RESOURCES & MATERIALS	ASSESSMENTS
Lesson 92	**Add the Ones First**		
Teacher Edition 395–99 **Worktext** 178–80	**92.1** Join sets for 2-digit addends by using representations. **92.2** Solve a word problem by writing an addition equation. **92.3** Complete a 2-digit addition problem in vertical form by using representations. **92.4** Explain how 2-digit addition should be used to care for others. BWS Caring (explain)	**Visuals** • Place Value Kit: tens, ones **Student Manipulatives Packet** • Number Cards: 0–9 • Tens/Ones Mat **BJU Press Trove*** • Video: Ch 12 Intro • Web Link: Virtual Manipulatives: Base Ten Number Pieces • Games/Enrichment: Subtraction Flashcards • PowerPoint® presentation	**Reviews** • Pages 173–74
Lesson 93	**Adding Two-Digit Numbers**		
Teacher Edition 400–403 **Worktext** 181–82	**93.1** Join sets for 2-digit addends. **93.2** Solve a word problem by writing an addition equation. **93.3** Complete a 2-digit addition problem in vertical form.	**Visuals** • Place Value Kit: tens, ones **Student Manipulatives Packet** • Number Cards: 0–9 • Tens/Ones Mat **BJU Press Trove** • Video: Add Ones • Games/Enrichment: Subtraction Flashcards • PowerPoint® presentation	**Reviews** • Pages 175–76
Lesson 94	**Adding Dimes & Pennies**		
Teacher Edition 404–7 **Worktext** 183–84	**94.1** Join sets of dimes and pennies for 2-digit addends. **94.2** Solve a money word problem by writing an addition or subtraction equation. **94.3** Solve a money word problem by using the Problem-Solving Plan.	**Visuals** • Visual 21: *Problem-Solving Plan* • Money Kit: 9 pennies, 9 dimes **Student Manipulatives Packet** • Number Cards: 0–9 • Tens/Ones Mat • Money Kit: 9 pennies, 9 dimes **BJU Press Trove** • Games/Enrichment: Subtraction Flashcards • PowerPoint® presentation	**Reviews** • Pages 177–78

*Digital resources for homeschool users are available on Homeschool Hub.

<table>
<tr><th>PAGES</th><th>OBJECTIVES</th><th>RESOURCES & MATERIALS</th><th>ASSESSMENTS</th></tr>
<tr><th colspan="4">Lesson 95 Adding Money</th></tr>
<tr><td>Teacher Edition
408–11
Worktext
185–86</td><td>95.1 Join sets of dimes and pennies for 2-digit addends.
95.2 Solve a money word problem by writing an addition or subtraction equation.
95.3 Solve a money word problem by using the Problem-Solving Plan
95.4 Compose an addition word problem.
95.5 Evaluate adding money for a selfish purpose.
BWS Caring (evaluate)</td><td>Teacher Edition
• Instructional Aid 27: Pictograph
Visuals
• Visual 21: Problem-Solving Plan
Student Manipulatives Packet
• Number Cards: 0–9
• Tens/Ones Mat
• Money Kit: 9 pennies, 9 dimes
BJU Press Trove
• Games/Enrichment: Subtraction Flashcards
• PowerPoint® presentation</td><td>Reviews
• Pages 179–80</td></tr>
<tr><th colspan="4">Lessons 96–97 Rename 10 Ones</th></tr>
<tr><td>Teacher Edition
412–15
Worktext
187–88</td><td>96–97.1 Rename 10 ones as 1 ten.
96–97.2 Join sets for 2-digit addends with renaming.
96–97.3 Complete a 2-digit addition problem with renaming.</td><td>Visuals
• DQ Puppet
• Place Value Kit: tens, ones
Student Manipulatives Packet
• Tens/Ones Mat
BJU Press Trove
• Video: Rename 10 Ones
• Web Link: Virtual Manipulatives: Base Ten Number Pieces
• PowerPoint® presentation</td><td>Reviews
• Pages 181–82</td></tr>
<tr><th colspan="4">Lesson 98 Rename 10 Pennies</th></tr>
<tr><td>Teacher Edition
416–19
Worktext
189–90</td><td>98.1 Rename 10 pennies as 1 dime.
98.2 Join sets of dimes and pennies for 2-digit addends with renaming.
98.3 Solve money problems with sums to 99¢.
98.4 Solve a word problem by writing an addition equation.</td><td>Visuals
• Visual 21: Problem-Solving Plan
• Money Kit: pennies, dimes
Student Manipulatives Packet
• Tens/Ones Mat
• Money Kit: pennies, dimes
BJU Press Trove
• Video: Rename 10 Pennies
• Web Link: Virtual Manipulatives: Money Pieces
• PowerPoint® presentation</td><td>Reviews
• Pages 183–84</td></tr>
</table>

PAGES	OBJECTIVES	RESOURCES & MATERIALS	ASSESSMENTS
Lesson 99 **Chapter 12 Review**			
Teacher Edition 420–23 **Worktext** 191–92	**99.1** Recall concepts and terms from Chapter 12.	**Visuals** • Visual 21: *Problem-Solving Plan* • Place Value Kit: tens, ones **Student Manipulatives Packet** • Number Cards: 0–9 • Tens/Ones Mat • Sign Cards: Greater than, Less than **BJU Press Trove** • Games/Enrichment: Subtraction Flashcards • PowerPoint® presentation	**Worktext** • Chapter 12 Review **Reviews** • Pages 185–86
Lesson 100 **Coding: Guiding a Friend**			
Teacher Edition 424–27 **Worktext** 193–94	**100.1** Follow cardinal directions on a grid. **100.2** Debug an algorithm. **100.3** Follow the Engineering Design Process to solve a problem. **100.4** Compose an algorithm that uses cardinal directions. **100.5** Program a friend to help him or her find a destination. BWS Caring (apply)	**Teacher Edition** • Instructional Aid 5: *STEM Engineering Design Process* • Instructional Aid 28: *Compass Rose* • Instructional Aid 29: *Trampoline Park Destination* **BJU Press Trove** • PowerPoint® presentation	**Teacher Edition** • Instructional Aid 30: *Coding Rubric: Guiding a Friend*
Lesson 101 **Test, Cumulative Review**			
Teacher Edition 428–30 **Worktext** 195	**101.1** Demonstrate knowledge of concepts from Chapter 12 by taking the test.		**Assessments** • Chapter 12 Test **Worktext** • Cumulative Review **Reviews** • Pages 187–88

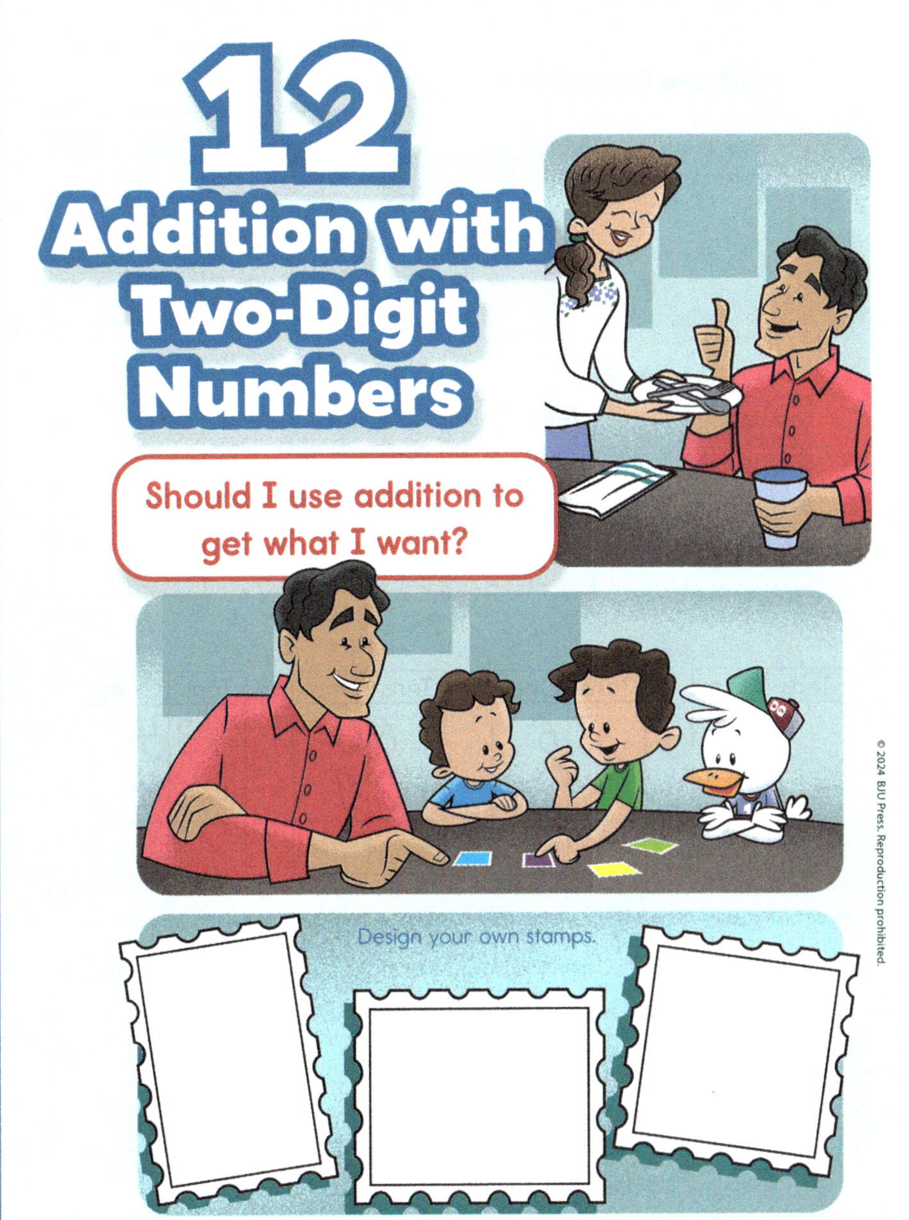

12
Addition with Two-Digit Numbers
Should I use addition to get what I want?
Design your own stamps.

Chapter 12

Should I use addition to get what I want?

Chapter Objectives

- Join sets to model 2-digit addition.
- Solve word problems by using the Problem-Solving Plan.
- Solve 2-digit addition problems in vertical form.
- Evaluate using addition to be selfish.
- Compose an algorithm that uses cardinal directions to help solve a problem.

92

Worktext pages 178–80
Reviews pages 173–74

Objectives

- **92.1** Join sets for 2-digit addends by using representations.
- **92.2** Solve a word problem by writing an addition equation.
- **92.3** Complete a 2-digit addition problem in vertical form by using representations.
- **92.4** Explain how 2-digit addition should be used to care for others. BWS

Biblical Worldview Shaping

- **Caring** (explain): Developing skill in adding 2-digit numbers helps us determine value. This knowledge can be used for selfish purposes, or it can be used to treat others the way we would like to be treated (92.4).

Printed Resources

- Visuals: Place Value Kit (tens, ones)
- Student Manipulatives: Number Cards (0–9)
- Student Manipulatives: Tens/Ones Mat

Digital Resources

- Video: Ch 12 Intro
- Web Link: Virtual Manipulatives: Base Ten Number Pieces
- Games/Enrichment: Subtraction Flashcards

Materials

- Subtraction flashcards
- 50 UNIFIX® Cubes (or Place Value Kit: tens, ones; for each student)

Add the Ones First

Add the ones first.
Add the tens next.

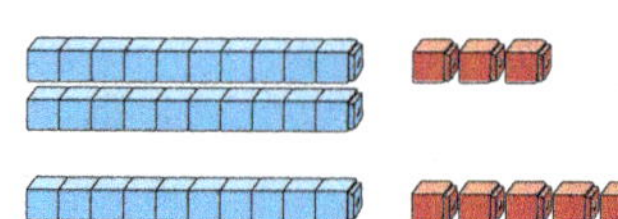

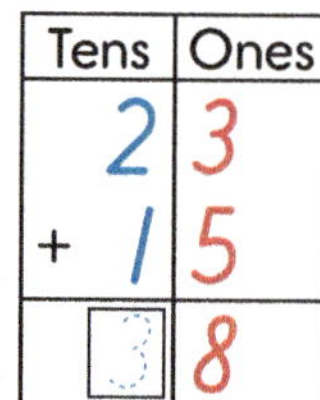

Add.

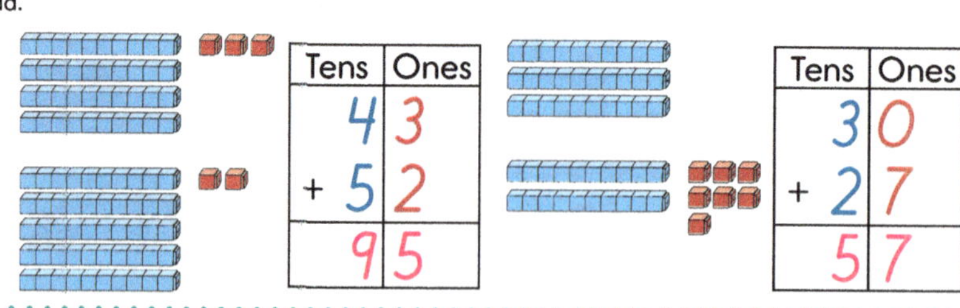

Tens	Ones
4	3
+ 5	2
9	5

Tens	Ones
3	0
+ 2	7
5	7

Add.

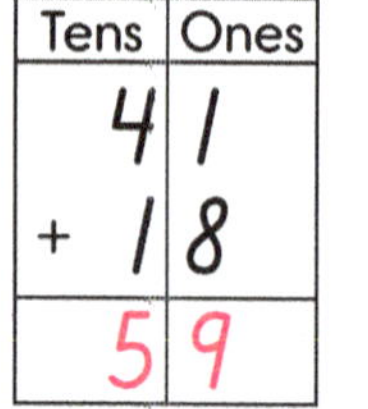

Tens	Ones
7	6
+	2
7	8

Tens	Ones
6	0
+ 2	9
8	9

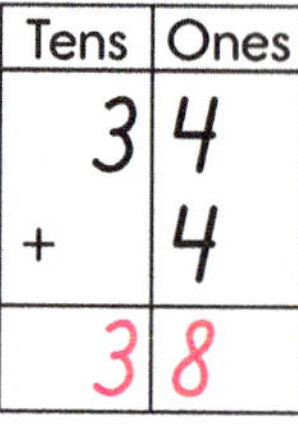

Write an equation for the word problem. Solve using the Tens/Ones frame.

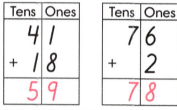

Jen needs 27 stamps.
Jeff needs 31 stamps.
How many stamps do they need in all?

Tens	Ones
2	7
+ 3	1
5	8

27 + 31 = 58 stamps

Chapter 12 • Lesson 92 one hundred seventy-nine 179

To demonstrate 2-digit addends by using manipulatives, display the tens horizontally in the Tens place and the ones individually in the Ones place. Display the first addend near the top of the Tens/Ones frame or Tens/Ones Mat and the second addend near the bottom.

Practice & Review

Memorize Subtraction Facts

Introduce the following facts.

8 – 6 8 – 4

Display each flashcard slowly. Invite students to give the answers.

Distribute Number Cards 0–9. Display each flashcard again. Direct the students to hold up the correct Number Card for each answer.

Practice these and the previously memorized subtraction facts.

Identify Tens & Ones in a Number

Display 5 tens and 4 ones in a Tens/Ones frame.

What number is shown? 54

How many tens are in 54? 5

How many ones are in 54? 4

Invite a student to use tens and ones to display the number 69.

How many tens are in 69? 6

How many ones are in 69? 9

Continue the activity for 82. 8 tens, 2 ones

Make 2-Digit Numbers

Distribute the Tens/Ones Mats and UNIFIX Cubes.

Write the number 43 for display. Direct the students to use their UNIFIX Cubes to show 43.

How many tens did you put in the Tens place? 4

How many ones did you put in the Ones place? 3

Repeat the activity for 26 and 17. 2 tens, 6 ones; 1 ten, 7 ones

Add.

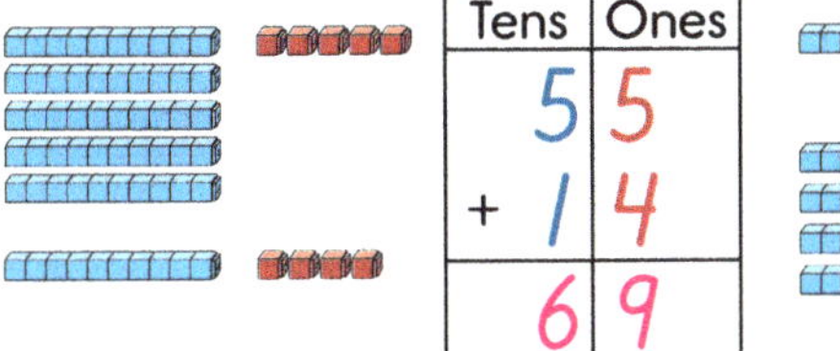

Tens	Ones
5	5
+ 1	4
6	9

Tens	Ones
1	5
+ 4	3
5	8

Add.

Tens	Ones
5	4
+ 2	5
7	9

Tens	Ones
2	2
+ 3	3
5	5

Tens	Ones
9	4
+	5
9	9

Tens	Ones
8	0
+ 1	7
9	7

Write the word to complete the sentence.

Two-digit addition helps me give to others.

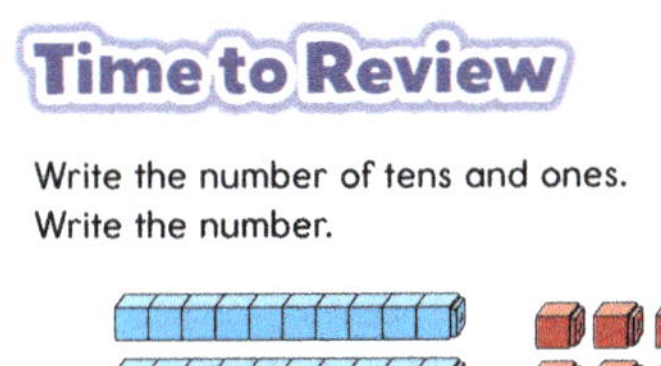

Write the number of tens and ones.
Write the number.

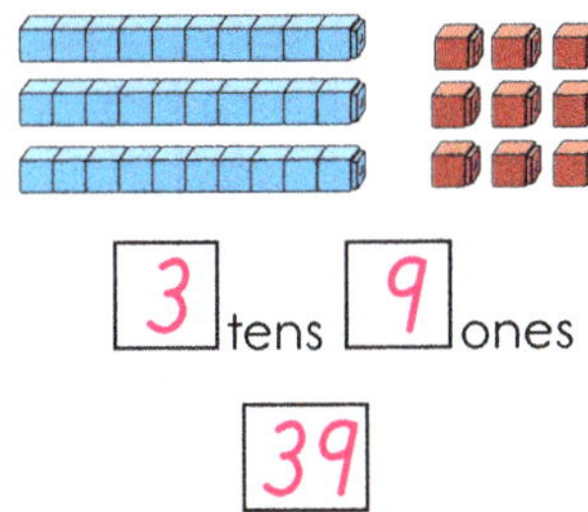

3 tens 9 ones

39

Draw a line to match.

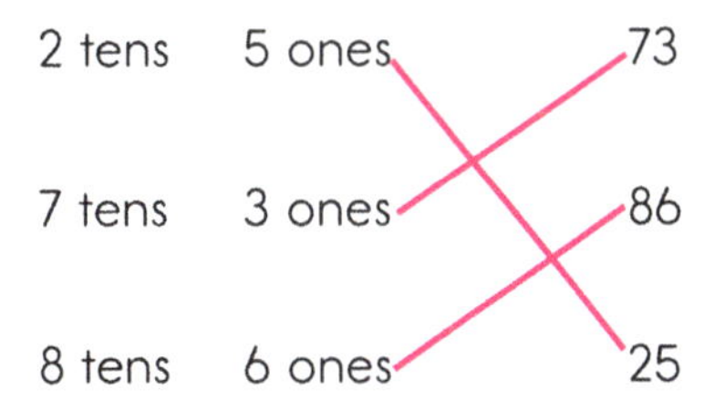

2 tens	5 ones	73
7 tens	3 ones	86
8 tens	6 ones	25

Engage

Essential Question

Direct attention to the chapter opener on Worktext page 178. Draw 4 postage stamp shapes and fill in their values as you **read aloud** the following scenario to introduce the chapter essential question.

Missionary Marc pushed back from the dinner table with a contented sigh. "That was a wonderful meal, Mrs. Tune. Thank you for having me for lunch today."

Smiling at the boys sitting at the table, Mr. Marc reached into his bag. "I brought some postage stamps from Spain for your collection." He arranged 4 brightly colored postage stamps on the table in front of Dan and his younger brother, Joel. "You boys can share them."

DQ, who was visiting his friends, looked over Dan's shoulder at the stamps. Because the stamps were from another country, Dan could not determine their value.

"How much is each one worth?" Dan asked.

Mr. Marc pointed to the bright blue stamp. "This one is worth 35¢ here in our country. The purple one is worth 54¢, the yellow one, 21¢, and the green one is worth 40¢."

Dan added the value of the purple and green stamps in his head. *Wow! They're worth 94¢ together*, he thought. *And I like those colors too.*

Dan quickly calculated the value of the other 2 stamps. *The blue and yellow stamps together are worth only 56¢*, he thought.

"Joel, I think you would like the blue one! It's your favorite color," Dan spoke encouragingly as he moved the blue stamp toward his brother.

Joel, who was only 4 years old, nodded. He agreed with mostly everything his big brother said.

"And you would probably like the yellow one to go with it, wouldn't you?"

Joel thought blue and yellow stamps were a great idea! Dan pulled the purple and green ones toward himself. "Thanks, Mr. Marc. These will be perfect in my collection!"

Dan smiled to himself. Being able to add 2-digit addends had made it easy for him to choose the stamps with the greatest value.

"My stamps are worth more than yours, Joel," Dan said with a laugh. Joel looked at his blue and yellow stamps with a hurt expression.

DQ caught Dan's gaze, but Dan looked away.

Invite a student to read aloud the essential question. Encourage the students to consider this question as you guide them to answer it by the end of the chapter.

Instruct

2-Digit Addend Word Problems

Model 2-digit addition by using **manipulatives**.

Distribute the Tens/Ones Mats and UNIFIX Cubes. Guide the students as they use the cubes to show a word problem on their mats.

Read aloud the following word problem. Model the problem in a Tens/Ones frame by using tens and ones.

Mrs. Hart's class has 14 boys and 12 girls. How many students are there in all?

How many boys are in the class? 14

Guide the students as they place 1 ten bar and 4 ones at the top of their mats.

How many girls are in the class? 12

Guide the students as they place 1 ten bar and 2 ones at the bottom of their mats.

92

How can you find how many students there are in all? join the sets; add

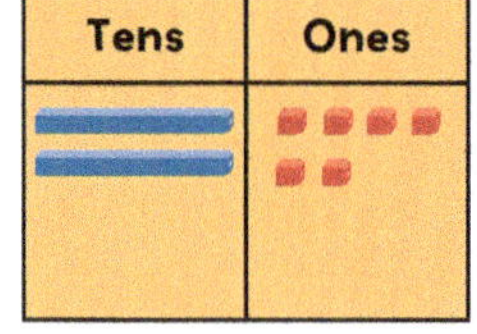

Explain that when they join sets with tens and ones, they should join the cubes in the Ones place first; then join the bars in the Tens place.

Instruct the students to move all the cubes in the Ones place together.

How many ones are there in all? 6

Instruct the students to move all the tens in the Tens place together.

How many tens are there in all? 2

What is the total number of students in the class? 26

What equation shows this word problem? 14 + 12 = 26

Display the equation and guide the students as they read it aloud.

Continue the activity by telling word problems for 23 + 16 = 39 and 11 + 31 = 42.

2-Digit Addition in Vertical Form

Model 2-digit addition in vertical form by using a **Tens/Ones frame**.

Write "24 + 15" vertically in a Tens/Ones frame as shown.

	Tens	Ones
	2	4
+	1	5

Remind the students that when they add 2-digit numbers, they always begin by adding the digits in the Ones place first.

What is 4 ones + 5 ones? 9 ones

Write "9" in the Ones place of the answer.

What is 2 tens + 1 ten? 3 tens

Write "3" in the Tens place of the answer.

Point to each addend and guide the students as they read the equation aloud. 24 plus 15 equals 39

Continue the activity with the following addition problems in a Tens/Ones frame. Emphasize adding the digits in the Ones place before adding the digits in the Tens place.

15 + 11 = 26

32 + 12 = 44

24 + 12 = 36

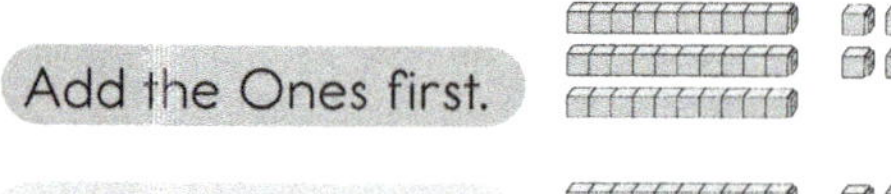

Add the Ones First

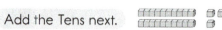

Add the Ones first.

Add the Tens next.

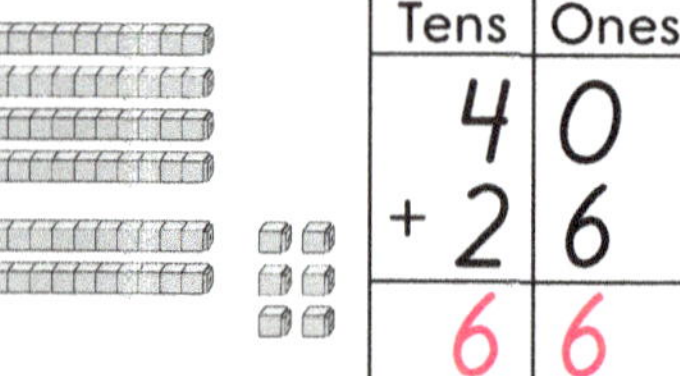

	Tens	Ones
	3	4
+	2	3
	5	7

Add.

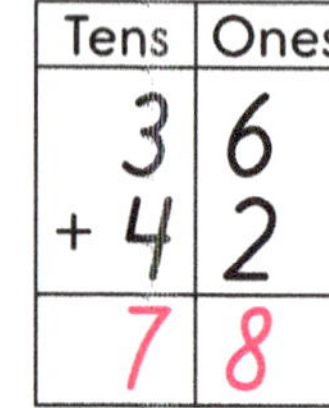

	Tens	Ones
	4	0
+	2	6
	6	6

	Tens	Ones
	6	5
+		2
	6	7

Add.

	Tens	Ones
	3	6
+	4	2
	7	8

	Tens	Ones
	5	3
+	3	5
	8	8

	Tens	Ones
	4	6
+	4	3
	8	9

	Tens	Ones
	4	4
+	2	2
	6	6

Write an equation for the word problem.
Solve using the Tens/Ones frame.

Ava picked up 23 seashells. Soren picked up 16 seashells. How many seashells did they pick up in all?

	Tens	Ones
	2	3
+	1	6
	3	9

23 + 16 = 39 seashells

Care for Others by Using Addition

Guide a **discussion** to help the students explain how 2-digit addition should be used to care for others.

Display the values of the stamps from the opening scenario (blue 35¢, purple 54¢, yellow 21¢, green 40¢). Invite students to add different combinations of the values of 2 stamps to compare their value. Include the 2 equations that have equal sums.

35¢ + 54¢ = 89¢; 35¢ + 21¢ = 56¢; 35¢ + 40¢ = 75¢; 54¢ + 21¢ = 75¢; 54¢ + 40¢ = 94¢; 21¢ + 40¢ = 61¢

Read aloud Matthew 22:39*b*.

What does it mean to love your neighbor as yourself? to treat others the way I would want to be treated

How could you use addition to share the stamps with someone else in a way that shows you care about them as much as you care about yourself? I could pair the stamps so that we both have the same value of stamps or so that the other person has a greater value of stamps than those I keep for myself. I could give the other person first pick of his or her favorite colors of stamps.

Point out to the students that learning to add 2-digit numbers helps them give to others by treating them as they would want to be treated.

Chapter 8 Review

Write the times under each clock.
Circle the time passed.

Activity	Start	Finish	Time Passed
	4:00	5:00	30 minutes or (1 hour)
	7:00	7:30	(30 minutes) or 1 hour
	3:00	4:00	30 minutes or (1 hour)

Addition Fact Review

Add.

3 + 2 = 5	6 + 1 = 7	5 + 2 = 7
1 + 8 = 9	5 + 4 = 9	6 + 6 = 12
9 + 3 = 12	6 + 5 = 11	8 + 2 = 10
4 + 3 = 7	7 + 2 = 9	6 + 3 = 9
8 + 0 = 8	4 + 4 = 8	9 + 2 = 11

174 one hundred seventy-four Math 1 Reviews

Apply

Worktext pages 179–80

Review the steps for adding 2-digit addends at the top of page 179. Read and guide completion of the remainder of the page.

Read and explain the directions for page 180. Assist the students as they complete the page independently.

Direct attention to the biblical worldview shaping statement and guide the students as they complete the sentence.

Assess

Reviews pages 173–74

Review telling time and time passed on page 174.

Extended Activity

Add 2-Digit Numbers by Using a Hundred Chart

Materials

- Instructional Aid 2: *Hundred Chart* for each student
- Visual 1: *Hundred Chart*
- Web Link: Virtual Manipulatives: Number Chart
- 1 UNIFIX® Cube for the teacher and for each student

Procedure

Display Visual 1. Distribute the *Hundred Chart* pages and the UNIFIX Cubes. Write the following addition problems in Tens/Ones frames for display.

13	21	49	61
+ 24	+ 52	+ 20	+ 36
37	73	69	97

Direct attention to the first problem.

What is the first addend? 13

Cover the number 13 on Visual 1 with a UNIFIX Cube and instruct the students to do the same on their pages.

Remind the students that they begin adding with the digits in the Ones place. Instruct the students to add 4 ones by moving their cubes 4 spaces to the right.

What number is your cube on now? 17

Explain that 3 ones plus 4 ones equals 7 ones and that the answer to this problem will be in the column that has a 7 in the Ones place. Write "7" in the Ones place of the answer.

Instruct the students to add 2 tens by moving their cubes 2 spaces down.

What number is your cube on now? 37

Write "3" in the Tens place of the answer.

Continue the activity for each addition problem.

2	3	4	5	6	7	8	9
12		14	15	16	17	18	19
22	23	24	25	26	27	28	29
32	33	34	35	36	37	38	39
42	43	44	45	46	47	48	49

Differentiated Instruction

The following activity provides additional practice for students experiencing difficulty with the concepts taught in this chapter.

Add 2-Digit Numbers

Provide students with UNIFIX Cubes, a Tens/Ones Mat, and a star cutout. Display the following problem:

$$\begin{array}{r} 31 \\ +\ 2 \\ \hline \end{array}$$

Instruct the students to use UNIFIX Cubes to make each addend on the Tens/Ones Mat. Guide them to place the star above the column that they should add first. Instruct them to add the Ones place first and then the Tens place. Emphasize that the star reminds them to start adding with the Ones place.

93

Worktext pages 181–82
Reviews pages 175–76

Objectives

- **93.1** Join sets for 2-digit addends.
- **93.2** Solve a word problem by writing an addition equation.
- **93.3** Complete a 2-digit addition problem in vertical form.

Printed Resources

- Visuals: Place Value Kit (tens, ones)
- Student Manipulatives: Number Cards (0–9)
- Student Manipulatives: Tens/Ones Mat

Digital Resources

- Video: Add Ones
- Games/Enrichment: Subtraction Flashcards

Materials

- Subtraction flashcards
- Different types of mail: letter, bill, birthday card, advertisement, magazine, a copy of an email
- 50 UNIFIX® Cubes (or Place Value Kit: tens, ones; for each student)

All addition problems in this chapter should be written vertically. All word-problem equations should be written horizontally.

Adding Two-Digit Numbers

Add.

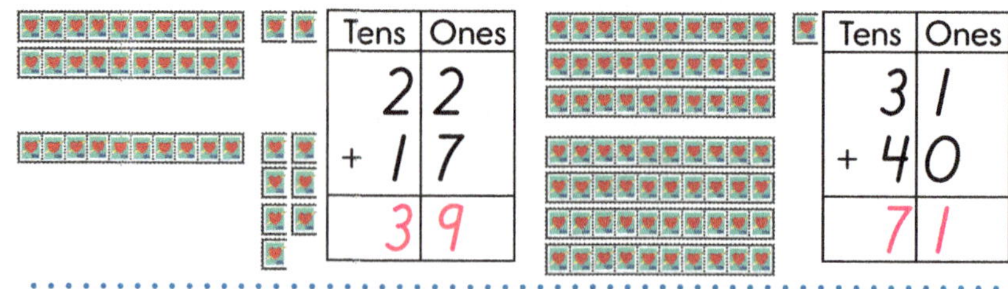

Tens	Ones
2	2
+ 1	7
3	9

Tens	Ones
3	1
+ 4	0
7	1

Write an equation for the word problem. Solve using the Tens/Ones frame.

Postman Carlos delivered 37 postcards in one week. The next week he delivered 32 postcards. How many did he deliver in all?

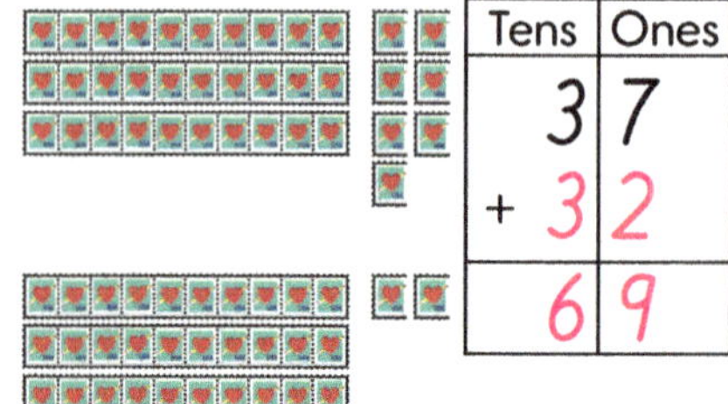

Tens	Ones
3	7
+ 3	2
6	9

37 + 32 = 69 postcards

Add.

Tens	Ones
3	4
+ 2	4
5	8

Tens	Ones
8	3
+ 1	2
9	5

$$53 + 24 = 77$$

$$70 + 17 = 87$$

Practice & Review

Memorize Subtraction facts

Introduce the following facts.

9 − 3 9 − 6

Display each flashcard slowly. Invite students to give the answers.

Distribute Number Cards 0–9. Display each flashcard again. Direct the students to hold up the correct Number Card for each answer.

Practice these and the previously memorized subtraction facts.

Identify Even & Odd Numbers

Display 5 tens and 8 ones. Ask a student to write the number. 58

What is an even number? a number that can be arranged in even pairs with nothing left over

What is an odd number? a number that does not make even pairs; there is an odd one left over

In which place do you look to tell whether the number 58 is even or odd? the Ones place

Which digit is in the Ones place in 58? 8

How can you find out whether 8 is even or odd? arrange the ones in pairs

Invite a student to display 8 ones and arrange them in pairs.

Can 8 be arranged into even pairs? yes

Is 58 an even number or an odd number? even number, because the 8 ones can be arranged in pairs with no ones left over

Continue the activity for 47 odd, 14 even, 36 even, and 21 odd.

Engage

Ministry of Mail

Discuss the ministry of sending mail.

Display different types of mail that are delivered by the postal service.

What are some kinds of mail you or your family receive? letters, bills, birthday cards, magazines, and advertisements

Invite students to tell about kinds of mail they like to receive.

Display an email. Explain that another type of mail that people send and receive is email, which is transmitted electronically on a computer, phone, or other electronic device. Explain that email can be delivered anytime and helps people stay in touch.

Write an equation for the word problem. Solve using the Tens/Ones frame.

Postman Dave has 23 letters to mail.
Postman Dan has 14 letters to mail.
How many letters will they mail in all?

Tens	Ones
2	3
+ 1	4
3	7

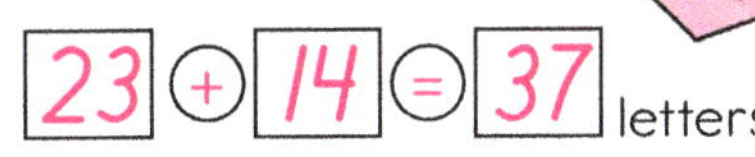

23 + 14 = 37 letters

Add.

Tens	Ones
4	4
+	2
4	6

Tens	Ones
1	4
+ 3	0
4	4

Tens	Ones
2	6
+ 5	1
7	7

Tens	Ones
7	1
+ 1	8
8	9

Tens	Ones
3	3
+ 5	4
8	7

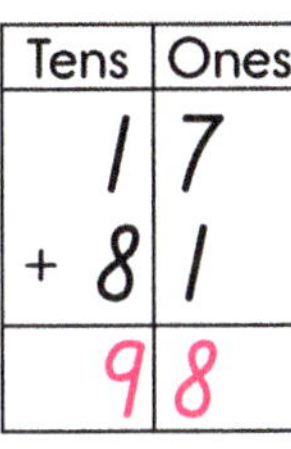

Tens	Ones
1	7
+ 8	1
9	8

Tens	Ones
4	3
+ 5	2
9	5

Tens	Ones
6	2
+ 3	0
9	2

Time to Review

Write the number. Circle the pairs of cubes.
Circle *even* if the number is even.
Circle *odd* if the number is odd.

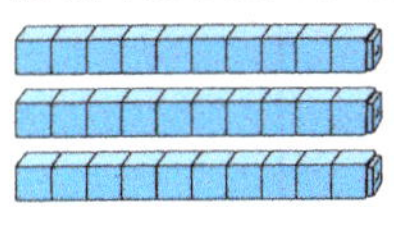

34

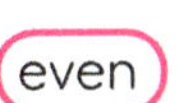

even odd

Subtract. Use the number line if needed.

0 1 2 3 4 5 6 7 8 9 10

8 − 6 = 2   9 − 6 = 3

8 − 4 = 4   9 − 3 = 6

Point out that missionaries like to receive mail. Remind the students that part of a Christian's ministry to missionaries is not only to pray for them but also to write to them. Explain that by writing to them, the students can find out the missionaries' prayer needs, encourage them in their work, and let them know that they are thinking about them.

Instruct

2-Digit Addend Word Problems

Model 2-digit addition by using **manipulatives**.

Distribute the Tens/Ones Mats and UNIFIX Cubes.

Read aloud the following word problem. Display manipulatives in a Tens/Ones frame for each step. Guide the students as they use UNIFIX Cubes on their mats.

Missionary Marc received 10 pieces of mail on Monday. He received 13 pieces on Tuesday. How many pieces of mail did he receive in all?

How many letters did Missionary Marc receive on Monday? 10

Guide the students as they show the sets for the first addend on their mats. 1 ten, 0 ones

How many letters did the missionary receive on Tuesday? 13

Guide the students as they show the sets for the second addend. 1 ten, 3 ones

How can you find out the total number of letters Missionary Marc received on Monday and Tuesday? join the sets; add

Remind the students that when joining sets with tens and ones, the cubes in the Ones place should always be joined first; then the bars in the Tens place are joined.

Direct the students to move all the cubes in the Ones place together on their mats.

How many ones are there in all? 3

Move all the tens in the Tens place together.

How many tens are there in all? 2

What is the total number of letters? 23

What is the equation that shows this word problem? 10 + 13 = 23

Write the equation for display and read it together.

Continue the activity with the following word problem.

Missionary Marc collects stamps. He had 25 stamps. Some friends gave him 23 more stamps. How many stamps does he have now? 25 + 23 = 48 stamps

2-Digit Addition in Vertical Form

Review 2-digit addition in vertical form by using a **Tens/Ones frame**.

Display 20 + 14 in a Ten/Ones frame.

What do you add first when you have 2-digit addends? the ones

What is 0 ones + 4 ones? 4 ones

What happens when a number is added to 0? The answer is the other number.

Write "4" in the Ones place of the answer.

What do you add next? the tens

What is 2 tens + 1 ten? 3 tens

Write "3" in the Tens place of the answer.

Guide the students as they read the problem aloud.

Write the following vertical addition problems for display and invite students to complete the problems. Emphasize adding the digits in the Ones place before adding the digits in the Tens place.

$$\begin{array}{r}15\\+21\\\hline 36\end{array}\qquad\begin{array}{r}32\\+17\\\hline 49\end{array}\qquad\begin{array}{r}40\\+6\\\hline 46\end{array}$$

93

Apply

Worktext pages 181–82

Review the steps for adding 2-digit addends at the top of page 181 and guide the students as they complete the equations. Guide the students as they write the equation for the word problem and then use the Tens/Ones frame to find the answer. Remind them to complete the equation. Read and guide completion of the remainder of the page.

Read and explain the directions for page 182. Assist the students as they complete the page independently.

Assess

Reviews pages 175–76

Review fractions on page 176.

Extended Activity

Bean Sticks for Adding 2-Digit Numbers

Materials

- Wax paper or aluminum foil
- 10 tongue depressors
- 120 dried lima beans
- Glue
- A resealable plastic bag

Procedure

Cover the work area with paper or foil. Students may work in pairs. Guide the students as they apply glue to the tongue depressors and place 10 beans in a row on top of the glue on each tongue depressor. Assist the students as they apply a final coat of glue over the top of the beans to make the beans adhere longer. Make 10 bean sticks and allow them to dry completely. Store the 10 bean sticks and remaining 20 lima beans in a resealable bag. Allow pairs of students to take turns using the bean sticks and beans to solve 2-digit addition problems.

Adding Two-Digit Numbers

Add.

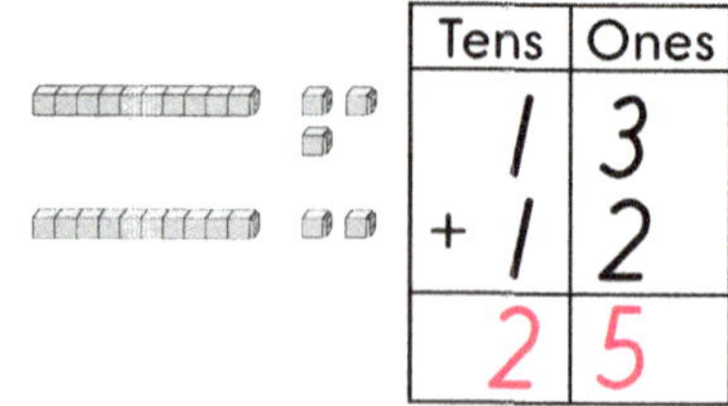

Tens	Ones
1	3
+ 1	2
2	5

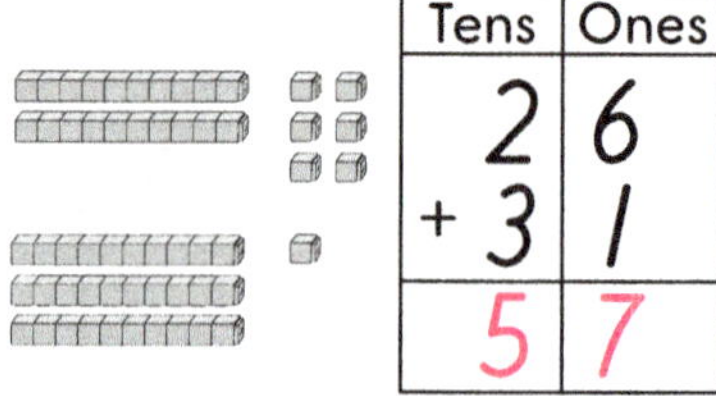

Tens	Ones
2	6
+ 3	1
5	7

Add.

Tens	Ones
6	4
+ 2	3
8	7

Tens	Ones
1	3
+ 5	2
6	5

Tens	Ones
2	4
+ 2	5
4	9

Tens	Ones
4	3
+ 2	0
6	3

$91 + 8 = 99$

$31 + 42 = 73$

$62 + 11 = 73$

$53 + 5 = 58$

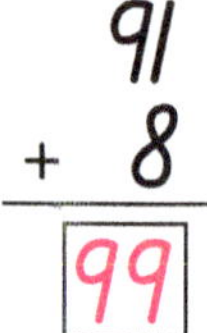

Write an equation for the word problem.
Solve using the Tens/Ones frame.

Iris sent 32 letters.
Mia sent 12 letters.
How many letters did they send in all?

Tens	Ones
3	2
+ 1	2
4	4

32 + 12 = 44 letters

Chapter 9 Review

Mark the fraction that names the shaded part.

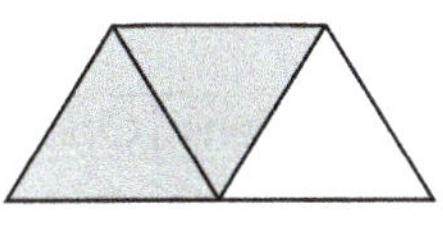
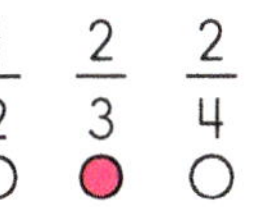

$\frac{1}{2}$ ○ $\frac{2}{3}$ ● $\frac{2}{4}$ ○

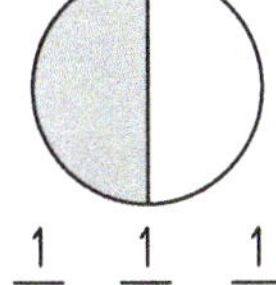

$\frac{1}{2}$ ● $\frac{1}{3}$ ○ $\frac{1}{4}$ ○

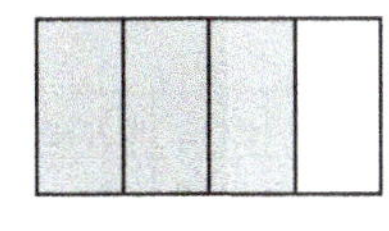
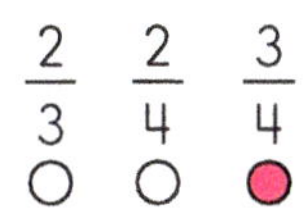

$\frac{2}{3}$ ○ $\frac{2}{4}$ ○ $\frac{3}{4}$ ●

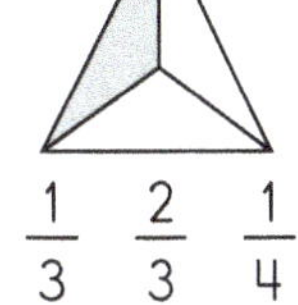

$\frac{1}{3}$ ● $\frac{2}{3}$ ○ $\frac{1}{4}$ ○

Color the shape to match the fraction.

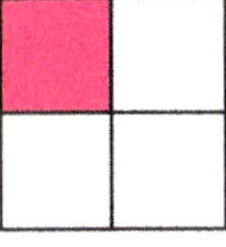
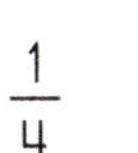

$\frac{1}{4}$

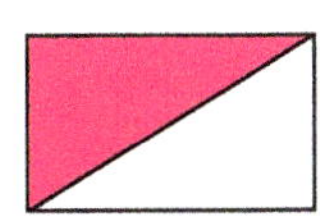
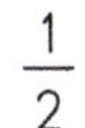

$\frac{1}{2}$

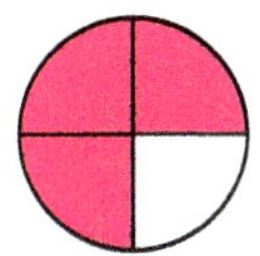

$\frac{3}{4}$

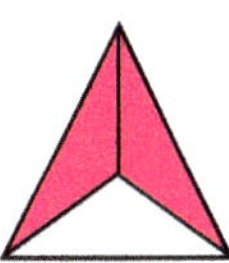

$\frac{2}{3}$

Subtraction Fact Review

Subtract.

10 - 1 = 9	5 - 3 = 2	6 - 2 = 4
5 - 0 = 5	7 - 5 = 2	9 - 8 = 1
6 - 6 = 0	8 - 3 = 5	7 - 6 = 1
5 - 4 = 1	9 - 0 = 9	6 - 4 = 2
4 - 2 = 2	8 - 8 = 0	5 - 2 = 3

94

Worktext pages 183–84
Reviews pages 177–78

Objectives

- **94.1** Join sets of dimes and pennies for 2-digit addends.
- **94.2** Solve a money word problem by writing an addition or subtraction equation.
- **94.3** Solve a money word problem by using the Problem-Solving Plan

Printed Resources

- Visual 21: *Problem-Solving Plan*
- Visuals: Money Kit (9 pennies, 9 dimes)
- Student Manipulatives: Number Cards (0–9)
- Student Manipulatives: Tens/Ones Mat
- Student Manipulatives: Money Kit (9 pennies, 9 dimes)

Digital Resources

- Games/Enrichment: Subtraction Flashcards

Materials

- Subtraction flashcards

The money addition lessons in this chapter use dimes and pennies to reinforce place value. Be sure to include the cent sign in the Tens/Ones frame.

All addition problems in this chapter should be written vertically. All word-problem equations should be written horizontally.

Adding Dimes & Pennies

Add.

Tens	Ones	
3	2	¢
+ 1	5	¢
4	7	¢

Tens	Ones	
2	3	¢
+ 2	1	¢
4	4	¢

Add.

Tens	Ones	
1	4	¢
+ 2	2	¢
3	6	¢

Tens	Ones	
2	1	¢
+	6	¢
2	7	¢

Tens	Ones	
3	0	¢
+ 4	2	¢
7	2	¢

Tens	Ones	
6	1	¢
+ 2	4	¢
8	5	¢

Read the word problem. Underline the question.
Circle the information. Write an addition or subtraction equation. Solve.

Problem-Solving Plan
Solve by thinking
1. What is the question?
2. What information is given?
3. Do I add or subtract?
4. Does my answer make sense?

DQ has 45¢. The postal worker gives DQ 14¢. How much money does DQ have?

45 + 14 = 59 ¢

Tens	Ones	
4	5	¢
+ 1	4	¢
5	9	¢

Emma had 7¢. She lost 5¢. How much money does Emma have now?

7 − 5 = 2 ¢

Tens	Ones	
	7	¢
−	5	¢
	2	¢

Practice & Review

Memorize Subtraction Facts

Introduce the following facts.

9 − 7 9 − 5 9 − 4

Display each flashcard slowly. Invite students to give the answers.

Distribute Number Cards 0–9. Display each flashcard again. Direct the students to hold up the correct Number Card for each answer.

Practice these and the previously memorized subtraction facts.

Engage

Associate 2-Digit Numbers with Dimes & Pennies

Review associating the value of dimes and pennies with place value in 2-digit numbers.

Display 5 dimes and 2 pennies in a Tens/Ones frame.

What is the value of 1 dime? 10¢

What is the value of 1 penny? 1¢

Guide the students as they count the dimes by 10s and the pennies by 1s.

What is the value of 5 dimes and 2 pennies? 52¢

Write "52¢" above the Tens/Ones frame.

How many tens are in 52? 5

How many ones are in 52? 2

Continue the activity for 68¢ and 46¢.

Distribute the Tens/Ones Mats and the coins.

Write "35¢" for display. Instruct the students to use their dimes and pennies to make 35¢.

How many dimes did you put in the Tens place? 3

How many pennies did you put in the Ones place? 5

Continue the activity for 13¢ and 49¢.
1 dime, 3 pennies; 4 dimes, 9 pennies

Add.

Tens	Ones
4	2 ¢
+ 1	5 ¢
5	7 ¢

Tens	Ones
3	3 ¢
+ 2	2 ¢
5	5 ¢

Add.

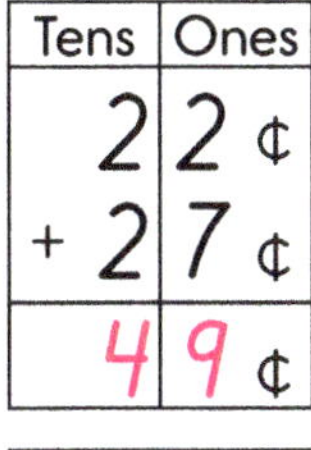

Tens	Ones
3	4 ¢
+ 3	4 ¢
6	8 ¢

Tens	Ones
2	2 ¢
+ 2	7 ¢
4	9 ¢

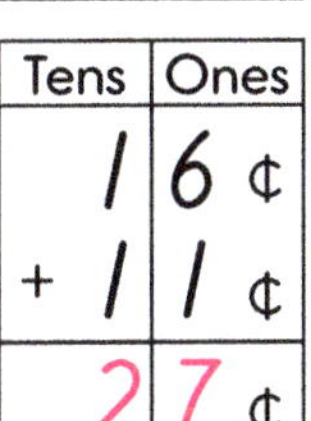

Tens	Ones
1	6 ¢
+ 1	1 ¢
2	7 ¢

Tens	Ones
1	0 ¢
+ 5	0 ¢
6	0 ¢

Tens	Ones
4	2 ¢
+ 3	3 ¢
7	5 ¢

Tens	Ones
6	2 ¢
+ 1	4 ¢
7	6 ¢

Time to Review

Write the total value.

43 ¢

54 ¢

82 ¢

26 ¢

Instruct

Join Sets of Coins for 2-Digit Addends

Use **coin manipulatives** to help the students solve 2-digit addition problems.

Distribute the Tens/Ones Mats and the coins if the students do not already have them out.

Write "32¢ + 24¢" in a Tens/Ones frame.

What is the first addend? 32¢

Instruct the students to place 3 dimes and 2 pennies at the top of their mats.

What is the second addend? 24¢

Instruct the students to place 2 dimes and 4 pennies at the bottom of their mats.

What do you add first when you have a 2-digit addend? the ones

Guide the students as they move all the pennies in the Ones place together on their mats.

How many pennies are there in all? 6

Ask the students to move all the dimes in the Tens place together.

How many dimes are there in all? 5

What is the sum of the addition problem? 56¢

Invite a student to complete the addition problem by writing the answer.

Continue the activity with the following problems.

14¢	23¢
+ 31¢	+ 5¢
45¢	28¢

Problem-Solving Plan

Use the **Problem-Solving Plan** to help the students write equations for solving money word problems.

Display Visual 21 and read aloud the questions. Explain that solving word problems can be made easier by following a plan.

Display the following word problem. Explain that the students need to think about the information given in the word problem so they can develop a plan to find the answer.

Read aloud the problem several times as you guide the students to identify the question and the information the problem gives.

DQ went to the post office to mail a letter and a postcard. It cost DQ 58¢ to mail the letter and 40¢ to mail the postcard. How much money did DQ pay to mail the letter and the postcard?

What is the question? How much money did DQ pay to mail the letter and the postcard?

Ask a student to underline the question in the displayed word problem.

What information is given? It cost 58¢ to mail the letter and 40¢ to mail the postcard.

Invite a student to circle the information given in the word problem.

Do you add or subtract to find the answer? add

What is the equation that you can use to solve this word problem?
58¢ + 40¢ = __¢

Ask another student to write the equation for display.

Point out that it is easier to add 2-digit numbers when the equation is rewritten in vertical form. Invite a student to write the problem in vertical form for display.

What do you add first when there are 2-digit addends? the ones

Direct a student to complete the equation. 98¢

Does the answer make sense? yes

94

Review Visual 21. Continue the activity by displaying the following word problem and guiding the students as they identify the question and the information given. Point out that it is not necessary that equations that are facts be rewritten in vertical form.

After DQ went to the post office, he had 9¢. DQ stopped by the store to buy a piece of gum for 2¢. How much money does DQ have now? 9¢ – 2¢ = 7¢

Apply

Worktext pages 183–84

Guide the students as they complete the problems in the top 2 sections of page 183.

Direct attention to the Problem-Solving Plan box. Review the plan. Read the directions and guide completion of the page.

Read and explain the directions for page 184. Assist the students as they complete the page independently.

Assess

Reviews Pages 177–78

Review fact families on page 178.

Extended Activity

Compare Sets of Dimes & Pennies by Using > and <

Materials

- A can containing 9 dimes
- A can containing 9 pennies
- Number Cards: 0–9 for each student
- Sign Cards: Greater than, Less than, Cent for each student

Procedure

Pair the students. Ask the first student to take 2 dimes and 3 pennies from the cans. Instruct him or her to use the Number Cards to write the value of the coins. 23¢ Instruct the second student to take 3 dimes and 2 pennies from the cans. Ask this student to use the Number Cards to write the value of his or her coins. 32¢ Instruct the pair of students to determine which set of coins has the greater value and then make the number sentence by using the correct sign. 23¢ < 32¢ or 32¢ > 23¢ Ask the students to place the coins back in the cans. Continue the activity by inviting different pairs of students to choose new sets of coins, compare the values, and write a number sentence.

Adding Dimes & Pennies

Add.

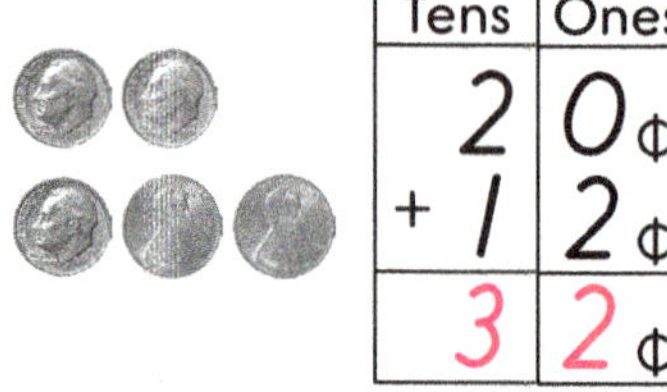

Tens	Ones
2	0¢
+ 1	2¢
3	2¢

Tens	Ones
1	2¢
+ 1	2¢
2	4¢

Add.

Tens	Ones
1	3¢
+ 2	5¢
3	8¢

Tens	Ones
3	2¢
+ 4	1¢
7	3¢

Tens	Ones
5	2¢
+ 2	6¢
7	8¢

Tens	Ones
4	4¢
+ 5	1¢
9	5¢

Tens	Ones
3	2¢
+ 2	3¢
5	5¢

Tens	Ones
4	6¢
+ 2	1¢
6	7¢

Tens	Ones
8	1¢
+ 1	4¢
9	5¢

Tens	Ones
6	3¢
+ 3	6¢
9	9¢

Write an equation for the word problem. Solve.

The baker baked 23 muffins.
Jill baked 16 muffins.
How many muffins did they bake in all?

 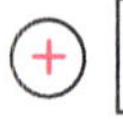   39 muffins

Tens	Ones
2	3
+ 1	6
3	9

Chapter 10 Review

Add or subtract.
Write the numbers for each fact family.

3 + 5 = 8
5 + 3 = 8
8 − 3 = 5
8 − 5 = 3

(4) (7) (11)

4 + 7 = 11
7 + 4 = 11
11 − 4 = 7
11 − 7 = 4

Write each fact family.

4 + 5 = 9
5 + 4 = 9
9 − 4 = 5
9 − 5 = 4

(3) (7) (10)

3 + 7 = 10
7 + 3 = 10
10 − 3 = 7
10 − 7 = 3

Subtraction Fact Review

0 1 2 3 4 5 6 7 8 9 10 11 12

Subtract. Use the number line if needed.

9	9	8	9	8	9	9
− 3	− 5	− 6	− 4	− 4	− 6	− 7
6	4	2	5	4	3	2

95 Worktext pages 185–86
Reviews pages 179–80

Objectives

- **95.1** Join sets of dimes and pennies for 2-digit addends.
- **95.2** Solve a money word problem by writing an addition or subtraction equation.
- **95.3** Solve a money word problem by using the Problem-Solving Plan.
- **95.4** Compose an addition word problem.
- **95.5** Evaluate adding money for a selfish purpose. BWS

Biblical Worldview Shaping

- **Caring** (evaluate): Doing anything we can to get everything we want is often encouraged and applauded in the world. But as believers, we must judge our actions by God's words. This includes loving others as we love ourselves (95.5).

Printed Resources

- Instructional Aid 27: *Pictograph*
- Visual 21: *Problem-Solving Plan*
- Student Manipulatives: Number Cards (0–9)
- Student Manipulatives: Tens/Ones Mat
- Student Manipulatives: Money Kit (9 pennies, 9 dimes)

Digital Resources

- Games/Enrichment: Subtraction Flashcards

Materials

- Subtraction flashcards

Practice & Review

Memorize Subtraction Facts

Distribute Number Cards 0–9. Display each flashcard. Ask the students to hold up the correct Number Card to indicate each answer.

Count by 5s to 150

Explain that you and the students will take turns as you count together by 5s to 150. You will begin by saying the number 5, and then they will say 10. Continue taking turns as they count to 150.

Adding Money

Add.

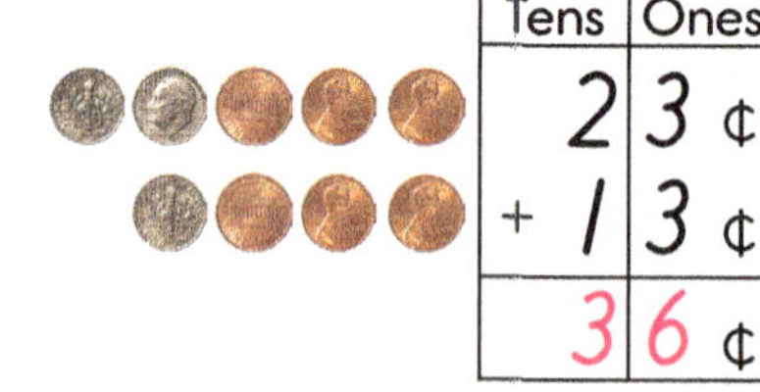

Tens	Ones	
2	3	¢
+ 1	3	¢
3	6	¢

Tens	Ones	
3	1	¢
+	4	¢
3	5	¢

Add.

Tens	Ones	
1	8	¢
+ 6	1	¢
7	9	¢

Tens	Ones	
5	3	¢
+ 1	6	¢
6	9	¢

14¢ + 72¢ = 86¢

21¢ + 34¢ = 55¢

Read the word problem. Underline the question. Circle the information. Write an addition or subtraction equation. Solve.

Finn had (7) stamps. He lost (3) stamps.
How many stamps does he have left?

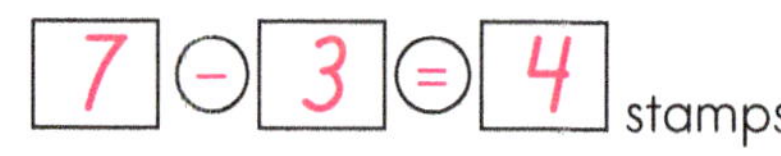

7 − 3 = 4 stamps

Tens	Ones
	7
−	3
	4

DQ has (14) stamps. Greg has (12) stamps.
How many stamps do they have in all?

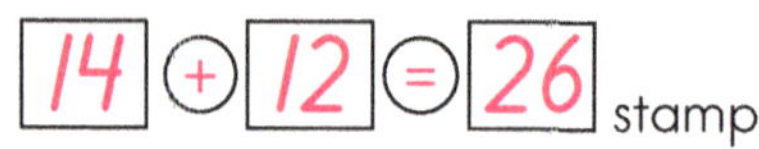

14 + 12 = 26 stamps

Tens	Ones
1	4
+ 1	2
2	6

Engage

Read a Pictograph

Guide a **review** of reading a pictograph.

Display the *Pictograph* page. Explain that this pictograph shows how many stamps each person has in his or her stamp collection. Direct attention to the stamp below the graph. Explain that each stamp on the graph represents 5 stamps and that the students will need to count by 5s to find out how many stamps each person has in his or her collection.

How many stamps does Ken have in his collection? 25

How many stamps does Meg have in her collection? 20

Continue until the students have identified how many stamps are in each person's collection.

Liam 15 stamps
Sam 15 stamps
Jill 30 stamps
Lily 5 stamps
Rosa 35 stamps

Instruct

Join Sets of Coins for 2-Digit Addends

Use **coin manipulatives** to help the students solve 2-digit addition problems.

Distribute the Tens/Ones Mats and the coins. Write "21¢ + 24¢" in vertical format for display.

95

Add.

Tens	Ones
1	9 ¢
+ 1	0 ¢
2	9 ¢

Tens	Ones
1	6 ¢
+ 8	1 ¢
9	7 ¢

Tens	Ones
2	4 ¢
+	4 ¢
2	8 ¢

Tens	Ones
1	3 ¢
+ 1	5 ¢
2	8 ¢

Circle the word to complete the sentence.

I should / (should not) add money to make a selfish choice.

Time to Review

Answer each question. Read the key below the pictograph.

How many letters did DQ send? 20

How many letters did Hank send? 10

How many letters did Carmen send? 10

How many letters did Josie send? 15

Who sent the most letters? DQ

Name	Letters Mailed
DQ	✉ ✉ ✉ ✉
Hank	✉ ✉
Carmen	✉ ✉
Josie	✉ ✉ ✉

✉ = 5 letters

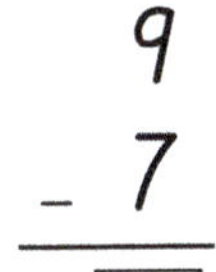
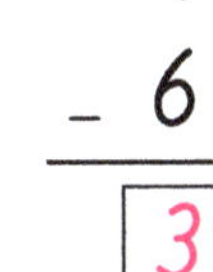

Subtract.
Use the number line if needed.

0 1 2 3 4 5 6 7 8 9 10

$9 - 7 = 2$ $9 - 6 = 3$ $9 - 5 = 4$ $9 - 4 = 5$ $9 - 3 = 6$

186 one hundred eighty-six Math 1

Ask the students to show the problem by using their dimes and pennies on their mats.

How did you make 21¢? 2 dimes, 1 penny

How did you make 24¢? 2 dimes, 4 pennies

What do you add first when you have 2-digit addends? the ones (pennies)

Instruct the students to move all the pennies in the Ones place together on their mats.

How many pennies are there in all? 5

Instruct the students to move all the dimes in the Tens place together.

How many dimes are there in all? 4

What is the sum of the addition problem? 45¢

Invite a student to complete the addition problem. 45¢

Continue the activity with the following problems.

46¢ + 33¢ = 79¢   53¢ + 32¢ = 85¢

Problem-Solving Plan

Use the **Problem-Solving Plan** to help the students write equations for solving money word problems.

Display Visual 21 and read aloud the questions. Remind the students that they need to think about the information given in a word problem so that they can develop a plan to find the answer.

Display the following word problem. Read aloud the problem several times as you guide the students to identify the question and the information the problem gives.

DQ's orange marble cost 8¢. His green marble cost 5¢. How much more did DQ's orange marble cost than his green marble?

What is the question? How much more did DQ's orange marble cost than his green marble?

Ask a student to underline the question in the displayed word problem.

What information is given? DQ's orange marble cost 8¢, and his green marble cost 5¢.

Invite a student to circle the information given in the word problem.

Do you add or subtract to find the answer? subtract

What is the equation that you can use to solve this word problem? 8¢ – 5¢ = __¢

Ask another student to write the equation for display.

Do you need to rewrite the equation in vertical form? No, it is not necessary because the equation is a fact.

Direct a student to complete the equation. 3¢

Does the answer make sense? yes

Review the Problem-Solving Plan. Continue the activity by displaying the following word problem.

DQ bought some supplies at the store. He purchased an envelope for 35¢ and some colored paper for 42¢. How much money did DQ spend at the store?

What is the question? How much money did DQ spend at the store?

Ask a student to underline the question.

What information is given? He spent 35¢ for an envelope and 42¢ for paper.

Invite a student to circle the information given.

Do you add or subtract to find the answer? add

What is the equation that you can use to solve this word problem? 35¢ + 42¢ = __¢

95

Invite a student to write the equation for display.

Should you rewrite the equation in vertical form? Yes, because it is easier to add 2-digit numbers in vertical form.

Ask a student to write the problem in vertical form.

What do you add first when there are 2-digit addends? the ones

Invite a student to complete the equation. 77¢

Does the answer make sense? yes

Compose an Addition Word Problem

Guide a **discussion** to help the students compose a word problem for a given equation.

Write "16¢ + 13¢ = __¢" for display. Ask a student to compose a word problem for the equation. Invite a student to rewrite the equation in vertical form and another student to solve it.

Write a new equation and ask students to compose other word problems.

Selfishness with Addition

Guide a **discussion** to help the students evaluate a situation from the perspective of biblical teaching.

Review the chapter opener scenario.

How many stamps did Mr. Marc give the boys to share between them? 4

What did Dan want to know about the stamps? how much they were worth; their value

Why do you think Dan wanted to know the value of the stamps? He wanted to choose the 2 stamps whose sum was the greatest for himself.

"Joel, I think you would like the blue one! It's your favorite color," Dan spoke encouragingly as he moved the blue stamp toward his brother.

Joel, who was only 4 years old, nodded. He agreed with mostly everything his big brother, Dan, said.

"And you would probably like the yellow one to go with it, wouldn't you?"

Adding Money

Add.

Tens	Ones
2	3¢
+ 1	3¢
3	6¢

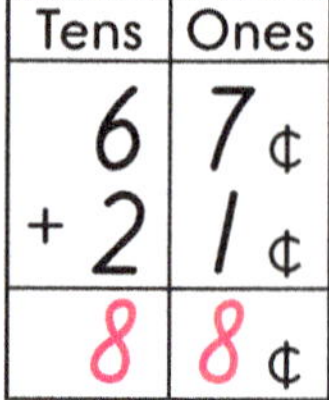

Tens	Ones
4	0¢
+ 2	2¢
6	2¢

Add.

Tens	Ones
6	5¢
+ 1	3¢
7	8¢

Tens	Ones
4	2¢
+ 2	4¢
6	6¢

Tens	Ones
2	6¢
+ 1	3¢
3	9¢

Tens	Ones
6	7¢
+ 2	1¢
8	8¢

34¢ + 52¢ = 86¢

14¢ + 25¢ = 39¢

23¢ + 34¢ = 57¢

75¢ + 11¢ = 86¢

Write an equation for the word problem. Solve.

Alice got 20 jellybeans.
John got 15 jellybeans.
How many jellybeans did they get in all?

20 + 15 = 35 jellybeans

Tens	Ones
2	0
+ 1	5
3	5

How did Dan make sure he would get the most valuable stamps? He encouraged his little brother Joel to take the 2 least valuable stamps.

"My stamps are worth more than yours, Joel," Dan said with a laugh. Joel looked at his blue and yellow stamps with a hurt expression.

Was Dan acting kindly or selfishly? selfishly

Point out that Dan's words make it clear that he used his skill with addition in a selfish way that hurt his little brother.

Remind the students of Matthew 22:39*b*, "Thou shalt love thy neighbour as thyself."

Was Dan treating Joel the way Dan would want to be treated? no

Encourage the students to remember Jesus' command and obey Him when they are tempted to use addition for a selfish purpose.

Chapter 11 Review

Use a ruler to draw each length. Start at the star.

3 inches ★

4 inches ★

5 inches ★

Subtraction Fact Review

Subtract.

5 − 4 = 1	9 − 2 = 7	7 − 4 = 3	6 − 5 = 1	10 − 2 = 8
7 − 3 = 4	6 − 4 = 2	3 − 3 = 0	8 − 2 = 6	9 − 3 = 6
6 − 2 = 4	5 − 2 = 3	9 − 8 = 1	10 − 1 = 9	7 − 1 = 6

180 one hundred eighty Math 1 Reviews

95

Apply

Worktext pages 185–86

Read and guide completion of page 185.

Read and explain the directions for page 186. Assist the students as they complete the page independently.

Direct attention to the biblical worldview shaping statement and guide the students as they complete the sentence.

Assess

Reviews pages 179–80

Review drawing a line to a specified length on page 180.

Extended Activity

Join Sets of Dimes & Pennies

Materials

- A menu for display: hamburger 55¢, hot dog 50¢, French fries 33¢, baby carrots 31¢, juice box 40¢, cookie 14¢
- Money Kit: 9 pennies, 9 dimes for each pair of students

Procedure

Pair the students. Distribute the coins and display the menu. Instruct 1 student in each pair to be the waiter and the other student to be the customer. Explain that the waiter will ask the customer what he or she would like to order. The customer may order 2 items only. The waiter will record the cost of each item ordered and find the total cost. The customer will count out the coins needed to pay for the order and give them to the waiter. Repeat the activity with the students exchanging roles.

96–97

Worktext pages 187–88
Reviews pages 181–82

Objectives

- **96–97.1** Rename 10 ones as 1 ten.
- **96–97.2** Join sets for 2-digit addends with renaming.
- **96–97.3** Complete a 2-digit addition problem with renaming.

Printed Resources

- Visuals: DQ Puppet
- Visuals: Place Value Kit (tens, ones)
- Student Manipulatives: Tens/Ones Mat

Digital Resources

- Video: Rename 10 Ones
- Web Link: Virtual Manipulatives: Base Ten Number Pieces

Materials

- 50 UNIFIX® Cubes (or Place Value Kit: tens, ones; for each student)

Lessons 96–98 are included as enrichment instruction. Present these lessons based on your schedule and the readiness of your students. You may spend 2 lessons covering Lessons 96–97, or you may present both lessons in 1 presentation.

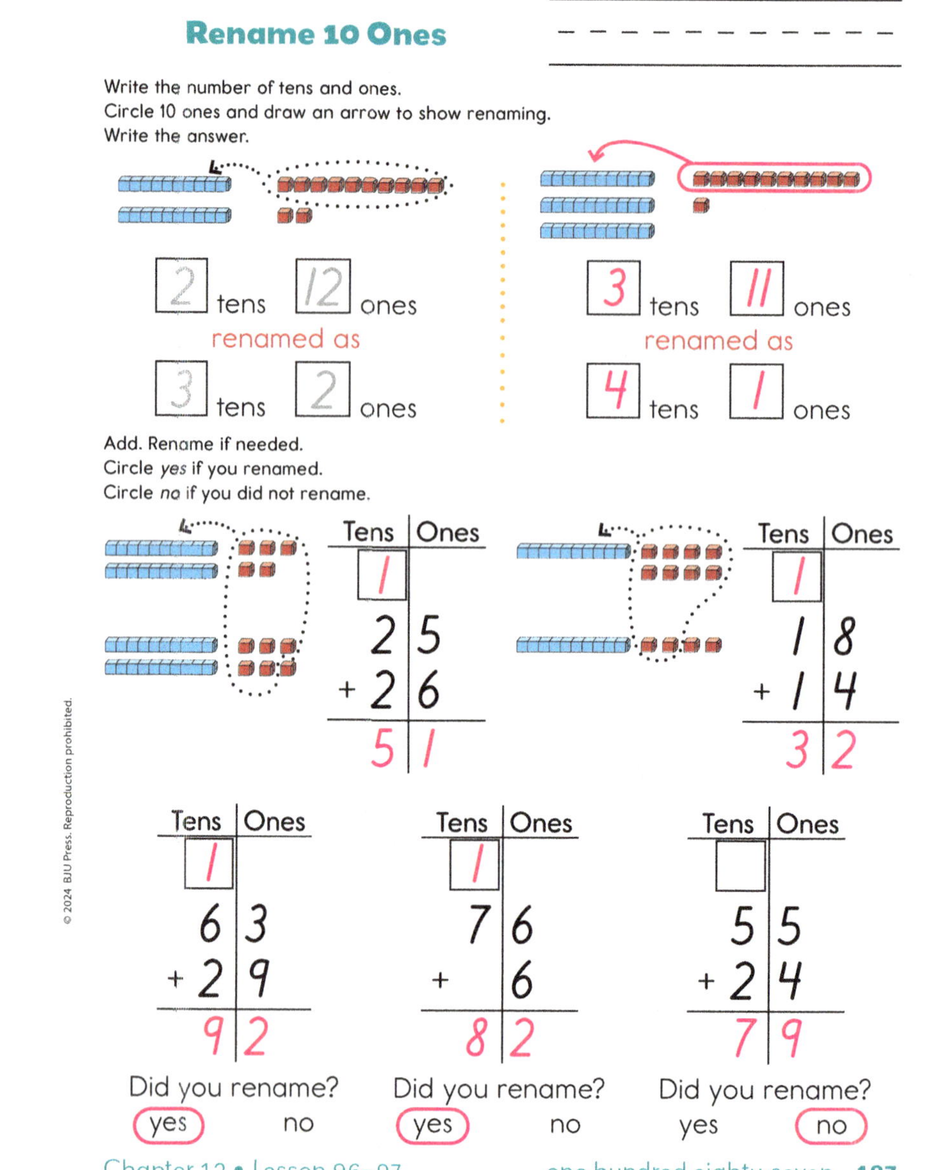

Rename 10 Ones

Write the number of tens and ones.
Circle 10 ones and draw an arrow to show renaming.
Write the answer.

2 tens 12 ones renamed as 3 tens 2 ones

3 tens 11 ones renamed as 4 tens 1 ones

Add. Rename if needed.
Circle *yes* if you renamed.
Circle *no* if you did not rename.

Tens	Ones
1	
2	5
+ 2	6
5	1

Tens	Ones
1	
1	8
+ 1	4
3	2

Tens	Ones
1	
6	3
+ 2	9
9	2

Did you rename? yes no

Tens	Ones
1	
7	6
+	6
8	2

Did you rename? yes no

Tens	Ones
5	5
+ 2	4
7	9

Did you rename? yes no

Chapter 12 • Lesson 96–97 one hundred eighty-seven 187

Engage

Names

Guide a **discussion** to prepare the students to rename 10 ones as 1 ten.

Display the DQ Puppet.

What animal is this a picture of? a duck

What name do you call this duck? DQ

Remind the students that DQ is a nickname, usually a short name that stands for a longer name. Invite the students to give DQ's full name. Duncan Quackerly Duckington

Does calling DQ by his longer name make him someone else? No, he is still the same duck whether he is called DQ or Duncan Quackerly Duckington.

Explain that in math a number is sometimes called by another name, or *renamed*, to make it easier to work with. Tell the students that renaming a number does not change its value. It is still worth the same amount.

Point out that in this lesson they will see how renaming a number helps them learn a new skill.

Instruct

Rename 10 Ones as 1 Ten

Use **manipulatives** to model renaming a set of 10 ones as 1 ten.

Display 14 ones in the Ones place of a Tens/Ones frame.

How many ones are there? 14

Point out that there are more ones in the Ones place than previously have been used there.

Are there more than 10 ones? Yes, 14 > 10.

Explain that 10 or more ones are too many for the Ones place. When there are 10 or more ones the students can *rename*, or give a new name to, a set of 10 ones. They can then trade the 10 ones for 1 ten because they both have the same value. They are different names for the same number.

Write "10 ones = 1 ten" for display.

Remove 10 ones from the Ones place and put 1 ten in the Tens place. Explain that the 10 ones received a new name and were moved to a new home in the Tens place.

What number is shown? 14

Explain that 14 ones was renamed as 1 ten and 4 ones.

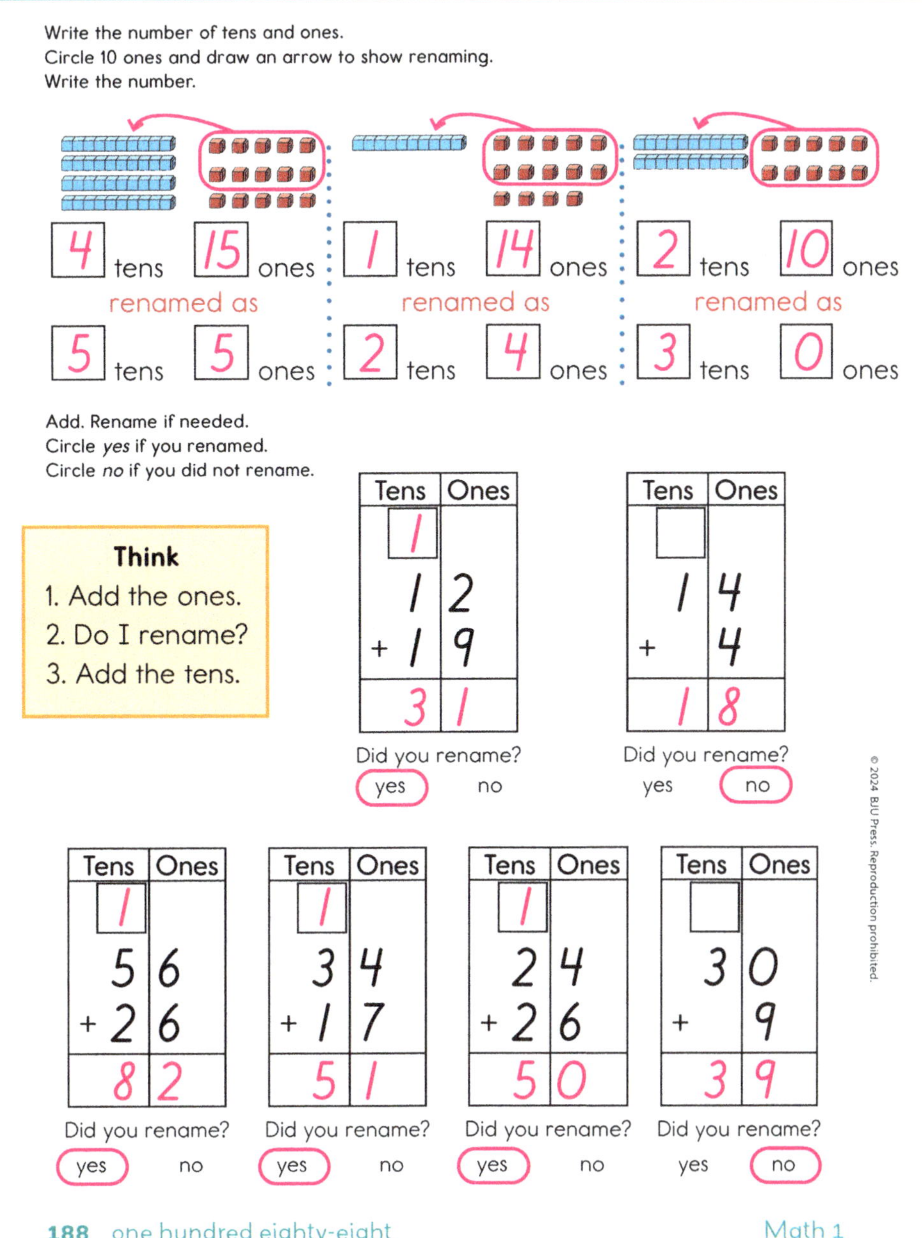

Write the number of tens and ones.
Circle 10 ones and draw an arrow to show renaming.
Write the number.
4 tens 15 ones
renamed as
5 tens 5 ones
1 tens 14 ones
renamed as
2 tens 4 ones
2 tens 10 ones
renamed as
3 tens 0 ones
Add. Rename if needed.
Circle yes if you renamed.
Circle no if you did not rename.
Think
1. Add the ones.
2. Do I rename?
3. Add the tens.
Tens Ones
Did you rename?
yes no
© 2024 BJU Press. Reproduction prohibited.
188 one hundred eighty-eight
Math 1

Point out that the value is the same, but the number looks different because 14 was made another way.

Continue the activity by renaming 12 ones and 18 ones.

Join Sets for 2-Digit Addends with Renaming

Guide the students as they use **manipulatives** to model joining sets for 2-digit addends.

Distribute the Tens/Ones Mats and UNIFIX Cubes. Instruct the students to show the problem on their mats as you model it by using manipulatives in a Tens/Ones frame.

Write "15 + 17" in a Tens/Ones frame as shown.

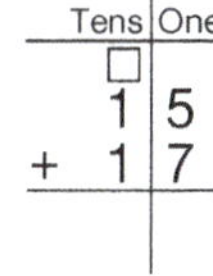

What is the first addend? 15

Instruct the students to place 1 ten bar and 5 ones at the top of their mats.

What is the second addend? 17

Instruct the students to place 1 ten bar and 7 ones at the bottom of their mats.

What do you add first when you have 2-digit addends? the ones

Instruct the students to move all the cubes in the Ones place together on their mats.

How many ones are there in all? 12

Are there enough ones to rename as 1 ten? yes

Instruct the students to group 10 ones by connecting 10 cubes together to make 1 ten bar.

96–97

Direct them to then place the new ten bar at the top of the Tens place on their mats. Model this procedure with your manipulatives.

If you are using pieces from the Place Value Kit to demonstrate, point out that since the ones in the Place Value Kit cannot connect together (like UNIFIX Cubes), you will remove 10 ones and trade them for 1 ten and then place the new ten at the top of the Tens place.

What do you add after you finish adding the ones? the tens

Direct the students to move all the tens in the Tens place together.

How many tens are there? 3

How many ones are there? 2

Direct attention to the problem written for display. Explain that when 5 ones are added to 7 ones the answer is 12 ones, but they cannot write 12 in the Ones place. Point out that when 10 ones were renamed as 1 ten there were 2 ones left in the Ones place, so they will write a 2 in the Ones place of the answer.

What happened to the 1 ten that was made from 10 ones? It was put in the Tens place.

Write "1" in the box above the Tens place to represent the 10 ones that were renamed as 1 ten.

What is 1 ten + 1 ten + 1 ten? 3 tens

Write "3" in the Tens place of the answer and read aloud the completed problem: "15 plus 17 equals 32."

Did you need to rename in this problem? Yes; when I added the ones together, there were 12 ones. The 12 ones are too many for the Ones place.

Continue the activity with the following problems. Point out that they do not need to rename the ones if there are fewer than 10.

12	21	39
+ 29	+ 28	+ 11
41	49	50

Apply

Worktext pages 187–88

Direct attention to the first problem on page 187. Read the directions and guide the students to write the number of tens and ones before renaming.

96–97

Point out the renaming of 10 ones as 1 ten. Explain that they will show the renaming by grouping 10 ones in a circle and drawing an arrow from the circle to show that the 10 ones now have a new name of 1 ten and will be counted with the tens. Guide the students as they complete the second problem.

Direct attention to the next set of problems on page 187. Remind the students to add the ones first and to decide if they need to rename 10 ones as 1 ten. Guide completion of the page.

Read and explain the directions and guide completion of page 188.

Assess

Reviews pages 181–82

Read the directions and guide completion of the pages.

Extended Activity

Celebrate Day 100

Procedure

Use one or more of the following activities to celebrate the 100th day of school.

- Make a 100-day chain. Cut out 100 strips of construction paper. Staple the strips together to form 100 links.
- Measure and cut a piece of string 100 inches in length.
- Exercise by doing 10 sets of different exercises: 10 toe-touches, 10 sit-ups, 10 jumping jacks, 10 hops on the left foot, 10 hops on the right foot, 10 knee-touches, 10 squats, 10 left arm circles, 10 right arm circles, 10 high stretches.
- Put together a 100-piece puzzle.
- Teach the number *one hundred* in Spanish (cien: See-EN), French (cent: sohn), and German (hundert: HOOND-airt).
- Place 100 footprints on the floor for the students to walk on.
- Make a "Thankful to God" list. Discuss the wonderful things that God has done. Number from 1 to 100. Ask students to tell things for which they are thankful to God.

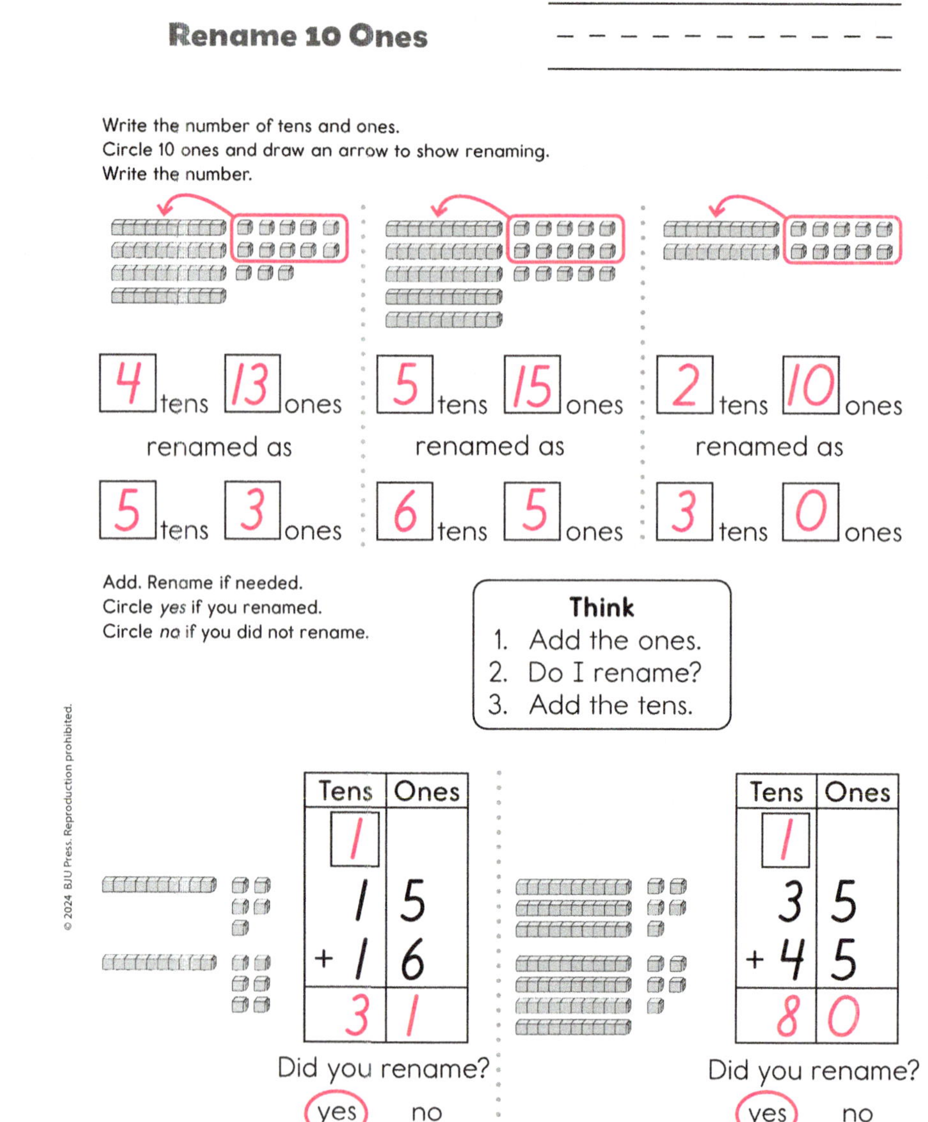

Add. Rename if needed.
Circle *yes* if you renamed.
Circle *no* if you did not rename.

Tens	Ones
☐	
1	1
+ 1	8
2	9

Did you rename?
yes (no)

Tens	Ones
☐	
1	0
+ 3	8
4	8

Did you rename?
yes (no)

Tens	Ones
1	
2	4
+ 4	8
7	2

Did you rename?
(yes) no

Tens	Ones
1	
1	6
+ 2	5
4	1

Did you rename?
(yes) no

Tens	Ones
☐	
3	0
+	9
3	9

Did you rename?
yes (no)

Tens	Ones
1	
2	4
+ 2	6
5	0

Did you rename?
(yes) no

Tens	Ones
☐	
3	3
+ 4	5
7	8

Did you rename?
yes (no)

Tens	Ones
1	
6	9
+ 2	3
9	2

Did you rename?
(yes) no

Tens	Ones
1	
7	1
+ 1	9
9	0

Did you rename?
(yes) no

98

Worktext pages 189–90
Reviews pages 183–84

Objectives

- **98.1** Rename 10 pennies as 1 dime.
- **98.2** Join sets of dimes and pennies for 2-digit addends with renaming.
- **98.3** Solve money problems with sums to 99¢.
- **98.4** Solve a word problem by writing an addition equation.

Printed Resources

- Visual 21: *Problem-Solving Plan*
- Visuals: Money Kit (pennies, dimes)
- Student Manipulatives: Tens/Ones Mat
- Student Manipulatives: Money Kit (pennies, dimes)

Digital Resources

- Video: Rename 10 Pennies
- Web Link: Virtual Manipulatives: Money Pieces

Engage

Using a Bank

Guide a **discussion** to acquaint the students with using a bank.

Invite the students to tell about their experience in visiting a bank.

What do people do at a bank? put money in or take money out of an account, save money, borrow money, or cash a check

Explain that a bank is a place where people store money so they do not have to carry it around with them all the time. Banks make it convenient for people to use their money safely.

Explain to the students that in this lesson they will pretend to use a bank for some of their money.

Instruct

Rename 10 Pennies as 1 Dime

Use **coin manipulatives** to help the students rename 10 pennies as 1 dime.

Distribute the Tens/Ones Mats and the coins. Instruct the students to place 4 dimes at the top of their desks to be used as their banks.

Rename 10 Pennies

Write the number of dimes and pennies.
Circle the number of pennies that can be traded for 1 dime.
Write the total value.

2 dimes 12 pennies
traded for
3 dimes 2 pennies
32 ¢

Add.
Circle *yes* if you renamed.
Circle *no* if you did not rename.

	Tens	Ones
(renamed)	1	
	5	5 ¢
+	3	6 ¢
	9	1 ¢

Did you rename? **yes** no

	Tens	Ones
(renamed)		
	6	1 ¢
+	2	4 ¢
	8	5 ¢

Did you rename? yes **no**

	Tens	Ones
(renamed)	1	
	7	8 ¢
+	1	2 ¢
	9	0 ¢

Did you rename? **yes** no

	Tens	Ones
(renamed)		
	3	0 ¢
+	2	9 ¢
	5	9 ¢

Did you rename? yes **no**

Write an equation for the word problem. Solve.

DQ stopped at a yard sale.
He bought a toy car for 25¢.
He also bought a toy bus for 35¢.
How much did DQ spend?

25 + 35 = 60 ¢

	Tens	Ones
(renamed)	1	
	2	5 ¢
+	3	5 ¢
	6	0 ¢

Guide the students to show 15¢ by putting 15 pennies in the Ones place of their mats.

How many pennies are the same as 1 dime? 10

Do you have enough pennies to rename as 1 dime? yes

Instruct the students to trade 10 pennies for 1 dime from their "banks" and to put the dime in the Tens place.

Are 15 pennies the same as 1 dime and 5 pennies? yes

Continue the activity by renaming 17¢ and 13¢.

Join Sets of Dimes & Pennies for 2-Digit Addends with Renaming

Use **coin manipulatives** to help the students solve money problems.

Instruct the students to use their dimes and pennies to show the following problem on their mats. Model the steps by using your manipulatives in a Tens/Ones frame.

Write "38¢ + 14¢" in a Tens/Ones frame as shown.

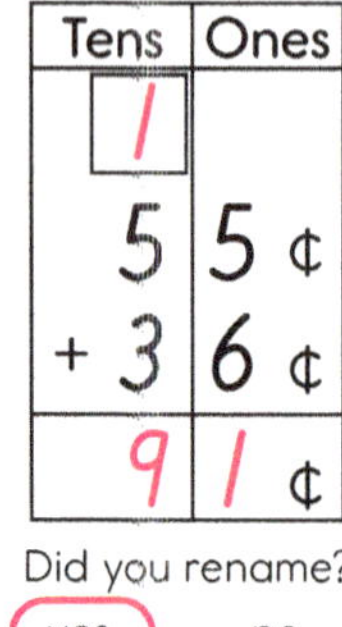

	Tens	Ones
	□	
	3	8 ¢
+	1	4 ¢
		¢

What is the first addend? 38¢

Instruct the students to place 3 dimes and 8 pennies at the top of their mats, putting the dimes in the Tens column and the pennies in the Ones column.

What is the second addend? 14¢

Instruct the students to place 1 dime and 4 pennies in the proper columns at the bottom of their mats.

Write the number of dimes and pennies.
Circle the number of pennies that can be traded for 1 dime.
Write the total value.

3 dimes 14 pennies
traded for
4 dimes 4 pennies
44 ¢

Add. Rename if needed.
Circle *yes* if you renamed.
Circle *no* if you did not rename.

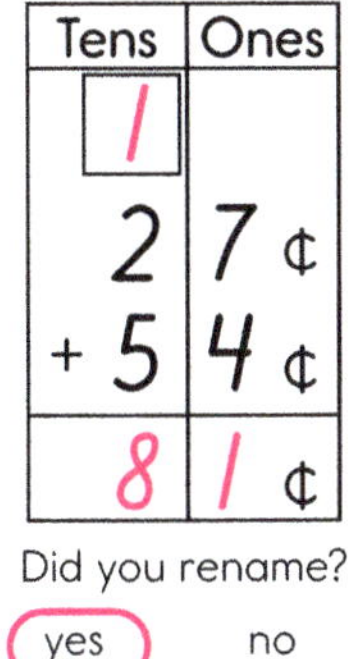

Did you rename?
yes no

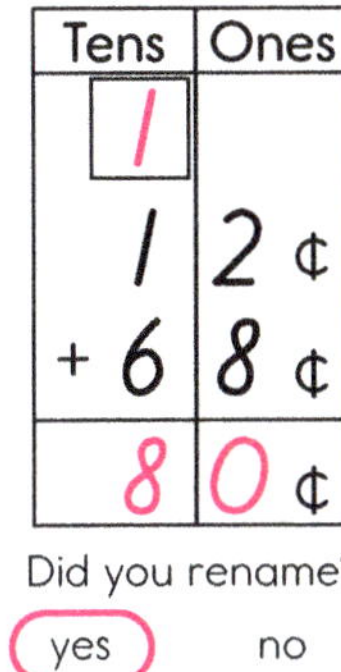

Did you rename?
yes no

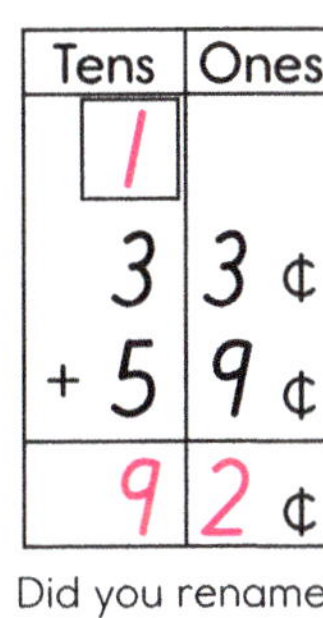

Did you rename?
yes no

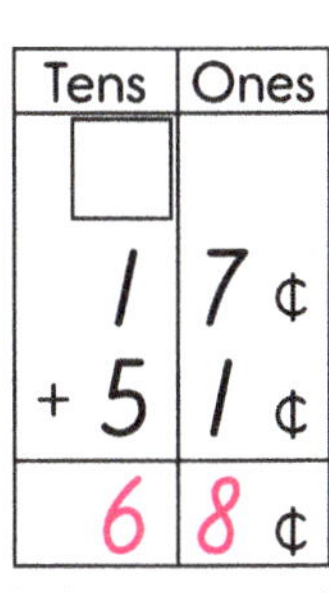

Did you rename?
yes no

Write an equation for the word problem. Solve.

Rita paid 32¢ for an orange fish.
She paid 29¢ for a brown fish.
How much did Rita pay for her pet fish?

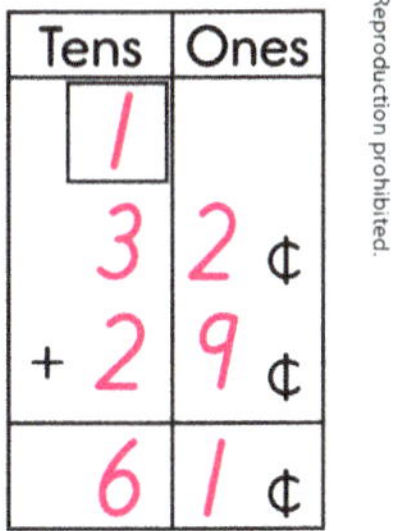

What do you add first when you have 2-digit addends? the ones (pennies)

Instruct the students to move all the pennies in the Ones place together on their mats.

How many pennies are there in all? 12

Are there enough pennies to rename as 1 dime? yes

Ask the students to trade 10 pennies for 1 dime from their "banks" and to put the dime in the Tens place.

What do you do after you add the ones (pennies)? add the tens (dimes)

Ask the students to move all the dimes in the Tens place together.

How many dimes are there? 5

How many pennies are there? 2

Direct attention to the Ones place in the vertical problem. Explain that 8¢ + 4¢ equals 12¢.

Can you write 12 in the Ones place? No; only 1 digit can be written in the Ones place; only numbers less than 10 are written in the Ones place.

What did you do with the 12 pennies? renamed 10 pennies as 1 dime

Write "2" in the Ones place of the answer. Explain that 12¢ was renamed as 1 dime and 2 pennies and that the 2 pennies remain in the Ones place, but the 1 dime is written in the box above the Tens place. Write "1" in the box.

What is 1 dime + 3 dimes + 1 dime? 5 dimes

Write "5" in the Tens place of the answer and read aloud the completed problem: "38¢ plus 14¢ equals 52¢."

Continue the activity with the following problems.

12¢	35¢
+ 9¢	+ 26¢
21¢	61¢

Write an Addition Equation to Solve a Problem

Guide the students to use the **Problem-Solving Plan** to solve a word problem.

Display Visual 21 and read aloud the questions. Direct the students to refer to the plan as they solve the following word problem. Read the problem aloud.

DQ went to the store to buy a box of crayons and a pencil. The crayons cost 61¢, and the pencil cost 19¢. How much did DQ pay for both items?

What is the question? How much did DQ pay for both items?

What information is given? The crayons cost 61¢, and the pencil cost 19¢.

How can you find the total cost of the 2 items? add

What equation can you use to solve the word problem? 61¢ + 19¢ = __¢

Invite a student to write the equation for display. Guide the students to determine that it is easier to add 2-digit addends when the equation is rewritten in vertical form. Ask a student to write the equation in vertical form.

Instruct the students to use their Tens/Ones Mats, dimes, and pennies to show the problem.

How will you show the problem? 6 dimes, 1 penny and 1 dime, 9 pennies

What is 1¢ + 9¢? 10¢

Is renaming needed? yes

Instruct the students to rename 10 pennies as 1 dime on their mats. Invite a student to write the renaming in the problem.

Why is a 1 written in the renaming box? 10 pennies are renamed as 1 dime.

What is the answer? 80¢

Ask a student to complete the equation.

Tell similar word problems for 45¢ + 17¢ = 62¢ and 13¢ + 8¢ = 21¢.

98

Apply

Worktext pages 189–90

Read and guide completion of page 189.

Read and explain the directions for page 190. Assist the students as they complete the page independently.

Assess

Reviews pages 183–84

Read the directions and guide completion of the pages.

Extended Activity

Add the Prices

Materials

- General store items, school items, or snacks labeled with price tags
- Student calculators

Procedure

Instruct the students to choose 2 items and to find the total price by writing both prices and adding them together. You may allow students to check their work by using a calculator.

If some of the prices are greater than 6 in the Ones place, some of the problems may include facts above 12.

Rename 10 Pennies

Write the number of dimes and pennies.
Circle the number of pennies that can be traded for 1 dime.
Write the total value.

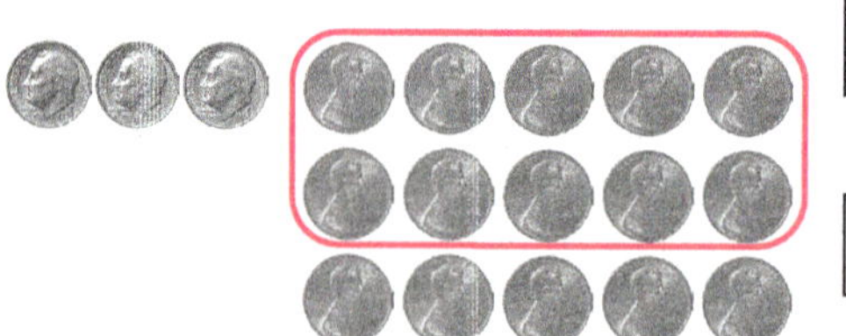

3 dimes 15 pennies
traded for

4 dimes 5 pennies
45¢

2 dimes 13 pennies
traded for

3 dimes 3 pennies
33¢

Write an equation for the word problem. Solve.

At the bake sale, a cupcake is 33¢ and a slice of pie is 38¢. How much will a cupcake and a slice of pie cost?

33 + 38 = 71¢

	Tens	Ones	
	1		
	3	3	¢
+	3	8	¢
	7	1	¢

Add. Rename if needed.
Circle *yes* if you renamed.
Circle *no* if you did not rename.

Tens	Ones
☐	
1	3 ¢
+ 1	5 ¢
2	8 ¢

Did you rename?
yes (no)

Tens	Ones
1	
3	4 ¢
+ 1	8 ¢
5	2 ¢

Did you rename?
(yes) no

Tens	Ones
1	
6	1 ¢
+ 1	9 ¢
8	0 ¢

Did you rename?
(yes) no

Tens	Ones
1	
4	6 ¢
+ 2	5 ¢
7	1 ¢

Did you rename?
(yes) no

Tens	Ones
☐	
6	1 ¢
+ 3	5 ¢
9	6 ¢

Did you rename?
yes (no)

Tens	Ones
1	
1	7 ¢
+ 5	5 ¢
7	2 ¢

Did you rename?
(yes) no

Tens	Ones
1	
1	2 ¢
+ 1	9 ¢
3	1 ¢

Did you rename?
(yes) no

Tens	Ones
1	
1	6 ¢
+ 2	5 ¢
4	1 ¢

Did you rename?
(yes) no

Tens	Ones
☐	
5	3 ¢
+ 1	1 ¢
6	4 ¢

Did you rename?
yes (no)

99

Worktext pages 191–92
Reviews pages 185–86

Should I use addition to get what I want?

Chapter Concept Review

- Practice concepts from Chapter 12 to prepare for the test.

Printed Resources

- Visual 21: *Problem-Solving Plan*
- Visuals: Place Value Kit (tens, ones)
- Student Manipulatives: Number Cards (0–9)
- Student Manipulatives: Tens/Ones Mat
- Student Manipulatives: Sign Cards (Greater than, Less than)

Digital Resources

- Games/Enrichment: Subtraction Flashcards

Materials

- UNIFIX® Cubes (or Place Value Kit: tens, ones)

Add.

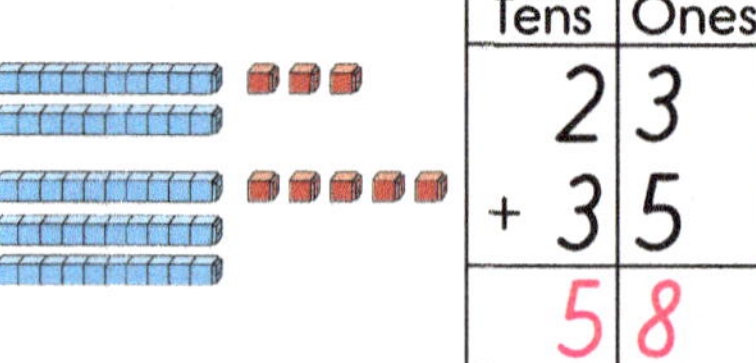

Tens	Ones
2	3
+ 3	5
5	8

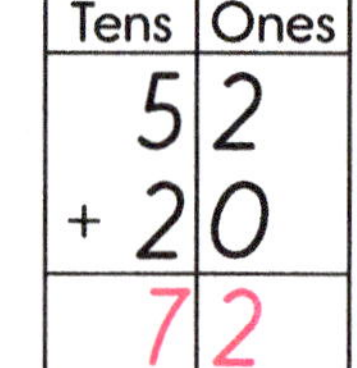

Tens	Ones
5	2
+ 2	0
7	2

Add.

Tens	Ones
2	3 ¢
+ 4	1 ¢
6	4 ¢

Tens	Ones
3	2 ¢
+ 1	4 ¢
4	6 ¢

Add.

Tens	Ones
6	0 ¢
+ 1	8 ¢
7	8 ¢

Tens	Ones
4	2 ¢
+ 3	5 ¢
7	7 ¢

Write an addition or subtraction equation. Solve.

The New Testament has many books.
The apostle Paul wrote 13 books.
Other men wrote 14 books.
How many books are in the New Testament?

13 + 14 = 27 books

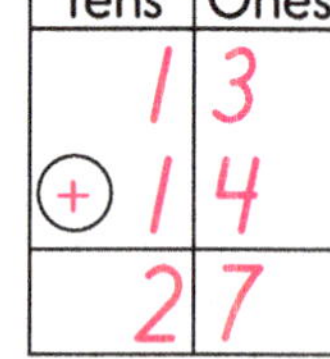

Tens	Ones
1	3
(+) 1	4
2	7

Practice & Review

Memorize Subtraction Facts

Distribute Number Cards 0–9. Practice all the previously memorized subtraction facts.

Compare 2-Digit Numbers by Using > and <

Distribute Number Cards 0–9 and the Greater-than and Less-than Signs. Display 5 tens and 4 ones. Instruct the students to use their Number Cards to write the number. 54

Display 6 tens and 7 ones to the right of the 5 tens and 4 ones. Instruct the students to use their Number Cards to write the new number. 67

Is 54 greater than or less than 67? less than

Instruct the students to place their Less-than Signs between the numbers.

Remind the students that this type of number sentence is read, "54 is less than 67." Guide the students as they read the number sentence together.

Continue the activity to compare and write number sentences for 76 > 73 and 14 < 41.

Instruct

Join Sets for 2-Digit Addends

Divide the students into teams. Explain that the teams will receive a point for each correct answer.

Distribute the Tens/Ones Mats and UNIFIX Cubes. Instruct the students to show the problem and find the answer by using the cubes on their mats. Write "25 + 13 = __" for display. 38 (3 tens, 8 ones)

Continue the game with the following addition problems.

11	41	26	24	12
+ 23	+ 7	+ 12	+ 23	+ 7
34	48	38	47	19

Problem-Solving Plan

Display Visual 21 and read aloud the questions. Read aloud the following word problem several times.

Mr. Block, a missionary, emailed 14 relatives and friends. Then Mr. Block emailed 10 churches that prayed and sent money each month. How many emails did Mr. Block send in all?

What is the question? How many emails did Mr. Block send in all?

What information is given? Mr. Block emailed 14 relatives and friends and 10 churches.

Do you add or subtract to find the answer? add

Write an addition or subtraction equation. Solve.

Paul wrote 9 letters to churches.
He wrote 2 letters to Timothy.
How many more letters did Paul write to churches?

9 − 2 = 7 letters

Tens	Ones
	9
−	2
	7

Add.

Tens	Ones
1	2
+ 1	6
2	8

Tens	Ones
4	0
+ 2	2
6	2

Tens	Ones
3	4
+ 3	1
6	5

Tens	Ones
2	9
+ 3	0
5	9

Add.

$13 + 11 = 24$ $42 + 46 = 88$ $94 + 5 = 99$ $75 + 12 = 87$

82¢ + 17¢ = 99¢ 41¢ + 31¢ = 72¢ 62¢ + 12¢ = 74¢ 53¢ + 30¢ = 83¢

Circle the word to complete the sentence.

Dan added to get the stamps he wanted. He was selfish with Joel. I should not use addition to be (selfish) kind.

What is the equation that you can use to solve this word problem? 14 + 10 = __

Write "14 + 10 = __" for display.

Should you rewrite the equation in vertical form? Yes, because it is easier to add 2-digit numbers in vertical form.

Invite a student to write the problem in vertical form.

What do you add first when there are 2-digit addends? the ones

Ask another student to solve the problem. 24

Does the answer make sense? yes

Write "24 emails" to complete the equation.

Review the Problem-Solving Plan and then read aloud the following word problem several times.

Mrs. Block had 9 stamps. She used 8 stamps to mail letters. How many stamps does Mrs. Block have now?

What is the question? How many stamps does Mrs. Block have now?

What information is given? Mrs. Block had 9 stamps, and she used 8 stamps to mail letters.

Do you add or subtract to find the answer? subtract

What is the equation that you can use to solve this word problem? 9 − 8 = __

Write "9 − 8 = __" for display.

Invite a student to solve the problem. 1

Does the answer make sense? yes

Do you need to rewrite the equation in vertical form? No, it is not necessary because the equation is a fact.

Write "1 stamp" to complete the equation.

Continue the activity with the following word problem.

Sam wrote to 2 people. The letter to his grandmother cost 58¢ to mail, and the postcard to his missionary friend cost 40¢ to mail. How much money did Sam spend to mail these letters? 58¢ + 40¢ = 98¢

Apply

Worktext pages 191–92

Read and explain the directions for the pages. Assist the students as they complete the pages independently.

Direct attention to the biblical worldview shaping section on page 192. Display the picture from the chapter opener and read aloud the essential question. Remind the students that Missionary Marc gave Dan and Joel some stamps to share for their collection. Dan used his skill with 2-digit addition to choose the most valuable stamps for himself. Guide the students as they complete the response on page 192.

Assess

Reviews pages 185–86

Use the Reviews pages to provide additional preparation for the chapter test.

Extended Activity

Write a Letter to a Missionary

Materials

- Stationery
- Postage stamp(s)
- A scale

99

Procedure

Choose a missionary family for the students to write a letter to. Explain that a letter is a way of showing friendliness. Draft the letter with the class; encourage the students to suggest some things to include in the letter. Write the letter for display. Rewrite the letter on stationery. Weigh the letter to determine the cost of the postage before mailing it to the missionary.

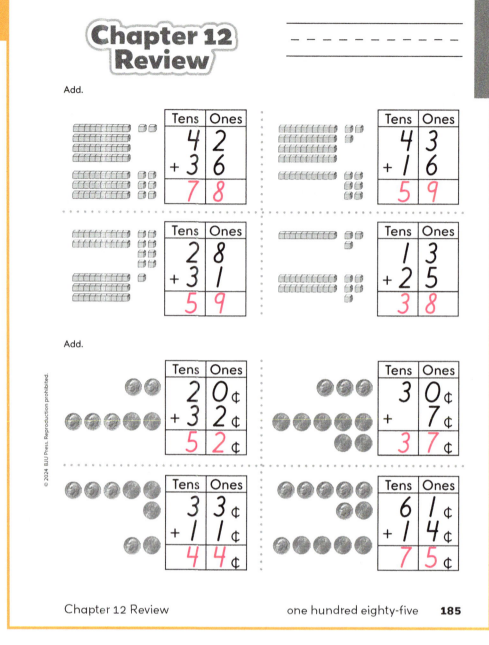

Add.

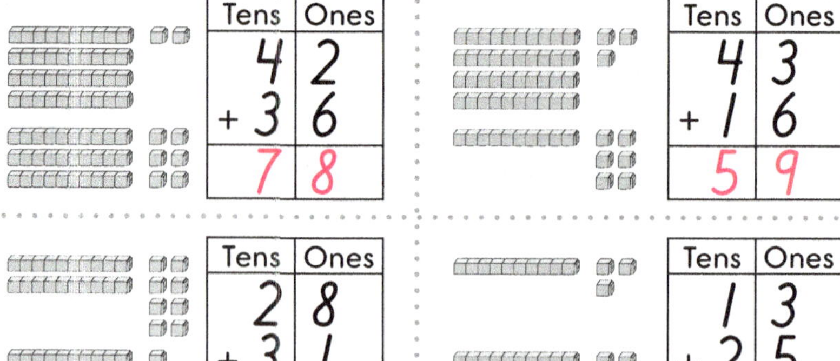

Tens	Ones
4	2
+ 3	6
7	8

Tens	Ones
4	3
+ 1	6
5	9

Tens	Ones
2	8
+ 3	1
5	9

Tens	Ones
1	3
+ 2	5
3	8

Add.

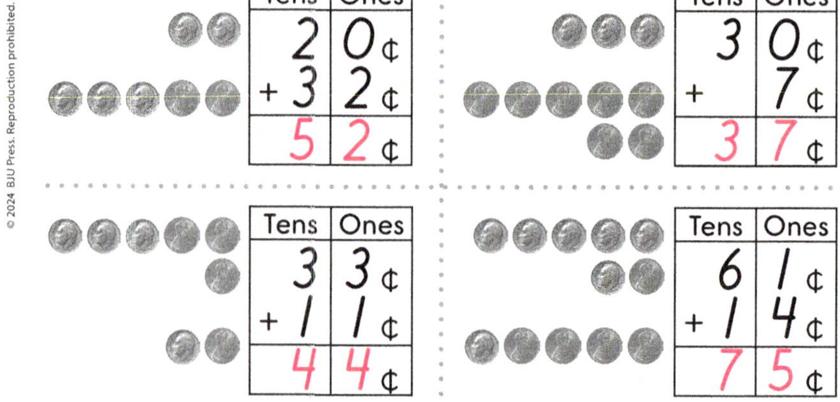

Tens	Ones
2	0¢
+ 3	2¢
5	2¢

Tens	Ones
3	0¢
+	7¢
3	7¢

Tens	Ones
3	3¢
+ 1	1¢
4	4¢

Tens	Ones
6	1¢
+ 1	4¢
7	5¢

Add.

Tens	Ones
9	3¢
+	5¢
9	8¢

Tens	Ones
7	1¢
+ 1	4¢
8	5¢

Tens	Ones
6	3¢
+ 3	2¢
9	5¢

Tens	Ones
5	4¢
+ 3	3¢
8	7¢

Add.

Tens	Ones
2	3
+ 4	1
6	4

Tens	Ones
4	3
+ 2	3
6	6

Tens	Ones
5	2
+ 2	5
7	7

Tens	Ones
4	4
+	2
4	6

$$\begin{array}{r} 33 \\ +54 \\ \hline 87 \end{array} \qquad \begin{array}{r} 15 \\ +\ 4 \\ \hline 19 \end{array} \qquad \begin{array}{r} 25 \\ +12 \\ \hline 37 \end{array} \qquad \begin{array}{r} 82 \\ +13 \\ \hline 95 \end{array}$$

Write an equation for each word problem. Solve.

Aubrey had 9 rocks in a bag.
The bag had a hole, and 2 rocks fell out.
How many rocks were left in the bag?

9 − 2 = 7 rocks

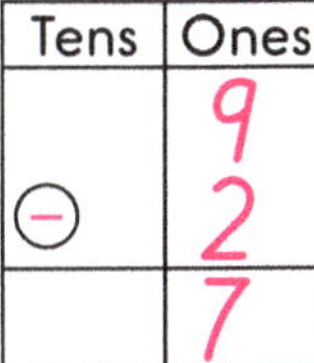

Tens	Ones
	9
−	2
	7

Matt has 23 rocks.
Tim has 15 rocks.
How many rocks do they have in all?

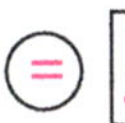

23 + 15 = 38 rocks

Tens	Ones
2	3
+ 1	5
3	8

100 Worktext pages 193–94

STEM

Objectives

- **100.1** Follow cardinal directions on a grid.
- **100.2** Debug an algorithm.
- **100.3** Follow the Engineering Design Process to solve a problem.
- **100.4** Compose an algorithm that uses cardinal directions.
- **100.5** Program a friend to help him or her find a destination. BWS

Biblical Worldview Shaping

- **Caring** (apply): Encourage students to recognize and even seek out opportunities to help others (101.5).

Printed Resources

- Instructional Aid 5: *STEM Engineering Design Process*
- Instructional Aid 28: *Compass Rose* (for the teacher and for each group)
- Instructional Aid 29: *Trampoline Park Destination* (for the teacher and for each group)
- Instructional Aid 30: *Coding Rubric: Guiding a Friend*

Materials

- 4 blank sheets of paper (same paper type/size/color as that used for the *Compass Rose* and *Trampoline Park Destination* pages; for the teacher and for each group)
- 3 × 5 card (for each student)

Preparation

Copy the *Compass Rose* page onto the same type of paper as the blank sheets. Make a copy of the *Trampoline Park Destination* page and cut out the image along the light gray frame border. Then copy only the light gray image onto the same type of paper as the blank sheets so the image is concealed and the page looks like the blank sheets when the page is face-down.

Clear spaces on the floor where the groups can position their 6 sheets of paper in grids that are 2 rows of 3 sheets.

Debug. Cross out the mistake.

Move east →.
~~Move east →.~~
Move south ↓.

~~Move west ←.~~
Move north ↑.

Move east →.
Move north ↑.
Move east →.
~~Move north ↑.~~

Engage

Engineering Design Process

Sing a **song** to help the students recall the steps of the Engineering Design Process.

Display the *STEM Engineering Design Process* page. Guide the students as they sing "STEM Engineering Design Process Song" to the tune of "Are You Sleeping?" Point out the steps of the Engineering Design Process.

I solve problems. I solve problems.
Yes, I do! Yes, I do!
I ask myself a question,
then think about the answer.
I plan to solve a problem,
then make a skillful model.
I test my model first,
and then I make it better.
I'm an engineer! Learning STEM this year!

Instruct

Cardinal Directions

Use a **visual display** to introduce following cardinal directions on a grid.

Draw a 2 × 4 grid for display. Draw a simple compass rose in the first cell of the second row and point out the names and locations of the cardinal directions on the rose.

Add a star to the last cell of the top row. Explain that the compass rose marks the cell where they will start, and the star marks their ending point or destination.

Ask. Think. Plan. Make. Test. Make Better.

Mark the tasks you completed.

- ○ I built a grid.
- ○ I planned an algorithm.
- ○ I programmed a friend.
- ○ I followed an algorithm.

Write the words to complete the sentence.

I will guide ______________________________
[friend's name]

to find ______________________________ [goal].

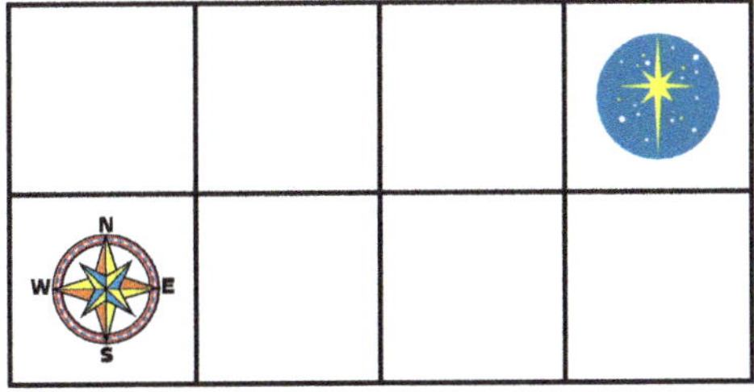

What compass directions could you give to guide me to the star? sample answer: east, east, east, north

Write the directions the students share for display. Invite a student to show the path on your displayed grid by moving his or her finger from cell to cell in order.

What is a list of steps that helps me finish my task called? an algorithm

Choose a student to write the algorithm in code. sample answers: →→→↑, EEEN

What is an algorithm called that has been written in code so that someone or something, such as a machine, can follow it? a program

Point out repeating symbols in the program and invite a student to write the program in a shorter way. sample answer: 3→ ↑

What do you call a repeated step in an algorithm or program? a loop

Debugging

Guide an **interactive activity** to debug an algorithm. Draw a 2 × 3 grid for display with the following layout and display the algorithm near it.

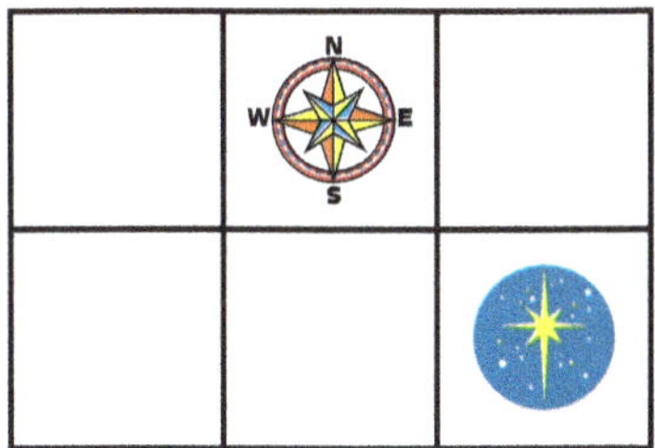

Move east →.
~~Move east →.~~
Move south ↓.

Invite a student to follow the algorithm on your displayed grid.

Does the algorithm work? no

What is wrong with it? It does not take me to the star; it takes me off the grid.

How could you fix it? I could leave out one of the "Move east →" steps.

Cross out the wrong step in the algorithm.

What is finding and fixing the mistakes in an algorithm or program called? debugging

Move the star to the cell below the compass rose and write a new algorithm.

Move east →.
Move south ↓.
Move west ←.
~~Move west ←.~~

Guide a **think-pair-share** to debug the algorithm. Invite students to share their answers and to correct the algorithm on the display. Cross out a "Move west ←" step.

Coding Game

Introduce a **coding game** to help the students program a friend to follow an algorithm.

Note: The game could be played outside.

Explain that the game requires 3 players who perform different tasks in the game.

1. A **builder/debugger** builds a grid and silently debugs the programmer's algorithm.
2. A **programmer** plans an algorithm and signals it to a stepper.

3. A **stepper** follows the signals to move to the destination on a grid.

Explain that the goal of the game is for the students (programmers) to program (give directions to) a friend who is new to the area (stepper) to help him or her find a destination (the trampoline park). Invite the students to share their experiences at a trampoline park.

Tell the students that they will be switching places in the game to try each task, so they need to listen carefully to what each player does.

Cardinal Directions Signals

Guide a **coding activity** to help the students signal cardinal (compass) directions.

Explain that during the game, the programmer will guide the stepper by giving him or her signals for the direction he or she should move for each step of the programmer's algorithm.

Invite the students to give ideas for how they could signal the cardinal directions with their arms. Sample answer: I could raise my hands above my head for north, point to the right for east, point left for west, and point straight down with both hands to my sides for south.

Practice signaling the 4 directions together as a class.

Explain that the programmers also need a signal for *stop* to let the steppers know that they have reached their destination.

Invite the students to choose a way of signaling *stop*. Sample answer: I could cross my arms above my head.

The following signals are shown as they would be viewed from behind the programmer.

Signal an Algorithm

Direct a **demonstration** to prepare the students to play the coding game.

Position the students so they can observe the demonstration. Invite a student to be a **builder** and help you lay out the 6 game pages on the floor in a 2 × 3 grid, according to an arrangement of the builder's choice. The *Compass Rose* page should face up, and the other 5 pages should face down.

Direct attention to the position of the *Trampoline Park Destination* page. Emphasize that you are letting the students see where the destination page is placed during the explanation of the game, but that when they play their game, only the builder and the programmer should see its location. The **stepper** will turn away so that he or she cannot see where the destination page is located.

Turn the *Trampoline Park Destination* page face-down in the grid before the next step.

Explain that after the builder has prepared the grid on the floor, the **programmer** will plan an algorithm to direct the stepper from the compass rose to the destination by signaling the steps, using cardinal directions.

Encourage the programmer to plan an algorithm with as few steps as possible to make it easier to remember.

Ask volunteers to be the programmer and the stepper. Ask the programmer to plan his or her algorithm.

Explain that the programmer must have the steps of his or her algorithm in mind before the game begins. Allow the programmer to write his or her algorithm on a 3 × 5 card as a reminder, if needed; but remind the programmer that the stepper should not see the algorithm. Encourage the programmer to memorize the algorithm so he or she can signal it smoothly.

Instruct the stepper to stand on the compass rose facing north.

Instruct the programmer to move to the north of the grid, facing north with his or her back to the stepper.

Instruct the students observing to be debuggers who silently watch to see if the algorithm works or if it has bugs. Ask them to stay silent until they are asked for their help.

Guide the programmer as he or she signals the steps, allowing time for the stepper to follow the signals. When the programmer signals *stop*, instruct the stepper to pick up the page he or she is standing on to check and see if it is the destination.

If the stepper did not reach the destination, ask a debugger to suggest a change to the algorithm or to the stepper's performance of the steps.

Allow as many students to participate as time permits.

Engineering Design Process

Guide the students as they **solve** a problem by using the Engineering Design Process.

Invite the students to state the problem they are trying to solve. I (programmer) am programming (giving directions to) a friend who is new to the area (stepper) to help him or her find a destination (the trampoline park).

Organize the students in groups of a builder, a programmer, and a stepper. Distribute the grid pages to the builder in each group. Builders should have prepared *Compass Rose* and *Trampoline Park Destination* pages and 4 blank sheets of paper.

Instruct the steppers to turn away from the builders while the builders get the grids ready.

Remind the builders to leave the compass rose facing up. Instruct the programmers to watch their builders carefully to see where the destination page is placed.

Allow the programmers time to develop and memorize the algorithm. Remind the builders that they are also the debuggers to check the algorithm as it is signaled.

Program a Friend

Guide a **physical game** to help the students program a friend.

Guide the steppers to position themselves on the compass roses facing north. Instruct the programmers to position themselves to the north of the grid with their backs to the steppers.

Ask the programmers to slowly signal their algorithms one signal at a time to the steppers, allowing each stepper time to read each signal and move before their programmers give another signal.

Remind the builders/debuggers to watch carefully to see that the programmers' algorithms work to guide the steppers to the destination.

Assist the groups as needed. Remind the students to debug kindly.

Tailor the game to meet the students' abilities. You may instruct 1 team at a time to play the game as the other students watch.

After the students have had a turn at each task in the group, **discuss** how they could use an algorithm to guide someone to an actual destination.

Assign cardinal directions in the classroom. Ask a student to stand facing north. Explain that the class will program the student to reach the exit door of the classroom.

Use questions such as the following to guide the class as they develop the algorithm to program the student. Display the program as the students dictate.

Which direction should (name of student) move first?

How many steps should (name of student) take?

Which direction should (name of student) move next?

How many steps?

Continue until the student arrives at the destination.

Help a Friend Find a Destination

Guide a **discussion** to help the students apply a biblical worldview shaping truth.

How would you feel if you were new at school and did not know how to find something? I might be unsure or afraid and want help.

Read aloud Galatians 6:10.

How does God tell us to treat others? to do good to all people

Explain that helping someone find the way when he or she needs help is a way that the students can show that they care for that person. Point out that the Worktext page provides an opportunity to put their coding skills to work by helping someone find the way to a destination.

Apply

Worktext pages 193–94

Read the directions and guide the students as they debug the first algorithm on page 193.

Assist the students as they complete the page independently.

Read the directions and guide the students as they mark the tasks on page 194 that they completed during the activity.

Direct attention to the biblical worldview shaping statement and guide the students as they complete the sentence. Encourage them to think of someone they could show care for by using the skills they practiced during the lesson to guide them to a destination.

Assess

Rubric

Use the *Coding Rubric: Guiding a Friend* page to assess the project. The rubric may be customized to include your chosen criteria.

101 Worktext page 195 Reviews pages 187–88

Chapter 12 Test

Administer the Chapter 12 Test.

Cumulative Concept Review

Worktext page 195

Review the following concepts. Adapt instructions and activities and provide reteaching as needed to meet the specific needs of your students.

- Extending a shape pattern: *abab, abbabb, abcabc* (Lesson 38)
- Extending a number pattern: *abab, abbabb, abcabc* (Lesson 39)
- Extending a letter pattern: *abab, abbabb, abcabc* (Lesson 39)
- Determining the value of a set of dimes and pennies (Lesson 49)
- Comparing 2-digit numbers by using > and < (Lesson 24)

Retain a copy of Worktext page 196 for the discussion of the Chapter 13 essential question during Lessons 102 and 106.

Reviews pages 187–88

Use the Reviews pages to help students retain previously learned skills.

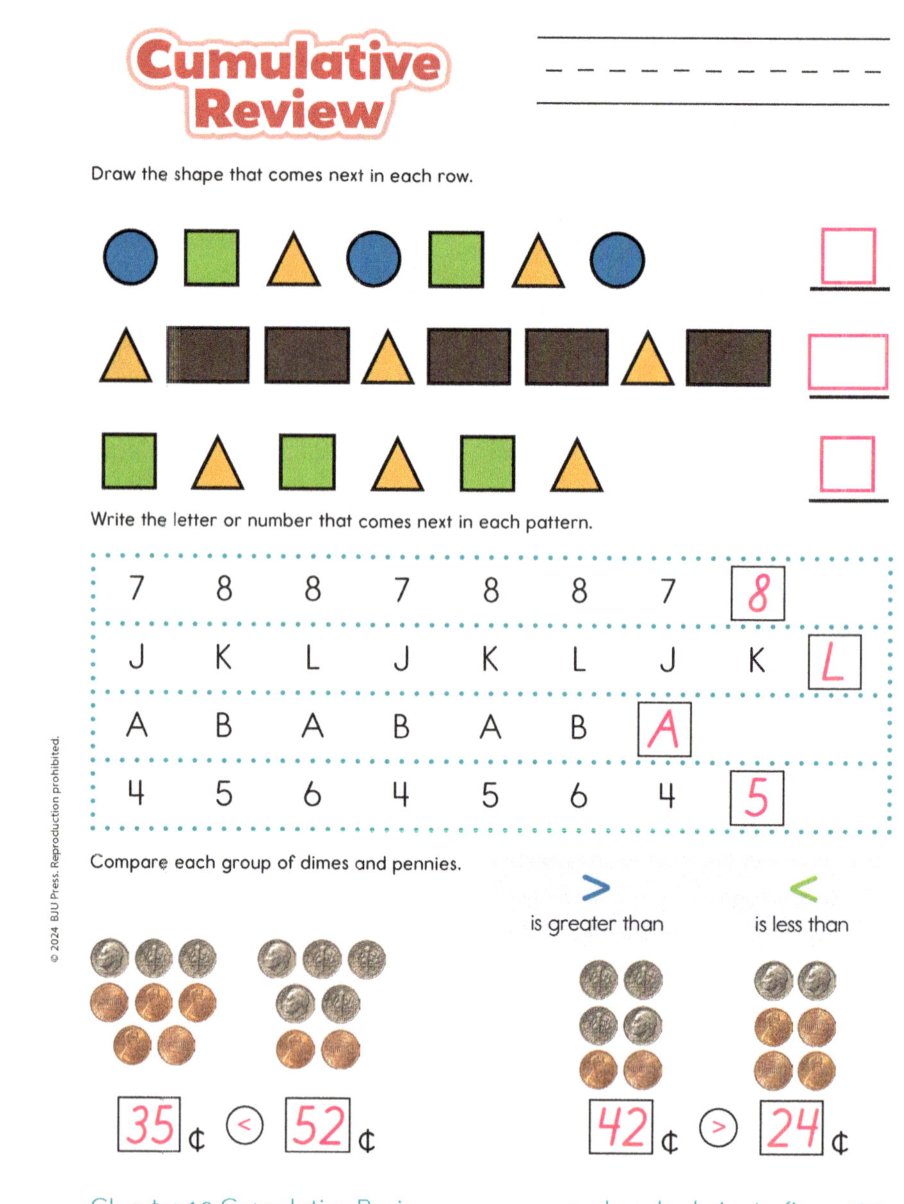

Cumulative Review

Write the number. Circle the single cubes to make pairs.
Circle *even* if the number is even.
Circle *odd* if the number is odd.

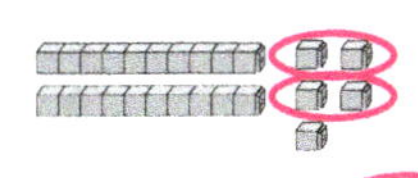

25

even odd

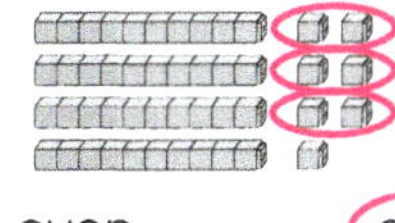

47

even odd

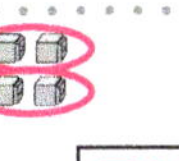

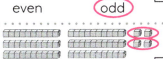

74

even odd

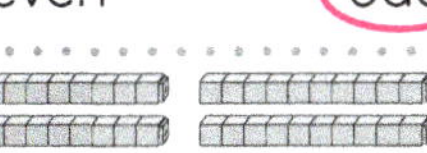

68

even odd

Write the time.

9:00 4:00 2:30 10:00 8:00

Write each fact family.

4 7 11

4 + 7 = 11
7 + 4 = 11
11 − 4 = 7
11 − 7 = 4

3 9 12

3 + 9 = 12
9 + 3 = 12
12 − 3 = 9
12 − 9 = 3

101

Addition Fact Review

Add.

$2 + 4 = 6$	$1 + 1 = 2$	$9 + 1 = 10$	$1 + 5 = 6$	$3 + 8 = 11$
$3 + 3 = 6$	$2 + 5 = 7$	$4 + 1 = 5$	$6 + 3 = 9$	$2 + 9 = 11$
$6 + 0 = 6$	$2 + 3 = 5$	$7 + 2 = 9$	$5 + 4 = 9$	$6 + 6 = 12$
$3 + 1 = 4$	$2 + 2 = 4$	$6 + 1 = 7$	$7 + 3 = 10$	$6 + 5 = 11$
$1 + 2 = 3$	$4 + 2 = 6$	$6 + 2 = 8$	$8 + 0 = 8$	$5 + 5 = 10$
$4 + 4 = 8$	$3 + 2 = 5$	$2 + 8 = 10$	$4 + 3 = 7$	$7 + 5 = 12$

CHAPTER 13: SUBTRACTION WITH TWO-DIGIT NUMBERS

PAGES	OBJECTIVES	RESOURCES	ASSESSMENTS
Lesson 102 Subtracting the Ones First			
Teacher Edition 435–39 **Worktext** 196–98	**102.1** Separate a set to subtract a 2-digit number by using representations. **102.2** Write a subtraction equation to solve a word problem. **102.3** Complete a 2-digit subtraction problem in vertical form. **102.4** Solve a word problem by using the Problem-Solving Plan. **102.5** Explain how people can use the Problem-Solving Plan to accomplish a task. BWS Working (explain)	**Visuals** • Visual 21: *Problem-Solving Plan* • Place Value Kit: tens, ones **Student Manipulatives Packet** • Tens/Ones Mat **BJU Press Trove*** • Video: Ch 13 Intro • Games/Enrichment: Fact Reviews (Subtraction Facts to 8) • Games/Enrichment: Subtraction Flashcards • Games/Enrichment: Fact Fun Activities • PowerPoint® presentation	**Reviews** • Pages 189–90
Lesson 103 Subtracting 2-Digit Numbers			
Teacher Edition 440–43 **Worktext** 199–200	**103.1** Separate a set to subtract a 2-digit number. **103.2** Write an addition or subtraction equation to solve a word problem. **103.3** Complete a 2-digit subtraction problem in vertical form. **103.4** Follow the Problem-Solving Plan to solve a word problem.	**Visuals** • Visual 21: *Problem-Solving Plan* • Place Value Kit: tens, ones **Student Manipulatives Packet** • Tens/Ones Mat **BJU Press Trove** • Video: Subtract 2 Digits • Games/Enrichment: Fact Reviews (Subtraction Facts to 8) • Games/Enrichment: Subtraction Flashcards • Games/Enrichment: Fact Fun Activities • PowerPoint® presentation	**Reviews** • Pages 191–92

*Digital resources for homeschool users are available on Homeschool Hub.

PAGES	OBJECTIVES	RESOURCES	ASSESSMENTS
Lesson 104 Subtracting Dimes & Pennies			
Teacher Edition 444–47 **Worktext** 201–2	**104.1** Separate a set of dimes and pennies to subtract a 2-digit number. **104.2** Compose a subtraction word problem. **104.3** Complete a 2-digit subtraction problem in vertical form. **104.4** Write a subtraction equation to solve a word problem.	**Visuals** • Visual 16: *12-Month Calendar* • Money Kit: 9 dimes, 9 pennies **Student Manipulatives Packet** • Tens/Ones Mat • Money Kit: 9 dimes, 9 pennies **BJU Press Trove** • Web Link: Virtual Manipulatives: Base Ten Money Pieces • Games/Enrichment: Fact Reviews (Subtraction Facts to 8) • Games/Enrichment: Subtraction Flashcards • Games/Enrichment: Fact Fun Activities • PowerPoint® presentation	**Reviews** • Pages 193–94
Lesson 105 Subtracting Money			
Teacher Edition 448–451 **Worktext** 203–4	**105.1** Separate a set of dimes and pennies to subtract a 2-digit number. **105.2** Complete a 2-digit subtraction problem in vertical form. **105.3** Choose the correct equation by reading a pictograph. **105.4** Write an addition or subtraction equation to solve a word problem. **105.5** Explain why a person should solve an equation even if the problem seems hard. BWS Working (formulate)	**Teacher Edition** • Instructional Aid 27: *Pictograph* **Visuals** • Money Kit: 9 dimes, 50 pennies **Student Manipulatives Packet** • Tens/Ones Mat • Money Kit: 9 dimes, 9 pennies **BJU Press Trove** • Web Link: Virtual Manipulatives: Base Ten Money Pieces • Games/Enrichment: Fact Reviews (Subtraction Facts to 8) • Games/Enrichment: Subtraction Flashcards • Games/Enrichment: Fact Fun Activities • PowerPoint® presentation	**Reviews** • Pages 195–96

PAGES	OBJECTIVES	RESOURCES	ASSESSMENTS
Lesson 106 Chapter 13 Review			
Teacher Edition 452–55 **Worktext** 205–6	**106.1** Recall concepts and terms from Chapter 13.	**Visuals** • Visual 21: *Problem-Solving Plan* **Student Manipulatives Packet** • Tens/Ones Mat **BJU Press Trove** • Games/Enrichment: Fact Fun Activities • Games/Enrichment: Subtraction Flashcards • PowerPoint® presentation	**Worktext** • Chapter 13 Review **Reviews** • Pages 197–98
Lesson 107 Test, Cumulative Review			
Teacher Edition 456–58 **Worktext** 207	**107.1** Demonstrate knowledge of concepts from Chapter 13 by taking the test.		**Assessments** • Chapter 13 Test **Worktext** • Cumulative Review **Reviews** • Pages 199–200

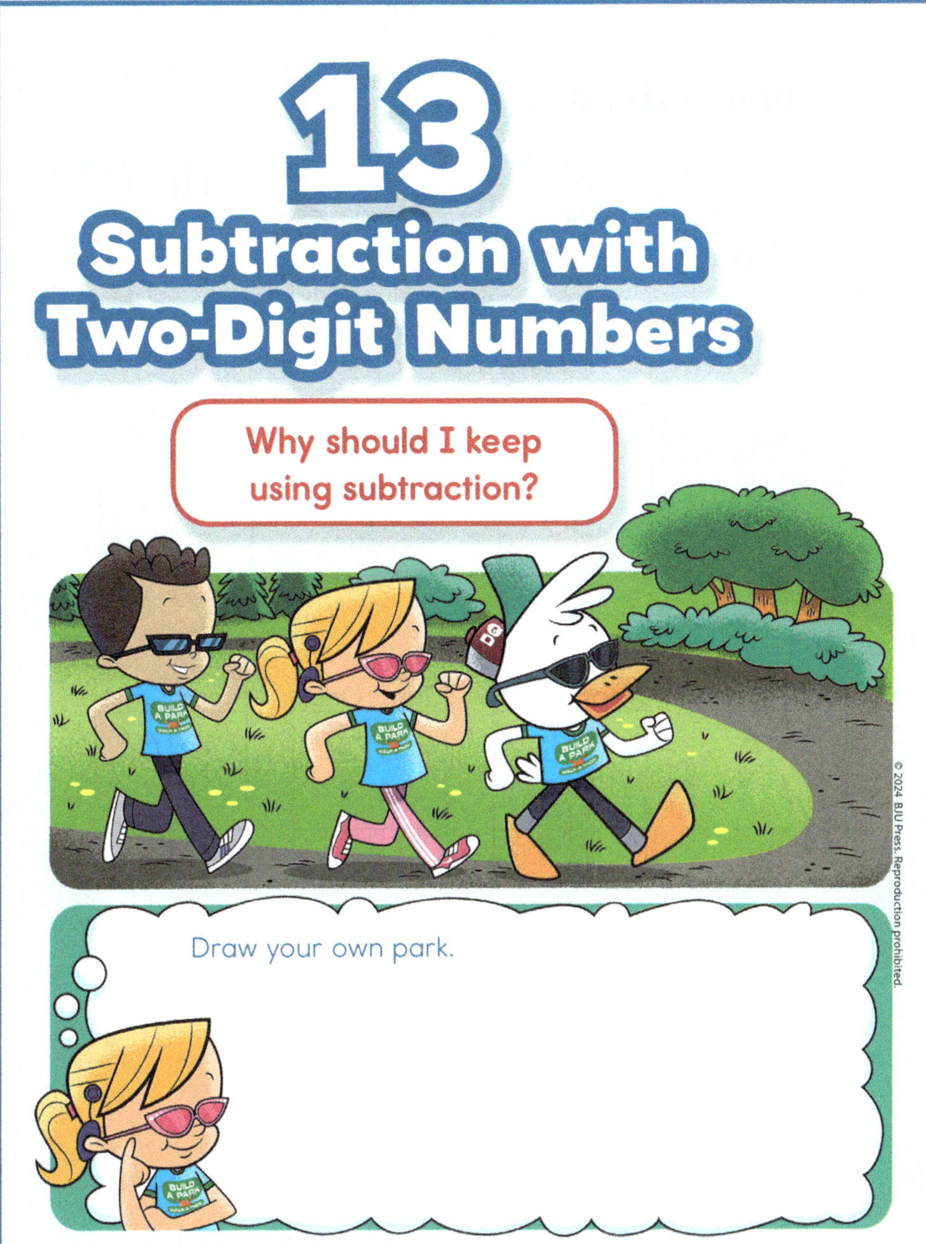
13
Subtraction with Two-Digit Numbers
Why should I keep using subtraction?
Draw your own park.

Chapter 13

Why should I keep using subtraction?

Chapter Objectives

- Separate sets to model 2-digit subtraction.
- Solve word problems by using the Problem-Solving Plan.
- Solve 2-digit subtraction problems in vertical form.
- Express a right view of using subtraction to do work.

102

Worktext pages 196–98
Reviews pages 189–90

Objectives

- **102.1** Separate a set to subtract a 2-digit number by using representations.
- **102.2** Write a subtraction equation to solve a word problem.
- **102.3** Complete a 2-digit subtraction problem in vertical form.
- **102.4** Solve a word problem by using the Problem-Solving Plan.
- **102.5** Explain how people can use the Problem-Solving Plan to accomplish a task. BWS

Biblical Worldview Shaping

- **Working** (explain): Real-world math problems are often stated in words. The Problem-Solving Plan helps us translate words to equations so we can solve the problems and complete the tasks (102.5).

Printed Resources

- Visual 21: *Problem-Solving Plan*
- Visuals: Place Value Kit (tens, ones)
- Student Manipulatives: Tens/Ones Mat

Digital Resources

- Video: Ch 13 Intro
- Games/Enrichment: Fact Reviews (Subtraction Facts to 8, page 41)
- Games/Enrichment: Subtraction Flashcards
- Games/Enrichment: Fact Fun Activities

Materials

- 50 UNIFIX® Cubes (or Place Value Kit: tens, ones; for each student)

Subtracting the Ones First

First: Subtract the Ones.
Next: Subtract the Tens.

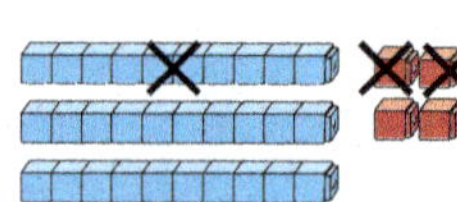

	Tens	Ones
	3	4
−	1	2
	2	2

Cross out the cubes to subtract. Write the answer.

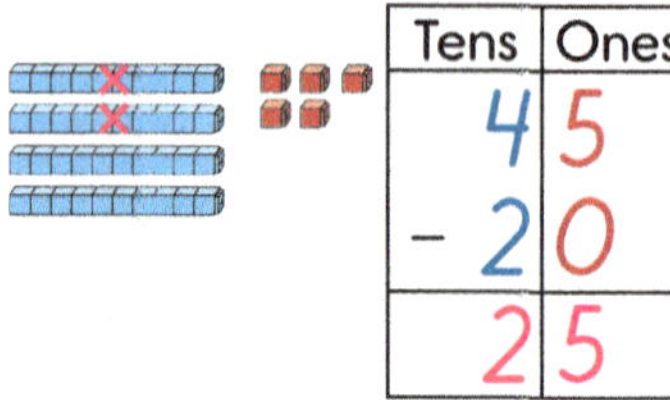

	Tens	Ones
	4	5
−	2	0
	2	5

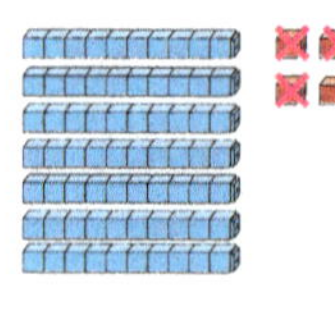

	Tens	Ones
	7	4
−		3
	7	1

Subtract.

	Tens	Ones
	5	6
−	1	3
	4	3

	Tens	Ones
	3	5
−	2	2
	1	3

	Tens	Ones
	8	5
−	6	3
	2	2

	Tens	Ones
	4	6
−	2	1
	2	5

Write an equation for the word problem.
Solve using the Tens/Ones frame.

There were 48 T-shirts made for the walkers. 36 walkers picked up their T-shirts. How many T-shirts are left for DQ to hand out?

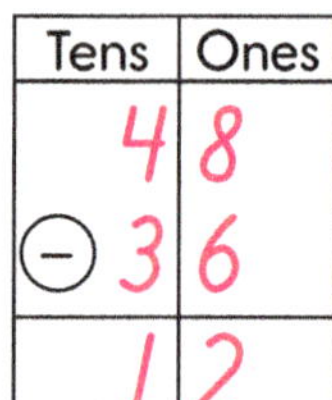

48 − 36 = 12 T-shirts

Direct the students to use their cubes to show the problem on their mats.

How did you show 32? 3 tens 2 ones

How did you show 15? 1 ten 5 ones

What will you add first? the ones

Direct the students to move all the cubes in the Ones place together on their mats.

How many ones are there in all? 7

Direct the students to move all the ten bars in the Tens place together.

How many tens are there in all? 4

What is the sum of the addition problem? 47

Ask a student to complete the problem for display. 47

Continue the activity, using the following problems.

$$\begin{array}{r} 24 \\ +\,15 \\ \hline 39 \end{array} \qquad \begin{array}{r} 12 \\ +\,31 \\ \hline 43 \end{array}$$

Practice & Review

Study Subtraction Facts

Choose a game or activity from the "Fact Fun Activities" available in Trove.

Join Sets to Add 2-Digit Numbers

Distribute a Tens/Ones Mat and a set of UNIFIX Cubes to each student. Write "32 + 15" for display.

Engage

Essential Question

Direct attention to the chapter opener on Worktext page 196 and **read aloud** the following scenario to introduce the chapter essential question.

"Don't forget your sunglasses and sunscreen," Liam's mom reminded him as he pulled his T-shirt over his head.

Cross out the cubes to subtract. Write the answer.

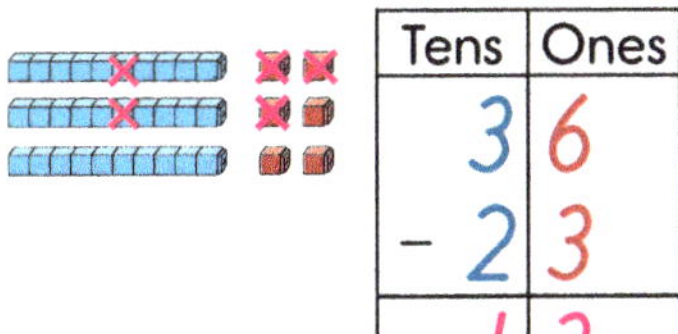

Tens	Ones
3	6
- 2	3
1	3

Tens	Ones
4	2
- 2	1
2	1

Subtract.

Tens	Ones
9	6
- 5	4
4	2

Tens	Ones
7	0
- 3	0
4	0

Tens	Ones
5	7
- 2	5
3	2

Tens	Ones
2	9
- 1	3
1	6

Write an equation for the word problem. Solve using the Tens/Ones frame.

On Monday, 33 people wore hats on their walk. That same day, 21 people wore sunglasses on their walk.
How many more people wore hats than sunglasses?

Tens	Ones
3	3
− 2	1
1	2

33 − 21 = 12 people

Trace the word to complete the sentence.

Solving a word problem helps me finish a task.

Time to Review

Add.

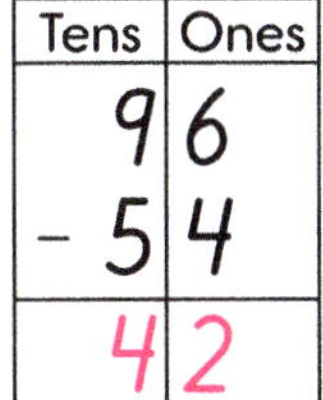

$41 + 18 = 59$ $63 + 24 = 87$ $16 + 13 = 29$ $52 + 25 = 77$ $76 + 21 = 97$

"I have them, Mom. Thanks for the reminder!" It was Saturday morning, and Liam was excited to be taking part in a walk-a-thon to help raise money for a community park.

Liam, Lily, and DQ joined others heading for the start of the walk, wearing their sunglasses and the matching shirts that DQ had helped hand out the day before.

Several local businesses had made donations, and the children had gotten pledges for each mile that they walked. The day, sunny and warm, promised to be a lot of fun.

"It will be so fun to help raise money for a park," said Lily. "We get to spend time with our friends too."

Liam agreed. "I'm hoping the new park will have a bike trail and a playground with climbing equipment."

"I'd like to have a water feature too," said DQ with a quack. He admired the shirts his friends were wearing. "It was a lot of fun handing out these T-shirts yesterday. Miss Markle helped me subtract to see how many shirts I had left to give out. Math really helped me get my work done."

A loud whistle called everyone to the starting line. DQ took his place between Liam and Lily, and soon they heard the starting signal. The walk had begun!

Invite a student to read aloud the essential question. Encourage the students to consider this question as you lead them to answer it by the end of the chapter.

Instruct

Separate a Set to Subtract a 2-Digit Number; Write a Subtraction Equation

Model separating a set to relate this skill to subtraction.

Distribute a Tens/Ones Mat and a set of UNIFIX Cubes to each student. Draw a Tens/Ones frame for display. Use your tens and ones to model the problem in the Tens/Ones frame while the students use their UNIFIX Cubes on their mats.

Display Visual 21 and read aloud the questions. Instruct the students to think about the information given in the word problem you will read so they can develop a plan to find the answer. Read aloud the following word problem several times.

Jenna took 38 cupcakes to a party at church. At the party, the children ate 25 cupcakes. How many cupcakes were not eaten at the party?

What is the question? How many cupcakes were not eaten?

What information is given? Jenna took 38 cupcakes, and the children ate 25.

Do you add or subtract to find the answer? subtract because the answer should be less than 38 since there are fewer cupcakes left, not more

What is the equation that represents this word problem? 38 – 25 = __

Write "38 – 25 = __" for display.

Direct the students to use their cubes to show 38 on their mats. Explain that when subtracting tens and ones, the cubes in the Ones place should always be subtracted first; then the tens can be subtracted. Instruct the students to remove 5 cubes from the Ones place.

Tens | Ones

Tens | Ones

Tens | Ones

How many ones are there now? 3

Explain that after subtracting the ones, the next step is to subtract the tens.

How many tens should you subtract? 2

Remove 2 ten bars from the Tens place.

How many tens are there now? 1

How many cubes remain after you subtracted 25 from 38? 13

How many cupcakes were not eaten? 13 cupcakes

Write "13 cupcakes" to complete the equation and read it aloud.

How is solving a subtraction problem different from solving an addition problem on your mat? For addition problems, both numbers are placed on the mat and the groups of cubes are joined. For subtraction problems, only the first number is placed on the mat and then cubes are removed from the mat.

	Tens	Ones
	2	5
–	1	3
	1	2

Continue the activity, telling word problems for 46 – 12 = 34 and 37 – 26 = 11.

Complete a 2-Digit Subtraction Problem

Demonstrate subtracting a 2-digit number.

Write "25 – 13" in a Tens/Ones frame.

Remind the students that when they subtract a 2-digit number, they should always begin by subtracting the digit in the Ones place.

Subtracting the Ones First

Cross out the cubes to subtract.
Write the answer.

First, subtract the Ones.
Next, subtract the Tens.

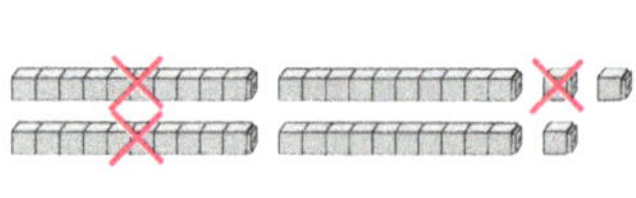

	Tens	Ones
	4	3
–	2	1
	2	2

Subtract.

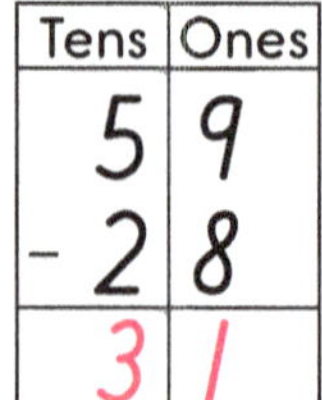

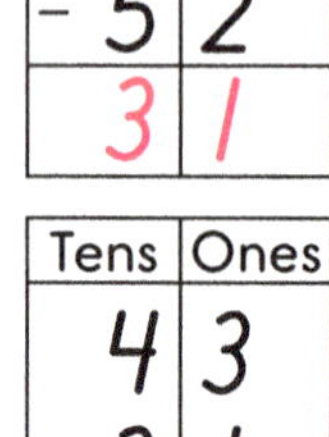

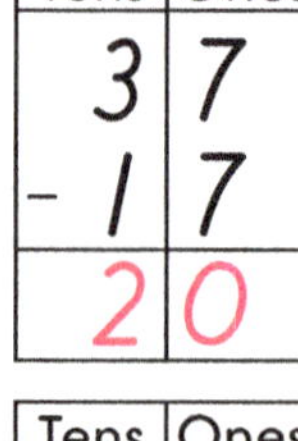

	Tens	Ones
	3	7
–	1	7
	2	0

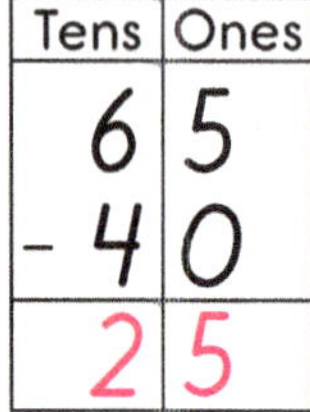

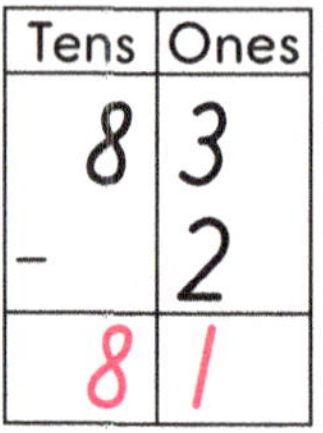

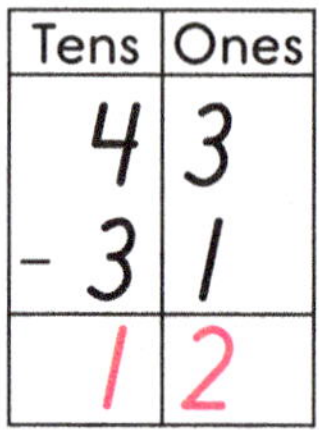

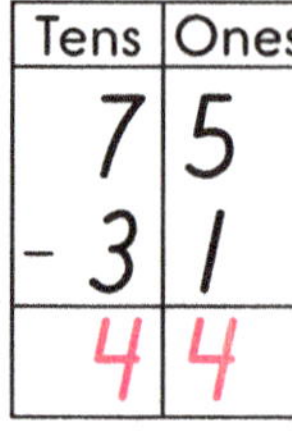

Write an equation for the word problem.
Solve using the Tens/Ones frame.

Jude had 35 bags of popcorn.
He gave 22 bags away.
How many bags of popcorn does Jude have left?

bags

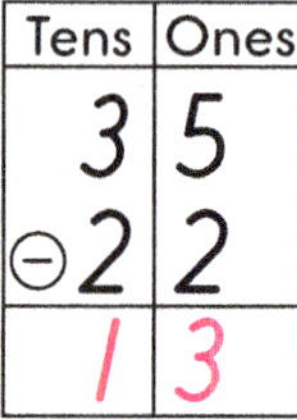

What is 5 ones minus 3 ones? 2 ones

Write "2" in the Ones place of the answer.

What is 2 tens minus 1 ten? 1 ten

Write "1" in the Tens place of the answer.

Guide the students as they read the problem while you point to each number and sign: *25 minus 13 equals 12.*

Continue the activity by writing each problem in a Tens/Ones frame and asking students to subtract.

Emphasize subtracting the digit in the Ones place before subtracting the digit in the Tens place.

75	52	38
– 11	– 32	– 14
64	20	24

Using the Problem-Solving Plan to Accomplish a Task

Guide a **discussion** to help the students explain a biblical worldview shaping truth.

Read aloud the following word problem.

There were 35 water bottles to set out on tables for the walk-a-thon. Lily set out the first 15 bottles. How many bottles were left? 20 bottles were left.

Guide the students as they determine the question, needed information, and equation.

Write "35 – 15" in a Tens/Ones frame for display. Guide the students as they solve the problem.

Remind the students that in this word problem Lily is accomplishing a task. She will find out how many bottles are left to set out.

Chapter 4 Review

Mark the number that matches the description.

3 in the Tens place	34 ●	43 ○
5 in the Ones place	56 ○	65 ●
7 in the Tens place	72 ●	27 ○
9 in the Tens place	19 ○	90 ●
2 tens, 5 ones	25 ●	52 ○
3 tens, 6 ones	63 ○	36 ●
7 tens, 4 ones	74 ●	47 ○

Subtraction Fact Review

Subtract.

$5 - 3 = 2$ $9 - 2 = 7$ $8 - 7 = 1$ $10 - 9 = 1$ $4 - 4 = 0$

$6 - 1 = 5$ $4 - 3 = 1$ $9 - 9 = 0$ $6 - 3 = 3$ $9 - 6 = 3$

$8 - 0 = 8$ $7 - 5 = 2$ $3 - 2 = 1$ $8 - 3 = 5$ $10 - 2 = 8$

What task is Lily accomplishing? setting out bottles of water

Explain that God made people to do work with math. Accomplishing a task such as bringing water is doing work. The Problem-Solving Plan is a way of thinking through a problem so that work can be done.

How can you use the Problem-Solving Plan to accomplish a task? I can use it to understand and solve a problem so that I can finish a task.

Apply

Worktext pages 197–98

Read and guide completion of page 197.

Read and explain the directions for page 198. Assist the students as they complete the page independently.

Direct attention to the biblical worldview shaping statement. Review the discussion in the lesson about serving in the walk-a-thon, and guide the students as they complete the sentence.

Assess

Reviews pages 189–90

Review matching a digit to its place on page 190.

Fact Reviews

Use "Subtraction Facts to 8" page 41 in Trove.

Differentiated Instruction

The following activity provides additional practice for the students experiencing difficulty with the concepts taught in this chapter.

Subtract a 2-Digit Number

Provide the students with UNIFIX Cubes, a Tens/Ones Mat, and a small paper star. Write "45 – 21" in vertical form for display. Instruct the students to use UNIFIX Cubes to represent the number 45. Remind them to subtract the Ones place first and then the Tens place. Direct them to place the star above the column that they should subtract first. Emphasize that the star will remind them to subtract the Ones place first.

103

Worktext pages 199–200
Reviews pages 191–92

Objectives

- **103.1** Separate a set to subtract a 2-digit number.
- **103.2** Write an addition or subtraction equation to solve a word problem.
- **103.3** Complete a 2-digit subtraction problem in vertical form.
- **103.4** Follow the Problem-Solving Plan to solve a word problem.

Printed Resources

- Visual 21: *Problem-Solving Plan*
- Visuals: Place Value Kit (tens, ones)
- Student Manipulatives: Tens/Ones Mat

Digital Resources

- Video: Subtract 2 Digits
- Games/Enrichment: Fact Reviews (Subtraction Facts to 8, page 42)
- Games/Enrichment: Subtraction Flashcards
- Games/Enrichment: Fact Fun Activities

Materials

- A large classroom calendar
- 50 UNIFIX® Cubes (or Place Value Kit: tens, ones; for each student)

Subtracting 2-Digit Numbers

Cross out the cubes to subtract. Write the answer.

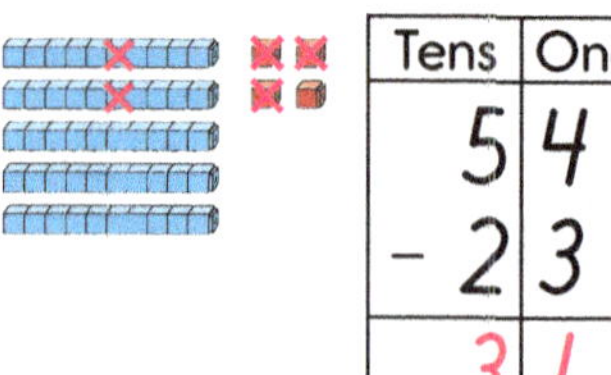

Tens	Ones
5	4
− 2	3
3	1

Tens	Ones
4	7
− 1	3
3	4

Subtract.

$$93 - 71 = 22 \qquad 53 - 42 = 11 \qquad 41 - 11 = 30 \qquad 57 - 31 = 26$$

Read the word problem.
Underline the question.
Circle the information.
Write an addition or subtraction equation. Solve.

Jade walked 23 laps.
DQ waddled 16 laps.
How many laps did they complete in all?

23 + 16 = 39 laps

Tens	Ones
2	3
+ 1	6
3	9

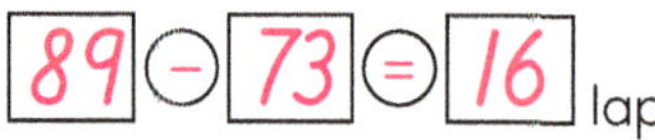

Gabe walked 89 laps.
Sam walked 73 laps.
How many more laps did Gabe walk than Sam?

89 − 73 = 16 laps

Tens	Ones
8	9
− 7	3
1	6

Practice & Review

Identify Days of the Week

Direct attention to the large calendar. Guide the students as they say the days of the week in order as you point to the days on the calendar. Ask the following questions.

What day of the week is the 21st of this month?

What day of the week is the 14th of this month?

How many Sundays are there in this month?

What day of the week is today?

What day of the week was yesterday?

What day of the week is tomorrow?

Study Subtraction Facts

Choose a game or activity from the "Fact Fun Activities" available in Trove.

Engage

Problem-Solving Plan

Review the Problem-Solving Plan to prepare the students to solve word problems.

Remind the students that the information on Visual 21 will help them solve word problems. Read aloud and discuss each question on the visual, using the following questions.

Why is it important to read the question in a word problem? It tells what I am trying to find out.

Why do you need to think about the information given in a word problem? I will use it to solve the problem.

How do you know whether to add or subtract? If I am joining 2 sets, I add; if I am separating objects from a set or comparing the number of objects in 2 sets, I subtract.

Why is it important that the answer make sense? If the answer doesn't make sense, I probably have made a mistake in solving the problem.

Instruct

Separate a Set to Subtract; Write an Equation for a Word Problem

Model subtraction by separating a set.

Subtract.

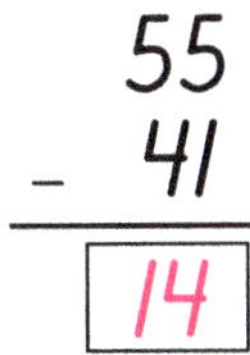

55 − 41 = 14	33 − 12 = 21	96 − 4 = 92	42 − 31 = 11
46 − 22 = 24	64 − 53 = 11	75 − 25 = 50	26 − 5 = 21

Time to Review

Color all the Fridays blue. Find December 15. Mark an *X* on it. Find the day after December 15. Mark a *T* on it. Find the day before December 15. Mark a *Y* on it.

Mark the correct answer.

1. How many Mondays are there in December? 3 ○ 4 ● 5 ○

2. What day is December 22? Monday ○ Tuesday ● Wednesday ○

Distribute a Tens/Ones Mat and a set of UNIFIX Cubes to each student. Use your tens and ones to model the problem in a Tens/Ones frame while the students use UNIFIX Cubes on their mats.

Read aloud the following word problem several times. Instruct the students to think about the information so that they can develop a plan to find the answer.

Rosa set out 48 snacks. Walkers ate 36 snacks. How many snacks are left?

What is the question? How many snacks are left?

What information is given? Rosa set out 48 snacks and walkers ate 36 snacks.

Do you add or subtract to find the answer? subtract

What equation would you use to solve the problem? 48 – 36 = __

Ask a student to write "48 – 36 = __" for display. Direct the students to represent the number 48 on their mats and subtract 6 as you demonstrate removing 6 cubes from the Ones place. Conclude together that 3 ten bars should be removed from the Tens place.

What is the answer to this problem? 1 ten 2 ones; 12 snacks

Does the answer make sense? yes

Write "48 – 36" in a Tens/Ones frame.

Invite a student to solve the problem. 12 Write "12 snacks" to complete the equation; read the equation aloud.

Read aloud the following word problem several times, allowing time for the students to think about the information and develop a plan for solving the word problem.

Sam walked for 15 minutes on Monday and for 30 minutes on Tuesday. How many minutes did Sam walk on those two days?

What is the question? How many minutes did Sam walk?

What information is given? Sam walked for 15 minutes on Monday and 30 minutes on Tuesday.

Do you add or subtract to find the answer? add

What equation would you use to solve the problem? 15 + 30 = __

Invite a student to write the equation "15 + 30 = __" for display. Direct the students to represent the addends 15 and 30 on their mats and find the sum.

What is the answer to this problem? 4 tens 5 ones; 45 minutes

Does the answer make sense? yes

Write "15 + 30" in the Tens/Ones frame and ask a student to solve the problem. 45 Write "45 minutes" to complete the equation; read the equation aloud.

Continue the activity, telling similar word problems for the following equations. 46 – 25 = 21, 39 – 12 = 27, and 16 + 32 = 48

2-Digit Subtraction in Vertical Form

Demonstrate 2-digit subtraction in vertical form.

Read aloud the following word problem.

The park is 28 miles from Ken's house. The library is 7 miles from his house. How much farther is the park from Ken's house than the library is?

What is the question? How much farther is the park from Ken's house than the library is?

Remind the students that this is a comparison problem.

Do you add or subtract to compare 2 numbers? subtract

What equation would you use to solve the problem? 28 – 7 = __

Write "28 – 7" for display in vertical form.

Explain that students will solve this problem without using the Tens/Ones frame or their Tens/Ones Mats.

103

Remind students that when they subtract a 2-digit number, they should always begin by subtracting the digit in the Ones place.

What is 8 ones minus 7 ones? 1 one

Write "1" in the Ones place of the answer.

What is 2 tens minus 0 tens? 2 tens

Write "2" in the Tens place.

Guide the students as they read the problem while you point to each number and sign: *28 minus 7 equals 21.*

Write the following problems and ask students to subtract. Emphasize subtracting the digit in the Ones place before subtracting the digit in the Tens place.

85	47	56	78
– 12	– 25	– 33	– 35
73	22	23	43

Apply

Worktext pages 199–200

Read and guide completion of page 199.

Read and explain the directions for page 200. Assist the students as they complete the page independently.

Assess

Reviews pages 191–92

Review writing the time and time equivalents on page 192.

Fact Reviews

Use "Subtraction Facts to 8" page 42 in Trove.

Extended Activity

Subtract a 2-Digit Number

Materials

- Visual 1: *Hundred Chart*
- Instructional Aid 2: *Hundred Chart* for each student
- 1 UNIFIX Cube for the teacher and for each student

Procedure

Display Visual 1. Distribute a *Hundred Chart* page and a UNIFIX Cube to each student. Write for display the following subtraction problems in Tens/Ones frames.

Subtracting 2-Digit Numbers

Cross out the cubes to subtract. Write the answer.

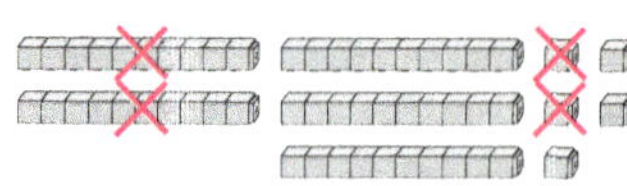

Tens	Ones
5	5
- 2	2
3	3

Subtract.

65	44	26	67
- 43	- 32	- 13	- 47
22	12	13	20

56	45	13	68
- 42	- 30	- 2	- 42
14	15	11	26

Write an equation for the word problem. Solve.

There were 36 trucks in the parking lot.
The boss sent 22 trucks out on jobs.
How many trucks are left in the parking lot?

Tens	Ones
3	6
⊖2	2
1	4

     trucks

67	34	49	56
– 22	– 12	– 30	– 36
45	22	19	20

Direct attention to the first problem.

What is the first number? 67

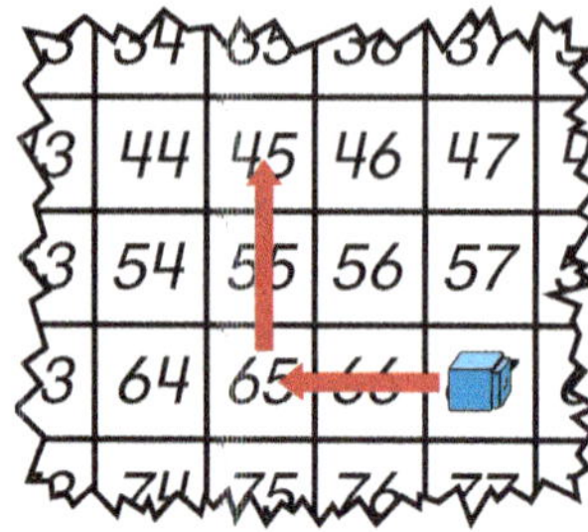

Place your cube on the 67 of Visual 1; direct the students to do the same on their charts.

Remind the students to subtract beginning with the digit in the Ones place. Demonstrate as you direct the students to subtract 2 ones by moving their cubes to the left 2 spaces.

What number is your cube on now? 65

Explain that 7 ones minus 2 ones is 5 ones and that the answer to the problem 67 – 32 will be in the chart column that has a 5 in the Ones place. Write "5" in the Ones place in the Tens/Ones frame.

Direct the students to subtract 2 tens by moving their cubes up 2 spaces.

What number is your cube on now? 45

Write "4" in the Tens place of the frame.

Continue the activity for each subtraction problem.

Chapter 8 Review

Write each time.

4:00

8:00

10:30

1:30

3:30

6:00

7:30

5:00

Complete each sentence.

1 hour = 60 minutes

a half hour = 30 minutes

Addition Fact Review

Add.

$6 + 2 = 8$ | $1 + 5 = 6$ | $6 + 4 = 10$ | $5 + 5 = 10$ | $9 + 3 = 12$

$8 + 1 = 9$ | $5 + 0 = 5$ | $7 + 5 = 12$ | $5 + 2 = 7$ | $8 + 4 = 12$

$2 + 9 = 11$ | $3 + 3 = 6$ | $6 + 5 = 11$ | $3 + 4 = 7$ | $7 + 3 = 10$

104

Worktext pages 201–2
Reviews pages 193–94

Objectives

- **104.1** Separate a set of dimes and pennies to subtract a 2-digit number.
- **104.2** Compose a subtraction word problem.
- **104.3** Complete a 2-digit subtraction problem in vertical form.
- **104.4** Write a subtraction equation to solve a word problem.

Printed Resources

- Visual 16: *12-Month Calendar*
- Visuals: Money Kit (9 dimes, 9 pennies)
- Student Manipulatives: Tens/Ones Mat
- Student Manipulatives: Money Kit (9 dimes, 9 pennies)

Digital Resources

- Web Link: Virtual Manipulatives: Base Ten Money Pieces
- Games/Enrichment: Fact Reviews (Subtraction Facts to 8, page 43)
- Games/Enrichment: Subtraction Flashcards
- Games/Enrichment: Fact Fun Activities

Lessons 104 and 105 use dimes and pennies to reinforce place value. Remember to include the cent sign in the Tens/Ones frames.

Subtracting Dimes & Pennies

Cross out the coins to subtract.
Write the answer.

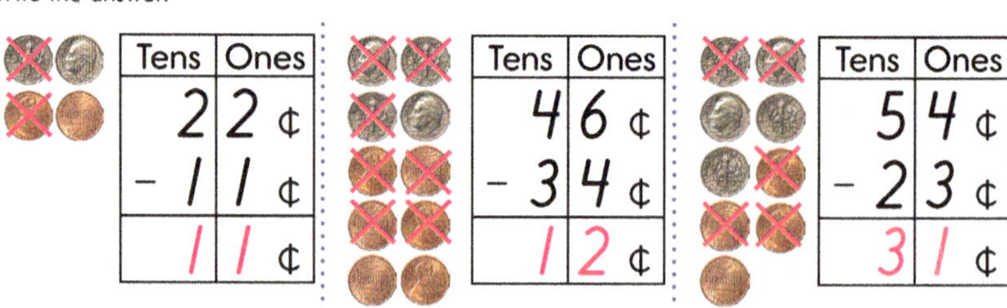

Tens	Ones	
2	2	¢
− 1	1	¢
1	1	¢

Tens	Ones	
4	6	¢
− 3	4	¢
1	2	¢

Tens	Ones	
5	4	¢
− 2	3	¢
3	1	¢

Subtract.

Tens	Ones	
4	5	¢
− 1	5	¢
3	0	¢

Tens	Ones	
7	4	¢
− 2	2	¢
5	2	¢

Tens	Ones	
6	9	¢
− 3	0	¢
3	9	¢

Tens	Ones	
3	5	¢
− 1	3	¢
2	2	¢

Tens	Ones	
5	3	¢
− 2	1	¢
3	2	¢

Tens	Ones	
6	6	¢
− 3	2	¢
3	4	¢

Tens	Ones	
4	4	¢
− 2	4	¢
2	0	¢

Tens	Ones	
5	5	¢
− 3	4	¢
2	1	¢

Write an equation for the word problem. Solve.

DQ had 36¢.
He spent 24¢ for ice to put in his drink.
How much money does DQ have left?

Tens	Ones	
3	6	¢
− 2	4	¢
1	2	¢

Practice & Review

Months of the Year

Display Visual 16. Guide the students as they say the months in order as you point to each one.

Choose groups of students to say the months of the year together.

Study Subtraction Facts

Choose a game or activity from the "Fact Fun Activities" available in Trove.

Engage

Add Money

Guide a **review** of addition with money to prepare the students for the content of the lesson.

Distribute a Tens/Ones Mat and set of coins to each student. Write "34¢ + 25¢" for display. Direct the students to show the problem on their mats.

How did you show 34¢? 3 dimes 4 pennies

How did you show 25¢? 2 dimes 5 pennies

What should you add first? the ones (pennies)

Instruct the students to move all the pennies in the Ones place together on the mats.

How many pennies are there in all? 9

Direct the students to move all the dimes in the Tens place together.

How many dimes are there in all? 5

What is the sum of the addition problem? 59¢

Ask a student to complete the problem and read it aloud.

Continue the activity for the following problems.

42¢	64¢
+ 36¢	+ 4¢
78¢	68¢

Instruct

Separate a Set to Subtract

Model subtraction to separate a set of money.

Use your dimes and pennies to demonstrate the problem in a Tens/Ones frame. Read aloud the following word problem.

Cross out the coins to subtract. Write the answer.

Tens	Ones
4	3 ¢
- 1	2 ¢
3	1 ¢

Tens	Ones
3	5 ¢
- 2	3 ¢
1	2 ¢

Subtract.

Tens	Ones
9	6 ¢
- 1	4 ¢
8	2 ¢

Tens	Ones
3	1 ¢
- 2	0 ¢
1	1 ¢

Tens	Ones
6	5 ¢
- 5	5 ¢
1	0 ¢

Tens	Ones
7	9 ¢
- 4	3 ¢
3	6 ¢

Tens	Ones
4	8 ¢
- 3	5 ¢
1	3 ¢

Tens	Ones
8	9 ¢
-	8 ¢
8	1 ¢

Tens	Ones
7	3 ¢
- 4	2 ¢
3	1 ¢

Tens	Ones
2	4 ¢
- 1	2 ¢
1	2 ¢

Time to Review

Number the months of the year in order.

2	February		March		December
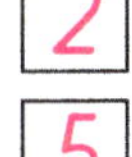	May		April		September
1	January		October		August
	June	7	July	11	November

Ryan had 87¢ when he went to the park to exercise. He bought a sports drink that cost 56¢. How much money did Ryan have left?

What is the question? How much money did Ryan have left?

What information is given? Ryan had 87¢, and he spent 56¢ on a drink.

Do you add or subtract to find the answer? subtract; because Ryan used part of his money to buy the drink

What equation would you use to solve the problem? 87¢ – 56¢ = __

Invite a student to write "87¢ – 56¢ = __" for display.

Direct the students to show 87¢ on their mats.

How many dimes and pennies are on your mat? 8 dimes 7 pennies

Explain that when subtracting tens and ones, the pennies in the Ones place should always be subtracted before the dimes in the Tens place. Direct the students to remove 6 pennies from the Ones place.

How many pennies are there now? 1

How many tens should you subtract? 5

Direct the students to remove 5 dimes from the Tens place.

How many dimes are there now? 3

What is the answer to this problem? 3 dimes 1 penny; 31¢

Write "31¢" to complete the equation and read it aloud.

Does the answer make sense? yes

How is solving a subtraction problem different from solving an addition problem on your mat? For addition problems, both numbers are placed on the mat and the groups of coins are joined. For subtraction problems, only the first number is placed on the mat and then coins are removed from the mat.

Compose a Subtraction Word Problem

Guide the students as they **model** a subtraction equation for a word problem.

Write "68¢ – 25¢ = __" for display. Ask several students to compose word problems for this equation.

Direct the students to use their dimes and pennies on their mats to find the answer. Ask a student to write the answer and to read the completed equation aloud. 43¢

Continue the activity for 54¢ – 13¢ = 41¢.

Complete a 2-Digit Subtraction Problem

Demonstrate subtracting 2-digit numbers in vertical form.

Write "47¢ – 32¢" in a Tens/Ones frame. Remind the students that when they subtract a 2-digit number, they should always begin by subtracting the digits in the Ones place.

What is 7 minus 2? 5

Write "5¢" in the Ones place of the answer.

What is 4 minus 3? 1

Write "1" in the Tens place of the answer.

Guide the students as they read the problem while you point to each number and sign: *47¢ minus 32¢ equals 15¢.*

Continue the activity by writing the following problems and inviting volunteers to subtract. Emphasize subtracting the digits in the Ones place before subtracting the digits in the Tens place.

95¢	67¢	56¢	78¢
– 12¢	– 20¢	– 46¢	– 4¢
83¢	47¢	10¢	74¢

Apply

Worktext pages 201–2

Read and guide completion of page 201.

Read and explain the directions for page 202. Assist the students as they complete the page independently.

104

Assess

Reviews pages 193–94

Review customary capacity on page 194.

Fact Reviews

Use "Subtraction Facts to 8" page 43 in Trove.

Extended Activity

Determine the Amount of Change from a Purchase

Materials

- General store items, school items, or snacks labeled with price tags less than $1.00
- Money Kits: 9 dimes and 9 pennies for a pair of students and 9 dimes and 9 pennies for the storekeeper

Procedure

Display the priced items on a table. Invite 1 student to be the storekeeper and 2 other students to be a team of shoppers.

Explain that the 2 shoppers will decide together which item to purchase and then will give the storekeeper 99¢ (9 dimes and 9 pennies) rather than the actual price on the item. Then 1 of the shoppers will write for display the subtraction problem needed to determine how much change they will receive from 99¢ from the storekeeper; the other shopper will solve the problem. After the other students decide whether the problem has been solved correctly, the storekeeper will give the shoppers the correct change.

Subtracting Dimes & Pennies

Cross out the coins to subtract.
Write the answer.

	Tens	Ones
	5	4¢
−	3	2¢
	2	2¢

Subtract.

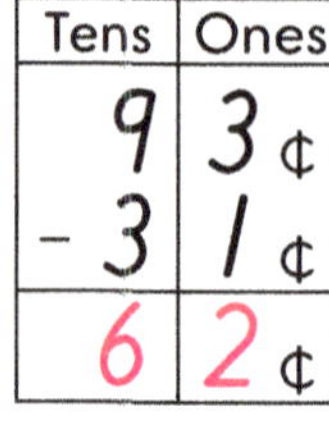

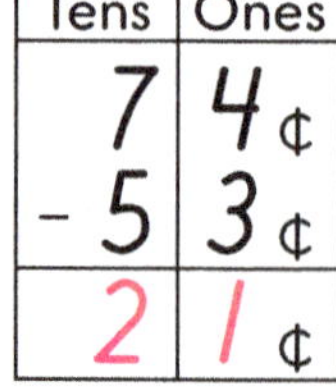

	Tens	Ones
	1	7¢
−		6¢
	1	1¢

	Tens	Ones
	9	3¢
−	3	1¢
	6	2¢

	Tens	Ones
	8	2¢
−	4	0¢
	4	2¢

	Tens	Ones
	7	4¢
−	5	3¢
	2	1¢

	Tens	Ones
	6	6¢
−	4	5¢
	2	1¢

	Tens	Ones
	9	8¢
−	7	6¢
	2	2¢

	Tens	Ones
	4	6¢
−	2	2¢
	2	4¢

	Tens	Ones
	2	7¢
−	1	4¢
	1	3¢

Write an equation for the word problem. Solve.

Hugo and Miles are going camping.
Hugo got a jug of water for 92¢.
Miles got a bag of ice for 71¢.
How much more did Hugo spend than Miles?

92 − 71 = 21¢

	Tens	Ones
	9	2¢
−	7	1¢
	2	1¢

Chapter 11 Review

Circle cups for each pint. Fill in each box.

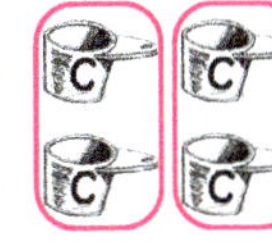
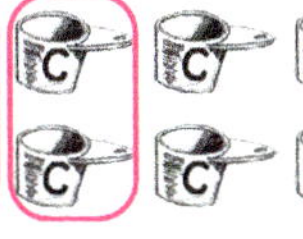

3 pints = 6 cups

Circle pints for each quart. Fill in each box.

2 quarts = 4 pints

Circle quarts for each gallon. Fill in each box.

1 gallon = 4 quarts

Subtraction Fact Review

Subtract.

8 - 6 = 2	9 - 6 = 3	7 - 7 = 0
7 - 1 = 6	7 - 2 = 5	9 - 4 = 5
5 - 4 = 1	9 - 3 = 6	6 - 3 = 3
9 - 7 = 2	8 - 2 = 6	9 - 0 = 9
8 - 4 = 4	9 - 5 = 4	11 - 2 = 9

105

Worktext pages 203–4
Reviews pages 195–96

Objectives

- **105.1** Separate a set of dimes and pennies to subtract a 2-digit number.
- **105.2** Complete a 2-digit subtraction problem in vertical form.
- **105.3** Choose the correct equation by reading a pictograph.
- **105.4** Write an addition or subtraction equation to solve a word problem.
- **105.5** Explain why a person should solve an equation even if it seems hard. BWS

Biblical Worldview Shaping

- **Working** (formulate): Any kind of work requires problem solving. Learning to persevere in solving math problems helps us develop perseverance in all areas of our life so that we can be faithful servants of God (105.5).

Printed Resources

- Instructional Aid 27: *Pictograph*
- Visuals: Money Kit (9 dimes, 50 pennies)
- Student Manipulatives: Tens/Ones Mat
- Student Manipulatives: Money Kit (9 dimes, 9 pennies)

Digital Resources

- Web Link: Virtual Manipulatives: Base Ten Money Pieces
- Games/Enrichment: Fact Reviews (Subtraction Facts to 8, page 44)
- Games/Enrichment: Subtraction Flashcards
- Games/Enrichment: Fact Fun Activities

You may place real pennies in a jar for the students to estimate and count instead of displaying pennies from the Visual Packet.

Practice & Review

Study Subtraction Facts

Choose a game or activity from the "Fact Fun Activities" available in Trove.

Subtracting Money

Cross out the coins to subtract. Write the answer.

Tens	Ones	
3	4	¢
− 2	2	¢
1	2	¢

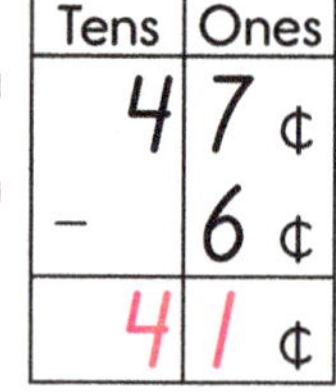

Tens	Ones	
4	7	¢
−	6	¢
4	1	¢

Subtract.

$78¢ - 15¢ = 63¢$ $86¢ - 72¢ = 14¢$ $68¢ - 54¢ = 14¢$ $88¢ - 6¢ = 22¢$

$96¢ - 25¢ = 71¢$ $54¢ - 13¢ = 41¢$ $82¢ - 50¢ = 32¢$ $53¢ - 32¢ = 21¢$

Write an equation for a word problem. Solve.

Dad spent 33¢ on nails and 45¢ on bolts. How much money did he spend in all?

(work space) 33¢ + 45¢ = 78¢

33 + 45 = 78 ¢

The red socks cost 96¢. The blue socks cost 82¢. How much more do the red socks cost?

(work space) 96¢ − 82¢ = 14¢

96 − 82 = 14 ¢

Complete Missing Addend Equations

Write "7 + __ = 10" for display. Remind the students that this is a missing addend equation: the sum is given, but 1 of the addends is missing. Display a set of 7 pennies.

Ask a student to add pennies to the set until there is a total of 10 pennies.

How many pennies did (name of student) add? 3

Write "3" to complete the missing addend equation.

Write "8 + __ = 12" and display a set of 8 pennies. Ask a student to add more pennies to the set until there is a total of 12 pennies.

How many pennies did (name of student) add? 4

Write "4" to complete the equation.

Continue the activity for 6 + 5 = 11, 4 + 4 = 8, 5 + 5 = 10. Invite students to complete the equations.

Point out that if the students know their facts, they can complete the missing addend equations without the use of manipulatives.

Engage

Estimate & Count Pennies

Guide an **activity** to help the students count pennies. Display 25 pennies. Invite the students to guess (without counting) how many pennies are displayed.

Are there more than 10 or fewer than 10?

Tom's Exercise Chart	
Monday	5 shoes
Tuesday	2 shoes
Wednesday	5 shoes
Thursday	3 shoes
Friday	2 shoes
Saturday	5 shoes

Key: 1 shoe = 5 laps

Use the key to read the graph. Mark the correct equation.

How many more laps did Tom run on Monday than on Tuesday?

- ● 25 – 10 = 15 laps
- ○ 25 + 10 = 35 laps

How many laps in all did Tom run on Thursday and Friday?

- ○ 15 – 10 = 5 laps
- ● 15 + 10 = 25 laps

Subtract.

37¢ – 10¢ = 27¢

85¢ – 44¢ = 41¢

28¢ – 17¢ = 11¢

49¢ – 25¢ = 24¢

Circle the phrase to complete the sentence.

I should (keep trying) / give up when subtraction seems hard.

Time to Review

Write the missing addend.

2 + 2 = 4 6 + 6 = 12 5 + 3 = 8

5 + 2 = 7 4 + 4 = 8 5 + 1 = 6

3 + 3 = 6 5 + 5 = 10 5 + 0 = 5

204 two hundred four Math 1

Are there more than 30 or fewer than 30?

Count the pennies together as you point to each one.

As time permits, add more pennies to the group and continue the activity.

Instruct

Separate a Set of Dimes & Pennies

Guide the students as they **model** subtracting money.

Distribute a Tens/Ones Mat and a set of coins to each student. Write "65¢ – 30¢" for display.

What do you subtract first when you are subtracting 2-digit numbers? the ones (pennies)

Direct the students to use their dimes and pennies to show the problem and find the answer.

What number did you make on your mat? 65¢

How many ones did you subtract? 0

How many pennies are there now? 5

How many tens did you subtract? 3

How many dimes are there now? 3

What is the answer to this subtraction problem? 35¢

Ask a student to complete the problem.

Continue the activity for the following problems.

46¢ – 32¢ = 14¢ 58¢ – 16¢ = 42¢ 84¢ – 31¢ = 53¢

Read a Pictograph; Write an Equation to Solve a Word Problem

Guide a **discussion** to help the students interpret a pictograph.

Display the *Pictograph* page. Explain that this pictograph shows how many stamps each person has in his or her stamp collection. Point to the stamp below the graph and explain that each stamp on the graph represents 5 stamps in a collection. Conclude together that the students will count by 5s to find out how many stamps each person has in his or her collection.

How many stamps does Ken have in his collection? 25

How many stamps does Meg have in her collection? 20

Would you add or subtract to find out how many fewer stamps Meg has than Ken? subtract

What equation would you write to find out how many fewer stamps Meg has than Ken? 25 – 20 = __

Write "25 – 20 = __ stamps" for display.

Remind the students that it is easier to solve a 2-digit subtraction problem when it is written in vertical form.

Ask a student to write the problem in vertical form and solve it. 5

Write the answer to complete the equation and then read it aloud: "25 – 20 = 5 stamps."

Direct attention to Meg's and Jill's stamp collections.

Would you add or subtract to find out how many stamps Meg and Jill would have if they put their collections together? add

What equation would you write to find the total number of stamps they would have? 20 + 30 = __

Write "20 + 30 = __ stamps" for display.

Invite a student to write the problem in vertical form and solve it. 50

Write the answer to complete the equation and then read it aloud: "20 + 30 = 50 stamps."

105

Continue the activity with the following problems.

How many more stamps does Ken have than Sam? 25 – 15 = 10 stamps

How many fewer stamps does Liam have than Sam? 15 – 15 = 0 stamps

How many stamps would Lily and Rosa have if they put their collections together? 5 + 35 = 40 stamps

Persevere in Solving an Equation

Share a **scenario** to help the students explain a biblical worldview shaping truth.

Read the following scenario aloud.

Lily and Liam were at school one day when they saw their friend Sam working on his math. He had a confused look on his face.

"I just can't seem to figure out this subtraction problem," Sam said. "I just want to quit and forget about this page."

"Sometimes we need to keep working even when subtraction seems hard," said Lily.

"That's right," agreed Liam. "You should keep working at it. You are probably close to finding the answer!"

"Let me look at it and see if I can help you understand," offered Lily.

Reinforce the idea that God made people to work with math. Explain that students should keep working at subtraction even when it seems hard. They should not lose courage and give up easily when work becomes difficult. Point out that school is the work that God has for students to do right now.

Why should you solve an equation even if it seems hard? I should keep trying so that I can do the work God gives me to do.

Apply

Worktext pages 203–4

Read and guide completion of page 203.

Read and explain the directions for page 204. Assist the students as they complete the page independently.

Direct attention to the biblical worldview shaping statement and guide the students as they complete the response.

Subtracting Money

Cross out the coins to subtract. Write the answer.

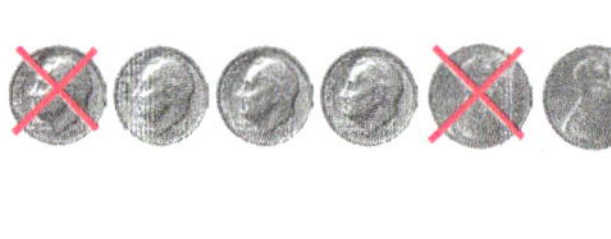

Tens	Ones
4	2¢
– 1	1¢
3	1¢

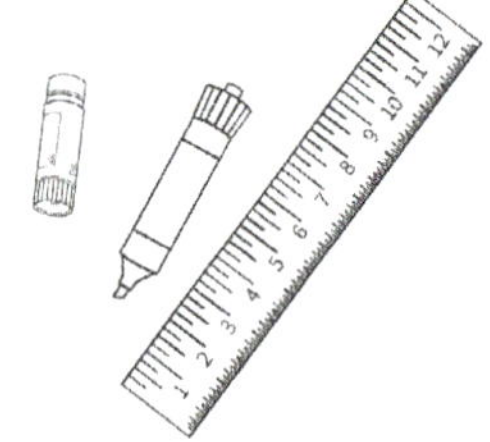

Subtract.

87¢ – 32¢ = 55¢	96¢ – 54¢ = 42¢	18¢ – 6¢ = 12¢	88¢ – 77¢ = 11¢
46¢ – 23¢ = 23¢	59¢ – 37¢ = 22¢	63¢ – 12¢ = 51¢	80¢ – 40¢ = 40¢

Write an equation for each word problem. Solve.

Lucy got a glue stick for 42¢ and a set of markers for 40¢. How much did Lucy spend in all?

work space

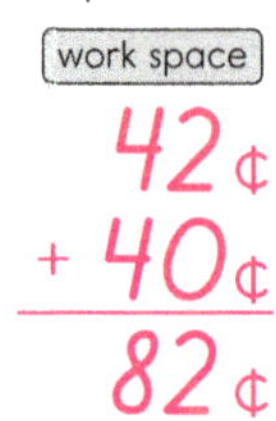

42¢ + 40¢ = 82¢

42 + 40 = 82¢

Finn had 75¢. He spent 43¢ on a ruler. How much money does Finn have left?

work space

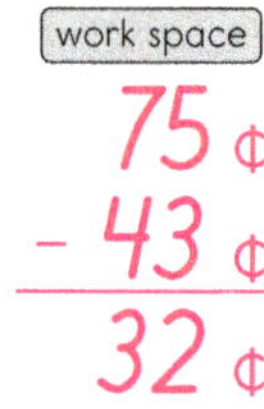

75¢ – 43¢ = 32¢

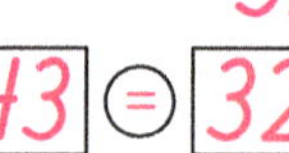

75 – 43 = 32¢

Chapter 13 • Lesson 105 one hundred ninety-five **195**

Assess

Reviews pages 195–96

Review measuring in inches on page 196.

Fact Reviews

Use "Subtraction Facts to 8" page 44 in Trove.

Extended Activity

Complete a "Magic" Square by Subtracting 2-Digit Numbers

Preparation

Draw a grid for display; write the 4 black numbers as shown. Do not write the answers in the grid.

78	34	44
55	23	32
23	11	12

Procedure

Distribute a sheet of paper to each student. Instruct the students to copy the grid on their papers. Explain that this is a special square that contains 6 different subtraction problems.

Point to the 2 numbers on the top row and direct the students to write and solve the problem 78 – 34 at the bottom of their papers and then to write the answer in the empty square on the top row. 44

Chapter 11 Review

Use an inch ruler to measure.

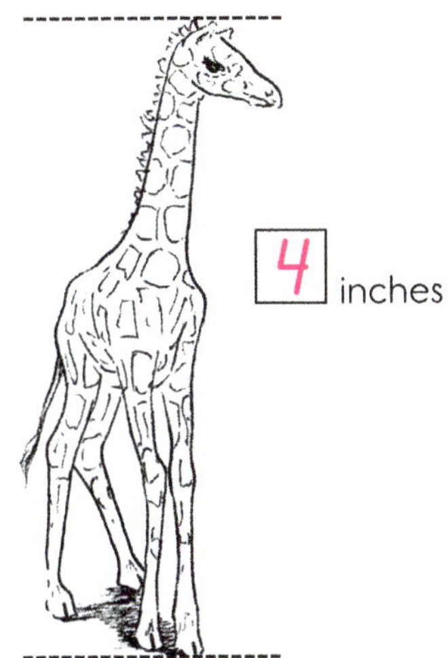

4 inches

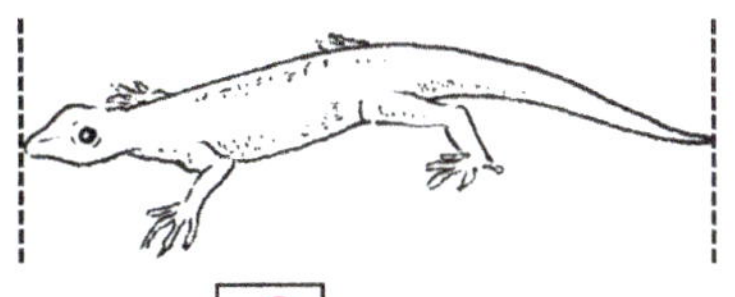

3 inches

2 inches

Addition Fact Review

Add.

9 + 1 = 10	4 + 2 = 6	9 + 2 = 11
7 + 0 = 7	3 + 3 = 6	3 + 2 = 5
6 + 3 = 9	8 + 4 = 12	3 + 9 = 12
5 + 4 = 9	5 + 5 = 10	7 + 2 = 9
5 + 6 = 11	5 + 2 = 7	7 + 5 = 12

Direct the students to write and solve the problem 55 – 23 and then to write their answers in the empty square on the second row. 32

Continue the activity for the problems in the first and second columns: 78 – 55 = 23 and 34 – 23 = 11.

Then direct the students to solve the problem in the third row (23 – 11 = 12) and the problem in the third column (44 – 32 = 12). Explain that they will know that all their answers are correct if the answer to the problem in the third row is the same as the answer to the problem in the third column.

106 Worktext pages 205–6 Reviews pages 197–98

Why should I keep using subtraction?

Chapter Concept Review

- Practice concepts from Chapter 13 to prepare for the test.

Printed Resources

- Visual 21: *Problem-Solving Plan*
- Student Manipulatives: Tens/Ones Mat

Digital Resources

- Games/Enrichment: Subtraction Flashcards
- Games/Enrichment: Fact Fun Activities

Materials

- 50 UNIFIX® Cubes (or Place Value Kit: tens, ones; for each student)

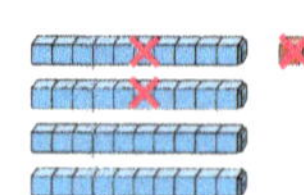

Chapter Review

Cross out cubes to subtract.
Write the answer.

Tens	Ones	
4	3	¢
− 2	1	¢
2	2	¢

Subtract.

Tens	Ones	
8	3	¢
− 4	1	¢
4	2	¢

Tens	Ones	
9	5	¢
− 7	2	¢
2	3	¢

Subtract.

$99 - 57 = 42$ $34 - 12 = 22$ $68 - 45 = 23$ $84 - 31 = 53$

Write an equation for each word problem. Solve.

DQ waddled 45 steps.
Jo walked 43 steps.
How many steps did they waddle and walk in all?

45 + 43 = 88 steps

(work space)
$45 + 43 = 88$

Sid passed out 64 water bottles.
Kevin passed out 33 water bottles.
How many more water bottles did Sid pass out than Kevin?

64 − 33 = 31 water bottles

(work space)
$64 - 33 = 31$

Practice & Review

Study Subtraction Facts

Choose a game or activity from the "Fact Fun Activities" available in Trove.

Count by 5s and 10s to 150

Ask a student to demonstrate a jumping jack. Guide the students as they count together by 10s to 150 as they do jumping jacks.

Invite a student to demonstrate touching his or her toes.

Guide the students as they count by 5s to 150 as they do toe touches.

Instruct

Separate a Set to Subtract a 2-Digit Number

Guide the students in a review game. Group the students into teams. Explain that teams will receive 1 point for each team member with the correct answer on his or her mat.

Distribute a Tens/Ones Mat and a set of UNIFIX Cubes to each student. Direct the students to show each problem by using their cubes on their mats. Write 1 of the following problems at a time for display.

$25 - 13 = 12$ $46 - 23 = 23$ $38 - 7 = 31$ $26 - 10 = 16$ $34 - 23 = 11$ $48 - 21 = 27$

Follow the Problem-Solving Plan

Display Visual 21 and read aloud the questions. Read aloud the following word problem several times.

Emily swam in her pool for 59 minutes on Friday. She swam for 45 minutes on Saturday. How many more minutes did Emily swim on Friday than she did on Saturday?

What is the question? How many minutes longer did Emily swim on Friday?

What information is given? She swam 59 minutes on Friday and 45 minutes on Saturday.

Do you add or subtract to find the answer? subtract

What is the equation you can use to find the answer? 59 – 45 = __

Write "59 – 45 = __" for display.

Should you rewrite the equation in vertical form? Yes; it will be easier to subtract in vertical form since the problem has 2-digit numbers.

Invite a student to write the equation in vertical form.

What do you subtract first when there are 2-digit numbers? the ones

Invite a student to solve the problem. 14

Does the answer make sense? yes

Write "14 minutes" to complete the equation.

Write an equation for each word problem. Solve.

Kelly has 75¢.
She bought a bottle of water for 50¢.
How much money does Kelly have left?

(work space)

$$\begin{array}{r} 75¢ \\ -\ 50¢ \\ \hline 25¢ \end{array}$$

75 − 50 = 25 ¢

Cross out the coins to subtract.
Write the answer.

Tens	Ones	
4	3	¢
− 2	2	¢
2	1	¢

Subtract.

Tens	Ones	
7	8	¢
− 1	5	¢
6	3	¢

Tens	Ones	
8	6	¢
− 7	6	¢
1	0	¢

Subtract.

54¢ − 23¢ = 31¢

48¢ − 20¢ = 28¢

42¢ − 31¢ = 11¢

67¢ − 35¢ = 32¢

78¢ − 53¢ = 25¢

97¢ − 24¢ = 73¢

89¢ − 46¢ = 43¢

76¢ − 30¢ = 46¢

Write the word to complete the sentence.

DQ subtracted T-shirts to do his work.

I should keep using subtraction to do the work God gives me.

Assess

Reviews pages 197–98

Use the Reviews pages to provide additional preparation for the chapter test.

Extended Activity

Determine the Amount Left after a Purchase

Materials

- A list of the following food items and prices:
  milk 42¢
  chips 31¢
  grapes 20¢
  hot dog 72¢
  nuts 23¢
- Money Kit: 9 dimes and 9 pennies for each student

Procedure

Explain that the students will have 99¢ to spend and may buy any of the items on the menu as long as they have enough money. Instruct the students to write the items they plan to purchase and the prices. Direct them to first add the prices of the items together to find the total cost and then to subtract the total cost from 99¢ to find out how much money they will have left over.

After the students have prepared the list and the total cost and money left over, give the students each 9 dimes and 9 pennies. Encourage them to count out the money needed for their purchases and to state the amount of money they have left. Check each problem for accuracy. If a problem is incorrect, encourage a student to ask a partner to help him or her find the error, and then allow the student to try to make his or her purchases again.

Continue the activity with the following word problems.

On Saturday Jack rode his exercise bike for 15 minutes in the morning and 30 minutes in the evening. How long did he ride his exercise bike on Saturday? 15 + 30 = 45 minutes

Josh has been lifting weights. This spring Josh can lift 95 pounds. Last fall Josh was able to lift 75 pounds. How many more pounds can Josh lift in the spring than he lifted in the fall? 95 − 75 = 20 pounds

Apply

Worktext pages 205–6

Read and explain the directions for the pages. Assist the students as they complete the pages independently.

Direct attention to the biblical worldview shaping statement on page 206. Display the picture from the chapter opener and remind the students that DQ helped subtract T-shirts to do his work.

Provide prompts as necessary to guide the students to the answer.

Why should I keep using subtraction?
I should keep using subtraction to do the work God gives me.

Chapter 13 Review

Cross out the cubes to subtract.
Write the answer.

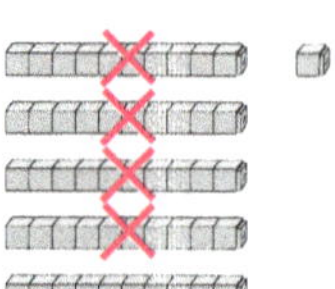

Tens	Ones
5	1
- 4	0
1	1

Subtract.

Tens	Ones
4	7¢
- 1	1¢
3	6¢

Tens	Ones
2	5¢
- 1	3¢
1	2¢

Subtract.

$49 - 36 = \boxed{13}$ $76 - 33 = \boxed{43}$ $64 - 52 = \boxed{12}$ $93 - 41 = \boxed{52}$

Write an equation for each word problem. Solve.

The train had 43 cars. Then 22 cars were taken off. How many cars does the train have now?

work space

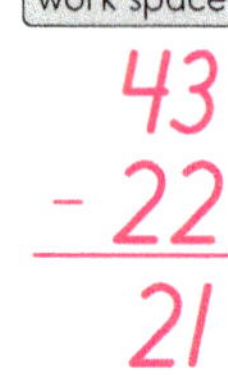

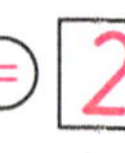

43 − 22 = 21 cars

There are 60 little boxes and 35 big boxes. How many boxes are there in all?

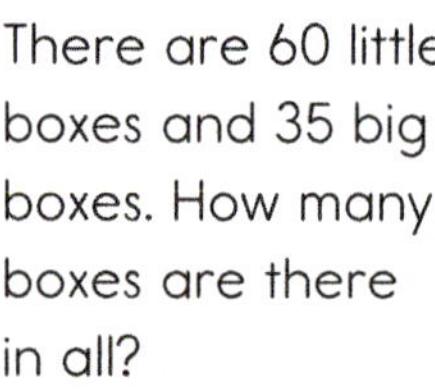

work space

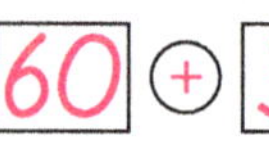

60 + 35 = 95 boxes

Cross out the coins to subtract.
Write the answer.

Tens	Ones
3	5¢
- 1	4¢
2	1¢

Subtract.

Tens	Ones
5	6¢
- 4	2¢
1	4¢

Tens	Ones
6	4¢
- 5	3¢
1	1¢

Subtract.

72¢ − 50¢ = 22¢

83¢ − 21¢ = 62¢

96¢ − 43¢ = 53¢

26¢ − 4¢ = 22¢

46¢ − 22¢ = 24¢

42¢ − 31¢ = 11¢

94¢ − 71¢ = 23¢

83¢ − 10¢ = 73¢

Write an equation for the word problem. Solve.

Ella got a ride on the train for 95¢. Ruby got a ride for 70¢. How much more did Ella spend than Ruby?

95 − 70 = 25¢

work space

95¢
− 70¢
25¢

107 Worktext page 207 Reviews pages 199–200

Chapter 13 Test

Administer the Chapter 13 Test.

Cumulative Concept Review

Worktext page 207

Review the following concepts. Adapt instructions and activities and provide reteaching as needed to meet the specific needs of your students.

- Counting by 5s and 10s to 100 (Lesson 27)
- Identifying even and odd numbers (Lesson 26)

Retain a copy of Worktext page 208 for the discussion of the Chapter 14 essential question during Lessons 108 and 112.

Reviews pages 199–200

Use the Reviews pages to help students retain previously learned skills.

Count by 5s. Write the missing numbers.

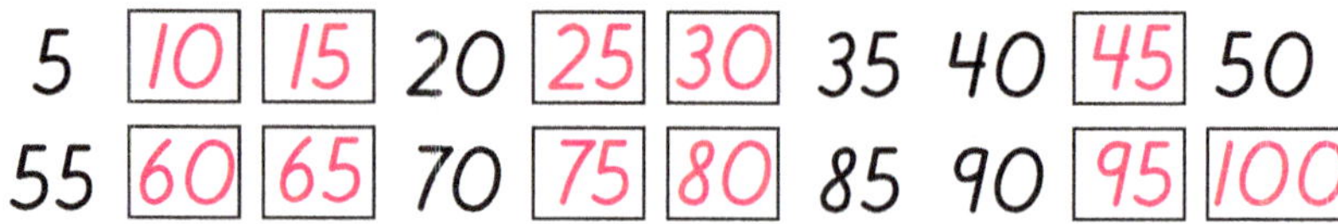

Count by 10s. Write the missing numbers.

Write the number. Circle the pairs of cubes.
Circle *even* if the number is even.
Circle *odd* if the number is odd.

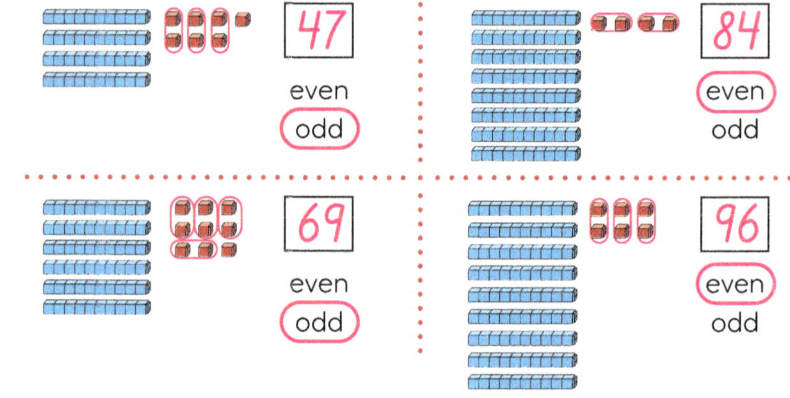

Add.

4	8	5	2	5
1	2	1	2	4
+ 3	+ 2	+ 6	+ 2	+ 3
8	12	12	6	12

Write each missing number.

60 

62   43 44 

   86 87 88

78 

80   

99 100   64 65 

Circle the third child in line.
Draw a line under the ninth child in line.
Draw a box around the fifth child in line.

107

Subtraction Fact Review

Subtract.

4 − 0 = 4	7 − 5 = 2	5 − 2 = 3	7 − 1 = 6	3 − 1 = 2
7 − 3 = 4	6 − 0 = 6	5 − 4 = 1	6 − 5 = 1	4 − 2 = 2
5 − 1 = 4	4 − 4 = 0	9 − 0 = 9	2 − 2 = 0	8 − 1 = 7
3 − 2 = 1	5 − 3 = 2	8 − 5 = 3	4 − 1 = 3	8 − 8 = 0
6 − 4 = 2	4 − 3 = 1	2 − 0 = 2	5 − 5 = 0	6 − 3 = 3
2 − 1 = 1	6 − 6 = 0	9 − 2 = 7	10 − 9 = 1	11 − 2 = 9

CHAPTER 14: *COUNT ON* WITH COINS

PAGES	OBJECTIVES	RESOURCES & MATERIALS	ASSESSMENTS
Lesson 108	**Counting Dimes, Nickels & Pennies**		
Teacher Edition 463–67 **Worktext** 208–10	**108.1** Identify the value of a penny, a nickel, and a dime. **108.2** Determine the value of a set of dimes and pennies and a set of nickels and pennies by *counting on*. **108.3** Determine whether there is enough money to purchase an item. BWS Working (explain)	**Teacher Edition** • Instructional Aid 9: *Price Tags* **Visuals** • Visual 14: *Money* • Money Kit: 7 pennies, 6 nickels, 6 dimes **Student Manipulatives Packet** • Number Cards: 0–9 • Red Mat • Money Kit: 4 pennies, 6 nickels, 4 dimes **BJU Press Trove*** • Video: Ch 14 Intro • Web Link: Virtual Manipulatives: Money Pieces • Web Link: U.S. Mint Coin Classroom • Games/Enrichment: Fact Reviews (Addition Facts to 12) • Games/Enrichment: Subtraction Flashcards • PowerPoint® presentation	**Reviews** • Pages 201–2
Lesson 109	**Solving Money Word Problems**		
Teacher Edition 468–71 **Worktext** 211–12	**109.1** Determine the value of a set of dimes and nickels and a set of dimes, nickels, and pennies by *counting on*. **109.2** Determine whether there is enough money to purchase an item. **109.3** Solve a money word problem by using the Problem-Solving Plan.	**Visuals** • Visual 21: *Problem-Solving Plan* • Money Kit: 7 pennies, 6 nickels, 7 dimes **Student Manipulatives Packet** • Number Cards: 0–9 • Red Mat • Money Kit: 6 pennies, 6 nickels, 6 dimes **BJU Press Trove** • Video: Nickels, Dimes, Pennies • Games/Enrichment: Fact Reviews (Addition Facts to 12) • Games/Enrichment: Subtraction Flashcards • PowerPoint® presentation	**Reviews** • Pages 203–4

*Digital resources for homeschool users are available on Homeschool Hub.

PAGES	OBJECTIVES	RESOURCES & MATERIALS	ASSESSMENTS
Lesson 110 **Money & Probability**			
Teacher Edition 472–75 **Worktext** 213–14	**110.1** Determine the value of a set of dimes, nickels, and pennies by *counting on*. **110.2** Identify equivalent amounts of money. **110.3** Predict the probability of choosing a dime, nickel, or penny. **110.4** Tally the results of a probability activity.	**Visuals** • Money Kit: 10 pennies, 7 nickels, 12 dimes **Student Manipulatives Packet** • Number Cards: 0–9 • Sign Cards: Greater than, Less than, Cent • Money Kit: 7 pennies, 7 nickels, 7 dimes • Red Mat **BJU Press Trove** • Games/Enrichment: Fact Reviews (Addition Facts to 12) • Games/Enrichment: Subtraction Flashcards • PowerPoint® presentation	**Reviews** • Pages 205–6
Lesson 111 **Quarters**			
Teacher Edition 476–79 **Worktext** 215–16	**111.1** Evaluate counting money in a careless way. BWS Working (evaluate) **111.2** Identify the value of a quarter. **111.3** Identify sets of pennies, nickels, and dimes equivalent to 1 quarter.	**Visuals** • Visual 14: *Money* • Money Kit: 25 pennies, 5 nickels, 6 dimes, 1 quarter **Student Manipulatives Packet** • Number Cards: 0–9 • Money Kit: 25 pennies, 5 nickels, 2 dimes • Red Mat **BJU Press Trove** • Video: Quarters • Web Link: Learning Coins • Web Link: U.S. Mint Coin Classroom • Games/Enrichment: Fact Reviews (Addition Facts to 12) • Games/Enrichment: Subtraction Flashcards • PowerPoint® presentation	**Reviews** • Pages 207–8

PAGES	OBJECTIVES	RESOURCES & MATERIALS	ASSESSMENTS
Lesson 112 **Chapter 14 Review**			
Teacher Edition 480–83 **Worktext** 217–18	**112.1** Recall concepts and terms from Chapter 14.	**Teacher Edition** • Instructional Aid 31: *Grocery Store Items* **Visuals** • Money Kit: 6 pennies, 6 nickels, 8 dimes, 1 quarter **Student Manipulatives Packet** • Tens/Ones Mat • Money Kit: 6 pennies, 6 nickels, 6 dimes, 1 quarter • Red Mat **BJU Press Trove** • Games/Enrichment: Fact Reviews (Addition Facts to 12) • Games/Enrichment: Subtraction Flashcards • Games/Enrichment: Fact Fun Activities • PowerPoint® presentation	**Worktext** • Chapter 14 Review **Reviews** • Pages 209–10
Lesson 113 **Test, Cumulative Review**			
Teacher Edition 484–86 **Worktext** 219	**113.1** Demonstrate knowledge of concepts from Chapter 14 by taking the test.		**Assessments** • Chapter 14 Test **Worktext** • Cumulative Review **Reviews** • Pages 211–12

14 Count on with Coins

What should I do if counting coins takes a lot of time?

DQ wants one more thing for water games. Circle the item he can afford.

Chapter 14

What should I do if counting coins takes a lot of time?

Chapter Objectives

- Identify individual coins and their value.
- Determine the value of a set of coins by *counting on*.
- Compare the value of a set of coins to a purchase price.
- Identify equivalent amounts of money.
- Solve a money word problem by using the Problem-Solving Plan.
- Predict the probability of choosing a certain coin.
- Evaluate attitudes toward counting coins.

108 Worktext pages 208–10
Reviews pages 201–2

Objectives

- **108.1** Identify the value of a penny, a nickel, and a dime.
- **108.2** Determine the value of a set of dimes and pennies and a set of nickels and pennies by *counting on*.
- **108.3** Determine whether there is enough money to purchase an item. BWS

Biblical Worldview Shaping

- **Working** (explain): Counting coins can be tedious, but it is important work. Counting carefully helps us know what we can buy (108.3).

Printed Resources

- Instructional Aid 9: *Price Tags*
- Visual 14: *Money*
- Visuals: Money Kit (7 pennies, 6 nickels, 6 dimes)
- Student Manipulatives: Number Cards (0–9)
- Student Manipulatives: Red Mat
- Student Manipulatives: Money Kit (4 pennies, 6 nickels, 4 dimes)

Digital Resources

- Video: Ch 14 Intro
- Web Link: Virtual Manipulatives: Money Pieces
- Web Link: U.S. Mint Coin Classroom
- Games/Enrichment: Fact Reviews (Addition Facts to 12, page 19)
- Games/Enrichment: Subtraction Flashcards

Materials

- Subtraction flashcards
- 3 general store items with price tags 30¢, 32¢, and 42¢

A general store provides teacher-guided experiences in applying money skills. Provide small toys, various empty food containers, and school items for the students to "purchase" during the lessons in this chapter as well as Chapter 19. Prices needed for this chapter are 27¢, 30¢, 32¢, 35¢, 42¢, 47¢, 70¢, 75¢, and 95¢.

Counting Dimes, Nickels & Pennies

Write the value of each coin.

 10 ¢
 5 ¢
 1 ¢

Write the value as you *count on*. Write the total.

10 ¢ 20 ¢ 21 ¢ 22 ¢

5 ¢ 10 ¢ 15 ¢ 20 ¢ 21 ¢

10 ¢ 20 ¢ 30 ¢ 31 ¢ 32 ¢

Practice & Review

Identify the Number of Tens & Ones by Using Dimes & Pennies

Display 4 dimes and 4 pennies in a Tens/Ones frame.

What is the value of 1 dime? 10¢

What is the value of 1 penny? 1¢

Guide the students as they count the dimes by 10s and the pennies by 1s.

What is the value of 4 dimes and 4 pennies? 44¢

Write "44¢" in the Tens/Ones frame.

How many tens are in 44? 4

How many ones are in 44? 4

Continue the activity for 57¢, 32¢, and 23¢.

Memorize Subtraction Facts

Introduce the following facts.

10 – 3   10 – 7   10 – 8

Distribute Number Cards 0–9. Practice the new subtraction facts and previously memorized facts as in earlier lessons.

How are 10 – 3 and 10 – 7 related? They belong to the same fact family.

108

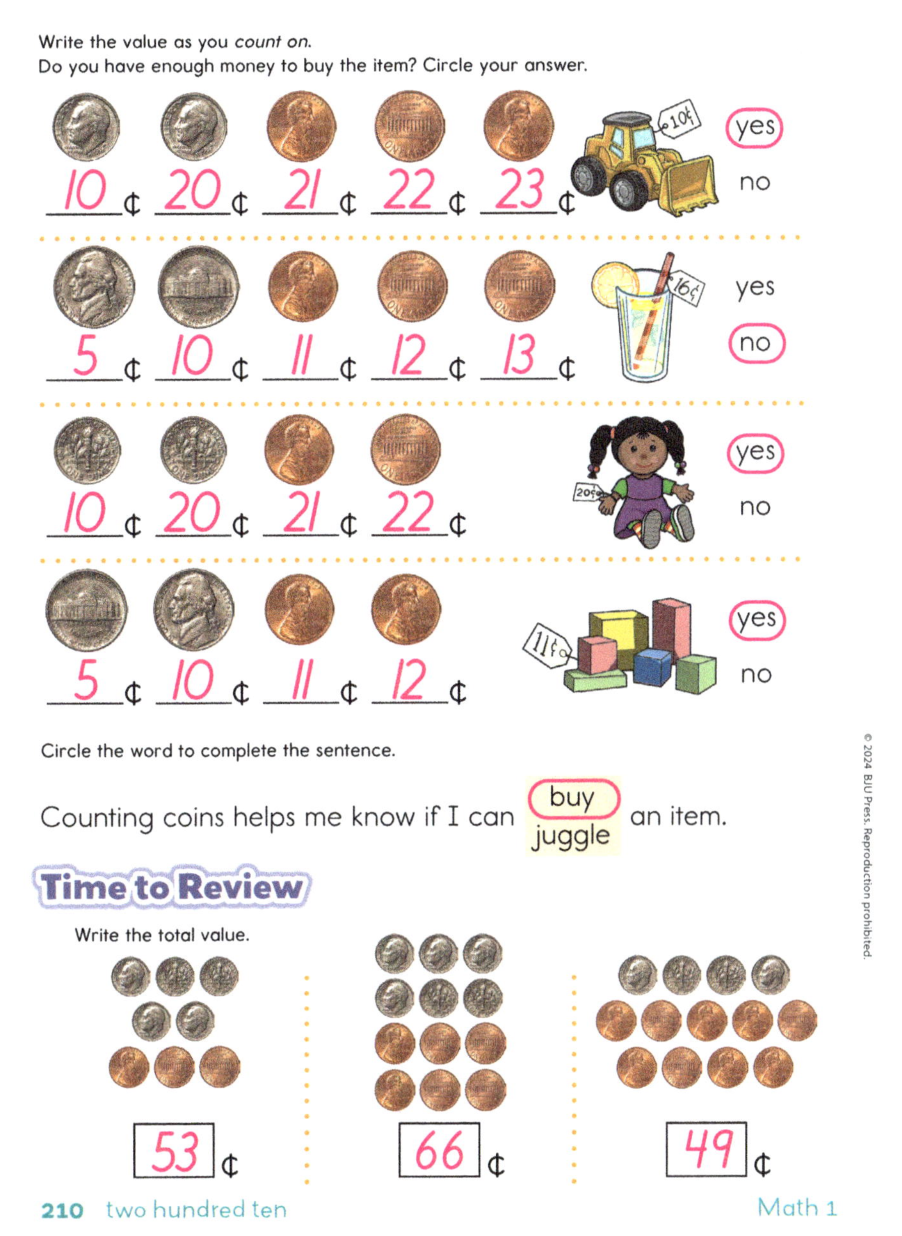
Write the value as you *count on*.
Do you have enough money to buy the item? Circle your answer.

10¢ 20¢ 21¢ 22¢ 23¢ — 10¢ — (yes) no

5¢ 10¢ 11¢ 12¢ 13¢ — 16¢ — yes (no)

10¢ 20¢ 21¢ 22¢ — 20¢ — (yes) no

5¢ 10¢ 11¢ 12¢ — 11¢ — (yes) no

Circle the word to complete the sentence.

Counting coins helps me know if I can (buy) / juggle an item.

Time to Review

Write the total value.

53¢ 66¢ 49¢

Engage

Essential Question

Direct attention to the chapter opener on Worktext page 208 and **read aloud** the following scenario to introduce the chapter essential question.

Oh, wow! thought DQ as he scanned the tables crowded with yard sale items. *I want that water blaster and these splash bombs. These will be so much fun this summer!* DQ's head feathers twitched with excitement as he thought about the water games ahead.

DQ collected his items and carried them to the line at the payment table where his friend Remy sat counting change. "Thank you, Mrs. Hunt. I hope you enjoy the whisk. My mother has a lot of those. I'm glad she's selling a few." Remy grinned as he plinked the coins into the metal money box. It was fun to see *and hear* the collection of coins growing with each purchase.

DQ placed his items and a pile of coins on the table. "We can have a great water fight this summer, Remy!"

Remy chuckled at the thought of clobbering his friend with one of the soft, soaked bombs. He had been working since early in the morning, and DQ's visit added some cheer to the work of counting coins.

Remy checked the price tags on the items. "The water blaster is 65¢, and the splash bombs cost 20¢." Remy wrote a vertical equation for 65¢ + 20¢ and added the cost of the items. "Your total is 85¢."

Remy arranged the coins DQ had given him and counted them aloud, starting with the dimes. "10, 20, 30, 40, 50, 60 (deep breath), 65, 70, 75, 80 (deep breath), 81, 82, 83, 84, 85¢. You have just enough, DQ. Whew! Counting money takes time!" Remy sighed. It was 11:00, and he was tired of counting coins. He wished he could leave and play with DQ.

DQ was secretly thankful that he wasn't the one counting the coins. "See you later, Remy!" DQ waddled off with the water toys.

The next customer stepped up to the table. This time Remy quickly glanced at the price of the item and the coins that were paid for it without actually counting the coins to be sure they were the right amount. *Oh well. It looks about right.* Remy felt a little stab of guilt as he smiled and took the money. *What could it hurt?* he thought. *It's not that much money. It's too much trouble to count every single coin every single time.*

Invite a student to read aloud the essential question. Encourage the students to consider this question as you guide them to answer it by the end of the chapter.

Instruct

Value of Coins

Guide a **visual analysis** to help the students identify the value of different coins.

Point to the penny on Visual 14.

How do you know that this coin is a penny? the copper/brown color, the smooth edge, the profile of Lincoln on the front, the words *one cent*

What is the value of a penny? 1¢

How do you count pennies? by 1s

Point to the cent sign on the visual.

What does this symbol stand for? cents

Display 4 pennies in a row. Count the pennies together to determine their total value: 1, 2, 3, 4¢.

Ask a student to point to the nickel on the visual.

108

How do you know that this coin is a nickel? the silver color, larger than a dime and a penny, the smooth edge, the profile of Jefferson on the front, the words *five cents*

What is the value of a nickel? 5¢

How do you count nickels? by 5s

Display 5 nickels in a row. Count the nickels together to determine their total value. 5, 10, 15, 20, 25¢

Invite a student to point to the dime on the visual.

How do you know that this coin is a dime? the silver color, the smallest in size, the ridges on the edge, the profile of Roosevelt on the front, the words *one dime*

What is the value of a dime? 10¢

How do you count dimes? by 10s

Display 6 dimes in a row. Count the dimes together to determine their total value. 10, 20, 30, 40, 50, 60¢

Value of a Set of Dimes & Pennies

Guide a **counting activity** to help the students determine the value of a set of coins by *counting on*.

Display 5 dimes and 3 pennies in random order.

What should you do first before you count the value of this set of pennies and dimes? separate the dimes and pennies

Which coins should you count first? the dimes because I count coins in the order of their value, from greatest to least; the dimes have the greater value

Place the coins in a row with the dimes to the left.

How do you count a set of dimes and pennies? begin by counting the dimes by 10s, take a deep breath, and then *count on* the pennies by 1s

Why do you take a deep breath after you count the dimes and before you begin counting the pennies? to remind myself to change the way I am counting

Count the coins together to determine their total value. 10, 20, 30, 40, 50, (deep breath), 51, 52, 53¢

Counting Dimes, Nickels & Pennies

Color green the apples with the value of a penny.
Color red the apples with the value of a nickel.
Color yellow the apples with the value of a dime.

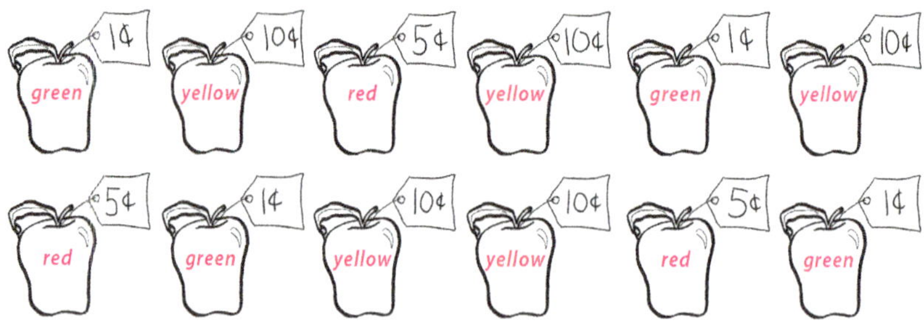

Write the value of Jason's coins as you *count on*. Does he have enough money to buy each fruit? Circle *yes* or *no*.

Continue the activity with a set of 6 dimes and 7 pennies and a set of 5 dimes and 5 pennies. 67¢; 55¢

Value of a Set of Nickels & Pennies

Guide a **counting activity** to help the students determine the value of a set of coins by *counting on*.

Display 3 nickels and 4 pennies in random order.

What should you do first? separate the nickels and pennies

Which coins should you count first? the nickels, because the nickels have the greater value

How do you count the nickels? by 5s

How do you count the pennies? by 1s

Count the coins together to determine their total value: 5, 10, 15, (deep breath), 16, 17, 18, 19¢.

What is the value of the coins? 19¢

Continue the activity with a set of 5 nickels and 4 pennies and a set of 6 nickels and 3 pennies. 29¢; 33¢

Determine Whether There Is Enough Money to Purchase an Item

Guide the students as they use **coin manipulatives** to determine whether they have enough money to make a purchase.

Distribute a Red Mat, 4 dimes, 6 nickels, and 4 pennies to each student. Direct them to place 4 dimes and 4 pennies on their mats.

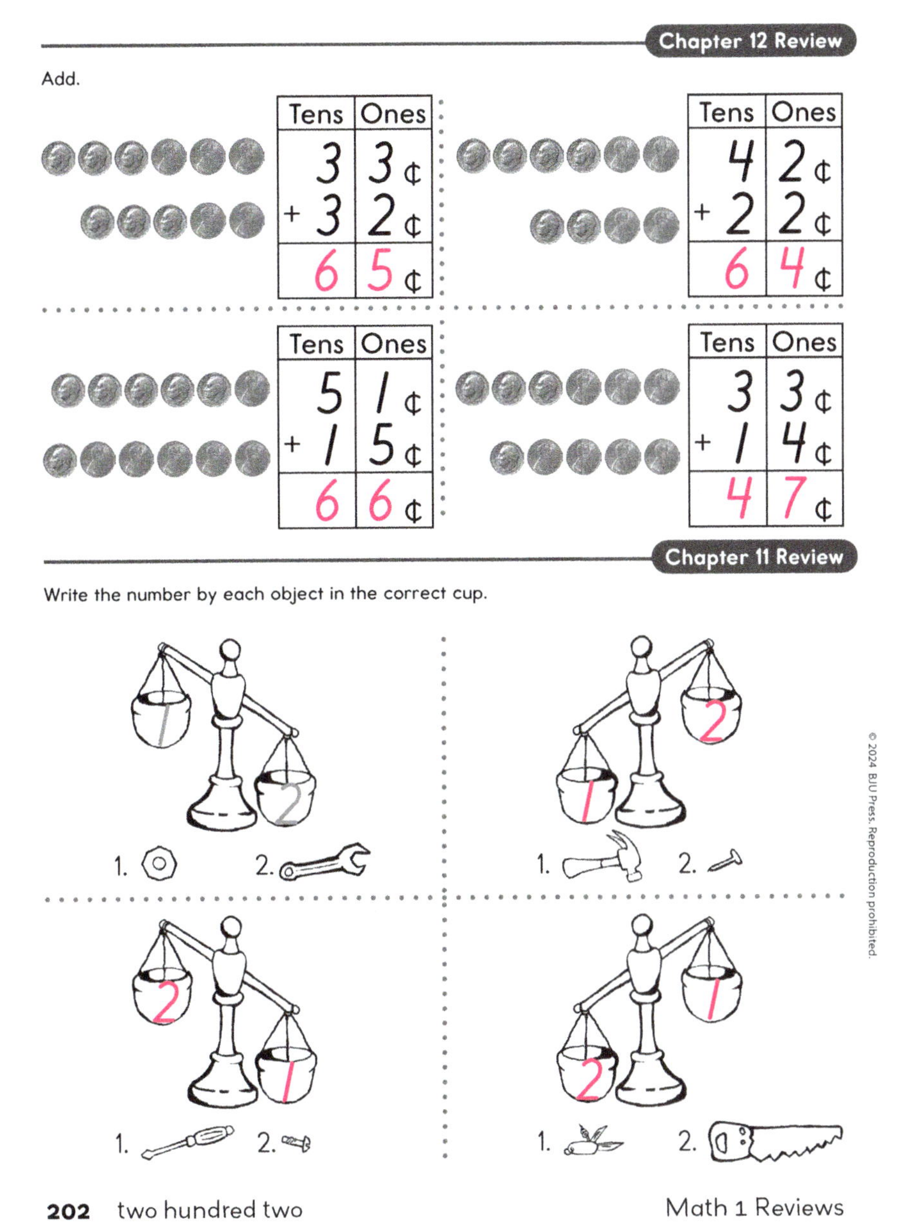

Chapter 12 Review
Add.
Tens
Ones
3 3¢
+ 3 2¢
6 5¢
4 2¢
+ 2 2¢
6 4¢
5 1¢
+ 1 5¢
6 6¢
3 3¢
+ 1 4¢
4 7¢
Chapter 11 Review
Write the number by each object in the correct cup.
202 two hundred two
Math 1 Reviews
© 2024 BJU Press. Reproduction prohibited.

108

Apply

Worktext pages 209–10

Read and guide completion of page 209.

Read and explain the directions for page 210. Assist the students as they complete the page independently.

Direct attention to the biblical worldview shaping statement and guide the students as they complete the sentence.

Assess

Reviews pages 201–2

Review adding 2-digit addends with money and comparing the weight of objects on page 202.

Fact Reviews

Use "Addition Facts to 12" page 19 in Trove.

How much is 4 dimes and 4 pennies? 44¢

Display a general store item with a 42¢ price tag.

What is the price of this item? 42¢

Do you have enough money to buy this item? yes; 42¢ < 44¢

Continue the activity with the following sets of coins and price tags.

5 nickels and 3 pennies 28¢; 30¢ price tag no

6 nickels and 4 pennies 34¢; 32¢ price tag yes

Point out that counting coins helps the students complete the task of buying an item.

If you tried to buy a snack at a store before you counted your money, what might happen? I might not have enough money.

Would you be able to complete the task of buying the snack if you did not have enough money? No; I would need to put the item back.

Explain that counting their money and determining whether they have enough money to buy something helps the students complete the tasks or work God gives them to do.

Read aloud Ecclesiastes 9:10*a*. Encourage the students to obey God by doing their best at their work.

Differentiated Instruction

Value of a Set of Coins

Distribute 35 pennies to the students. Instruct them to count the pennies and say the value. Ask the students to count out 10 pennies and replace them with 1 dime. Continue the activity by replacing the remaining pennies with 2 more dimes and 1 nickel. Ask the students to give the total value of the coins. Ask them to say whether 3 dimes and 1 nickel is the same value as the 35 pennies. Continue the activity, using different values of pennies and other coin combinations.

109

Worktext pages 211–12
Reviews pages 203–4

Objectives

- **109.1** Determine the value of a set of dimes and nickels and a set of dimes, nickels, and pennies by *counting on*.
- **109.2** Determine whether there is enough money to purchase an item.
- **109.3** Solve a money word problem by using the Problem-Solving Plan.

Printed Resources

- Visual 21: *Problem-Solving Plan*
- Visuals: Money Kit (7 pennies, 6 nickels, 7 dimes)
- Student Manipulatives: Number Cards (0–9)
- Student Manipulatives: Red Mat
- Student Manipulatives: Money Kit (6 pennies, 6 nickels, 6 dimes)

Digital Resources

- Video: Nickels, Dimes, Pennies
- Games/Enrichment: Fact Reviews (Addition Facts to 12, page 20)
- Games/Enrichment: Subtraction Flashcards

Materials

- Subtraction flashcards
- 3 general store items with price tags 70¢, 75¢, and 95¢

Solving Money Word Problems

Write the value as you count. Write the total.

10 ¢ 20 ¢ 25 ¢ — 25 ¢

10 ¢ 20 ¢ 25 ¢ 30 ¢ — 30 ¢

10 ¢ 15 ¢ 20 ¢ 21 ¢ 22 ¢ — 22 ¢

Write the total.

23 ¢

Write an equation for the word problem. Solve.

Kinsley has 2 dimes in her bank.
She put 3 nickels in her bank.
How much money is in her bank?

20 + 15 = 35 ¢

(work space)

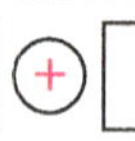  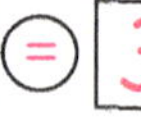

$$\begin{array}{r} 20¢ \\ +\ 15¢ \\ \hline 35¢ \end{array}$$

Practice & Review

Write Numbers to 100

Distribute the Number Cards. Instruct the students to use their cards to show the number 95. Invite a student to write "95" for display so the students can check their cards. Continue the activity for numbers 13, 48, 82, and 70.

Memorize Subtraction Facts

Introduce the following facts.

10 – 4 10 – 5 10 – 6

Distribute Number Cards 0–9. Practice the new subtraction facts and previously memorized facts as in earlier lessons.

How are 10 – 4 and 10 – 6 related? They belong to the same fact family.

Engage

Compare 2-Digit Numbers by Using > and <

Guide a **review activity** to help the students compare the value of 2 sets of coins.

Display a set of 2 dimes and 5 pennies. Display a set of 5 dimes and 2 pennies to the right of the first set of coins.

Invite a student to write the value of the set of coins on the left. 25¢ Invite another student to write the value of the set of coins on the right. 52¢ Invite a third student to write a > or < to complete the comparison sentence. 25¢ < 52¢

Continue the activity to compare 42¢ to 24¢ 42¢ > 24¢ and 23¢ to 32¢. 23¢ < 32¢

Instruct

Value of a Set of Dimes & Nickels

Guide a **counting activity** to help the students determine the value of a set of coins by *counting on*.

Display 4 dimes and 2 nickels in random order.

What should you do first? separate the nickels and dimes

Which coins should you count first? the dimes, because a dime has a greater value than a nickel

How do you count the dimes? by 10s

Write the total value.
Do you have enough money to buy the item for the missionary?
Circle the correct answer.

30¢ — yes / no (circled: yes)

17¢ — yes / no (circled: no)

36¢ — yes / no (circled: yes)

26¢ — yes / no (circled: yes)

Time to Review

Cross out to subtract.
Write the answer.

10 − 3 = 7
10 − 8 = 2
10 − 6 = 4
10 − 7 = 3
10 − 4 = 6
10 − 5 = 5

How do you count the nickels? by 5s

Place the coins in a row with the dimes to the left.

How do you count a set of dimes and nickels? begin by counting the dimes by 10s, take a deep breath, and then *count on* the nickels by 5s

Why do you take a deep breath after you count the dimes and before you begin counting the nickels? to remind myself to change the way I am counting

Guide the students as they count the value of the coins: 10, 20, 30, 40, (deep breath), 45, 50¢.

Distribute the Red Mats and the coins. Instruct the students to place 5 dimes and 2 nickels on their mats.

Guide the students as they count the value of the coins: 10, 20, 30, 40, 50, (deep breath), 55, 60¢.

Continue the activity with 3 dimes and 2 nickels. 10, 20, 30, (deep breath), 35, 40¢

Value of a Set of Dimes, Nickels & Pennies

Guide a **counting activity** to help the students determine the value of a set of coins by *counting on*.

Display 3 dimes, 2 nickels, and 2 pennies in random order.

What should you do first? separate the dimes, nickels, and pennies

In what order should you count the value of this set of coins? the dimes first, the nickels next, and the pennies last; I count coins in the order of their value, from greatest to least.

Guide the students as they count the 3 dimes by 10s, then *count on* the 2 nickels by 5s, and then *count on* the 2 pennies by 1s: 10, 20, 30, (deep breath), 35, 40, (deep breath), 41, 42¢.

Continue the activity with a set 4 dimes, 3 nickels, and 1 penny and a set of 2 dimes, 2 nickels, and 3 pennies. 56¢; 33¢

Determine Whether There Is Enough Money to Purchase an Item

Guide the students as they use **coin manipulatives** to determine whether they have enough money to make a purchase.

Direct the students to place 4 dimes and 4 nickels on their mats.

How much is 4 dimes and 4 nickels? 60¢

Display a general store item with a 70¢ price tag.

What is the price of this item? 70¢

Do you have enough money to buy this item? no; 60¢ < 70¢

Continue the activity with the following sets of coins and price tags.

5 dimes and 6 nickels 80¢; 75¢ price tag yes

6 dimes, 2 nickels, and 6 pennies 76¢; 95¢ price tag no

Problem-Solving Plan

Follow the **Problem-Solving Plan** to help the students solve a word problem.

Display Visual 21 and read aloud the following word problem.

DQ had 1 dime and 1 nickel in his pocket. The next day he found 3 dimes in the weeds. How much money does DQ have now?

What is the question? How much money does DQ have now?

What information is given? DQ had 1 dime and 1 nickel in his pocket, and he found 3 dimes.

Do you add or subtract to find the answer? add

Direct the students to place 1 dime and 1 nickel at the top of their mats.

How much money did DQ have in his pocket? 15¢

Direct the students to place 3 dimes at the bottom of their mats.

How much money did DQ find in the weeds? 30¢

What is the equation that you can use to solve this word problem? 15¢ + 30¢ = __¢

Write "15¢ + 30¢ = __" for display.

Should you rewrite the equation in vertical form? Yes, it will be easier to add in vertical form since the addends are 2-digit numbers.

Invite a student to write it in vertical form.

What do you add first when there are 2-digit addends? the ones

Ask a student to solve the problem. 45¢

Does the answer make sense? yes

Write "45¢" to complete the equation.

Continue the activity with the following word problem.

Alex has 6 dimes and 5 nickels. He needs new shoelaces that cost 70¢ for his soccer cleats. Does Alex have enough money to buy the shoelaces? yes; 85¢ – 70¢ = 15¢

Apply

Worktext pages 211–12

Read and guide completion of page 211.

Read and explain the directions for page 212. Assist the students as they complete the page independently.

Assess

Reviews pages 203–4

Review writing an addition or subtraction equation to solve a problem on page 204.

Fact Reviews

Use "Addition Facts to 12" page 20 in Trove.

Extended Activity

Add Sets of Coins

Materials

- Small clear container with lid for each student
- Assorted stickers
- A permanent marker
- Money Kit: pennies, nickels, dimes for each student

Procedure

Make coin openings in the lids to make banks. Guide the students as they decorate their containers with small stickers. Use the marker to label the containers with the students' names. Distribute the coins. Direct the students to put 2 dimes in their banks.

How much money is in your bank? 20¢

Instruct them to put in 3 more dimes and 2 nickels.

How much money did you add to what was already in the bank? 40¢

Write "20¢ + 40¢ = __" for display.

How much money do you have in your bank in all? 20¢ + 40¢ = 60¢

Continue the activity with other coin combinations.

Write the value as you *count on*. Write the total.

10¢ 20¢ 25¢ 30¢ 35¢ 36¢ 37¢ — 37¢

10¢ 20¢ 30¢ 40¢ 45¢ 46¢ 47¢ — 47¢

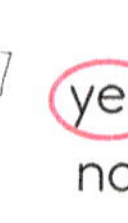

Write the total value. Do you have enough money to buy the item? Circle *yes* or *no*.

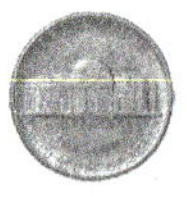

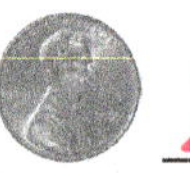

36¢

yes (circled) no

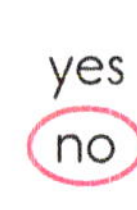

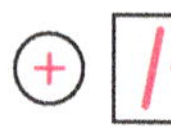

22¢

yes no (circled)

Write an equation for the word problem. Solve.

James has 3 dimes in his bank. He puts in 1 dime and 4 pennies. How much money is in the bank?

30¢ + 14¢ = 44¢

work space

$$\begin{array}{r} 30¢ \\ +\ 14¢ \\ \hline 44¢ \end{array}$$

Chapter 14 • Lesson 109 two hundred three **203**

Chapter 12 Review

Write an equation for each word problem. Solve.

Farmer Bob picked 42 baskets of corn.
His helper picked 25 baskets of corn.
How many baskets of corn did they pick in all?

42 + 25 = 67 baskets

work space

$$\begin{array}{r} 42 \\ +\,25 \\ \hline 67 \end{array}$$

Grandma had 67 pea plants.
In the storm, 43 pea plants died.
How many pea plants does Grandma have left?

67 − 43 = 24 pea plants

work space

$$\begin{array}{r} 67 \\ -\,43 \\ \hline 24 \end{array}$$

Mom canned 53 jars of string beans on Friday
and 36 jars on Saturday.
How many jars of beans did Mom can in all?

53 + 36 = 89 jars

work space

$$\begin{array}{r} 53 \\ +\,36 \\ \hline 89 \end{array}$$

Subtraction Fact Review

Cross out to subtract. Write the answer.

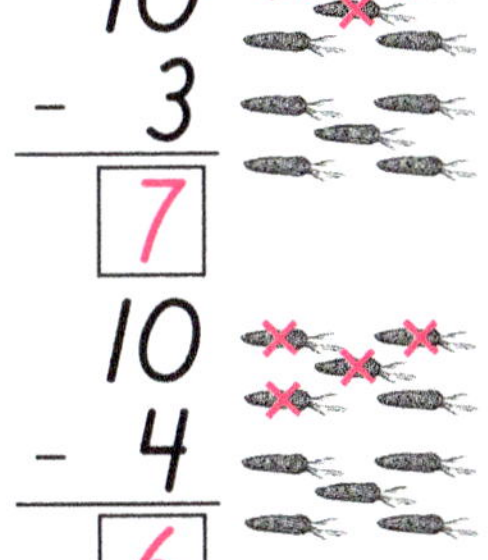  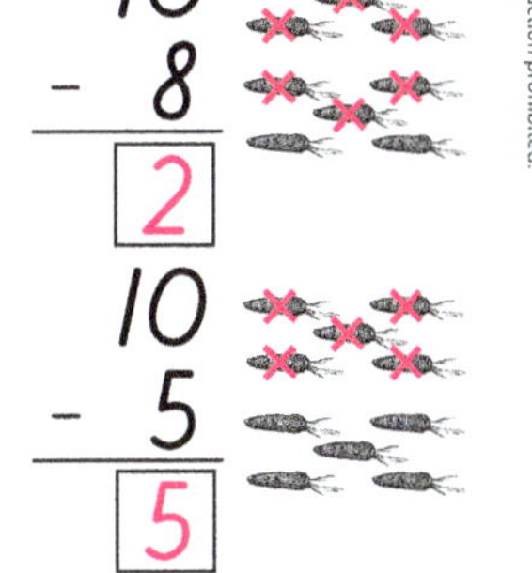

$$\begin{array}{r} 10 \\ -\,3 \\ \hline 7 \end{array} \qquad \begin{array}{r} 10 \\ -\,7 \\ \hline 3 \end{array} \qquad \begin{array}{r} 10 \\ -\,8 \\ \hline 2 \end{array}$$

$$\begin{array}{r} 10 \\ -\,4 \\ \hline 6 \end{array} \qquad \begin{array}{r} 10 \\ -\,6 \\ \hline 4 \end{array} \qquad \begin{array}{r} 10 \\ -\,5 \\ \hline 5 \end{array}$$

110

Worktext pages 213–14
Reviews pages 205–6

Objectives

- **110.1** Determine the value of a set of dimes, nickels, and pennies by *counting on.*
- **110.2** Identify equivalent amounts of money.
- **110.3** Predict the probability of choosing a dime, nickel, or penny.
- **110.4** Tally the results of a probability activity.

Printed Resources

- Visuals: Money Kit (10 pennies, 7 nickels, 12 dimes)
- Student Manipulatives: Number Cards (0–9)
- Student Manipulatives: Sign Cards (Greater than, Less than, Cent)
- Student Manipulatives: Money Kit (7 pennies, 7 nickels, 7 dimes; for each pair of students)
- Student Manipulatives: Red Mat

Digital Resources

- Games/Enrichment: Fact Reviews (Addition Facts to 12, page 21)
- Games/Enrichment: Subtraction Flashcards

Materials

- Subtraction flashcards
- 3 general store items with price tags 27¢, 35¢, and 47¢
- A small paper bag
- Real coins: 1 penny, 5 nickels, 5 dimes

Money & Probability

Circle the coin you think will be chosen the most. Answers may vary.

Mark a tally for each coin chosen during the activity.

Tallies	

Circle the coin that was chosen the most.

How much money is in the bag? Write the total value.

51 ¢

Practice & Review

Compare 2-Digit Numbers by Using > and <

Distribute the Number Cards and Sign Cards to the students. Display 5 dimes and 4 pennies. Direct the students to show the value of the coins on their desks by using their Number Cards and Cent Signs. 54¢

Display 7 dimes and 3 pennies to the right of the first set of coins. Direct the students to use their Number Cards and Cent Signs to show the value of the second set of coins to the right of the value of the first set. 73¢

Is 54¢ greater than or less than 73¢? less than

Instruct the students to place their Less-than Signs between the numbers. Guide the students as they read the number sentence: "54¢ is less than 73¢."

Continue the activity by comparing the values of coins and writing number sentences for 36¢ > 21¢ and 34¢ < 52¢.

Memorize Subtraction Facts

Introduce the following facts.

11 – 3 11 – 8 11 – 9

Distribute Number Cards 0–9. Practice the new subtraction facts and previously memorized facts as in earlier lessons.

How are 11 – 3 and 11 – 8 related? They belong to the same fact family.

Engage

Giving to God

Guide a **discussion** to help the students evaluate a monetary gift given to God from the perspective of biblical teaching.

Read aloud the Bible account in Luke 21:1–4. Explain that one day in the temple, rich men were giving great amounts of money to the Lord, but a poor widow gave fewer than 2 pennies to the Lord. Point out that Jesus said the widow gave more than the rich men.

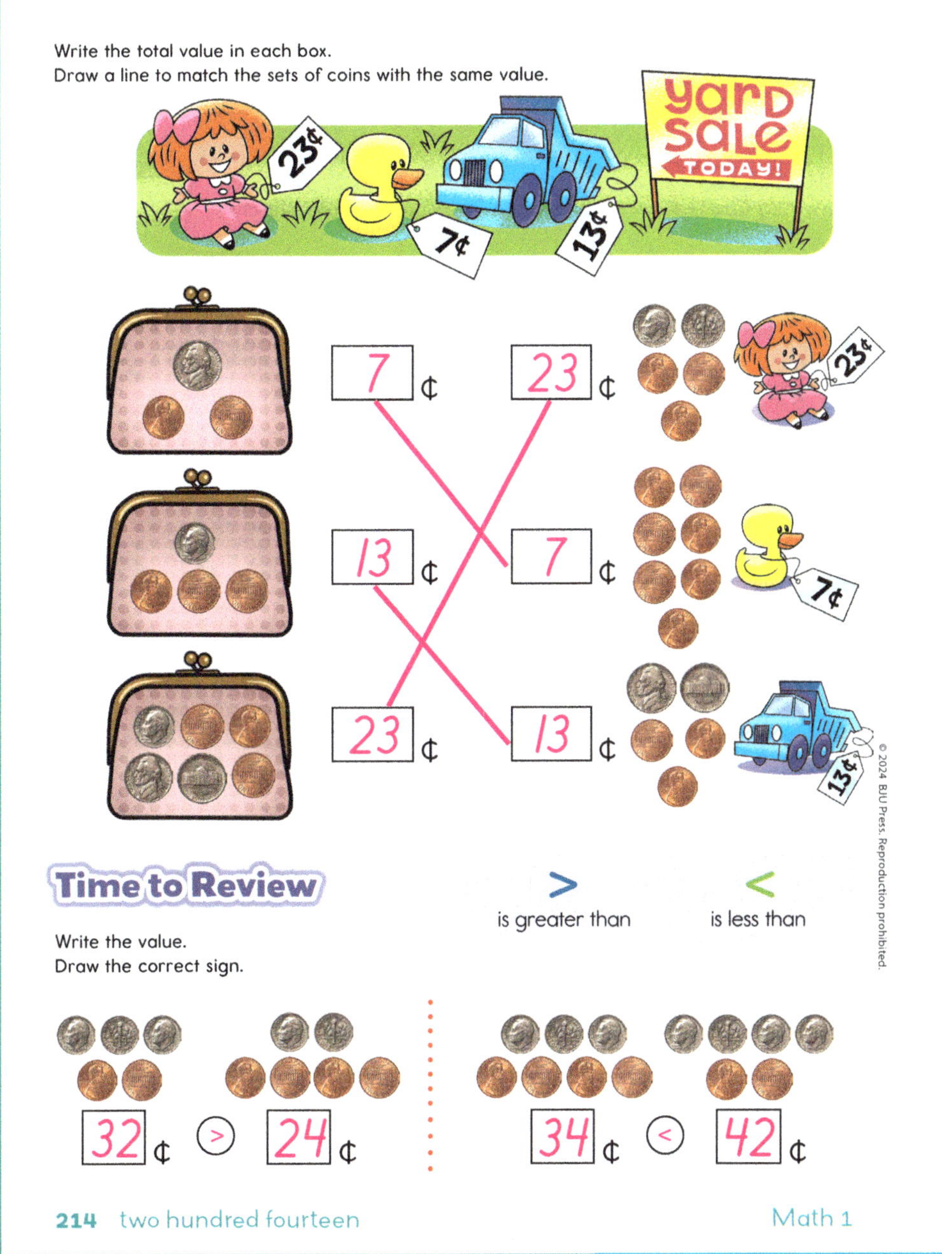

Write the total value in each box.
Draw a line to match the sets of coins with the same value.
yard sale
TODAY!
23¢
7¢
13¢
7 ¢
23 ¢
13 ¢
7 ¢
23 ¢
13 ¢
© 2024 BJU Press. Reproduction prohibited.
Time to Review
Write the value.
Draw the correct sign.
> is greater than
< is less than
32 ¢ > 24 ¢
34 ¢ < 42 ¢
214 two hundred fourteen
Math 1

How could the widow have given more than the rich men? The poor widow gave everything she could, but the rich men had much more money left over.

Explain that in God's eyes, the attitude with which we give is more important than how much we give. The widow's gift showed that she was not afraid to give so much of the little she had, because she trusted God to give her what she needed.

Instruct

Value of a Set of Dimes, Nickels & Pennies

Guide a **counting activity** to help the students determine the value of a set of coins by *counting on*.

Display 2 dimes, 1 nickel, and 2 pennies in random order.

What should you do first? separate the pennies, nickel, and dimes

In what order do you count the value of the set of coins? dimes, nickel, pennies; I count the greatest coin value first.

Count the coins together to determine their total value. 10, 20, (deep breath), 25, (deep breath), 26, 27¢

Continue the activity with 3 dimes, 3 nickels, and 1 penny and with 4 dimes, 3 nickels, and 2 pennies. 46¢; 57¢

Equivalent Amounts of Money

Guide a **discussion** to prepare the students to identify the coins needed for a purchase.

Display 1 nickel.

What is the value of the nickel? 5¢

Invite a student to show coins that are worth the same as 1 nickel. 5 pennies

Display 1 dime.

What is the value of the dime? 10¢

What other ways can you show 10¢? 2 nickels, 1 nickel and 5 pennies, 10 pennies

Invite students to display the coins.

Guide the students to **think-pair-share** to identify the coins needed for a purchase.

Pair the students. Distribute the coins and a Red Mat to each pair. Display the general store item that costs 35¢. Explain that the students will pretend to buy the item.

Instruct the students to place the coins needed to purchase the item on their mats. Direct them to quietly count their coins with their partners to check their answers. Invite students to display the coins they chose. Possibilities include 3 dimes and 1 nickel; 7 nickels; 2 dimes and 3 nickels; or 3 dimes and 5 pennies.

Continue the activity with the general store item that costs 27¢. Possibilities include 2 dimes and 7 pennies; 2 dimes, 1 nickel, and 2 pennies; or 1 dime, 3 nickels, and 2 pennies.

Continue with the item that costs 47¢. Possibilities include 4 dimes, 1 nickel, and 2 pennies; or 4 dimes and 7 pennies.

Direct the students to continue the think-pair-share activity, comparing the value of 2 sets of coins.

Display 1 nickel and 5 pennies on the left and 1 dime on the right. Direct the students to think and then pair with their partner to answer the following questions.

Is the set of 1 nickel and 5 pennies the same amount of money as the set of 1 dime? Yes, both equal 10¢.

Place 1 dime, 1 nickel, and 2 pennies on the left and 4 nickels and 2 pennies on the right.

Do the 2 sets of coins have the same value? no

Instruct the students to use their manipulatives to determine the answer to the question if needed. Invite students to count the value of each set of coins to verify the answer. 17¢; 22¢

110

Probability Activity

Guide a **probability activity** to help the students predict and tally the results of a probability experiment.

Invite a student to place 5 real dimes in a bag.

> **Which coin do you think is more likely to be chosen from the bag—a nickel or a dime?** a dime, because no nickels were placed in the bag

Remove the 5 dimes and ask a student to place 5 real nickels and 1 real dime in the bag.

> **Which coin do you think is more likely to be chosen from the bag?** a nickel, because there are more nickels than dimes

> **Which coin do you think is less likely to be chosen from the bag?** a dime, because there are fewer dimes than nickels

Display a nickel and a dime. Explain that the students will keep track of which coins they pull out of the bag by making tally marks. Shake the bag, then invite a student to reach into the bag and take out the first coin that his or her hand touches. Record the result with a tally next to the corresponding displayed coin.

Put the coin back into the bag, shake the bag, and invite another student to take out a coin and draw a tally next to the corresponding coin.

Continue the activity for a total of 10 times. Discuss the results and compare them with the predictions that were made.

Money & Probability

Write the total value in each box.
Draw a line from each present to the child with the same amount of money.

Present	Child
10¢	35¢
50¢	25¢
35¢	10¢
25¢	50¢

Chapter 14 • Lesson 110 two hundred five **205**

Apply

Worktext pages 213–14

Direct attention to the probability activity on page 213. Instruct the students to circle the coin that they think will be chosen the most.

Place 4 dimes, 2 nickels, and 1 penny in a bag. Shake the bag and invite a student to reach into the bag and take out the first coin that his or her hand touches. Direct the students to record the results with tallies next to the coins on page 213.

Put the coin back into the bag. Continue the activity for a total of 10 times. Guide the students as they complete the page.

Read and explain the directions for page 214. Assist the students as they complete the page independently.

Assess

Reviews pages 205–6

Review shapes and equal parts of shapes on page 206.

Fact Reviews

Use "Addition Facts to 12" page 21 in Trove.

Extended Activity

Determine the Number of Dimes or Nickels in a Set of Coins When Given the Value of the Set

Materials

- Money Kit: 5 nickels, 8 dimes for each student
- Red Mat for each student

Procedure

Distribute the coins and instruct the students to use their coins to show the answers to the following questions.

Explain that the students have 15¢. The coins are only nickels.

Chapter 9 Review

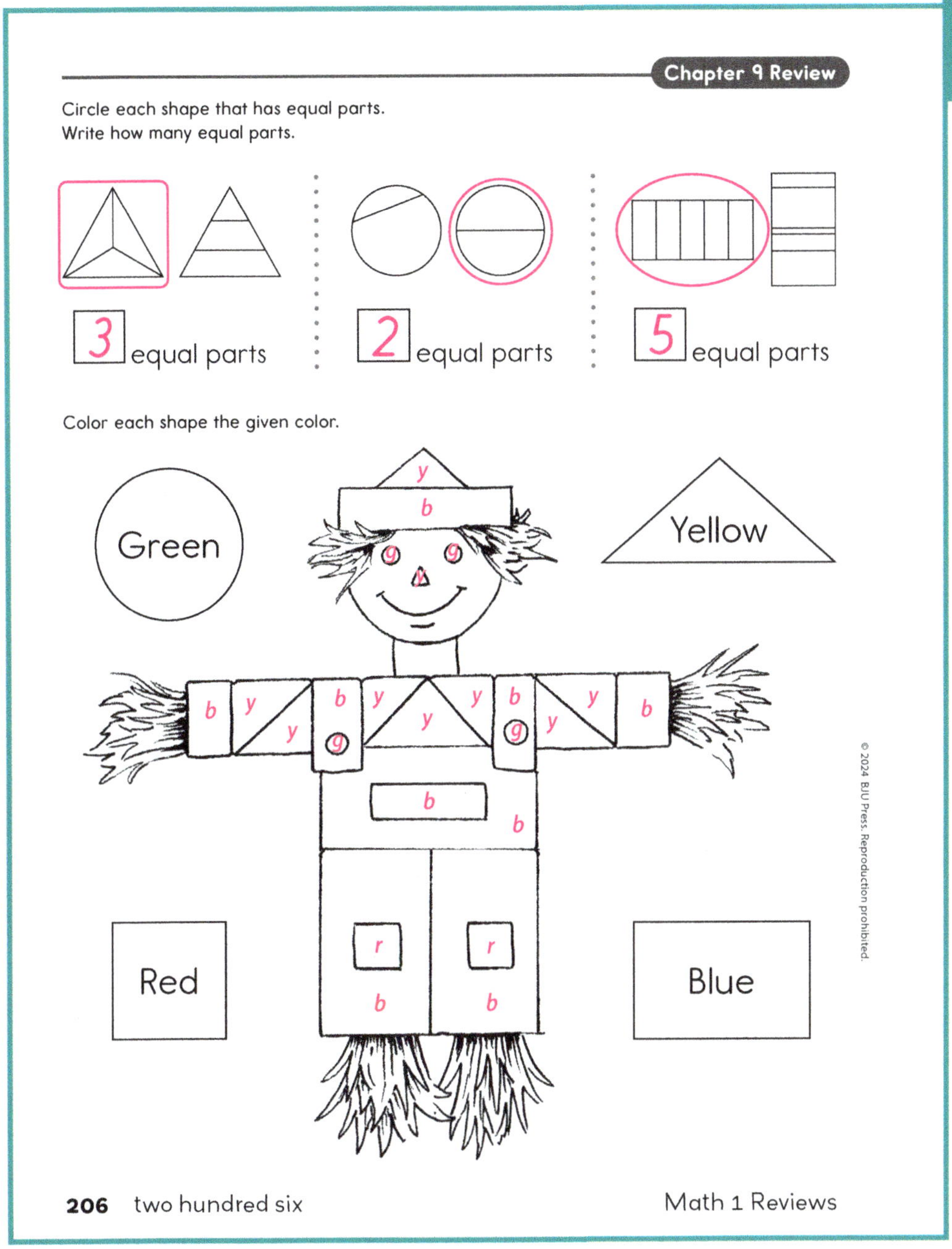

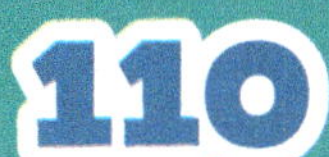

Circle each shape that has equal parts.
Write how many equal parts.

3 equal parts

2 equal parts

5 equal parts

Color each shape the given color.

Green

Yellow

Red

Blue

How many nickels do you need to make 15¢? 3

They have 60¢. The coins are only dimes.

How many dimes do you need to make 60¢? 6

They have 40¢. They have 3 dimes and some nickels.

How many nickels do you need to add to the dimes to make 40¢? 2

They have 50¢. They have 2 nickels and some dimes.

How many dimes do you need to add to the nickels to make 50¢? 4

111 Worktext pages 215–16 Reviews pages 207–8

Objectives
- **111.1** Evaluate counting money in a careless way. BWS
- **111.2** Identify the value of a quarter.
- **111.3** Identify sets of pennies, nickels, and dimes equivalent to 1 quarter.

Biblical Worldview Shaping
- **Working** (evaluate): Since carelessness in counting money may result in harm to others, we must complete this task carefully even if we do not want to (111.1).

Printed Resources
- Visual 14: *Money*
- Visuals: Money Kit (25 pennies, 5 nickels, 6 dimes, 1 quarter)
- Student Manipulatives: Number Cards (0–9)
- Student Manipulatives: Money Kit (25 pennies, 5 nickels, 2 dimes)
- Student Manipulatives: Red Mat

Digital Resources
- Video: Quarters
- Web Link: Learning Coins
- Web Link: U.S. Mint Coin Classroom
- Games/Enrichment: Fact Reviews (Addition Facts to 12, page 22)
- Games/Enrichment: Subtraction Flashcards

Materials
- Subtraction flashcards
- Real coins: pennies, nickels, dimes, quarters (enough for each student to have 1 coin)

Quarters

25¢ quarter

Write the total value of each set of coins.

25 ¢ 25 ¢ 25 ¢

Write the total value.

32 ¢

45 ¢

Practice & Review

Identify Even & Odd Numbers

Display 4 dimes and 6 pennies. Write "46¢" for display.

What is an even number? a number that can be arranged in even pairs with nothing left over

What is an odd number? a number that does not make even pairs; there is an odd one left over

In which place do you look to tell whether 46 is an even number or an odd number? the Ones place

Which digit is in the Ones place in 46? 6

How can you find out whether 6 is even or odd? put the pennies into pairs

Invite a student to group the pennies into pairs.

Can 6 be arranged into even pairs? yes

Is 46 an even number or an odd number? even number, because the 6 ones can be arranged in even pairs with no ones left over

Repeat the procedure for 35¢ odd, 62¢ even, and 53¢ odd.

Memorize Subtraction Facts

Introduce the following facts.

11 – 7   11 – 4

Distribute Number Cards 0–9. Practice the new subtraction facts and previously memorized facts as in earlier lessons.

How are 11 – 7 and 11 – 4 related? They belong to the same fact family.

Engage

Careless Work

Guide a **discussion** to help the students evaluate a situation from the perspective of biblical teaching.

Mark an *X* on the coins needed to make the same value as a quarter.

Circle the word to complete the sentence.

I should / should not finish my task of counting coins.

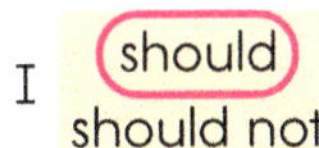

Subtract.
Use the number line if needed.

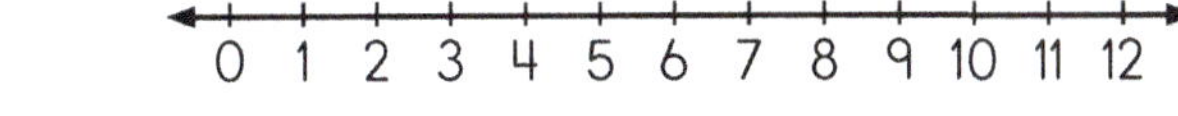

11 − 3 = 8   11 − 8 = 3   11 − 9 = 2

11 − 7 = 4   11 − 4 = 7   11 − 2 = 9

216 two hundred sixteen   Math 1

Display a copy of the Worktext chapter opener picture.

What was Remy's job? to count the coins for the purchases to be sure the right amount was paid

Why did Remy wish he could leave and play with DQ? He was tired of the work of counting coins.

Read aloud the following paragraph from the chapter opener scenario.

The next customer stepped up to the table. This time Remy quickly glanced at the price of the item and the coins that were paid for it without actually counting the coins to be sure they were the right amount. *Oh well. It looks about right.* Remy felt a little stab of guilt as he smiled and took the money. *What could it hurt?* he thought. *It's not that much money. It's too much trouble to count every single coin every single time.*

Invite students to tell whether they think Remy did the right thing.

Why did Remy feel guilty? He counted the coins carelessly. He knew he was doing the wrong thing.

What might happen if you count coins carelessly? I might accept the wrong amount of money for a purchase. If I take too much money, I am not being fair to the customer. If I take too little money, I am not being faithful to complete the work God gives me to do. I might cheat those to whom the money belongs. My carelessness could hurt others.

Read aloud Colossians 3:23.

Did Remy do his work heartily as to the Lord? no

What should Remy have done instead? He should have taken the time to count the coins carefully even though it was not easy.

Encourage the students to be faithful in their work, knowing that what they do is for the Lord. Remind them that God is their Helper when they are tempted to do their work carelessly (Hebrews 4:16).

Instruct

Value of a Quarter

Guide a **visual analysis** to help the students identify a quarter.

Display Visual 14. Point out the 2 quarters and read the 3 ways to refer to a quarter.

Explain that the first president of the United States is pictured on the front of the quarter.

Whose picture is on the quarter? George Washington

Encourage the students to discuss other things they notice about the quarter. the design on the back, the date on the front, the words *Liberty, In God We Trust, United States of America, and quarter dollar*

How can you tell the quarter apart from the other coins you have studied? the silver color, largest of the coins I have studied, ridges on the edge, the profile of George Washington on the front, the words *quarter dollar*

Identify Coins

Guide an **examination activity** to help the students identify different coins.

Distribute quarters and instruct the students to handle them and observe the features. Discuss the various symbols. Point out that quarters have had many different designs on the back through the years.

Distribute the remaining real coins to the students. Allow them to examine the coins. Say the name of a coin and ask the students who have that coin to hold it up.

Instruct the students to exchange coins with another student. Continue the activity.

Display a penny, a nickel, a dime, and a quarter from your Money Kit, one by one in random order, inviting the students to say the value of each.

Identify Sets of Coins Equivalent to 1 Quarter

Use **coin manipulatives** to enable the students to identify sets of coins equivalent to a quarter.

Distribute the coin manipulatives and Red Mats to the students. Display 1 quarter.

What is the value of a quarter? 25¢

Write "25¢" for display.

How many pennies equal the value of a quarter? 25

Instruct the students to place 25 pennies on their mats as you display 25 pennies. Count the pennies together by 1s.

What is the value of 25 pennies? 25¢

Which coin is the same value as 25 pennies? a quarter

Explain that it might be easier to count the pennies if they were placed in rows of 5 pennies. Demonstrate as the students show their work on their mats.

How can you count the pennies now? by 5s

Count the pennies together by 5s.

Which coin do 5 pennies equal? 1 nickel

Direct the students to replace each row of 5 pennies with 1 nickel as you demonstrate. Count the nickels together by 5s.

What is the value of 5 nickels? 25¢

What coin is the same value as 5 nickels? a quarter

Instruct the students to place 25 pennies on their mats in rows of 10 pennies as you demonstrate.

How many rows of 10 do you have? 2

How many pennies are in the last row? 5

Which coin can replace a row of 10 pennies? a dime

Instruct the students to replace each row of 10 pennies with 1 dime as you demonstrate. Count the coins together to determine their total value: 10, 20, (deep breath), 21, 22, 23, 24, 25¢.

Quarters

Help Mother find her way to the cashier.
Write the value of each set.
Draw lines to join all the coin sets that equal a quarter.

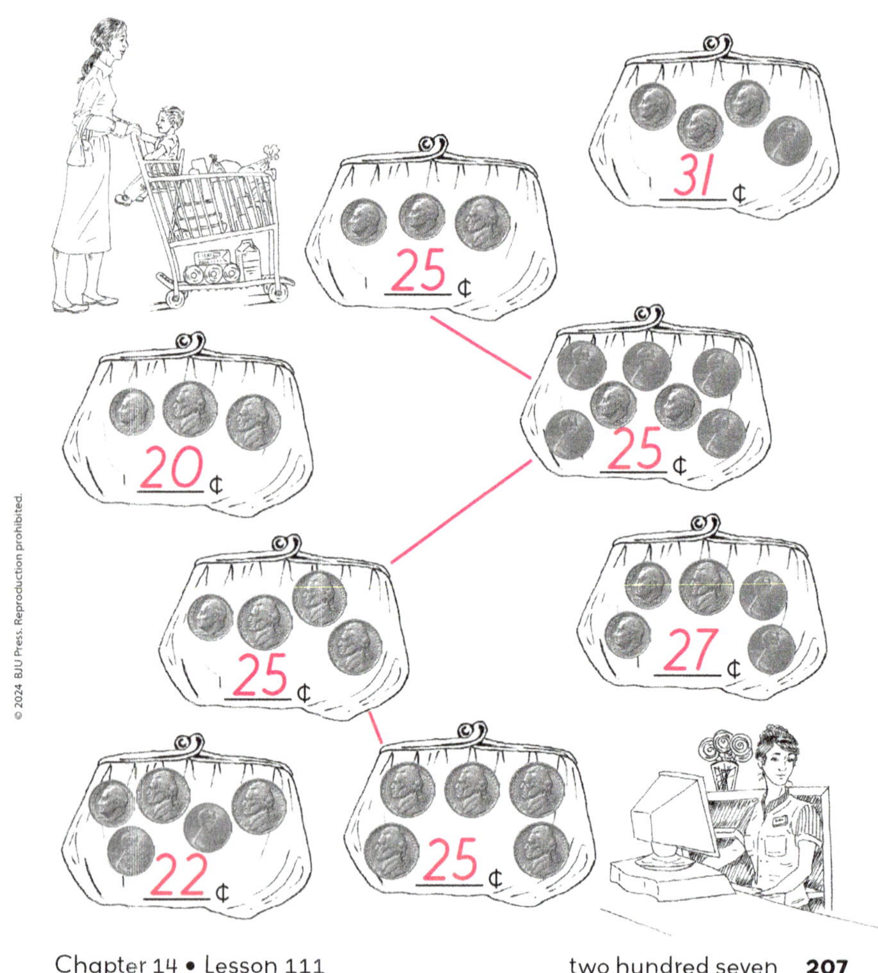

What is the value of 2 dimes and 5 pennies? 25¢

Which coin is the same as 2 dimes and 5 pennies? a quarter

Which coin can replace the 5 pennies? a nickel

Instruct the students to replace the 5 pennies with 1 nickel as you demonstrate. Guide the students as they count the value of the coins. 10, 20, (deep breath), 25¢

Ask the students to show other possible coin combinations that equal the value of 1 quarter, 25¢. Possibilities include 4 nickels and 5 pennies; 1 dime and 3 nickels; 1 dime, 2 nickels, and 5 pennies; or 1 dime and 15 pennies.

Apply

Worktext pages 215–16

Read and guide completion of page 215.

Read and explain the directions for page 216. Assist the students as they complete the page independently.

Direct attention to the biblical worldview shaping statement and guide the students as they complete the sentence.

Assess

Reviews pages 207–8

Review activities that take more or less than an hour on page 208.

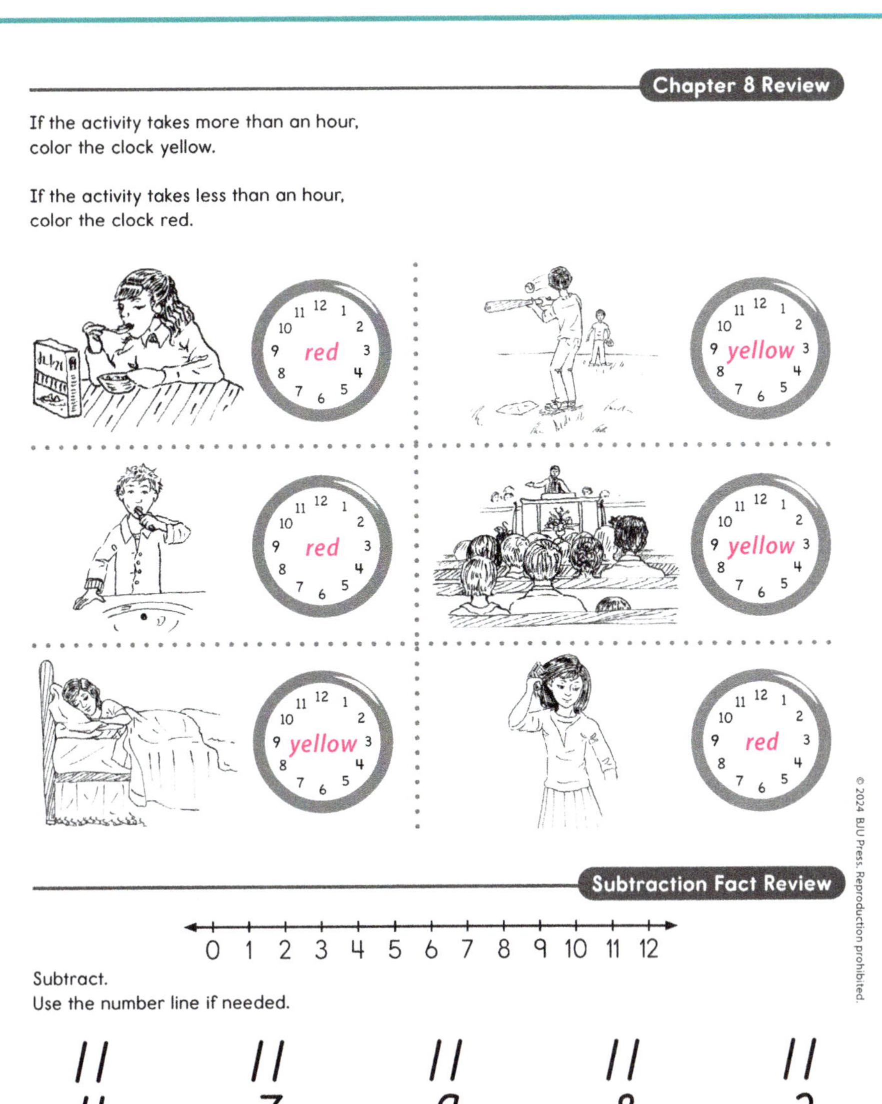

Chapter 8 Review

If the activity takes more than an hour, color the clock yellow.

If the activity takes less than an hour, color the clock red.

red | yellow
red | yellow
yellow | red

Subtraction Fact Review

0 1 2 3 4 5 6 7 8 9 10 11 12

Subtract.
Use the number line if needed.

$11 - 4 = 7$ $11 - 7 = 4$ $11 - 9 = 2$ $11 - 8 = 3$ $11 - 3 = 8$

208 two hundred eight — Math 1 Reviews

Fact Reviews

Use "Addition Facts to 12" page 22 in Trove.

Extended Activity

Go on a Money Field Trip

Materials

- A list of items costing less than $1 from a store
- A permission slip for each student

Procedure

Make arrangements with a local store for the students to take a tour. Explain to the management that the students have been learning about money and that each will be purchasing an item "on their own." When sending the permission slip home, include the list of items and request that the parent specify which item the child may buy.

What should I do if counting coins takes a lot of time?

Chapter Concept Review

- Practice concepts from Chapter 14 to prepare for the test.

Printed Resources

- Instructional Aid 31: *Grocery Store Items*
- Visuals: Money Kit (6 pennies, 6 nickels, 8 dimes, 1 quarter)
- Student Manipulatives: Tens/Ones Mat
- Student Manipulatives: Money Kit (6 pennies, 6 nickels, 6 dimes, 1 quarter)
- Student Manipulatives: Red Mat

Digital Resources

- Games/Enrichment: Fact Reviews (Addition Facts to 12, page 23)
- Games/Enrichment: Subtraction Flashcards
- Games/Enrichment: Fact Fun Activities

Write the value of each coin.

 25¢ quarter   5¢ nickel   1¢ penny   10¢ dime

Write the total value.
Write the letter of what you can buy with the exact amount.

A 27¢

B 65¢

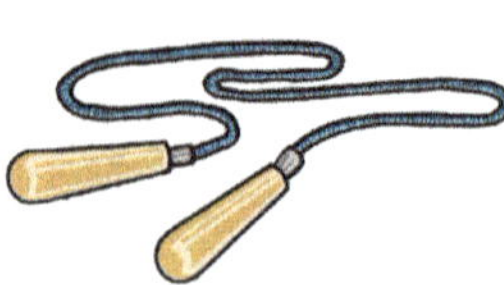
C 40¢

      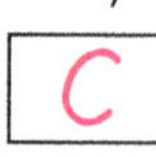

Total 40¢ Buy C

     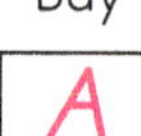

Total 27¢ Buy A

Mark an X on the coins needed to make the same value as a quarter.

Practice & Review

Subtract 2-Digit Numbers by Using Dimes & Pennies

Distribute the Tens/Ones Mats and the coins to the students. Demonstrate the following problem in a Tens/Ones frame as you read it aloud.

Aviel had 46¢ with him when he went to the grocery store. He bought a cookie for 23¢. How much money does Aviel have now?

What is the question? How much money does Aviel have now?

What information is given? Aviel had 46¢, and he spent 23¢ on a cookie.

Do you add or subtract to find the answer? subtract, because Aviel used part of his money to pay for the cookie

What equation would you use to solve the problem? 46¢ – 23¢ = __¢

Write "46¢ – 23¢ = __" for display.

What should you do first? subtract the ones

Remove 3 pennies from the Ones place.

How many pennies are there now? 3

How many tens should you subtract? 2

Remove 2 dimes from the Tens place.

How many dimes are there now? 2

What is the answer to this problem? 2 dimes and 3 pennies; 23¢

Write "23¢" to complete the equation.

Does the answer make sense? yes

Review Subtraction Facts

Choose a game or activity from the "Fact Fun Activities" available in Trove.

Instruct

Recognize & Identify the Value of Coins

Distribute the Red Mats and the coins. Direct the students to hold up the correct coin in response to the following questions:

Which coin's value is 25¢? quarter

Which coin's value is 1¢? penny

Which coin is the smallest in size? dime

Which coin's value is 5¢? nickel

Which coin has a picture of Thomas Jefferson? nickel

Which coin's value is 10¢? dime

Write the total value in each box.
Draw a line to match the items with the same price.

30 ¢ | 15 ¢
15 ¢ | 20 ¢
20 ¢ | 30 ¢

Circle the word to complete the sentence.

Remy counted coins to find the cost.
It took lots of time.
Counting coins helps me stop / finish the work God gives me to do.

Display a penny, a nickel, a dime, and a quarter from your Money Kit, one by one in random order, asking the students to say the name and the value of each one.

Determine the Value of a Set of Coins by *Counting on*

Instruct the students to place 2 dimes, 2 nickels, and 2 pennies on their mats as you display the coins in random order.

What is the first thing you do when counting the value of a set of coins? separate the different coins into groups

Which coins do you count first? the coins that have the greatest value

Instruct the students to separate the coins. Guide them as they *count on* the coins: 10, 20, (deep breath), 25, 30, (deep breath), 31, 32¢.

Continue the activity with 3 dimes, 3 nickels, and 1 penny and with 2 nickels and 6 pennies. 46¢; 16¢

Identify Sets of Coins Equivalent to 1 Quarter

What is the value of a quarter? 25¢

Write "25¢" for display. Direct the students to place the value of a quarter on their mats, using the following combinations.

only nickels 5 nickels

1 dime and some nickels 1 dime and 3 nickels

1 nickel and some dimes 1 nickel and 2 dimes

2 dimes and some pennies 2 dimes and 5 pennies

5 pennies and some nickels 5 pennies and 4 nickels

Determine Whether There Is Enough Money to Purchase an Item

Display 8 dimes, 6 pennies, and the *Grocery Store Items* page. Count the money together to find its total value: 10, 20, 30, 40, 50, 60, 70, 80, (deep breath), 81, 82, 83, 84, 85, 86¢.

What is the value of the money? 86¢

Is there enough money to purchase the soup? yes

What other items can you buy with 86¢? any of the items

Continue the activity with the following sets of coins.

1 quarter banana

5 dimes green beans, banana

4 dimes, 4 nickels, 6 pennies soup, crackers, green beans, banana, yogurt

Identify the Coins Needed to Purchase an Item; Identify Equivalent Amounts of Money

Direct attention to the *Grocery Store Items* page. Instruct the students to place coins on their mats to show the amount of money needed to purchase a cup of yogurt. Remind them that the amount can be shown in different ways.

Invite several students to demonstrate the different ways to purchase yogurt. Possibilities include 6 dimes and 4 pennies; 4 dimes, 4 nickels, and 4 pennies; or 3 dimes, 6 nickels and 4 pennies.

Continue the activity for the banana. Possibilities include 2 dimes and 3 pennies; 1 dime, 2 nickels, and 3 pennies; or 4 nickels and 3 pennies.

Continue the activity for the green beans. Possibilities include 4 dimes, 1 nickel, and 4 pennies; 3 dimes, 3 nickels, and 4 pennies; or 2 dimes, 5 nickels, and 4 pennies.

Apply

Worktext pages 217–18

Read and explain the directions for the pages. Assist the students as they complete the pages independently.

112

Direct attention to the biblical worldview shaping section on page 218. Display the picture from the chapter opener and read aloud the chapter essential question. Review what happened when Remy got tired of the work he was given to do and what choice he made. Remind the students that God's Word teaches us to work hard to finish the tasks God gives us to do, even if our work takes a lot of time (Colossians 3:23). Guide the students as they complete the response on page 218.

Assess

Reviews pages 209–10

Use the Reviews pages to provide additional preparation for the chapter test.

Fact Reviews

Use "Addition Facts to 12" page 23 in Trove.

Extended Activity

Purchase Items from the General Store

Materials

- Money Kit: 99¢ in coins for each pair of students
- General store items (See Lesson 108.)

Procedure

Pair the students. Distribute 99¢ in coins to each pair. Designate one student to purchase as many items as he or she can from the general store. Designate the other student as "cashier" to receive the money and to give change.

Write the value of each coin.

 1 ¢ penny   10 ¢ dime  5 ¢ nickel  25 ¢ quarter

Write the total value. Do you have enough money to buy the item? Circle *yes* or *no.*

     27 ¢   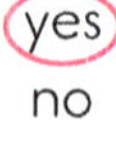

(yes) no

    18 ¢  

yes (no)

       46 ¢  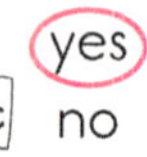

(yes) no

Mark an *X* on the coins needed to make the same value as a quarter.

   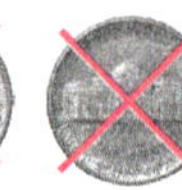    

Write the total value in each box.
Draw a line to match the items with the same price.

12 ¢

15 ¢

15 ¢

25 ¢

30 ¢

12 ¢

25 ¢

30 ¢

113 Worktext page 219 Reviews pages 211–12

Chapter 14 Test

Administer the Chapter 14 Test.

Cumulative Concept Review

Worktext page 219

Review the following concepts. Adapt instructions and activities and provide reteaching as needed to meet the specific needs of your students.

- Recognizing tools for measurement (Lesson 90)
- Completing a 2-digit addition problem in vertical form (Lesson 92)

Retain a copy of Worktext page 220 for the discussion of the Chapter 15 essential question during Lessons 114 and 122.

Reviews pages 211–12

Use the Reviews pages to help students retain previously learned skills.

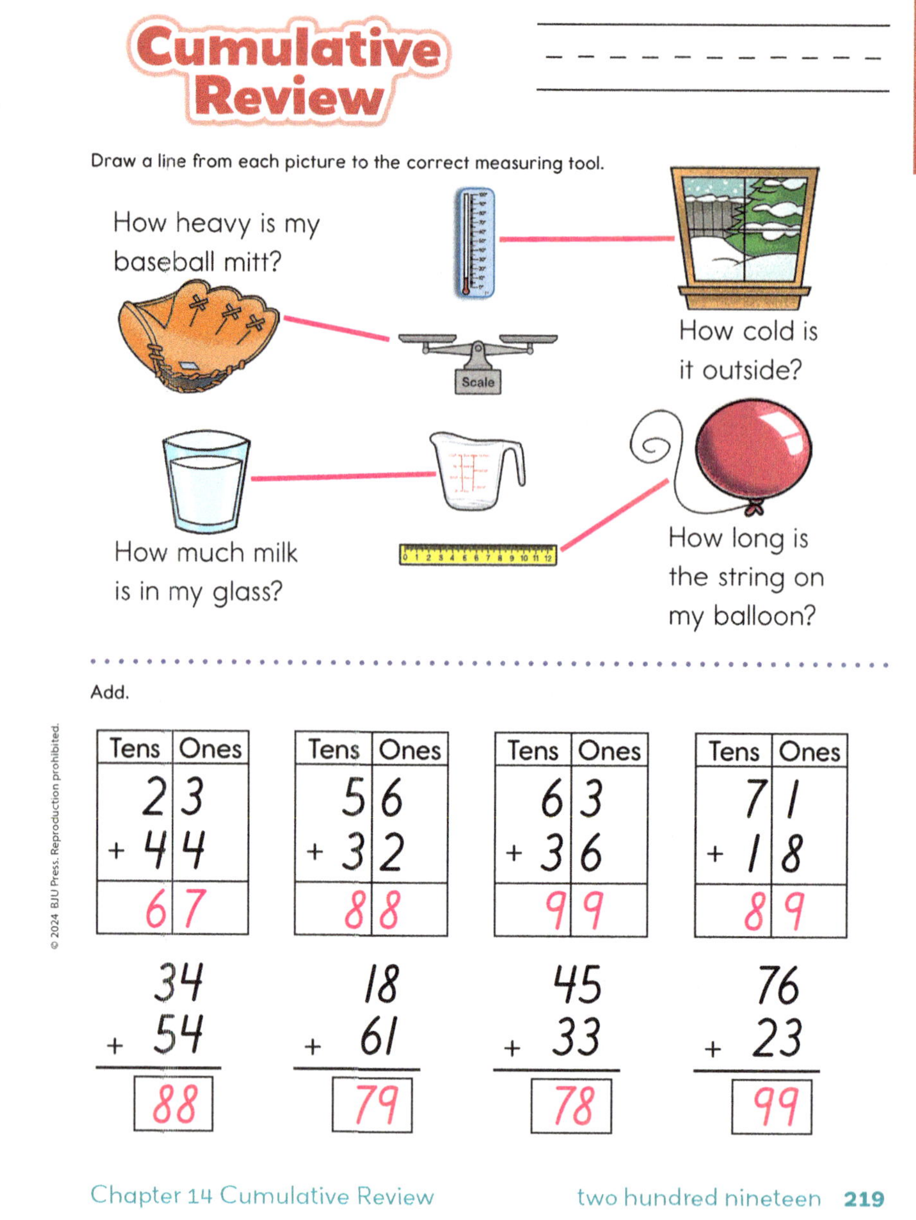

Subtract.

Tens	Ones
5	8
- 2	7
3	1

Tens	Ones
4	9
- 3	2
1	7

Tens	Ones
8	3
- 5	2
3	1

Tens	Ones
3	7
- 1	4
2	3

$67 - 40 = 27$

$95 - 3 = 92$

$26 - 13 = 13$

$75 - 33 = 42$

Draw a line to match each picture to the correct thermometer.

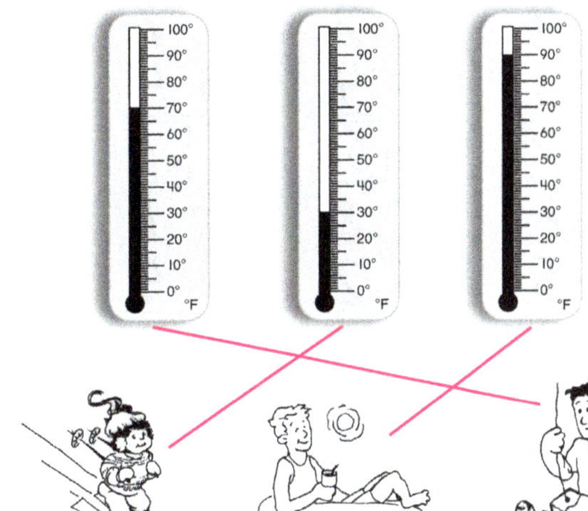

113

Addition Fact Review

Add.

7 + 2 = 9	5 + 4 = 9	9 + 2 = 11	8 + 1 = 9	3 + 3 = 6
8 + 0 = 8	6 + 4 = 10	5 + 5 = 10	7 + 3 = 10	8 + 4 = 12
3 + 6 = 9	6 + 6 = 12	5 + 3 = 8	5 + 6 = 11	7 + 0 = 7
7 + 5 = 12	9 + 3 = 12	2 + 4 = 6	7 + 4 = 11	3 + 4 = 7
4 + 4 = 8	6 + 5 = 11	8 + 2 = 10	1 + 9 = 10	2 + 7 = 9
0 + 9 = 9	4 + 7 = 11	6 + 3 = 9	5 + 7 = 12	3 + 2 = 5

CHAPTER 15: GEOMETRY

PAGES	OBJECTIVES	RESOURCES	ASSESSMENTS
Lesson 114 Solid Figures			
Teacher Edition 491–95 **Worktext** 220–22	**114.1** Distinguish 2-dimensional and 3-dimensional figures. **114.2** Identify a sphere, a cone, a cylinder, and a rectangular prism. **114.3** Describe the attributes of a sphere, a cone, a cylinder, and a rectangular prism. **114.4** Classify 3-dimensional objects as a sphere, a cone, a cylinder, or a rectangular prism. **114.5** Explain how classifying figures and objects helps people explore God's world. BWS Exploring (explain)	**Visuals** • Visual 13: *Solid Figures* • Shapes Kit: 1 circle, 1 rectangle, 1 square, 1 triangle **Student Manipulatives Packet** • Number Cards: 0–9 **BJU Press Trove*** • Video: Ch 15 Intro • Games/Enrichment: Fact Reviews (Subtraction Facts to 10) • Games/Enrichment: Subtraction Flashcards • PowerPoint® presentation	**Reviews** • Pages 213–14
Lesson 115 More Solid Figures			
Teacher Edition 496–99 **Worktext** 223–24	**115.1** Identify a cube and a triangular pyramid. **115.2** Distinguish between a cube and a rectangular prism. **115.3** Describe the attributes of a cube and a triangular pyramid. **115.4** Construct a triangular pyramid.	**Teacher Edition** • Instructional Aid 32: *Pyramid Pattern* **Visuals** • Visual 13: *Solid Figures* **Student Manipulatives Packet** • Number Cards: 0–9 **BJU Press Trove** • Games/Enrichment: Fact Reviews (Subtraction Facts to 10) • Games/Enrichment: Subtraction Flashcards • PowerPoint® presentation	**Reviews** • Pages 215–16
Lesson 116 Faces, Curves & Corners			
Teacher Edition 500–3 **Worktext** 225–26	**116.1** Describe the attributes of solid figures. **116.2** Draw faces of solid figures.	**Teacher Edition** • Instructional Aid 32: *Pyramid Pattern* **Visuals** • Visual 13: *Solid Figures* **Student Manipulatives Packet** • Number Cards 0–9 **BJU Press Trove** • Games/Enrichment: Fact Reviews (Subtraction Facts to 10) • Games/Enrichment: Subtraction Flashcards • PowerPoint® presentation	**Reviews** • Pages 217–18

*Digital resources for homeschool users are available on Homeschool Hub.

PAGES	OBJECTIVES	RESOURCES	ASSESSMENTS
Lesson 117 **Compose Solid Figures**			
Teacher Edition 504–7 **Worktext** 227–28	**117.1** Identify solid figures used to compose a picture. **117.2** Compose a solid figure by using solid figures. **117.3** Compose a picture by using solid figures.	**Visuals** • Money Kit: 1 quarter **Student Manipulatives Packet** • Red Mat • Money Kit: 5 dimes, 5 nickels, 5 pennies **BJU Press Trove** • Games/Enrichment: Fact Reviews (Subtraction Facts to 10) • Games/Enrichment: Fact Fun Activities • PowerPoint® presentation	
Lesson 118 **Plane Figures**			
Teacher Edition 508–11 **Worktext** 229–30	**118.1** Distinguish between a circle, a triangle, a square, and a rectangle. **118.2** Describe attributes of a triangle, a square, and a rectangle. **118.3** Predict the results of a probability activity. **118.4** Tally the results of a probability activity.	**Teacher Edition** • Instructional Aid 33: *Spinner* **Visuals** • Visual 12: *Plane Figures* • Number Word Cards: *zero* to *ten* • Shapes Kit: 3 red circles, 3 blue circles, 2 rectangles, 3 squares, 6 triangles **Student Manipulatives Packet** • Number Cards 0–9 **BJU Press Trove** • Web Link: Mathigon • Games/Enrichment: Subtraction Flashcards • PowerPoint® presentation	**Reviews** • Pages 219–20
Lesson 119 **Same Shape, Size, or Color**			
Teacher Edition 512–15 **Worktext** 231–32	**119.1** Identify plane figures that are the same or different shape, color, or size. **119.2** Make a Venn diagram. **119.3** Draw plane figures that are the same or different size.	**Teacher Edition** • Instructional Aid 34: *Dot Grid* **Visuals** • Shapes Kit: 5 rectangles—2 green, 1 red, 1 blue, 1 yellow; 5 triangles—2 red, 1 green, 1 blue, 1 yellow; 4 circles—1 of each color; 4 squares—1 of each color **Student Manipulatives Packet** • Shapes Kit: 1 blue circle, 4 squares, 4 triangles, 4 rectangles **BJU Press Trove** • Web Link: Mathigon • Games/Enrichment: Fact Fun Activities • PowerPoint® presentation	**Reviews** • Pages 221–22

PAGES	OBJECTIVES	RESOURCES	ASSESSMENTS
Lesson 120 Symmetry			
Teacher Edition 516–19 **Worktext** 233–34	**120.1** Identify symmetrical and asymmetrical figures. **120.2** Identify a line of symmetry. **120.3** Draw a line of symmetry on an object.	**Teacher Edition** • Instructional Aid 35: *Symmetrical Shapes* • Instructional Aid 36: *Asymmetrical Shapes* • Instructional Aid 37: *Line of Symmetry* **BJU Press Trove** • Games/Enrichment: Fact Fun Activities • PowerPoint® presentation	**Reviews** • Pages 223–24
Lesson 121 Patterns			
Teacher Edition 520–23 **Worktext** 235–36	**121.1** Copy an action pattern and a shape pattern. **121.2** Extend patterns of shapes, numbers, or letters. **121.3** Compose a picture by using plane figures. **121.4** Relate learning shapes to exploring God's world. BWS Exploring (explain)	**Visuals** • Shapes Kit (6 of each): yellow triangles, blue triangles, yellow circles, red circles, green squares, blue rectangles, orange hexagons, purple rhombuses, brown trapezoids **Student Manipulatives Packet** • Shapes Kit (6 of each): yellow triangles, blue triangles, yellow circles, red circles, green squares, blue rectangles, orange hexagons, purple rhombuses, brown trapezoids **BJU Press Trove** • Games/Enrichment: Fact Fun Activities • PowerPoint® presentation	**Reviews** • Pages 225–26
Lesson 122 Chapter 15 Review			
Teacher Edition 524–27 **Worktext** 237–38	**122.1** Recall concepts and terms from Chapter 15.	**Teacher Edition** • Instructional Aid 35: *Symmetrical Shapes* • Instructional Aid 36: *Asymmetrical Shapes* **Visuals** • Visual 13: *Solid Figures* • Visual 12: *Plane Figures* • Shapes Kit: 6 red rectangles, 6 blue squares • Money Kit: 6 pennies, 4 nickels, 4 dimes **BJU Press Trove** • Games/Enrichment: Fact Fun Activities • PowerPoint® presentation	**Worktext** • Chapter 15 Review **Reviews** • Pages 227–28
Lesson 123 Test, Cumulative Review			
Teacher Edition 528–30 **Worktext** 239	**123.1** Demonstrate knowledge of concepts from Chapter 15 by taking the test.		**Assessments** • Chapter 15 Test **Worktext** • Cumulative Review **Reviews** • Pages 229–30

How can shapes help me learn about God's world?

Can you find all the shapes?
- ▲ 1 triangle
- ▬ 5 rectangles
- ● 4 circles

Chapter 15

How can shapes help me learn about God's world?

Chapter Objectives

- Describe various 3-dimensional and 2-dimensional figures.
- Distinguish between a set of 3-dimensional figures and a pair of 2-dimensional figures.
- Predict the probability of choosing a certain shape.
- Classify shapes according to their attributes by using a Venn diagram.
- Identify symmetrical objects and a line of symmetry.
- Extend a pattern by using shapes.
- Compose a shape from 3-dimensional and 2-dimensional shapes.
- Explain how geometry helps people explore God's world.

114

Worktext pages 220–22
Reviews pages 213–14

Objectives

- **114.1** Distinguish 2-dimensional and 3-dimensional figures.
- **114.2** Identify a sphere, a cone, a cylinder, and a rectangular prism.
- **114.3** Describe the attributes of a sphere, a cone, a cylinder, and a rectangular prism.
- **114.4** Classify 3-dimensional objects as a sphere, a cone, a cylinder, or a rectangular prism.
- **114.5** Explain how classifying figures and objects helps people explore God's world. BWS

Biblical Worldview Shaping

- **Exploring** (explain): Classifying objects requires us to carefully consider all their characteristics. Examining objects closely enough to notice and compare sizes, shapes, and sounds helps us learn more about God's creation (114.5).

Printed Resources

- Visual 13: *Solid Figures*
- Visuals: Shapes Kit (1 circle, 1 rectangle, 1 square, 1 triangle)
- Student Manipulatives: Number Cards (0–9)

Digital Resources

- Video: Ch 15 Intro
- Games/Enrichment: Fact Reviews (Subtraction Facts to 10, page 45)
- Games/Enrichment: Subtraction Flashcards

Materials

- Subtraction flashcards
- Several spheres such as a baseball, marble, globe, balloon, red grape
- Several cones such as a party hat, funnel, toy top, megaphone, wooden cone-shaped block
- Several cylinders such as a soup can, roll of paper towels, cup, soft drink can, potato chip container
- Several rectangular prisms such as a cereal box, tissue box, pudding box, brick, book, wooden block

Solid Figures

Draw a line to match each solid figure with its name.

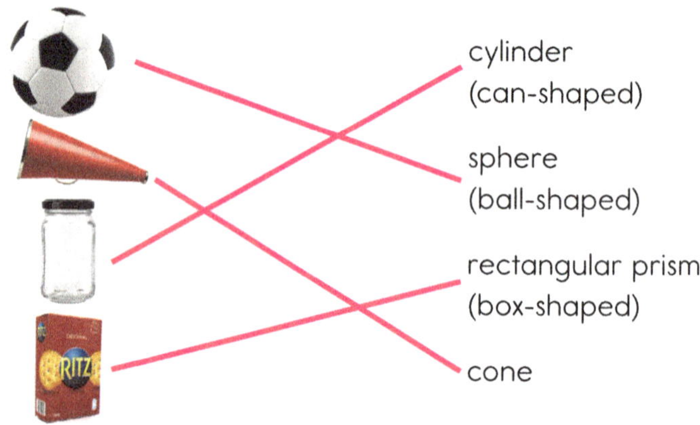

Do these solid figures have flat sides, curved sides, or both?
Mark the correct answer.

○ flat sides
○ curved sides
● both

○ flat sides
● curved sides
○ both

○ flat sides
○ curved sides
● both

● flat sides
○ curved sides
○ both

If an object you use to represent solid figures does not have one of the flat sides (faces), prepare the needed face from paper and attach it to the object. For example, prepare and attach a circle to the opening of a party hat.

Practice & Review

Memorize Subtraction Facts

Introduce the following facts.

11 – 5 11 – 6

Distribute Number Cards 0–9. Practice the new subtraction facts and previously memorized facts as in earlier lessons.

Identify a Circle, a Square, a Triangle & a Rectangle

Display a circle.

How many sides does this shape have? 0

How many corners does this shape have? 0

What is this shape? a circle

Display a rectangle.

How many sides does this shape have? 4

How many corners does this shape have? 4

What is this shape? a rectangle

Display a square.

How many corners does this shape have? 4

How many sides does this shape have? 4

What is special about the 4 sides? They are the same length.

What is this shape? a square

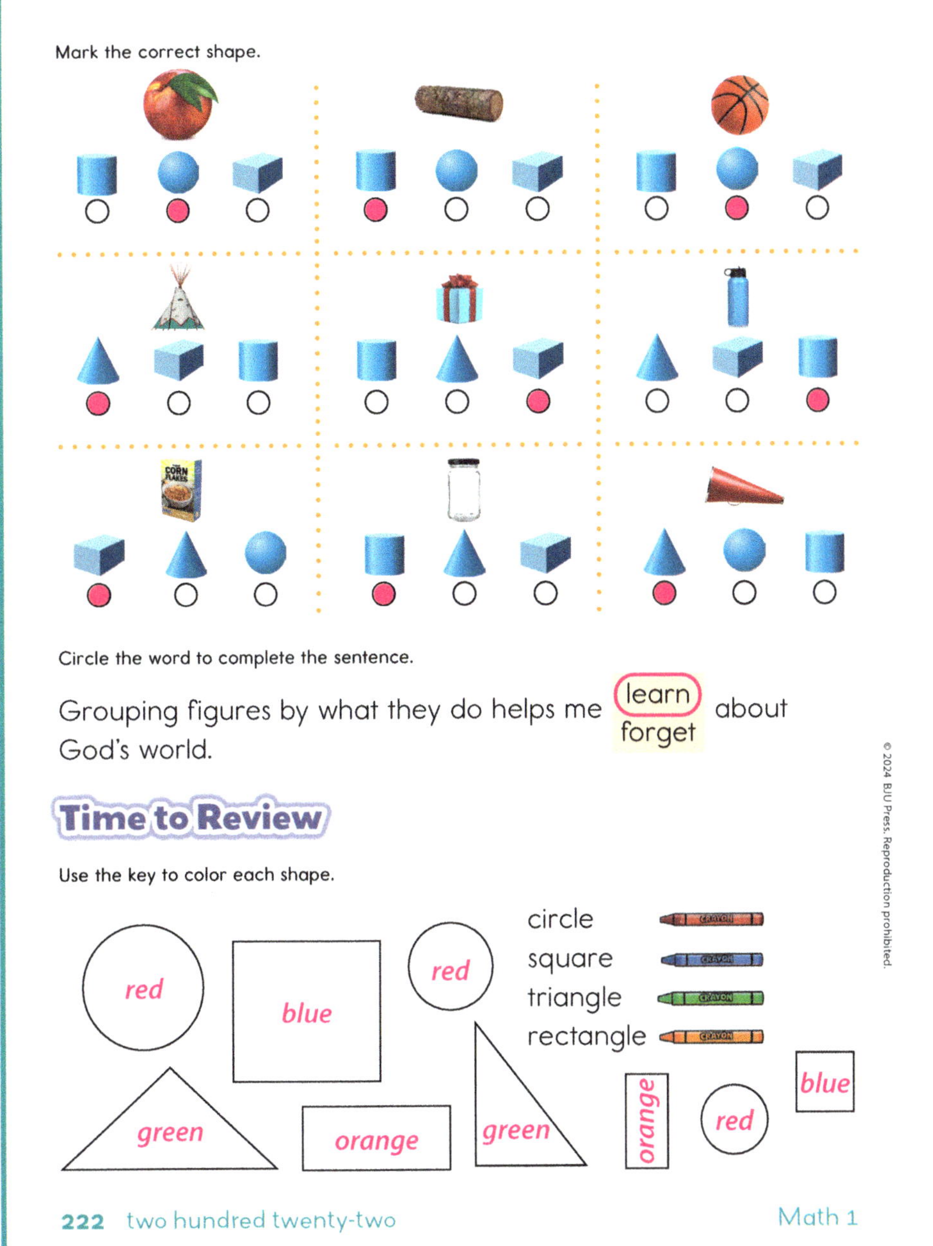
Mark the correct shape.

Circle the word to complete the sentence.

Grouping figures by what they do helps me learn / forget about God's world.

Time to Review

Use the key to color each shape.

circle
square
triangle
rectangle

red
blue
red
green
orange
green
orange
red
blue

DQ admired the triangle. It was named for its shape, and it seemed easy enough to play.

After the short concert, some students were trying out a few of the instruments.

Liam first tried the trumpet and trombone. Then he called Lily and DQ over to the large bass drum.

"Hear this low sound," said Liam, as he struck the drum with a "boom-boom-boom." "I would really like to play this someday!"

"With its curved surface and face like a circle, can you guess which shape it is?" asked DQ.

"It's another cylinder," replied Lily. "It sure is big compared to the flute. And it makes a deeper sound too."

Soon it was time for DQ to wave goodbye to his friends. As he waddled away, he thought about all the instruments, their interesting shapes, and their beautiful music.

Consider playing recordings of instruments for the students to hear the different sounds.

Direct a student to read aloud the essential question. Encourage the students to consider this question and develop a response by the end of the chapter.

Remind the students that the sides of a square are always the same length and that opposite sides of a rectangle are the same length.

Display a triangle.

How many sides does this shape have? 3

How many corners does this shape have? 3

What is this shape? a triangle

Engage

Essential Question

Direct attention to the chapter opener on Worktext page 220 and **read aloud** the following scenario to introduce the chapter essential question.

"Shhhh, DQ! The musicians are ready to play," said Lily. DQ shut his duckbill and turned his webbed feet toward the orchestra. DQ, Lily, Liam, and the other schoolchildren were gathered to hear the older students play some songs for them.

They listened to the different instrument families: the singing strings, the blaring brass, the melodious woodwinds, and the rhythmic percussion. Each instrument in a family had a different shape, size, and sound.

"I like the cymbals best," said Liam. "They have a shiny circle shape and a crashing sound!"

Lily appreciated the clear, light sound of the flute. She thought the instrument looked a bit like a cylinder.

Instruct

Distinguish 2-Dimensional & 3-Dimensional Figures

Use visual aids to illustrate 2D and 3D figures.

Display a square and a wooden block.

How are these figures different? The square is flat; the block is not flat.

Explain that the square is a *plane figure*; it has 2 dimensions: length and width. The block is a 3-dimensional object, or a *solid figure*, because it has length, width, and height. Direct attention to the 3 dimensions of the block, repeating the name of each dimension.

Identify a Sphere

Guide the students as they use **objects** to describe a sphere.

Distribute balls to various students. Explain that although these balls are different sizes and colors, they have the same shape. Direct a student to describe a ball. It is round and smooth; it can roll.

Display Visual 13. Point out that all these objects are solid figures and that they are not flat.

Which solid figures are ball-shaped objects? soccer ball, orange

Explain that all ball-shaped solid figures are called *spheres*. Point out that spheres have curved sides which allow them to roll.

Ask a volunteer to roll a ball.

Does a sphere have any flat sides? no

Does it have any corners? no

What other spheres do you see in the room? balls, a globe

Identify a Cone

Use **objects** to help the students describe a cone.

Display the cones. Display the party hat.

How would you describe this hat? It comes to a point at the top; it has a circle at the bottom.

What cone-shaped objects are on the visual? ice-cream cone, party hat

Explain that all these objects are called *cones* because they have a cone shape.

Explain that the flat sides of a solid figure are called *faces* and that faces can be circles, rectangles, squares, or triangles.

How many flat sides does a cone have? 1

Which shape is the face of a cone? a circle

Can a cone roll? Yes, it can roll on its curved side.

Ask a volunteer to roll the hat. Explain that a cone has a curved side or a *curve* which allows it to roll, but a cone does not roll as easily as a sphere since it has a flat side and a pointed end.

What other cones do you see in the room? a top, an ice-cream cone, a megaphone, a block

Identify a Cylinder

Use **objects** to help the students describe a cylinder.

Display the cylinders. Hold up a can.

How would you describe this can? It is curved on the sides; it has a circle on the top and bottom; it can roll.

What can-shaped objects are solid figures on the visual? the candle, the drum

Explain that all can-shaped objects are called *cylinders*.

How many flat sides, or faces, does a cylinder have? 2

Solid Figures

Draw a line to match each solid figure with its name.

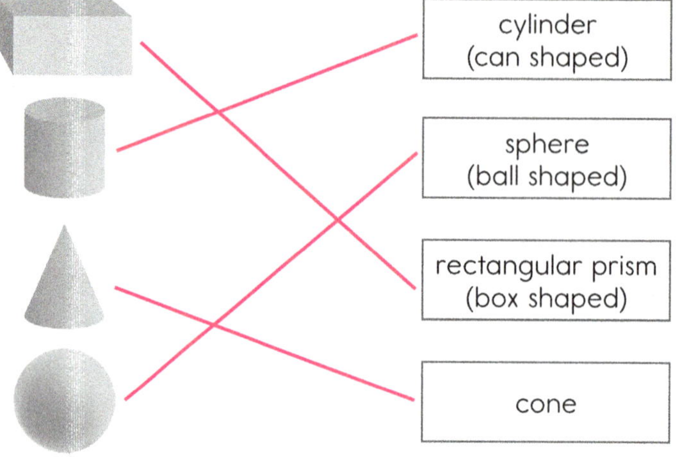

Does each solid figure have flat sides, curved sides, or both?
Mark the correct answer.

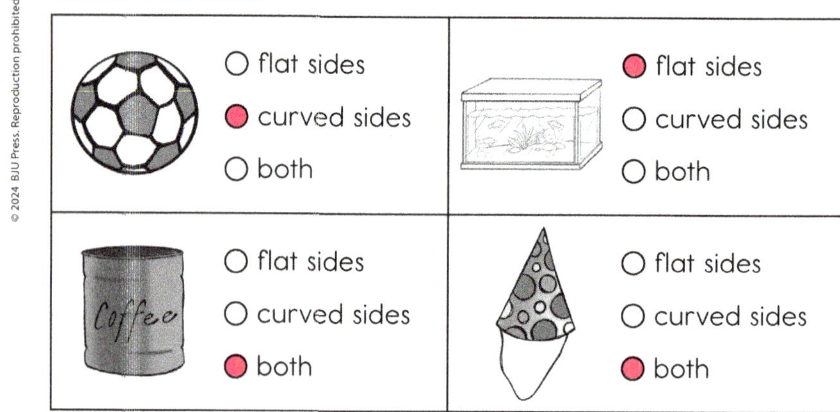

What shape are the faces of a cylinder? circle

Can a cylinder roll? yes, because it has a curved side or curve

Ask a volunteer to roll a cylinder.

How easily does the cylinder roll compared to the sphere and the cone? The cylinder does not roll as well as a sphere, because the cylinder has 2 flat sides. A cylinder rolls better than a cone since a cylinder does not have a pointed end.

What other cylinders do you see in the room? chalk, roll of paper towels, trash can, soft drink can

Identify a Rectangular Prism

Use **objects** to help the students describe a rectangular prism.

Display the rectangular prisms. Display a cereal box.

How would you describe this box of cereal? It has flat sides; it has edges and corners; it cannot roll.

What box-shaped objects are solid figures on the visual? tissue box, gift

Explain that box-shaped objects are called *rectangular prisms*. Count the faces of the box together.

How many faces does a rectangular prism have? 6

What shape are the faces of a rectangular prism? rectangle

Does a rectangular prism have any curved sides? no

Can a rectangular prism roll? no

Chapter 13 Review

Subtract.

56 − 42 = 14	74 − 43 = 31	89 − 37 = 52	48 − 38 = 10
65 − 34 = 31	83 − 51 = 32	58 − 12 = 46	77 − 26 = 51

Write an equation for the word problem. Solve.

The boy picked up 36 rocks on the riverbank. He lost 12 on his walk back to camp. How many rocks does the boy have left?

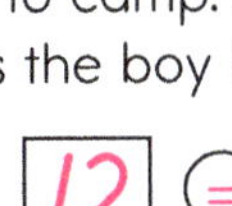

36 − 12 = 24 rocks

work space

36
−12
24

Addition Fact Review

Add.

2 + 8 = 10	4 + 2 = 6	5 + 6 = 11
5 + 3 = 8	9 + 2 = 11	4 + 3 = 7
7 + 4 = 11	8 + 3 = 11	4 + 6 = 10

214 two hundred fourteen Math 1 Reviews

Point to the corners of the box and explain that wherever 3 faces meet, they form a corner. Count the corners of the box together.

How many corners does a rectangular prism have? 8

What other rectangular prisms do you see in the room? books, blocks, boxes

Classify 3D Objects

Use a **visual** to help the students classify 3D objects.

Use Visual 13 to review the names of the first 4 solid figures. Display the words "sphere," "cone," "cylinder," and "rectangular prism." Direct attention to all the solid figures displayed earlier in the lesson. Ask volunteers to select objects and stand below the category that matches their chosen objects. Review the number of faces, curves, and corners of each object to help the students decide if the object is in the correct category.

Classifying Objects Helps People Explore God's World

Guide a **discussion** to help the students explain a biblical worldview shaping truth.

Display a cylinder.

Does it have a curved side? yes

Does it have flat sides, or faces? yes

Can it roll? yes

Display a rectangular prism.

What kind of sides does it have? flat sides

Does it have edges and corners? yes

Can it roll? no

Ask the students to think about the shape of a bicycle tire.

Would a round or box-shaped wheel work better for a tire? round because it rolls

Point out that understanding the features of a shape (such as whether it will roll) helps them learn more about how things work in God's world.

Grouping shapes by their features or attributes is called *classifying*. Explain that identifying and classifying shapes helps the students describe what they see so that they can use that knowledge to learn more about the world God has made.

How does classifying objects help you explore God's world? It helps me learn more about the world God has made.

Apply

Worktext pages 221–22

Guide completion of page 221.

Read and explain the directions for page 222. Assist the students as they complete the page independently.

Direct attention to the biblical worldview shaping statement and guide the students as they complete the sentence.

Assess

Reviews pages 213–14

Review subtracting 2-digit numbers on page 214.

Fact Reviews

Use "Subtraction Facts to 10" page 45 in Trove.

Extended Activity

Identify a Sphere, a Cone, a Cylinder & a Rectangular Prism

Materials

- 4 sheets of posterboard or chart paper
- Magazines

Procedure

Write a label at the top of each sheet: "sphere," "cone," "cylinder," and "rectangular prism." Instruct the students to cut out magazine pictures of objects that illustrate the shapes and then glue them to the correct sheet. Discuss together whether each picture has been placed in the correct category.

115 Worktext pages 223–24 Reviews pages 215–16

Objectives

- **115.1** Identify a cube and a triangular pyramid.
- **115.2** Distinguish between a cube and a rectangular prism.
- **115.3** Describe the attributes of a cube and a triangular pyramid.
- **115.4** Construct a triangular pyramid.

Printed Resources

- Instructional Aid 32: *Pyramid Pattern* (optional)
- Visual 13: *Solid Figures*
- Student Manipulatives: Number Cards (0–9)

Digital Resources

- Games/Enrichment: Fact Reviews (Subtraction Facts to 10, page 46)
- Games/Enrichment: Subtraction Flashcards

Materials

- Subtraction flashcards
- Gumdrop pyramid model, using 6 toothpicks and 4 gumdrops (See directions for model preparation at the end of this lesson.)
- 6 toothpicks and 4 gumdrops for model demonstration (for the teacher and for each student)
- Several cubes such as a wooden block, photo cube, tissue box, UNIFIX® Cube
- Several triangular pyramids such as a block, candle, paperweight
- Several rectangular prisms (from Lesson 114)

Other pyramids will be introduced in an upper grade level along with the term *base*. This lesson introduces basic recognition of pyramids as having triangular faces that meet at a point (vertex).

Practice & Review

Memorize Subtraction Facts

Introduce the following facts.

12 – 3   12 – 9

More Solid Figures

Draw a line to match each solid figure with its name.

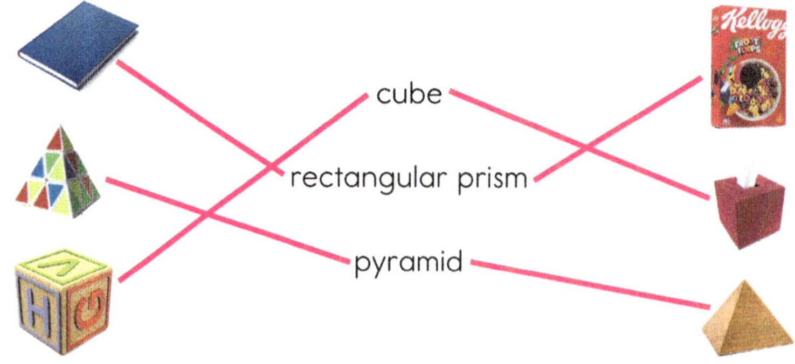

Do these solid figures have flat sides, curved sides, or both? Mark the correct answer.

- ● flat sides
- ○ curved sides
- ○ both

- ● flat sides
- ○ curved sides
- ○ both

Cross out the solid figure that does not belong. Mark the sentence that tells why.

- ● It does not have flat sides.
- ○ It does not have curved sides.

Distribute Number Cards 0–9. Practice the new subtraction facts and previously memorized facts as in earlier lessons.

Practice these and the previously memorized subtraction facts.

Engage

Solid Figures in Nature

Guide the students in a **review** of solid figures.

Remind the students that solid figures can be found everywhere. They can be found in God's creation, such as the shape of a tree trunk (cylinder), the shape of a pea (sphere), and the shape of a pinecone (cone). People have also used shapes in their inventions, such as bricks for buildings (rectangular prisms), pillars for bridges (cylinders), balls for play (spheres), and party hats (cones).

Display Visual 13. Review spheres, cones, cylinders, and rectangular prisms.

Discuss other examples of natural or man-made solid figures.

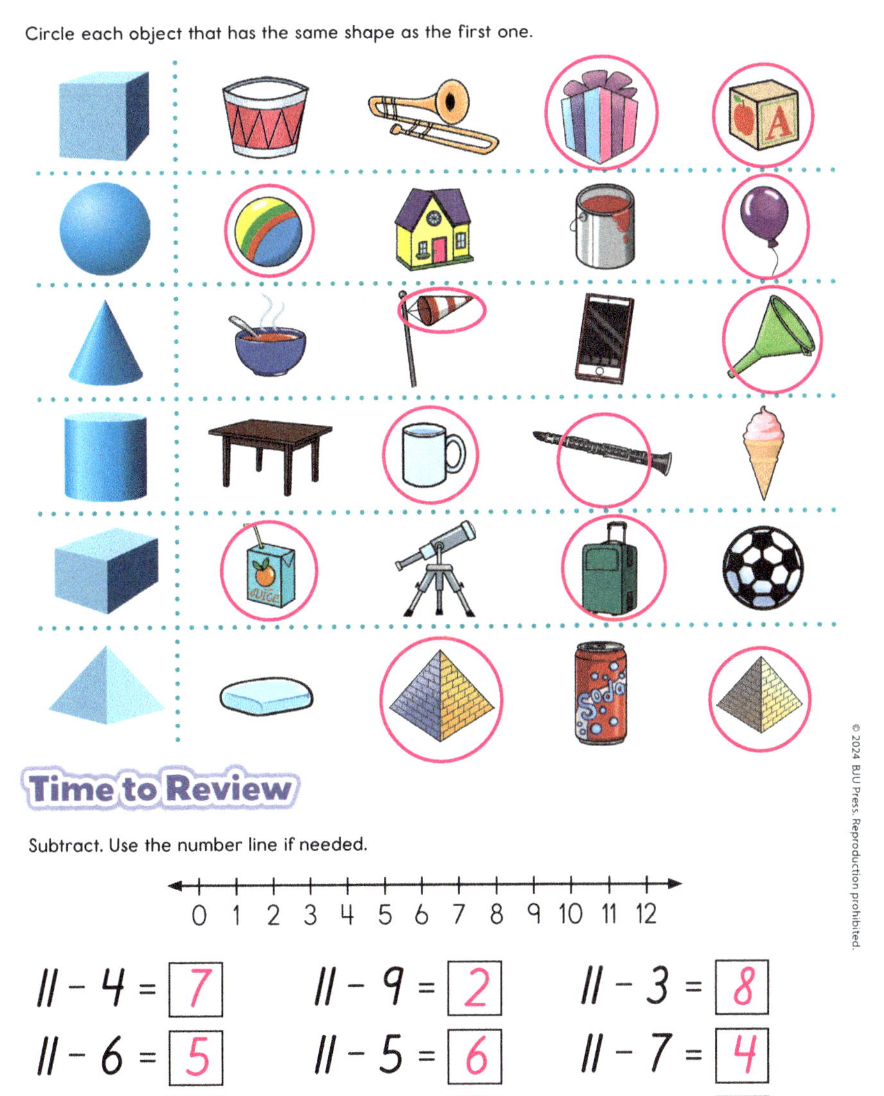

Circle each object that has the same shape as the first one.

Time to Review

Subtract. Use the number line if needed.

0 1 2 3 4 5 6 7 8 9 10 11 12

11 − 4 = 7	11 − 9 = 2	11 − 3 = 8
11 − 6 = 5	11 − 5 = 6	11 − 7 = 4
11 − 8 = 3	12 − 9 = 3	12 − 3 = 9

Instruct

Identify a Cube

Use **visuals** to help the students identify a cube.

Display the wooden block (cube).

How would you describe this block? It has flat sides; it has corners; the sides are the same size; the sides are squares.

Display Visual 13. Explain that this figure is called a *cube*.

Which of these objects is in the shape of a cube? planter

What other cubes do you see in the room? UNIFIX Cube, photo cube, tissue box

Distinguish between a Cube & a Rectangular Prism

Guide a **discussion** to help the students distinguish a cube and a rectangular prism.

Remind the students that solid figures are identified by their faces (flat sides), curves (curved sides), and corners.

Does the cube have any curved sides? no

Can a cube roll since it does not have any curves? no

Guide the students as they count the faces of the cube as you point to each face.

How many faces does the cube have? 6

Count together the corners of the cube as you point to each corner.

How many corners does the cube have? 8

Display the wooden block and a cereal box (or another rectangular prism). Explain that these 2 types of solid figures are similar. Draw 2 columns for display and label them "cube" and "rectangular prism." Write "faces" and "corners" in separate rows to the left of the 2 columns.

How many faces does the cube have? 6

How many corners does the cube have? 8

Record the answers in the column labeled "cube."

Guide the students as they count the faces and corners of the rectangular prism. Record the answers in the column labeled "rectangular prism." 6 faces, 8 corners

Do either of these figures have curved sides? no

Point out that both cubes and rectangular prisms have the same number of faces and corners and no curves.

Direct the students to look at the faces to see how the 2 figures are different.

How are the faces of the rectangular prism different from the faces of the cube? The faces on the rectangular prism are rectangles, and the faces on the cube are squares.

Display other cubes and rectangular prisms. Invite students to choose objects and to stand beside the display column that matches their chosen objects.

After all objects have been chosen, guide the students as they determine whether each object is in the correct category by looking at its faces.

Identify a Pyramid

Use a **visual aid** to help the students identify a pyramid.

More Solid Figures

blue red yellow green purple orange

Use the color key to color the objects.

Chapter 15 • Lesson 115 two hundred fifteen **215**

Show an example of a triangular pyramid. Explain that this solid figure is called a *pyramid*. Hold the pyramid in the palm of your hand for the students to observe.

What do you notice about the faces on this pyramid? They are all triangles.

Does this figure have any curves? no

Does this pyramid have corners? yes

Guide the students as they count the faces of the pyramid as you point to each face.

How many faces does this pyramid have? 4

Count together the corners of the pyramid as you point to each corner.

How many corners does this pyramid have? 4

Explain that a pyramid has triangles for faces on the sides and that the sides always make a point at the top.

Display the gumdrop pyramid model. Explain that this model is also a solid figure, even though you can see through it. Point out the 4 triangle faces and the 4 corners.

Direct attention to the pyramids on Visual 13.

Which of these objects is in the shape of a pyramid? candle

Explain that there are fewer solid figures that are pyramids than any of the other shapes. Conclude together that pyramids are usually made by people and not found in nature.

Construct a Triangular Pyramid

Guide a **hands-on activity** to construct a pyramid.

Distribute a set of toothpicks and gumdrops to each student.

Demonstrate each step, giving help as needed. Instruct the students to connect 3 toothpicks to 3 gumdrops to form a triangle. Next, direct them to lay their triangles flat on their desks and to place a toothpick in the top of each gumdrop so that the toothpicks point up. Finally, direct them to connect the 3 toothpicks together with the other gumdrop at the top.

Guide the students in identifying the triangular-shaped faces and the corners of the pyramid.

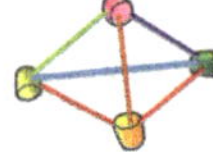

> **Differentiated Instruction**
>
> **Identify Solid Figures**
>
> Write the following labels on separate sheets of paper and display them: "sphere," "rectangular prism," "cylinder," "cone," "cube," and "pyramid." Place an example of each figure with the label. Provide several types of each figure and instruct the students to sort the items and place each item with the correct label and figure. Direct the students to explain their answers.

Apply

Worktext pages 223–24

Guide completion of page 223.

Read and explain the directions for page 224. Assist the students as they complete the page independently.

Assess

Reviews pages 215–16

Review adding 2-digit numbers on page 216.

Fact Reviews

Use "Subtraction Facts to 10" page 46 in Trove.

Chapter 12 Review

Add.

43 + 52 = 95	59 + 30 = 89	83 + 14 = 97	35 + 24 = 59
65 + 31 = 96	76 + 22 = 98	24 + 14 = 38	56 + 12 = 68

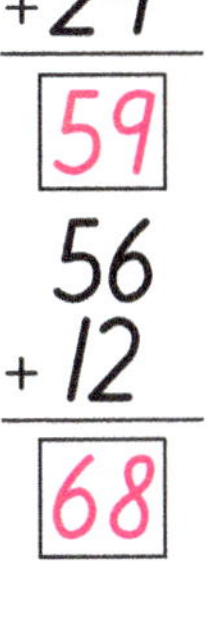

Write an equation for the word problem.
Solve.

In the morning 12 hummingbirds came to the feeder. In the afternoon 10 hummingbirds came to the feeder. How many hummingbirds came in all?

work space

12
+ 10
22

12 + 10 = 22 hummingbirds

Subtraction Fact Review

Subtract.

10 − 3 = 7	8 − 4 = 4	11 − 8 = 3	7 − 4 = 3	11 − 2 = 9
10 − 7 = 3	9 − 5 = 4	10 − 6 = 4	8 − 2 = 6	10 − 4 = 6

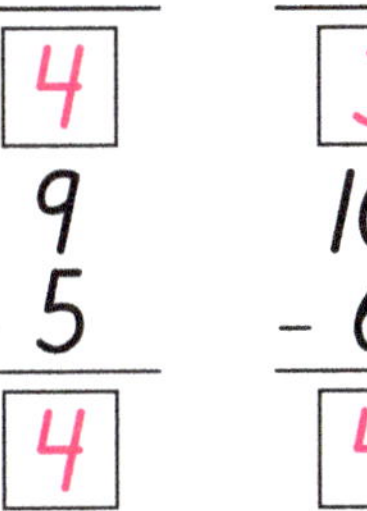
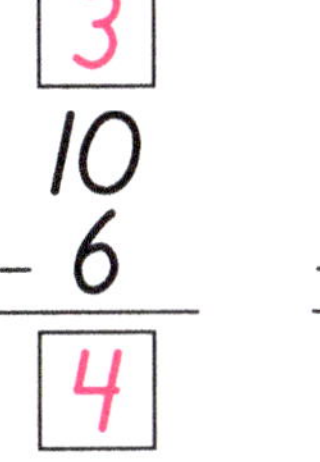
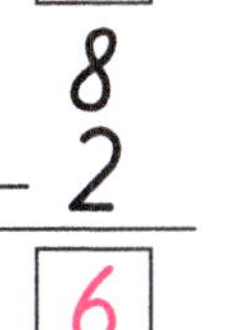

Extended Activity

Identify Solid Figures

Materials

- Cloth bag, paper sack, or cylindrical oatmeal container
- Assorted solid figures: spheres, cones, cylinders, rectangular prisms, cubes

Procedure

Place the solid figures in the container. Ask a student to reach inside without looking and to hold an object without taking it out of the container. Instruct the student to describe the figure (how many faces, corners, and curved sides). Encourage other students to guess the shape of the object; then direct the student to remove the object from the container to reveal the object's identity. Continue the activity; ask other students to describe an object until all the objects have been identified.

Variation: Identify plane figures by placing an assortment of squares, rectangles, triangles, and circles in the container.

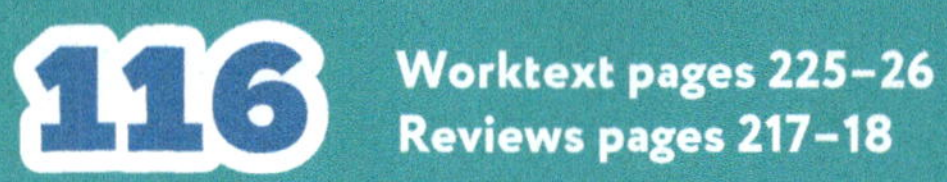

Objectives

- **116.1 Describe attributes of solid figures.**
- **116.2 Draw faces of solid figures.**

Printed Resources

- Instructional Aid 32: *Pyramid Pattern*
- Visual 13: *Solid Figures*
- Student Manipulatives: Number Cards (0–9)

Digital Resources

- Games/Enrichment: Fact Reviews (Subtraction Facts to 10, page 47)
- Games/Enrichment: Subtraction Flashcards

Materials

- Subtraction flashcards
- Gumdrop pyramid model (from Lesson 115)
- Solid figures: 1 cylinder, 1 cone, 1 sphere, 1 cube, 1 triangular pyramid, 1 rectangular prism (for each group of 3–4 students)

If you do not have enough pyramids, you may use the *Pyramid Pattern* to construct several paper pyramids.

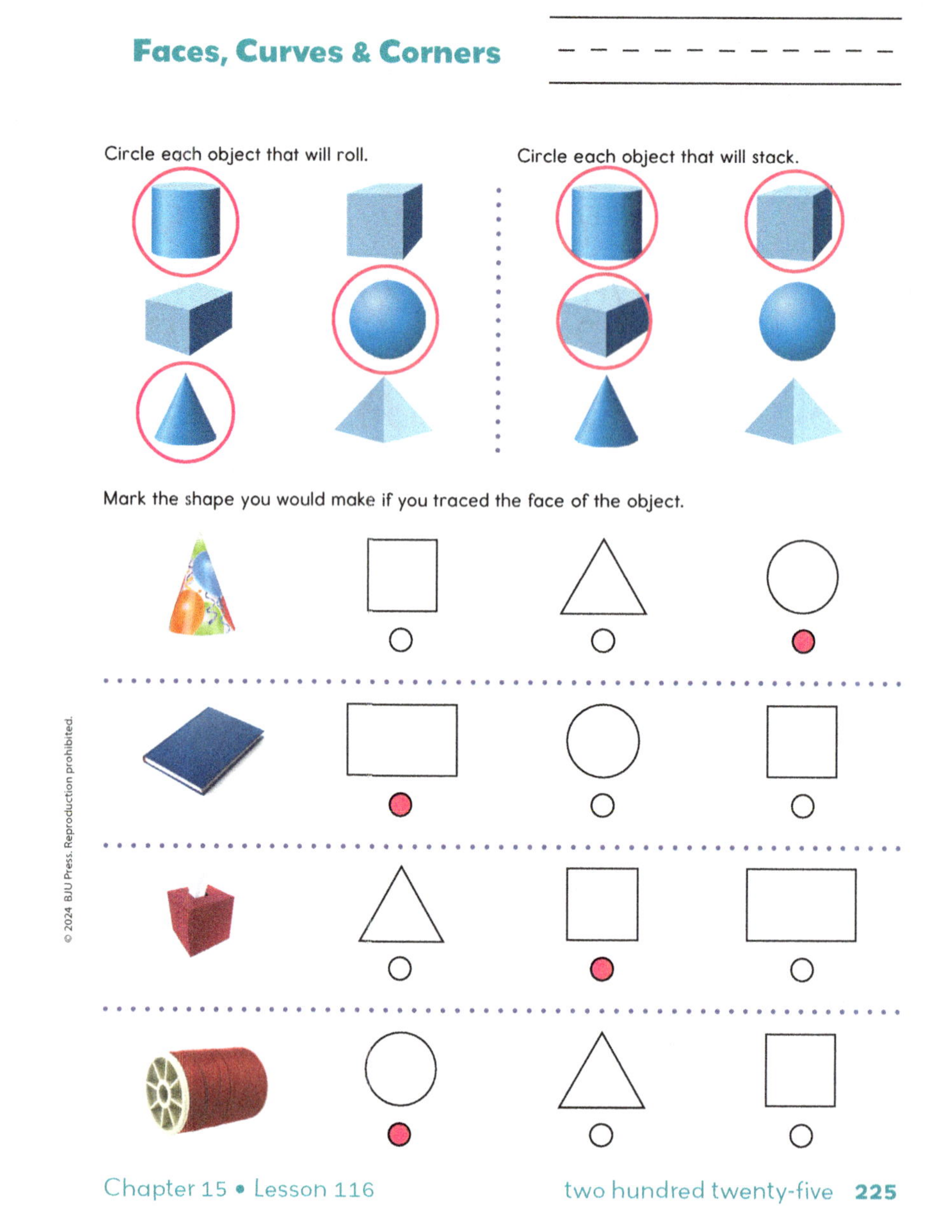

Faces, Curves & Corners
Circle each object that will roll.
Circle each object that will stack.
Mark the shape you would make if you traced the face of the object.
© 2024 BJU Press. Reproduction prohibited.
Chapter 15 • Lesson 116
two hundred twenty-five 225

Practice & Review

Memorize Subtraction Facts

12 – 4   12 – 8

Display each flashcard slowly. Invite students to give the answers.

Distribute Number Cards 0–9. Practice the new subtraction facts and previously memorized facts as in earlier lessons.

Write Numbers to 150

Direct the students to display their Number Cards to represent 134.

Ask a volunteer to show 134 for display for students to check their cards. Continue the activity for numbers 97, 102, 136, and 150.

Write the Number That Comes *before, after, or between*

Direct the students to use their Number Cards to show the number that matches each clue. Invite a student to write the correct number for display for students to check their cards. Give these and other similar clues.

Which number comes just *before* 73? 72

Which number comes just *after* 59? 60

Which number comes *between* 36 and 38? 37

Engage

Identify Solid Figures

Guide a **review** of solid figures.

Display Visual 13 with the names of figures covered. Direct attention to a row and ask the students to identify the solid figures in that row. Direct a student to remove the cover to check the name. Continue the activity until all the figures are identified.

Mark an *X* in the box if the solid figure has curves, faces, or corners.

	Curves (curved sides)	Faces (flat sides)	Corners
Sphere	X		
Cone	X	X	
Cylinder	X	X	
Rectangular prism		X	X
Cube		X	X
Pyramid		X	X

Time to Review

Add.

$33 + 54 = 87$ $52 + 24 = 76$ $27 + 31 = 58$ $75 + 24 = 99$

Subtract.

$67 - 43 = 24$ $88 - 24 = 64$ $93 - 32 = 61$ $77 - 65 = 12$

Instruct

Identify Curves

Guide a **collaborative activity** to help the students identify curves.

Group the students. Distribute a set of solid figures to each group.

Remind the students that solid figures are identified by their faces (flat sides), curves (curved sides), and corners.

Explain that in this lesson they will discover how solid figures are alike and different.

Direct each group of students to find the sphere in the set of solid figures.

Does a sphere have any corners? No; there are no points and no flat sides to make corners.

Does a sphere have any faces? No; there are no flat sides.

Does a sphere have any curves? yes

Explain that the students can test their answers by rolling the spheres. Conclude together that if a solid figure rolls, it has a curved side.

Did your sphere roll? yes

Instruct the students to test the other solid figures to see whether they will roll. Allow a few minutes for students to try each object in their sets.

Which other solid figures roll? a cone, a cylinder

Identify Faces

Guide a **hands-on activity** to identify faces.

Remind the students that a face is a flat side. Explain that most solid figures with faces can be stacked on top of each other. Direct each group of students to stack or try to stack all the figures that have faces.

Which figures can you stack? any figure with a flat surface

Allow time for the students to stack their figures. Conclude together that the flat end of the cone (or pyramid) will stack on any other flat surface; if they hold the pointed end of the cone in their hands, they can stack something on it, but they cannot stack anything on the pointed end.

Identify Corners

Guide a **discussion** to identify corners.

Remind the students that a corner is formed where 3 faces meet. Direct the students to look at the solid figures in their sets.

Which solid figures have corners? a cube, a rectangular prism, a pyramid

Which solid figures do not have any corners? a sphere, a cylinder, a cone

Direct the students in each group to count the corners on the cube and the rectangular prism in their set.

How many corners does a cube have? 8

How many corners does a rectangular prism have? 8

Direct the students to count the corners on the pyramid.

How many corners does this triangular pyramid have? 4

Draw Faces of Solid Figures

Guide a **hands-on activity** to help the students draw faces of solid figures.

Trace around the face of a pyramid for display. Direct the students to use paper and pencil and to take turns tracing around their pyramids.

What shape are the faces on this pyramid? triangles

Explain that faces on solid figures can also be circles, squares, or rectangles.

Instruct the students to choose a solid figure from their group's set of objects, lay the object on a face (flat side), and trace around the face.

Direct the students to trace all the solid figures in their group's set. Remind the students that they will need to take turns so that each person in their group has the opportunity to trace each solid figure.

Which figures have a face that is a circle? a cylinder and a cone

Do cylinders and cones have the same number of circular faces? No; the cone has 1 face, and the cylinder has 2 faces.

Which solid figure has only squares for faces? a cube

Which solid figure has rectangles for faces? a rectangular prism

Apply

Worktext pages 225–26

Guide completion of page 225.

Read and explain the directions for page 226. Assist the students as they complete the page independently.

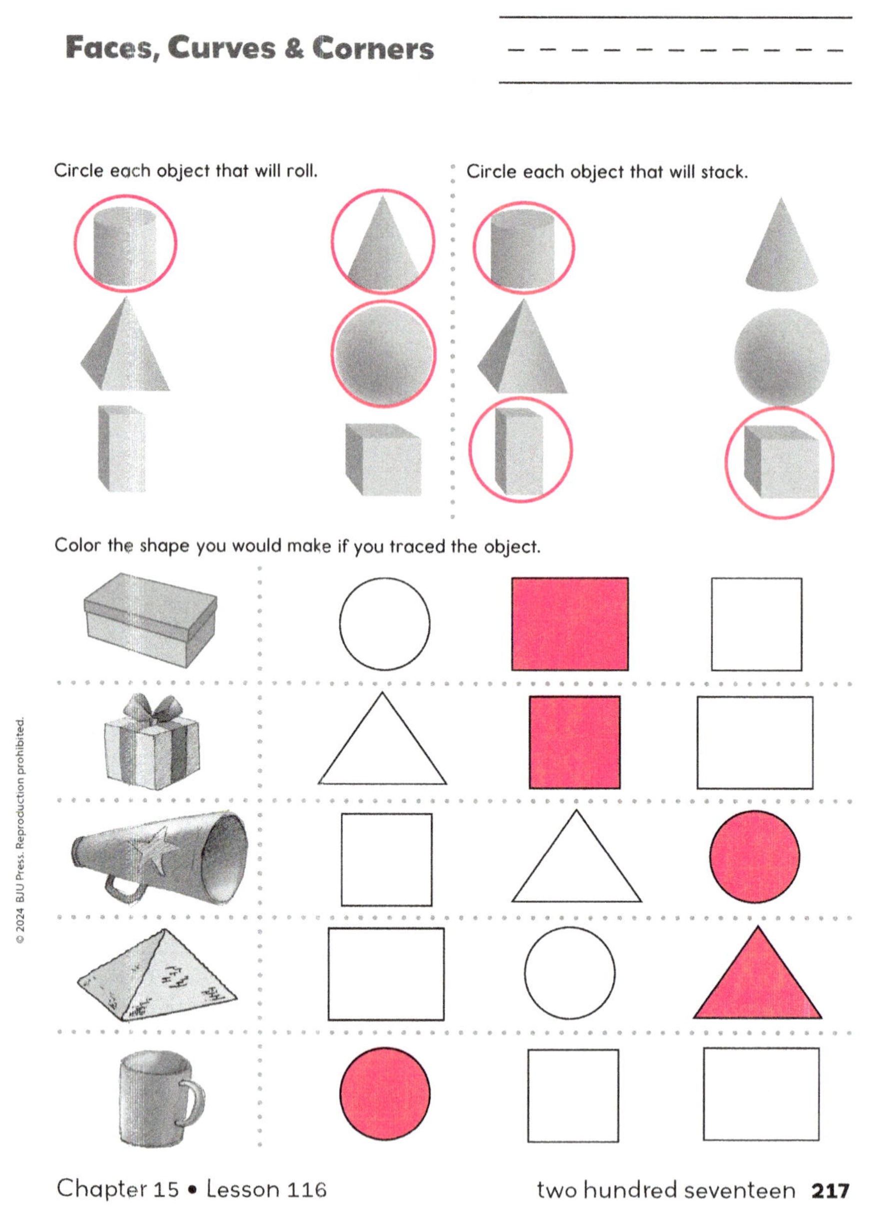

Faces, Curves & Corners
Circle each object that will roll.
Circle each object that will stack.
Color the shape you would make if you traced the object.
© 2024 BJU Press. Reproduction prohibited.
Chapter 15 • Lesson 116
two hundred seventeen 217

Chapter 11 Review

Use an inch ruler to measure.

Use an inch ruler to draw each length. Start at the star.

3 inches ★

2 inches ★

5 inches ★

Subtraction Fact Review

0 1 2 3 4 5 6 7 8 9 10 11 12

Subtract. Use the number line if needed.

11 - 6 = 5	12 - 3 = 9	12 - 4 = 8
11 - 5 = 6	12 - 9 = 3	12 - 8 = 4

Assess

Reviews pages 217–18

Review using an inch ruler to measure on page 218.

Fact Reviews

Use “Subtraction Facts to 10” page 47 in Trove.

Extended Activity

Match Solid Figures

Materials

- Index cards
- Pictures of a rectangular prism, a cube, a cone, a cylinder, and a pyramid (optional)

Preparation

Prepare a set of 5 face cards and 5 solid figure cards for each group of students. To make 5 face cards, draw a shape on 5 index cards: 1 triangle, 1 square, 1 rectangle, and 2 circles. To make 5 solid figure cards, draw (or glue a picture of) a solid figure on 5 index cards: 1 rectangular prism, 1 cube, 1 cone, 1 cylinder, and 1 triangular pyramid.

Procedure

Arrange the students in groups of 2 or 3. Direct the students to lay the cards face-down on a desk in 2 rows of 5 cards. Explain that they are to take turns turning over 2 cards. If the face card matches the kind of faces on the solid figure card, the student keeps both of the cards and the next student takes his or her turn. If the cards do not match, both cards are turned back over. The student with the most matches is the winner.

Worktext pages 227–28

Objectives

- **117.1 Identify solid figures used to compose a picture.**
- **117.2 Compose a solid figure by using solid figures.**
- **117.3 Compose a picture by using solid figures.**

Printed Resources

- Visuals: Money Kit (1 quarter)
- Student Manipulatives: Red Mat
- Student Manipulatives: Money Kit (5 dimes, 5 nickels, 5 pennies)

Digital Resources

- Games/Enrichment: Fact Reviews (Subtraction Facts to 10, page 48)
- Games/Enrichment: Fact Fun Activities

Materials

- A brown paper bag
- Solid figures: 1 cone, 1 sphere, 1 cylinder, 4 cubes that are all the same size, 2 rectangular prisms that are both the same size (for the teacher)
- Solid figures: cones, spheres, cylinders, 4 cubes that are all the same size, 2 rectangular prisms that are both the same size (for each small group of students)

Compose Solid Figures

Circle the combinations that have the same shape.

 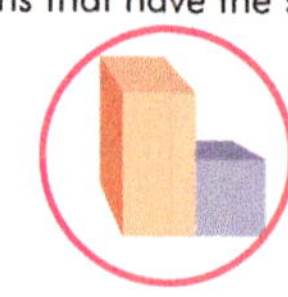  

   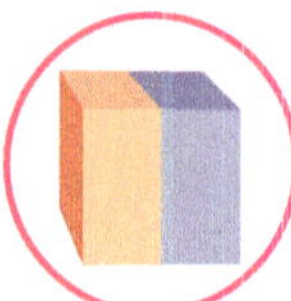

Circle each kind of figure used in the combination.

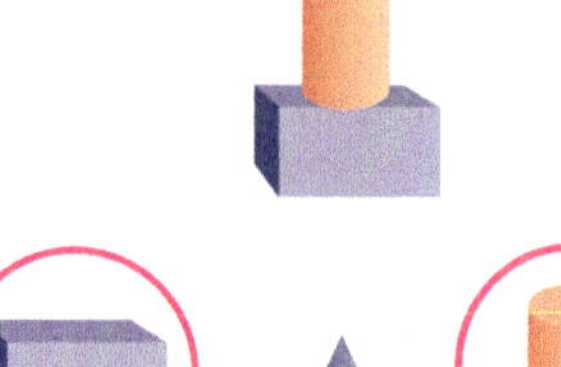

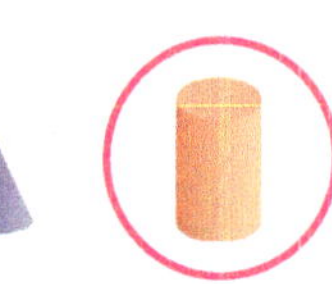

Practice & Review

Review Subtraction Facts

Select a game or activity from the “Fact Fun Activities” in Trove. Practice the previously memorized subtraction facts.

Identify Sets of Coins Equivalent to 1 Quarter

Display a quarter.

What coin is this? a quarter

What is the value of a quarter? 25¢

Write “25¢” for display. Distribute a Red Mat and set of coins to each student. Direct the students to place the value of a quarter on their mats, using the following combinations.

only nickels 5 nickels

1 dime and some nickels 1 dime and 3 nickels

1 nickel and some dimes 2 dimes and 1 nickel

2 dimes and some pennies 2 dimes and 5 pennies

5 pennies and some nickels 4 nickels and 5 pennies

Engage

Differentiate Shapes

Guide a **visual analysis** of shapes to help the students identify the one that does not belong.

Display 3 cubes and 1 rectangular prism.

Which one is different? the rectangular prism because its faces are not all the same size

Remove the rectangular prism.

How would you describe the remaining set? They are cubes.

Continue the activity with the following sets:

3 cones and 1 cylinder

3 spheres and 1 cone

Instruct

Identify Solid Figures

Use **visual aids** to identify solid figures.

Display a cone, point side down, with a sphere over it.

What 2 solid figures am I holding? a sphere and a cone

What real-life object does this new shape remind you of? an ice-cream cone

Turn the figure upside-down.

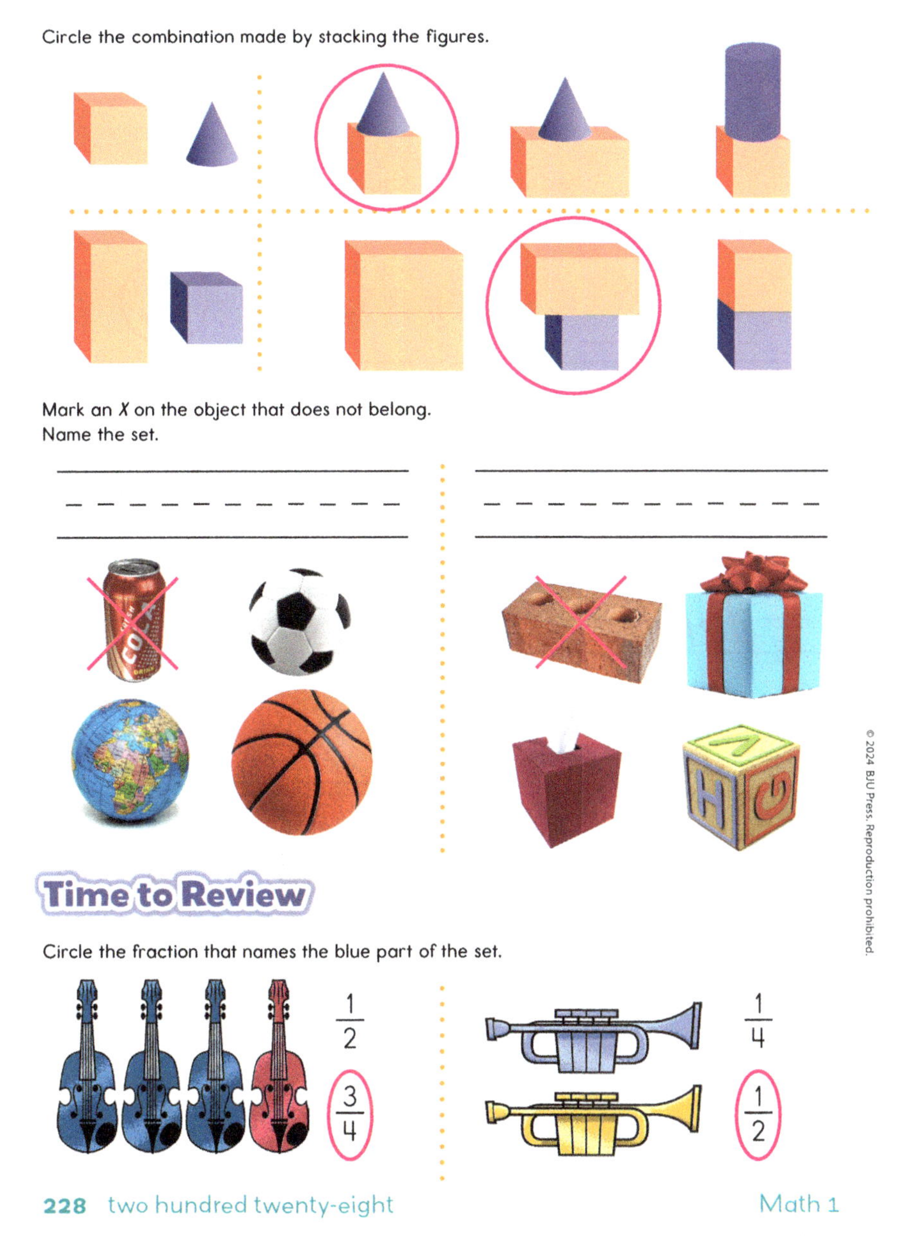
Circle the combination made by stacking the figures.

Mark an *X* on the object that does not belong.
Name the set.

Time to Review

Circle the fraction that names the blue part of the set.

$\frac{1}{2}$ $\frac{3}{4}$

$\frac{1}{4}$ $\frac{1}{2}$

228 two hundred twenty-eight Math 1

What real-life object does this shape remind you of? a person wearing a party hat

Display other combinations of 3D shapes and ask what they resemble.

a cylinder with a sphere over it a tree

a cylinder with a cone over it a pine tree

a cube with a cone over it a house or a circus tent

Display a tower made of a cube, a cylinder, and a cone.

What solid figures were used to make this tower? a cube, a cylinder, and a cone

Continue the activity by using other sets of stacked solid figures. Each time ask a student to identify the solid figures that were used to build the tower.

Compose a Solid Figure by Using Solid Figures

Guide a **hands-on activity** to compose a solid figure.

Group the students and distribute 4 cubes (all the same size) and 2 rectangular prisms (both the same size) to each group. Model each step with your solid figures.

Ask the students to identify each shape and its attributes. cube: all flat surfaces in the shape of a square; rectangular prism: all flat surfaces, some in the shape of a rectangle and some in the shape of a square

Direct the students to place 2 cubes next to each other, with the sides touching.

What shape did you make? a rectangular prism

How do you know it is a rectangular prism? All its surfaces are flat, some surfaces are in the shape of a rectangle, and 2 surfaces are in the shape of a square.

Allow time for the students to experiment with combining the 4 cubes in different ways to make longer and taller rectangular prisms.

Direct the students to place the 2 rectangular prisms next to each other, with the sides touching.

What shape did you make? a longer rectangular prism

How do you know it is a rectangular prism? All its surfaces are flat and in the shape of a rectangle.

Allow time for the students to experiment with combining the 2 rectangular prisms in different ways.

Conclude together that rectangular prisms and cubes can be joined to make rectangular prisms.

Compose a Picture by Using Solid Figures

Guide a **collaborative activity** to compose a picture.

Distribute additional solid figures (spheres, cylinders, and cones) to each group of students.

Direct each group to make a design or picture, using some or all their solid figures.

Invite a student from each group to explain which solid figures the group used to make its design.

Apply

Worktext pages 227–28

Guide completion of page 227.

Read and explain the directions for page 228. Assist the students as they complete the page independently.

Assess

Reviews

There are no Reviews pages planned for this lesson.

Fact Reviews

Use "Subtraction Facts to 10" page 48 in Trove.

118 Worktext pages 229–30 Reviews pages 219–20

Objectives

- **118.1** **Distinguish between a circle, a triangle, a square, and a rectangle.**
- **118.2** **Describe attributes of a triangle, a square, and a rectangle.**
- **118.3** **Predict the results of a probability activity.**
- **118.4** **Tally the results of a probability activity.**

Printed Resources

- Instructional Aid 33: *Spinner* (4 sections colored blue and 2 sections colored yellow)
- Visual 12: *Plane Figures*
- Visuals: Number Word Cards (*zero* to *ten*)
- Visuals: Shapes Kit (3 red circles, 3 blue circles, 2 rectangles, 3 squares, 6 triangles)
- Student Manipulatives: Number Cards (0–9)

Digital Resources

- Web Link: Mathigon
- Games/Enrichment: Subtraction Flashcards

Materials

- Subtraction flashcards
- 1 paper bag
- 1 paper clip (for use with the spinner)
- Blue and yellow markers

Plane Figures

circle rectangle square triangle

Trace the shape. Draw the shape. Write the name.

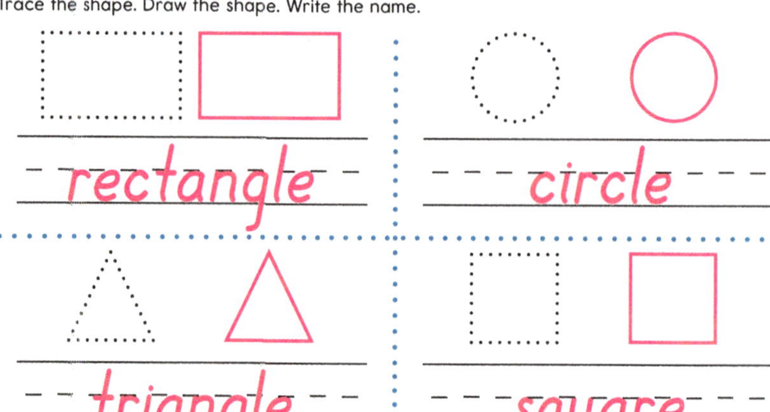

Mark the correct circle.

Which shape is someone more likely to pick from the bag?
- ● circle
- ○ rectangle

Which shape is impossible to pick?
- ○ rectangle
- ● triangle

Practice & Review

Identify Number Words & Draw Tally Marks

Display a Number Word Card. Invite a student to read the word. If the student is correct, direct him or her to draw for display the tally marks to represent that number.

Memorize Subtraction Facts

12 – 5 12 – 7 12 – 6

Distribute Number Cards 0–9. Practice the new subtraction facts and previously memorized facts as in earlier lessons.

Practice these and the previously memorized subtraction facts.

Engage

Distinguish between Plane Figures & Solid Figures

Use a **visual** to differentiate plane and solid figures.

Display a rectangle and a cereal box.

Which shape is flat? the rectangle

Conclude together that the rectangular prism is not flat. Explain that it is a solid figure because it has 3 dimensions: length, width, and height; the rectangle is flat because it has only 2 dimensions: length and width.

Instruct

Distinguish between a Circle, a Triangle, a Square & a Rectangle

Guide a **discussion** to help the students distinguish plane figures.

Display the shapes. Ask a volunteer to remove all the circles, counting each circle as he or she removes it.

How many circles were there? 6

Do all the circles look the same? no; some circles are red and other circles are blue

What is the same about all the circles? They have the same shape. They have 0 sides and 0 corners.

Color the star on each corner yellow.
Mark an X on each side. Fill in each box.

A rectangle has 4 sides and 4 corners.

A triangle has 3 sides and 3 corners.

A square has 4 sides and 4 corners.

What color is someone likely to spin?

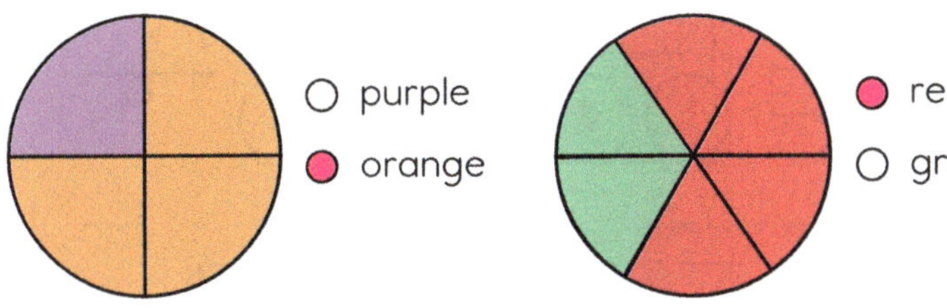

Time to Review

Subtract. Use the number line if needed.

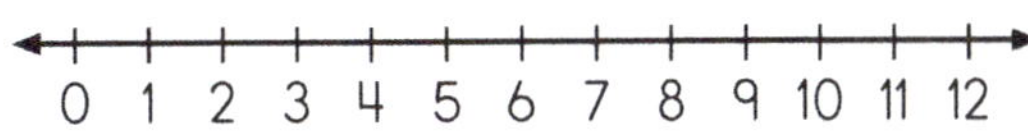

12 − 4 = 8 12 − 6 = 6 12 − 8 = 4

12 − 5 = 7 12 − 7 = 5 12 − 9 = 3

230 two hundred thirty Math 1

Display Visual 12. Explain that these objects are called plane figures since they are flat.

Which figures are shaped like a circle? doughnut, clock

What foods are shaped like a circle? hamburgers, pancakes, cookies, cakes, lollipops

Ask a volunteer to remove all the triangles, counting each triangle as he or she removes it. 6

How many sides does a triangle have? 3

Point out that corners are where the sides meet.

How many corners does a triangle have? 3

Which figures are shaped like a triangle? a pennant, a musical triangle

What foods are shaped like a triangle? a slice of pizza, a slice of pie or cake, a tortilla chip

Ask a volunteer to remove all the rectangles, counting each rectangle as he or she removes it. 2

How many sides does a rectangle have? 4

What do you notice about the sides? Two sides are longer, but the opposite sides are the same length.

How many corners does a rectangle have? 4

Are the corners like the corners of the triangle? no

Which figures are shaped like a rectangle? a dollar bill, a door

What foods are shaped like a rectangle? a candy bar, a wafer, a piece of nut bread, a graham cracker

Ask a student to remove all the squares, counting each square as he or she removes it. 3

How many sides does a square have? 4

How many corners does a square have? 4

Point out that the square is a special kind of rectangle whose sides are the same length. Conclude together that the corners of the square are like the corners of the rectangle.

Which objects are shaped like a square? a picture frame, a checkerboard

What foods are shaped like a square? brownies, waffles, a saltine cracker, a square piece of breakfast cereal

Predict & Tally the Results of a Probability Activity

Guide a **hands-on activity** to help the students predict and tally results.

Write the words "certain," "more likely," "less likely," and "impossible" for display.

Place 6 triangles and 2 squares in a bag. Explain that a student will pick a shape from the bag without looking.

If there are 6 triangles and 2 squares in the bag, will someone be more likely or less likely to pick a triangle? more likely, because there are more triangles in the bag

Will someone be more likely or less likely to pick a square? less likely, because there are fewer squares in the bag

Display a triangle and a square. Explain that the students can use tally marks to keep track of the shapes pulled out of the bag. Shake the bag; then direct a student to reach into the bag and pull out the first shape that his or her hand touches. Record the results with a tally next to the displayed shape.

Return the shape to the bag. Shake the bag. Ask another student to take out a shape and record the results.

Continue for a total of 10 times. Discuss the results.

Could someone pick a circle from this bag? no, because there are no circles in the bag

Explain that it would be impossible to pick a circle shape since none are in the bag.

Display the *Spinner*. Explain that spinners can also be used to show probability.

See Lesson 72 for an illustration of the spinner.

Which colors are on this spinner? blue and yellow

How many sections are colored blue? 4

How many sections are colored yellow? 2

What is the probability that the paper clip will land on red? impossible

What is the probability that it will land on blue? more likely

What is the probability that it will land on yellow? less likely

Spin the paper clip 10 times and record the results.

Apply

Worktext pages 229–30

Guide completion of page 229.

Read and explain the directions for page 230. Assist the students as they complete the page independently.

Assess

Reviews pages 219–20

Review completing expanded form on page 220.

Plane Figures

Color the circles green.
Color the squares red.
Color the rectangles brown.
Color the triangles yellow.

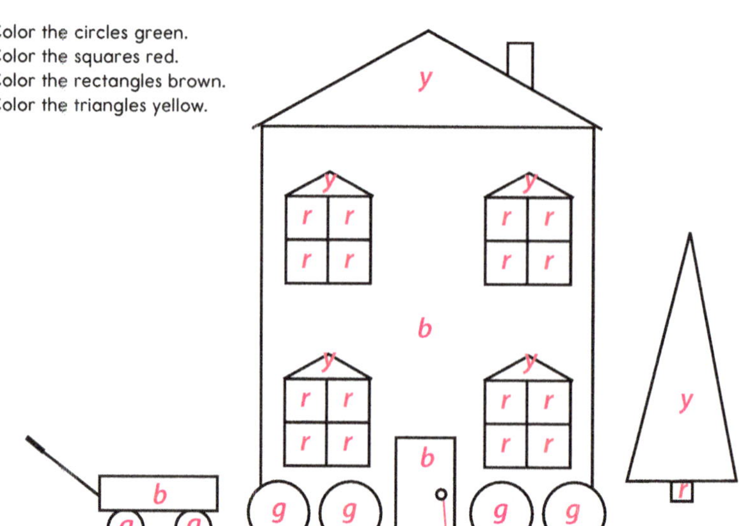

Write the number of corners and sides.

Shape	Corners	Sides
square	4	4
rectangle	4	4
triangle	3	3

Chapter 4 Review

Mark the correct answer.

60 + 8 =	63 ○	68 ●	58 ○	86 ○
30 + 6 =	36 ●	30 ○	63 ○	46 ○
90 + 3 =	39 ○	90 ○	93 ●	30 ○

Complete each expanded form.

56 = 50 + 6

34 = 30 + 4

85 = 80 + 5

98 = 90 + 8

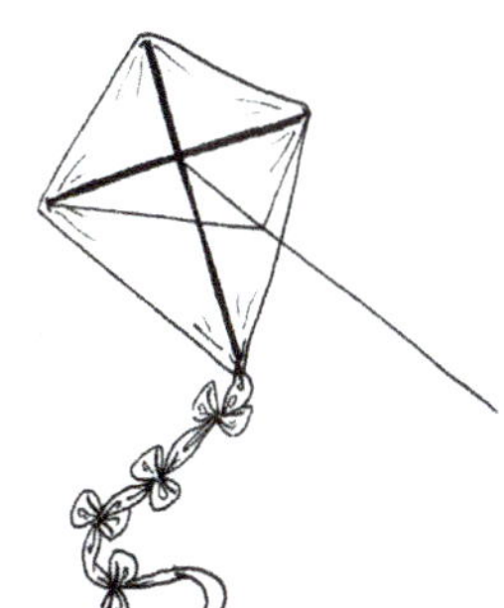

Addition Fact Review

Add.

8 + 2 = 10	3 + 7 = 10	6 + 6 = 12	7 + 2 = 9	3 + 6 = 9
5 + 3 = 8	7 + 4 = 11	9 + 3 = 12	8 + 0 = 8	5 + 6 = 11

119

Worktext pages 231–32
Reviews pages 221–22

Objectives

- **119.1** Identify plane figures that are the same or different shape, color, or size.
- **119.2** Make a Venn diagram.
- **119.3** Draw plane figures that are the same or different size.

Printed Resources

- Instructional Aid 34: *Dot Grid* (for the teacher and for each student)
- Visuals: Shapes Kit (5 rectangles —2 green, 1 red, 1 blue, 1 yellow; 5 triangles—2 red, 1 green, 1 blue, 1 yellow; 4 circles—1 of each color; 4 squares—1 of each color)
- Student Manipulatives: Shapes Kit (1 blue circle; for the teacher)
- Student Manipulatives: Shapes Kit (4 squares, 4 triangles, 4 rectangles)

Digital Resources

- Web Link: Mathigon
- Games/Enrichment: Fact Fun Activities

Practice & Review

Review Subtraction Facts

Select a game or activity from the "Fact Fun Activities" in Trove. Practice the previously memorized subtraction facts.

Identify the Fraction for Part of a Set

Display 3 circles and 1 square.

How many shapes are there in all? 4

How many shapes are square? 1

Which fraction of the set of shapes is square? $\frac{1}{4}$

Write "$\frac{1}{4}$" for display.

Display 2 circles and 2 squares.

How many shapes are there in all? 4

How many shapes are square? 2

Which fraction of the set of shapes is square? $\frac{2}{4}$ or $\frac{1}{2}$

Write "$\frac{2}{4}$" and "$\frac{1}{2}$" for display.

Continue the activity with sets of rectangles and triangles.

Same Shape, Size, or Color

Color the figure that is the same shape and size.

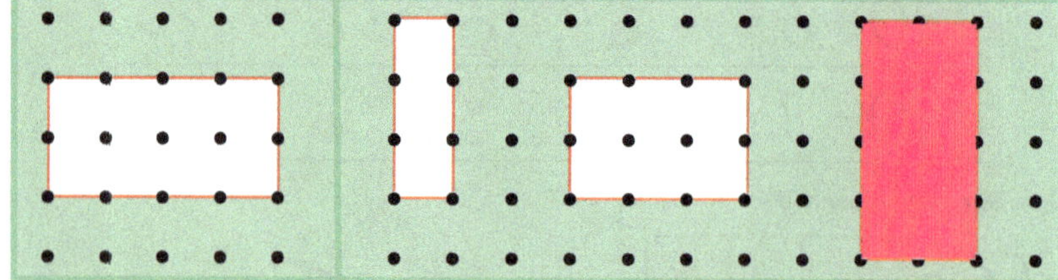

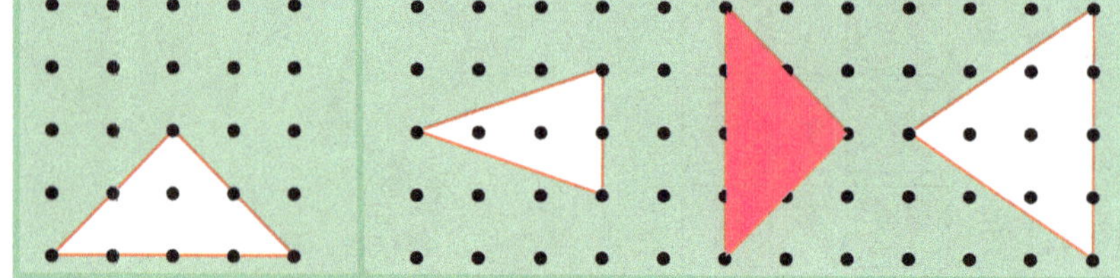

Draw a figure that is the same shape and size.

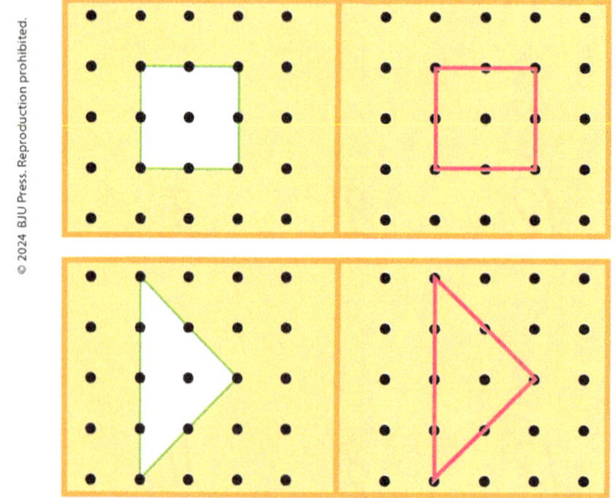

Engage

Size & Shape

Relate size and shape to a biblical worldview truth.

Ask a student to stand beside you.

Are we the same size? no

Are we the same shape? no

Read aloud Psalm 139:14. Explain that God created each person different from everyone else. God has a special plan for each person's life.

Instruct

Identify Plane Figures That Are the Same or Different Shape

Guide a **sorting activity** to identify figures that are the same or different shape.

Distribute the shapes. Direct the students to put all their shapes on the side of their desks.

Display 1 red square. Direct the students to hold up a shape that is the same shape as your red shape; then direct them to put their shapes in the middle of their desks. Instruct the students to sort through their other shapes and to place all their squares in a row in the middle of their desks. Students should have 4 squares in a row.

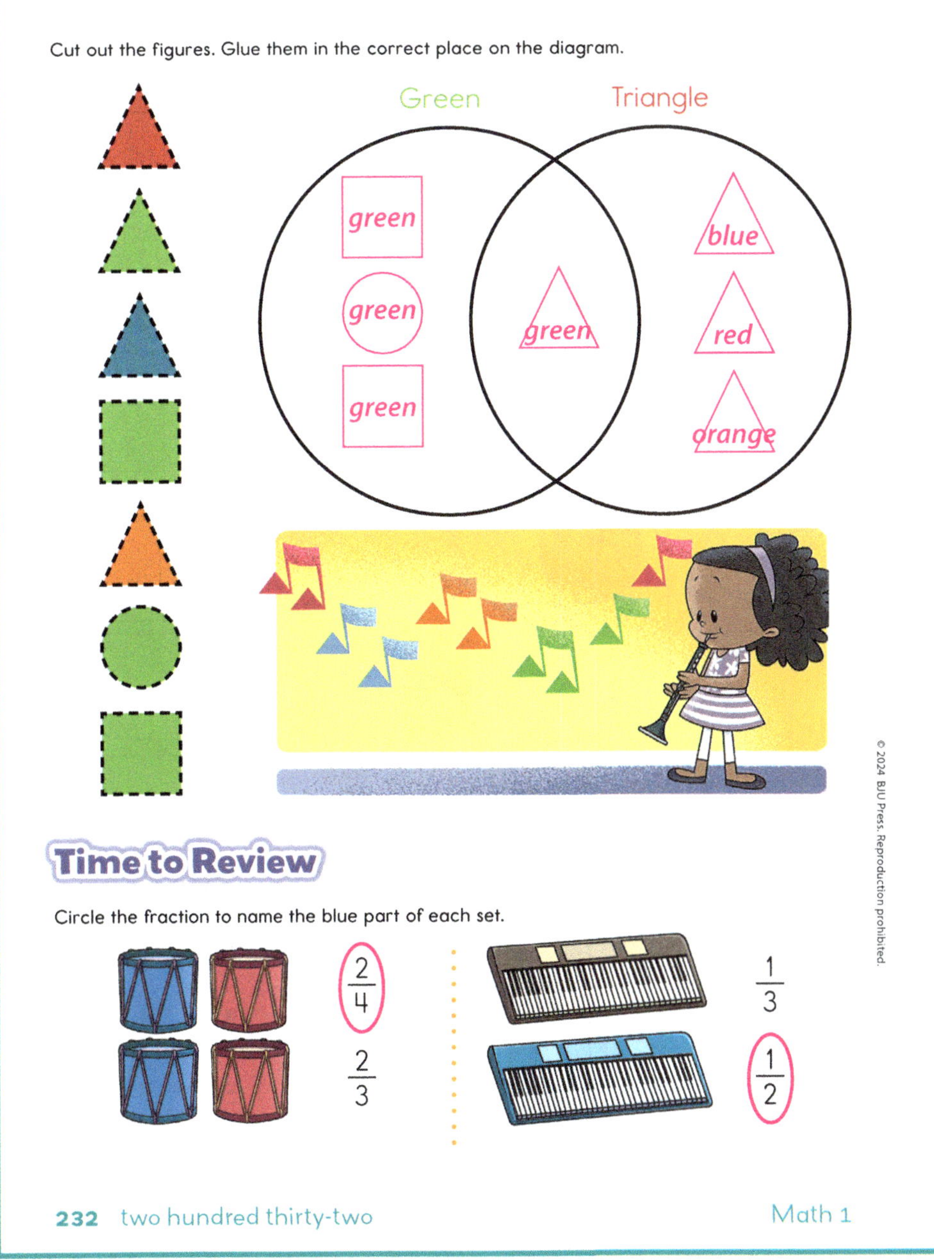

Direct the students to move their squares to the side of their desks. Display 1 blue circle. Direct the students to hold up a shape that is a different shape than your blue shape; then direct them to put their shapes in the middle of their desks.

Instruct the students to sort through their other shapes and to place all the shapes that are different from a circle in a row on their desks. Students will have 4 squares, 4 rectangles, and 4 triangles in a row.

Make a Venn Diagram

Guide the students as they complete a **Venn diagram**.

Draw a Venn diagram for display. Label one region *Red* and explain that all the red shapes belong inside that circle. Label the other region *Triangle* and explain that all the triangles belong inside that circle.

Display a red rectangle.

Which circle does the red rectangle belong in? the circle labeled *Red*

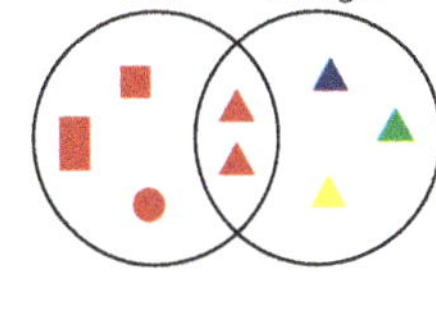

Display a blue triangle.

Which circle does the blue triangle belong in? the circle labeled *Triangle*

Follow the same procedure for a red square and a green triangle.

Display a red triangle.

Ask the students if they think the red triangle belongs in the circle for red shapes or in the circle for triangles.

Place the red triangle in the region formed by the intersection of the two circles. Explain that this region is shared by both circles, so any shape that is both red and a triangle belongs in the shared region.

Continue the activity with any remaining shapes that can be placed on the diagram. Explain that this is a *Venn diagram* and that Venn diagrams show how things belong together.

Remove the shapes and relabel the regions: *Blue* and *Square*. Continue the activity, directing the students to place the shapes in the correct regions.

Identify Plane Figures That Are the Same or Different Size

Use **visual aids** to help the students identify figures that are the same or different size.

Display 1 large blue circle (from Visuals) and 1 small blue circle (from a Student Manipulatives Packet).

Are these the same shape? yes

What shape are they? circles

Are these circles exactly the same, or are they different? Different; 1 circle is large and the other circle is small.

Place the small circle on top of the large circle for the students to observe.

Display 2 green rectangles.

What shape are these? rectangles

Are these rectangles exactly the same, or are they different? the same

Explain that these rectangles are the same shape and the same size. Place 1 rectangle on top of the other rectangle for the students to observe.

Draw Plane Figures That Are the Same or Different Size

Use **drawings** to distinguish figures that are the same or different size.

Distribute the *Dot Grid* page to each student and display your copy. Draw a large square on one of the grids.

What is this shape? a square

Draw a small square inside the large square, using only 4 dots (1 × 1).

What is this shape? a square

Is the second square the same size as the first square? No; the first square is larger than the second square.

Point out that these figures are the same shape, but they are not the same size.

Instruct the students to draw a rectangle on one of their grids and then to draw another rectangle that is the same size on the grid next to it. Provide help as needed.

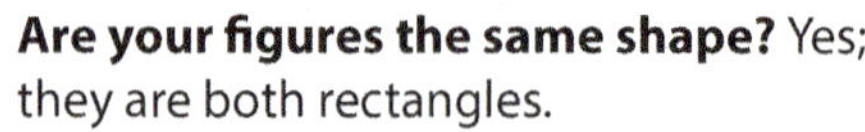

Are your figures the same shape? Yes; they are both rectangles.

Are your figures the same size? yes

Repeat the activity, drawing triangles that are different sizes and drawing squares that are the same size.

Apply

Worktext pages 231–32

Read and explain the directions for page 231. Direct the students to complete the page.

Instruct the students to cut out the shapes on the side of page 232. Explain that they should place all the green shapes in the circle labeled *Green* and all the triangles in the circle labeled *Triangle.*

Where will you place the green triangle? in the region that both circles share

Direct the students to glue their shapes after they have placed them in the circles.

Students may use shapes from their Student Manipulatives Packets to do the activity without cutting and gluing.

Read the directions for the last activity and guide its completion.

Same Shape, Size, or Color

Color the shape that is the same shape and size as the first one.

Draw a figure that is the same shape and size.

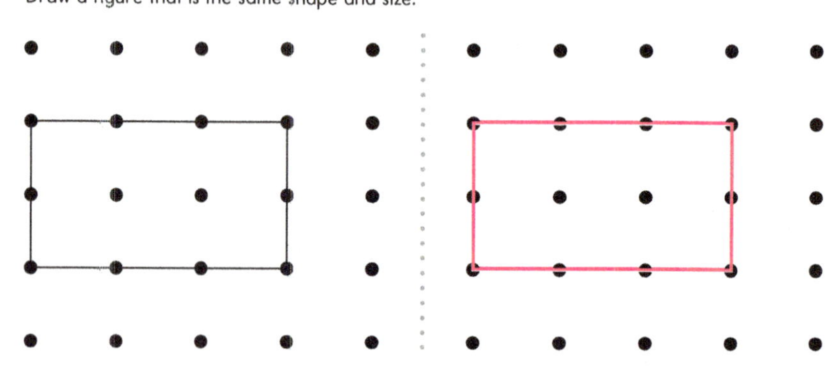

Chapter 8 Review

Complete each sentence.

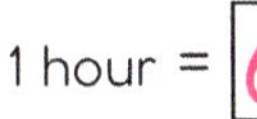

1 hour = 60 minutes half-hour = 30 minutes

Write each time.

6:00

9:30

1:30

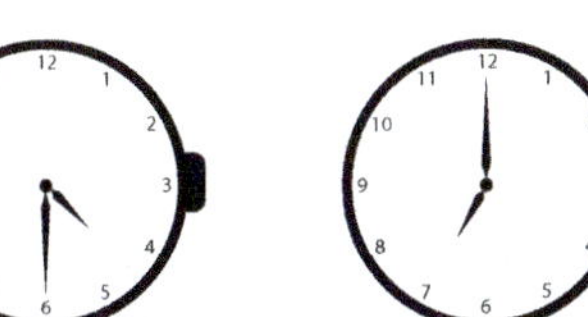

4:30

7:00

5:00

Subtraction Fact Review

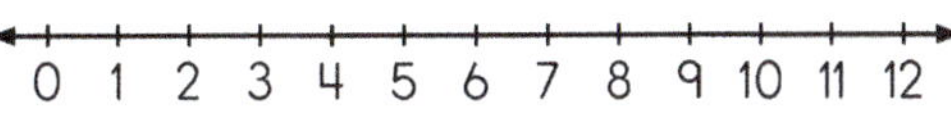

Subtract. Use the number line if needed.

11 − 6 = 5 11 − 5 = 6 12 − 3 = 9

12 − 9 = 3 12 − 4 = 8 12 − 8 = 4

12 − 5 = 7 12 − 7 = 5 12 − 6 = 6

Assess

Reviews pages 221–22

Review writing the time on page 222.

Extended Activity

Make a Venn Diagram

Materials

- A Venn diagram drawn on a sheet of paper for each student
- Colorful sticker dots, sponge shapes and paint, or stamps and ink pads

Procedure

Distribute a Venn diagram and a set of supplies to each student. Instruct the students to label the two circles of their Venn diagrams as directed (a specific color and a specific shape of the stickers, sponges, or stamps). Instruct the students to complete the Venn diagram.

120 Worktext pages 233–34 Reviews pages 223–24

Objectives

- **120.1** Identify symmetrical and asymmetrical figures.
- **120.2** Identify a line of symmetry.
- **120.3** Draw a line of symmetry on an object.

Printed Resources

- Instructional Aid 35: *Symmetrical Shapes* (2 copies)
- Instructional Aid 36: *Asymmetrical Shapes*
- Instructional Aid 37: *Line of Symmetry* (for each student)

Digital Resources

- Games/Enrichment: Fact Fun Activities

Materials

- Several leaves of different colors, sizes, and shapes

Preparation

Cut out the figures on the *Symmetrical Shapes* and *Asymmetrical Shapes* pages.

Practice & Review

Count 1–150

Arrange the students in 2 groups. Instruct the students in Group 1 to begin with number 1 and to count until you clap your hands.

Instruct the students in Group 2 to *count on* again until you clap your hands. Continue alternating groups until you reach 150.

Distinguish between *Left & Right*

Direct the students to stand and follow the directions for *left* and *right*.

Raise your right hand.
Show me your left thumb.
Stand on your right foot.
Touch your left shoulder.
Make a fist with your right hand.
Hop on your right foot.
Touch your left elbow.
Put your left hand on your head.

Review Subtraction Facts

Select a game or activity from the "Fact Fun Activities" in Trove. Study the previously memorized subtraction facts.

Symmetry

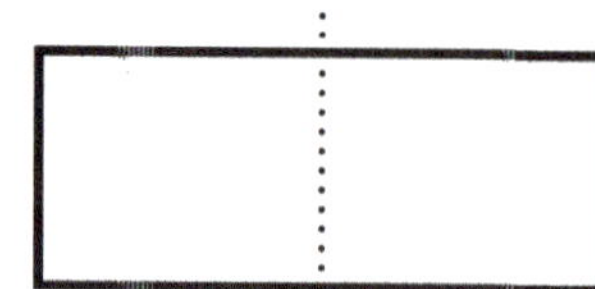

This is a line of symmetry.
The parts match.

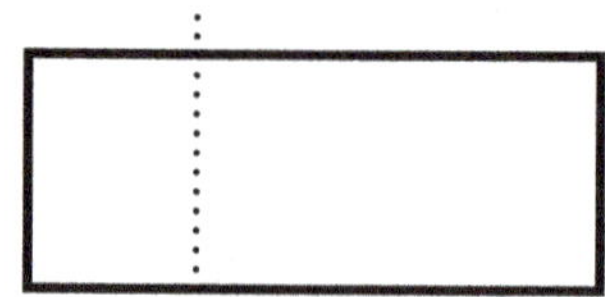

This is not a line of symmetry.
The parts do not match.

Look at the line on each object.
Mark an *X* on each object that has matching equal parts.

Draw a line of symmetry so each shape has matching equal parts.

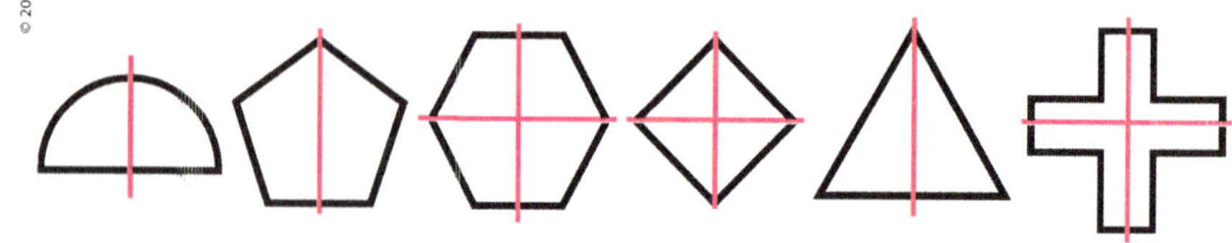

Engage

Introduce Symmetry

Guide a **visual analysis** to help the students understand symmetry.

Display the leaves for the students to observe.

Are the leaves the same or different? different

How are the leaves different? size, shape, and color

Explain that all things, even these small leaves, were created by God to show His wonderful design.

Choose one of the symmetrical leaves and fold the leaf in half.

What do you notice about the left and right sides of this leaf? They are both the same.

Point out that both sides of the leaf have matching equal parts.

Instruct

Identify Symmetrical & Asymmetrical Figures

Demonstrate folding figures to help the students identify symmetrical and asymmetrical figures.

Display a circle from the *Symmetrical Shapes* page.

What is this shape? a circle

Look at the line on each letter.
Circle each letter that has symmetry.

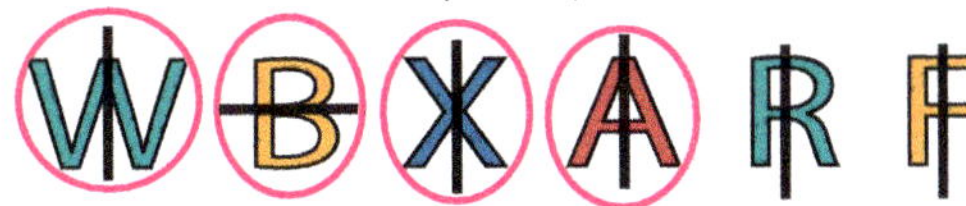

Look at the line on each object.
Circle each figure that has matching equal parts.

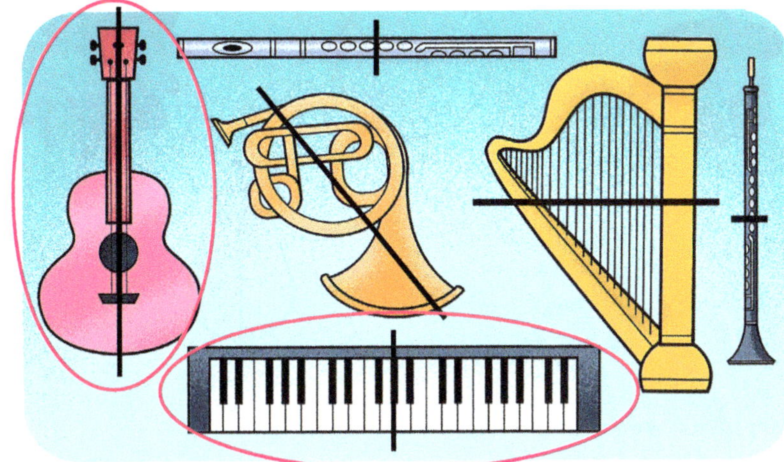

Time to Review

Color the music notes on the left of the girl yellow.
Color the music notes on the right of the girl blue.
Color the music notes above the girl orange.

Can I fold this circle so that the sides match? yes

Demonstrate folding the circle so that the sides match.

Explain that when a shape can be folded into 2 equal parts that match, it has symmetry. Write *symmetry* for display. Explain that a circle is an unusual shape because it can be folded in half in any direction and the parts will be equal. Demonstrate by folding the circle 2 or 3 ways.

Are the 2 parts equal? yes

Display a heart.

What is this shape? a heart

If I fold this heart in half, will the sides match? yes, if folded vertically, but no, if folded horizontally

Fold the heart vertically to show them that the left and right sides match, so the heart has symmetry.

Follow the same procedure for the remaining symmetrical shapes, folding the triangle and the letter *T* vertically so that the parts match.

Display the asymmetrical triangle (from the *Asymmetrical Shapes* page).

Invite students to tell whether they think there is any way to fold this shape into 2 equal parts.

Fold the shape as close to half as possible. Open up the shape so that the students can see the fold line.

Are the 2 parts equal? no

Does this shape have symmetry? no

Display the letter *P*.

Can I fold the letter *P* in half to show symmetry? no

Demonstrate that *P* does not have symmetry because it cannot be folded into 2 equal parts that match.

Follow the same procedure for the remaining asymmetrical shapes.

Identify a Line of Symmetry

Use **visual aids** to identify a line of symmetry.

Write "line of symmetry" for display. Display the heart again. Point out that the fold line divides the heart into 2 equal parts. Explain that this line is called a *line of symmetry*. (You may draw a line along the fold to make the fold more visible.)

Display the second heart and fold it horizontally.

Does this fold line divide the heart into 2 equal parts? no

Why is this fold line not a line of symmetry? The parts are not equal; they do not match.

Display the second circle. Fold down a small part of the circle.

Is this fold line a line of symmetry? No; the 2 parts are not equal, or the circle was not folded correctly in half.

Continue the activity, folding down a small part of the second symmetrical triangle and folding the second letter *T* horizontally. Point out that only a line that divides an object into 2 equal parts is a line of symmetry.

Draw a Line of Symmetry

Guide a **folding activity** to identify a line of symmetry.

Draw a rectangle for display.

What is this shape? a rectangle

How could I draw a line of symmetry to show 2 equal parts? Draw a vertical line or a horizontal line to divide the figure into 2 equal parts.

Demonstrate that a rectangle has 2 lines of symmetry.

First, draw a vertical line of symmetry. Then erase that line and draw a horizontal line of symmetry. Explain that both lines can be a line of symmetry for the rectangle.

Distribute a *Line of Symmetry* page to each student. Direct each student to carefully cut out the shapes on the solid lines and to fold the shapes on the dotted lines. Conclude together that if the 2 parts of a shape match, the dotted line is a line of symmetry; if they do not match, it is not a line of symmetry.

Direct the students to place each shape with a line of symmetry on the left of their desks and each shape without a line of symmetry on the right of their desks.

Which shapes show a line of symmetry? circle, square, and rectangle

Which shapes do not show a line of symmetry? heart and triangle

Apply

Worktext pages 233–34

Guide completion of page 233.

Read and explain the directions for page 234. Assist the students as they complete the page independently.

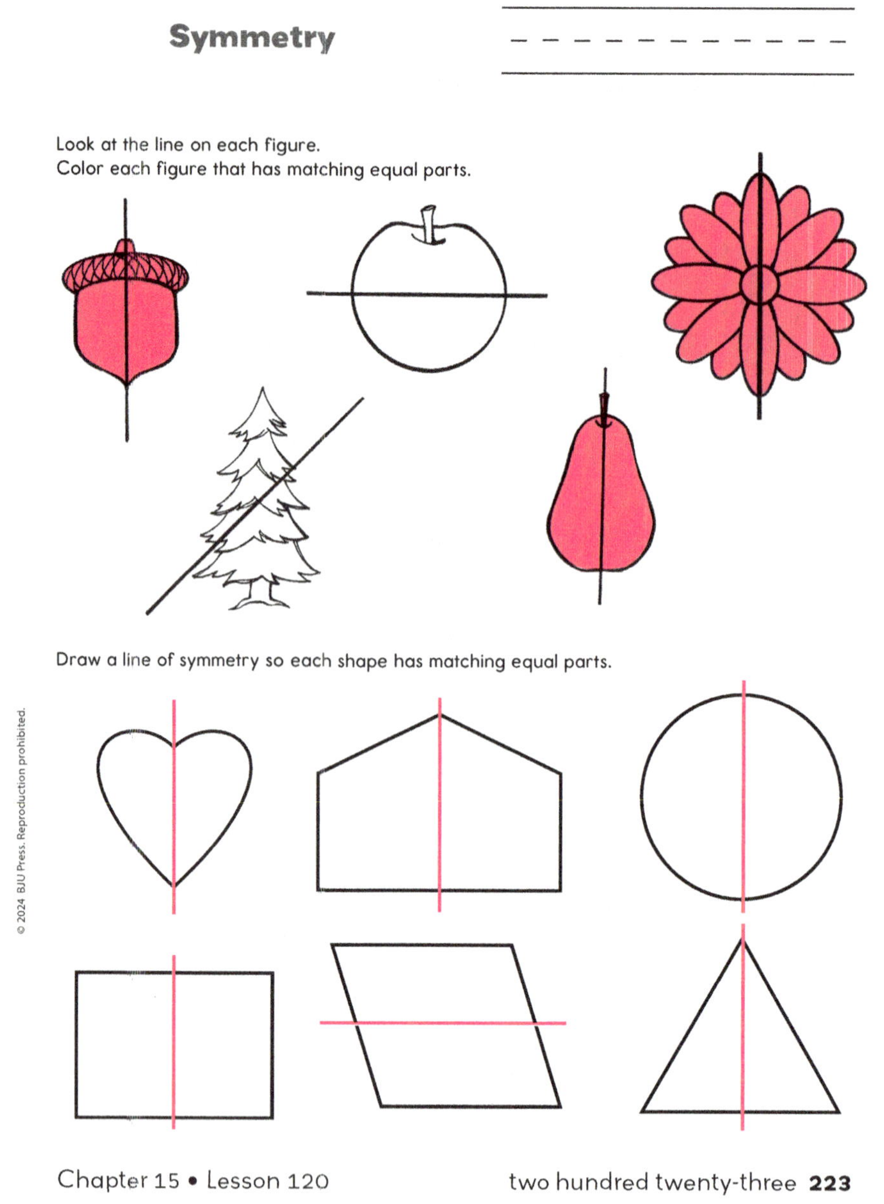

Symmetry
Look at the line on each figure.
Color each figure that has matching equal parts.
Draw a line of symmetry so each shape has matching equal parts.
© 2024 BJU Press. Reproduction prohibited.
Chapter 15 • Lesson 120
two hundred twenty-three 223

Chapter 12 Review

Dad's Garden	
green beans	
carrots	
corn	
peas	
red pepper	

= 5 seeds planted

Use the information in the graph.
Mark the correct equation.

How many more green bean seeds did Dad plant than carrots?

- ● $30 - 20 = 10$ seeds
- ○ $30 + 20 = 50$ seeds

How many corn and pea seeds did Dad plant in all?

- ○ $25 - 20 = 5$ seeds
- ● $25 + 20 = 45$ seeds

What kind of seed did Dad plant the most?

- ○ red pepper
- ● green beans

Subtraction Fact Review

Subtract.

$8 - 6 = 2$	$10 - 7 = 3$	$10 - 5 = 5$	$8 - 5 = 3$	$11 - 9 = 2$
$9 - 4 = 5$	$11 - 8 = 3$	$8 - 7 = 1$	$10 - 9 = 1$	$10 - 4 = 6$
$7 - 6 = 1$	$8 - 4 = 4$	$9 - 9 = 0$	$11 - 2 = 9$	$9 - 7 = 2$

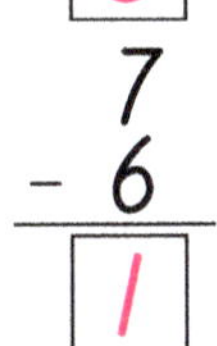

224 two hundred twenty-four Math 1 Reviews

Assess

Reviews pages 223–24

Review interpreting a pictograph on page 224.

Extended Activity

Make a Symmetrical Butterfly

Materials

- $\frac{1}{2}$ sheet of construction paper for each student
- 1 round top clothespin for each student
- 1 chenille wire for each student
- Glue

Procedure

Distribute the paper, clothespins, and chenille wires. Explain that the students will make symmetrical butterflies.

Demonstrate as you give directions for cutting the symmetrical shapes. Instruct the students to fold their papers in half, keeping the fold on their left sides, and to draw a butterfly wing by beginning and ending at the fold. Direct the students to cut along the lines to make butterfly wings. Explain that they should not cut on the folds.

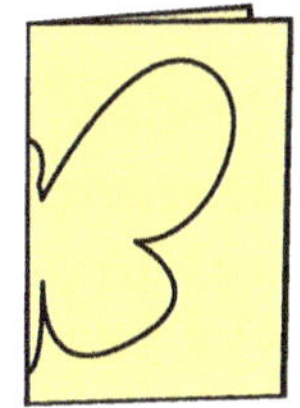

Direct the students to decorate the wings and add eyes with crayons or felt-tip pens; then glue the wings onto the clothespin. Instruct them to bend the chenille wire in half and then curl the ends of it with a pencil; glue the chenille wire onto the clothespin to resemble antennae.

121

Worktext pages 235–36
Reviews pages 225–26

Objectives

- **121.1** Copy an action pattern and a shape pattern.
- **121.2** Extend patterns of shapes, numbers, or letters.
- **121.3** Compose a picture by using plane figures.
- **121.4** Relate learning shapes to exploring God's world. BWS

Biblical Worldview Shaping

- **Exploring** (explain): Patterns are all around us. Recognizing patterns is one way we learn about and make sense of God's world. Learning shapes helps us recognize patterns (121.4).

Printed Resources

- Visuals: Shapes Kit (6 of each; yellow triangles, blue triangles, yellow circles, red circles, green squares, blue rectangles, orange hexagons, purple rhombuses, brown trapezoids)
- Student Manipulatives: Shapes Kit (6 of each; yellow triangles, blue triangles, yellow circles, red circles, green squares, blue rectangles, orange hexagons, purple rhombuses, brown trapezoids)

Digital Resources

- Games/Enrichment: Fact Fun Activities

Materials

- An article of clothing with a striped pattern or a piece of striped fabric

Patterns

Write the number of shapes used to make each picture.

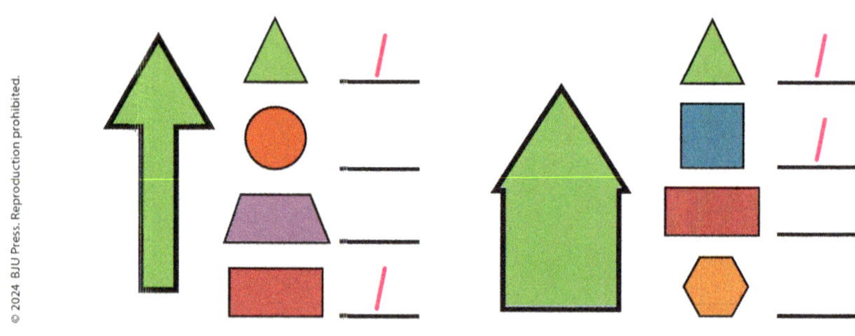

Practice & Review

Identify $\frac{1}{2}$ & $\frac{2}{2}$

Display an orange hexagon. Explain that this shape is called a hexagon since it has 6 sides. Place 2 brown trapezoids on the hexagon.

How many brown shapes does it take to make the same shape as the orange shape? 2

Remove 1 brown trapezoid from the hexagon and hold it up. Explain that the brown shape is 1 half of the orange shape. Write "$\frac{1}{2}$" for display.

How many halves does it take to equal 1 whole? 2

Place the brown trapezoid back on the orange shape.

Write for display "$\frac{2}{2}$ = 1 whole."

Identify $\frac{1}{3}$, $\frac{2}{3}$ & $\frac{3}{3}$

Place 3 purple rhombuses on the orange hexagon.

How many purple shapes does it take to make the same shape as the orange shape? 3

Display 1 of the purple shapes. Explain that the purple shape is 1 third of the orange shape. Write "$\frac{1}{3}$" for display.

Display 2 purple shapes.

Which fraction of the orange shape do the 2 purple shapes represent? 2 thirds

Write "$\frac{2}{3}$" for display.

Display 3 purple shapes.

Which fraction of the orange shape do 3 of these shapes represent? 3 thirds

How many thirds does it take to equal 1 whole? 3

Place the 3 purple rhombuses back on the orange hexagon. Write for display "$\frac{3}{3}$ = 1 whole."

Review Subtraction Facts

Select a game or activity from the "Fact Fun Activities" in Trove. Practice the previously memorized subtraction facts.

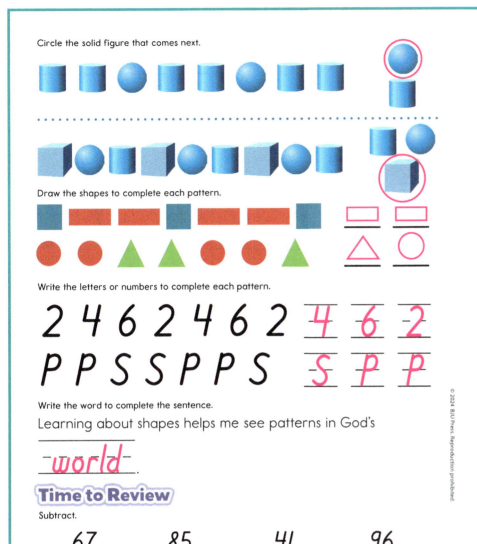

Circle the solid figure that comes next.

Draw the shapes to complete each pattern.

Write the letters or numbers to complete each pattern.

2 4 6 2 4 6 2 4 6 2

P P S S P P S S P P

Write the word to complete the sentence.

Learning about shapes helps me see patterns in God's world.

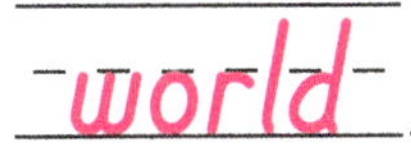

Subtract.

67	85	41	96
− 43	− 73	− 21	− 63
24	12	20	33

Engage

Introduce Patterns

Use a **visual** to illustrate a pattern.

Display the striped article of clothing. Explain that the fabric used to make this piece of clothing contains a pattern.

What is the pattern? Answers should give the order of the color of the stripes on the clothing.

Remind the students that a pattern shows that something is repeated in a certain order. Explain that patterns can be found in many places in addition to clothing; some other places that patterns are found include music, art, and in nature with plants and animals.

Instruct

Identify & Copy an Action Pattern: *aabaab, abbabb*

Guide a **movement activity** to help the students copy a pattern.

Direct the students to watch for a pattern in your actions.

Demonstrate the action pattern: *clap hands, clap hands, snap fingers, clap hands, clap hands, snap fingers.*

What is the pattern? clap, clap, snap, clap, clap, snap

Demonstrate the pattern again; then instruct the students to copy the pattern.

Direct the students to watch as you demonstrate a new pattern: *stomp foot, clap hands, clap hands, stomp foot, clap hands, clap hands.*

What is the new pattern? stomp, clap, clap, stomp, clap, clap

Instruct the students to copy this pattern.

Extend a Letter & a Number Pattern: *aabaab, abcabc*

Use **direct instruction** to guide the students to extend a letter and a number pattern.

Write for display the pattern *a a b a a b* as you say the letters aloud.

Which letter comes next? a

Explain that students can continue, or *extend*, the pattern. Underline the first 3 letters. Explain that these letters are the *base* of the pattern. The next 3 letters repeat the base of the pattern.

What would the pattern look like if you repeated the base 2 more times? a a b a a b a a b a a b

Invite a student to write the letters that will extend the pattern so that the base is repeated 2 more times.

Write for display the pattern *10 20 30 10 20 30* as you say the numbers aloud.

What is the base of this number pattern? 10 20 30

Ask a student to write the numbers that will extend the pattern so that the base is repeated 1 more time. 10 20 30 10 20 30 10 20 30

Copy & Extend a Shape Pattern: *abab, abcabc, abbabb*

Guide a **discussion** to help the students extend a shape pattern.

Distribute the set of shapes to each student. Use the yellow triangles and the yellow circles to display a shape pattern. Say the pattern aloud: *triangle, circle, triangle, circle.*

Which pattern do you see? triangle, circle, triangle, circle (all yellow)

Direct the students to copy this same pattern, using the shapes at their desks.

What is the base of this pattern? triangle, circle

How would you extend the pattern so that the base is repeated 1 more time? add a yellow triangle; then add a yellow circle

Instruct the students to extend their patterns so that the base is repeated 1 more time.

What does your pattern look like? triangle, circle, triangle, circle, triangle, circle

Continue the activity by using the following patterns. After the students copy the pattern, direct them to repeat the base 1 time. Point out that since these patterns include different colors, the student should make sure that the colors and shapes match your example.

green square, blue rectangle, red circle, green square, blue rectangle, red circle

orange hexagon, blue triangle, blue triangle, orange hexagon, blue triangle, blue triangle

Ask a volunteer to make up his or her own pattern. Let him or her display it for the other students. Direct students to copy the volunteer's pattern.

Compose a Picture by Using Plane Figures

Guide a **hands-on activity** to compose a picture.

Direct attention to Worktext page 235. Direct the students to use their shapes to make the first picture. After the students have made the picture, discuss which shapes they used.

Learning Shapes Helps People Explore God's World

Guide a **discussion** to help the students relate a biblical worldview shaping truth.

Direct attention to the first problem on Workext page 236.

What pattern do you see? cylinder, cylinder, sphere, cylinder, cylinder, sphere, cylinder, cylinder, ____

What figure would come next? sphere

If you did not recognize a figure's shape, could you find a pattern? probably not

Remind the students that God created this world and that math is an important part of it. When students learn about shapes, they can recognize patterns. The better they understand math, the better they can understand the world in which they live.

How can learning about shapes help you explore God's world? Learning about shapes helps me see patterns in God's world.

Patterns

Circle the note that comes next.

Draw the shapes to complete each pattern.

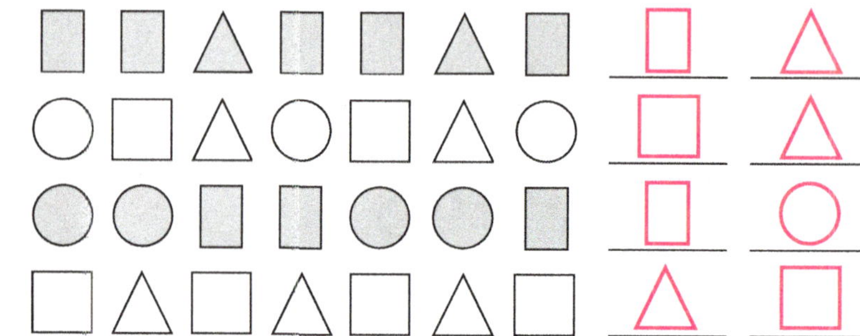

Write the letters or numbers to complete each pattern.

J C B J C B J C B

4 6 4 6 4 6 4 6 4

8 5 5 8 5 5 8 5 5

Chapter 12 Review

Our Pets

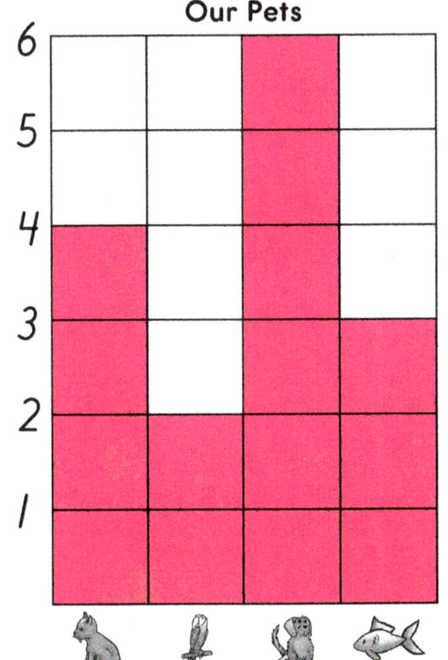

Use the information to make a bar graph.

 = 4
 = 2
 = 6
 = 3

Circle the pet the kids had the most.
Underline the pet the kids had the least.

cat bird dog fish

Mark the correct answer.

How many more kids had dogs than birds?

○ 6 + 2 = 8
● 6 − 2 = 4

Substraction Fact Review

Subtract.
Use the number line if needed.

0 1 2 3 4 5 6 7 8 9 10 11 12

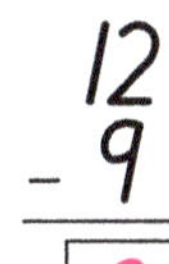
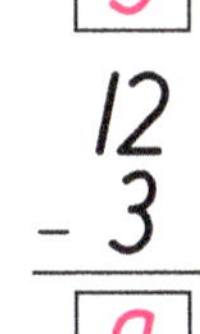

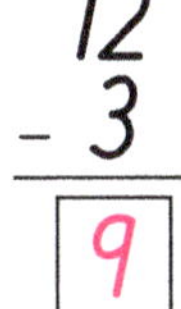
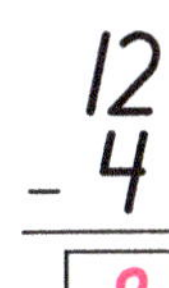

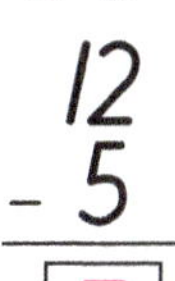
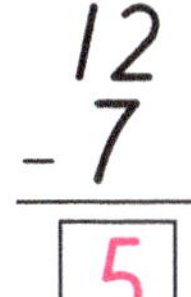

11 − 5 = 6	12 − 9 = 3	12 − 4 = 8	12 − 5 = 7	12 − 6 = 6
11 − 6 = 5	12 − 3 = 9	12 − 8 = 4	12 − 7 = 5	11 − 4 = 7

Apply

Worktext pages 235–36

Assist the students as they complete the pages independently.

Direct attention to the biblical worldview shaping statement and guide the students as they complete the sentence.

Assess

Reviews pages 225–26

Review making and interpreting a bar graph on page 226.

Extended Activity

Make a Pattern

Materials

- A 3″ × 12″ strip of paper for each student
- 2 or 3 different shapes of pasta, such as spirals, wheels, shells
- Glue
- Food coloring and rubbing alcohol (optional)

Preparation

If desired, color the pasta by using 2 tablespoons of rubbing alcohol mixed with a few drops of food coloring. Allow time for pasta to dry before proceeding.

Procedure

Distribute a strip of paper and 3 pieces of each pasta shape to each student.

Instruct the students to make a pattern on their desks, using the pasta; then direct them to glue the pattern onto the strip of paper.

After the patterns have dried, give the students an opportunity to identify each pattern and to decide which pasta shape comes next.

122 Worktext pages 237–38
Reviews pages 227–28

How can shapes help me learn about God's world?

Chapter Concept Review

- Practice concepts from Chapter 15 to prepare for the test.

Printed Resources

- Instructional Aid 35: *Symmetrical Shapes*
- Instructional Aid 36: *Asymmetrical Shapes*
- Visual 13: *Solid Figures*
- Visual 12: *Plane Figures*
- Visuals: Shapes Kit (6 red rectangles, 6 blue squares)
- Visuals: Money Kit (6 pennies, 4 nickels, 4 dimes)

Digital Resources

- Games/Enrichment: Fact Fun Activities

Materials

- A sphere, cone, cylinder, rectangular prism, cube, and pyramid (from Lessons 14–15)

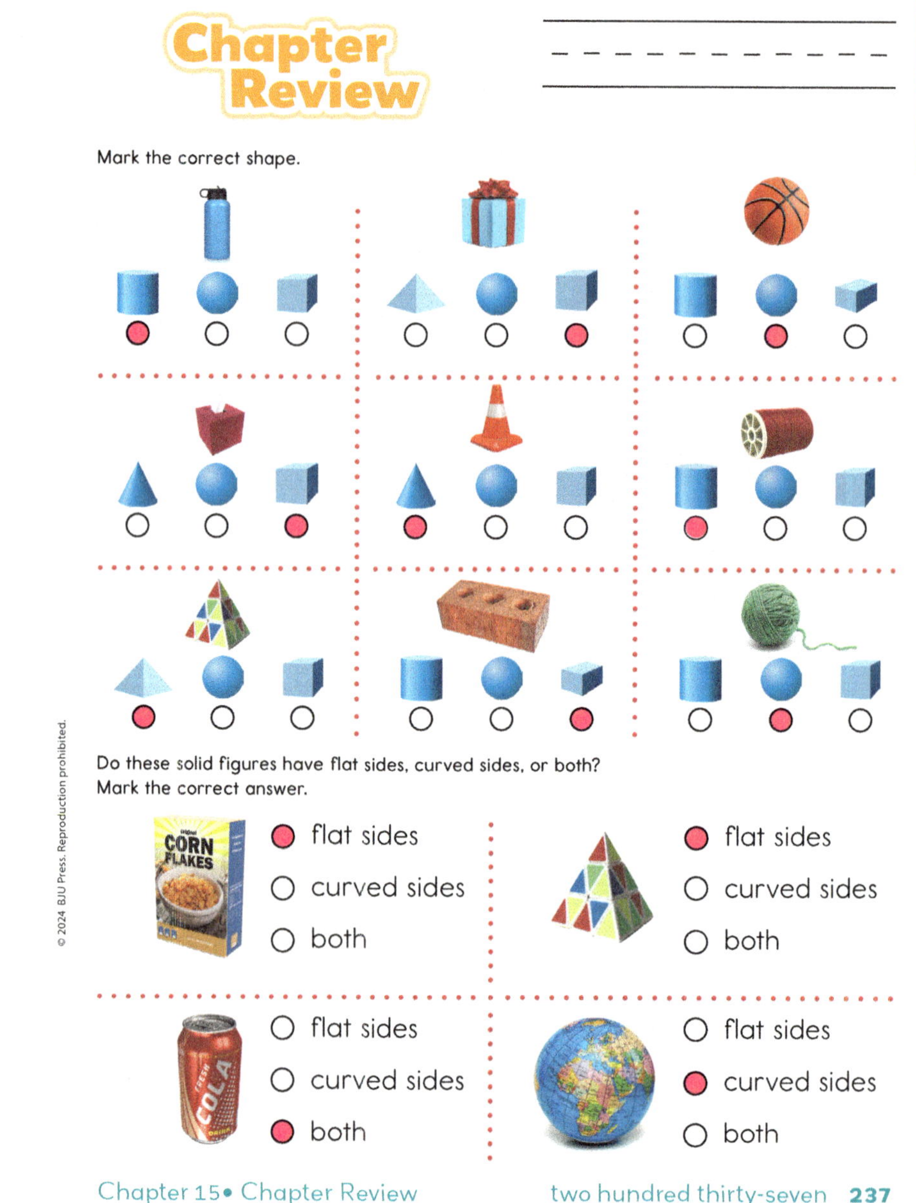
Chapter Review

Mark the correct shape.

Do these solid figures have flat sides, curved sides, or both? Mark the correct answer.

flat sides
curved sides
both

flat sides
curved sides
both

flat sides
curved sides
both

flat sides
curved sides
both

Practice & Review

Determine the Value of a Set of Coins by Counting on

Display randomly 4 dimes, 4 nickels, and 4 pennies.

What do you do first when counting the value of a set of coins? separate the different coins into groups

Which coins do you count first? coins with the greatest value

Separate the coins and guide the students as they count the value of them: 10, 20, 30, 40 *(deep breath)*, 45, 50, 55, 60 (deep breath), 61, 62, 63, 64¢.

Follow the same procedure with 3 dimes, 3 nickels, and 1 penny; and 1 dime, 3 nickels, and 6 pennies. 46¢; 31¢

Review Subtraction Facts

Select a game or activity from the "Fact Fun Activities" in Trove. Practice the previously memorized subtraction facts.

Instruct

Identify Solid Figures

Place the solid figures at the front of the room. Display Visual 13. Direct the students to listen to the clues and identify the figure.

Which solid figure has 2 faces (2 flat sides) and 1 curve (1 curved side)? a cylinder

Ask a volunteer to find the solid object that is a cylinder and to hold it up.

What shape are the faces? circles

Can a cylinder roll? yes

Can cylinders be stacked on top of each other? yes

Which cylinders are pictured on the visual? the candle and the drum

Continue the activity with the remaining solid figures. Discuss whether the object can roll and whether it can be stacked. Use the following clues.

Which solid figure has 1 circular face and 1 curved surface? a cone; It can roll, but it cannot be stacked on its pointed end.

Which solid figure has 0 faces and 1 curved surface? a sphere; It can roll, but it cannot be stacked.

Which figure has 6 square faces and 0 curves? a cube; It cannot roll, but it can be stacked.

Write the number of sides and corners.

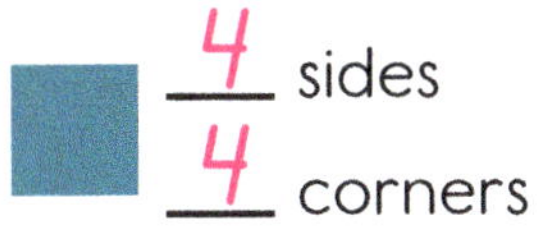
4 sides
4 corners

3 sides
3 corners

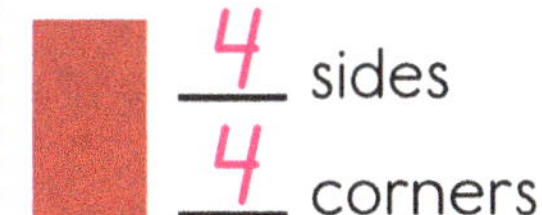
4 sides
4 corners

Mark an X on the figure that is the same size and shape.

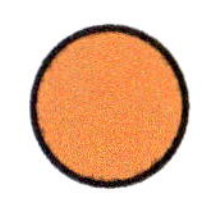   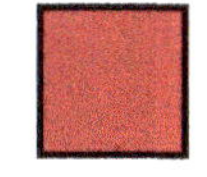 

Draw the shape to complete the pattern.

  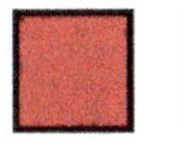    

Write the numbers to complete the pattern.

6 7 8 6 7 8 6 7 8 6

Mark the answer.

Which shape is someone more likely to pick?

- ● circle
- ○ rectangle

Draw a line of symmetry so each shape has matching parts.

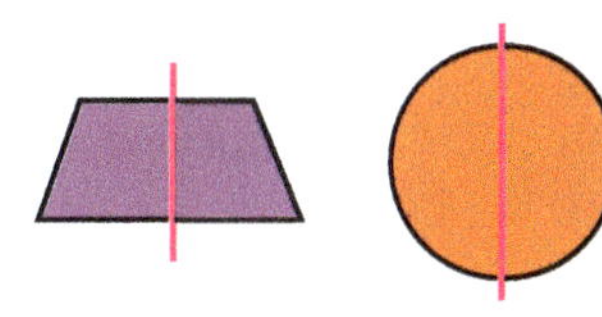

Trace the word to complete the sentence.

How can shapes help me?

Shapes help me learn about God's world.

Which figure has 6 rectangular faces and 0 curves? a rectangular prism; It cannot roll, but it can be stacked.

Which figure has triangular faces and 0 curves? a pyramid; It cannot roll, and it cannot be stacked on its pointed end.

Identify Plane Figures

Display Visual 12.

Which shape has 3 sides and 3 corners? a triangle

Ask a volunteer to point to the triangle on the visual.

What objects are shaped like a triangle? a pennant, a musical triangle

Continue with the remaining figures.

Which shape has 4 corners and 4 sides that are all the same length? a square

Which shape has no corners and no sides? a circle

Which shape has 4 corners and 4 sides with opposite sides being the same length? a rectangle

Predict & Tally the Results of a Probability Activity

Place 4 squares and 1 rectangle in a bag.

If there are 4 squares and 1 rectangle in the bag, will someone be more likely or less likely to choose a square? more likely, because there are more squares in the bag

Will someone be more likely or less likely to choose a rectangle? less likely, because there are fewer rectangles in the bag

Could someone choose a triangle from this bag? No; there are no triangles in the bag.

Complete a probability activity and tally the results. (See Lesson 118.)

Recognize & Extend a Pattern

Display the following shapes as you say the pattern aloud: *red rectangle, blue square, blue square, red rectangle, blue square, blue square.*

What is the base of this pattern? red rectangle, blue square, blue square

What comes next in this pattern? a red rectangle

Ask a volunteer to use your shapes to extend the pattern so that the base is repeated 1 more time.

Write the pattern *x y z x y z* for display.

What is the base of this letter pattern? x y z

Ask another student to extend the pattern so that the base is repeated 2 more times.

Write the pattern *1 2 2 1 2 2* for display.

What is the base of this number pattern? 1 2 2

Ask a volunteer to extend the pattern so that the base is repeated 2 more times.

Draw a Line of Symmetry

Draw a square for display.

What is this shape? a square

How do you know whether a figure has symmetry? if it can be divided into 2 equal parts; if you can fold the shape and the 2 parts match

Explain that you want the students to show that a square has symmetry. Ask a volunteer to draw a line of symmetry. The line of symmetry may be vertical, horizontal, or diagonal.

Is this the only way that a square can be divided to show symmetry? no

Erase the line drawn by the student and ask another volunteer to draw a different line of symmetry.

122

Draw or trace a trapezoid. Invite a student to draw a line of symmetry. The line may be vertical or diagonal. Write "A" for display. Ask a volunteer to draw a line of symmetry. The line should be vertical.

Identify Symmetrical & Asymmetrical Figures

Display several asymmetrical figures and symmetrical figures. As you display each shape, direct a student to tell whether the shape has symmetry and to explain his or her answer.

Apply

Worktext pages 237–38

Read and explain the directions for the pages. Assist the students as they complete the pages independently.

Direct attention to the biblical worldview shaping section on page 238. Display the picture from the chapter opener and remind the students that DQ and his friends recognized some shapes in the orchestral instruments. Reinforce the idea that learning about shapes helps them recognize and understand the world around them.

Provide prompts as necessary to guide the students to the answer.

How can shapes help me learn about God's world? Knowing shapes helps me recognize patterns. Recognizing patterns is one way I learn about God's world.

Assess

Reviews pages 227–28

Use the Reviews pages to provide additional preparation for the chapter test.

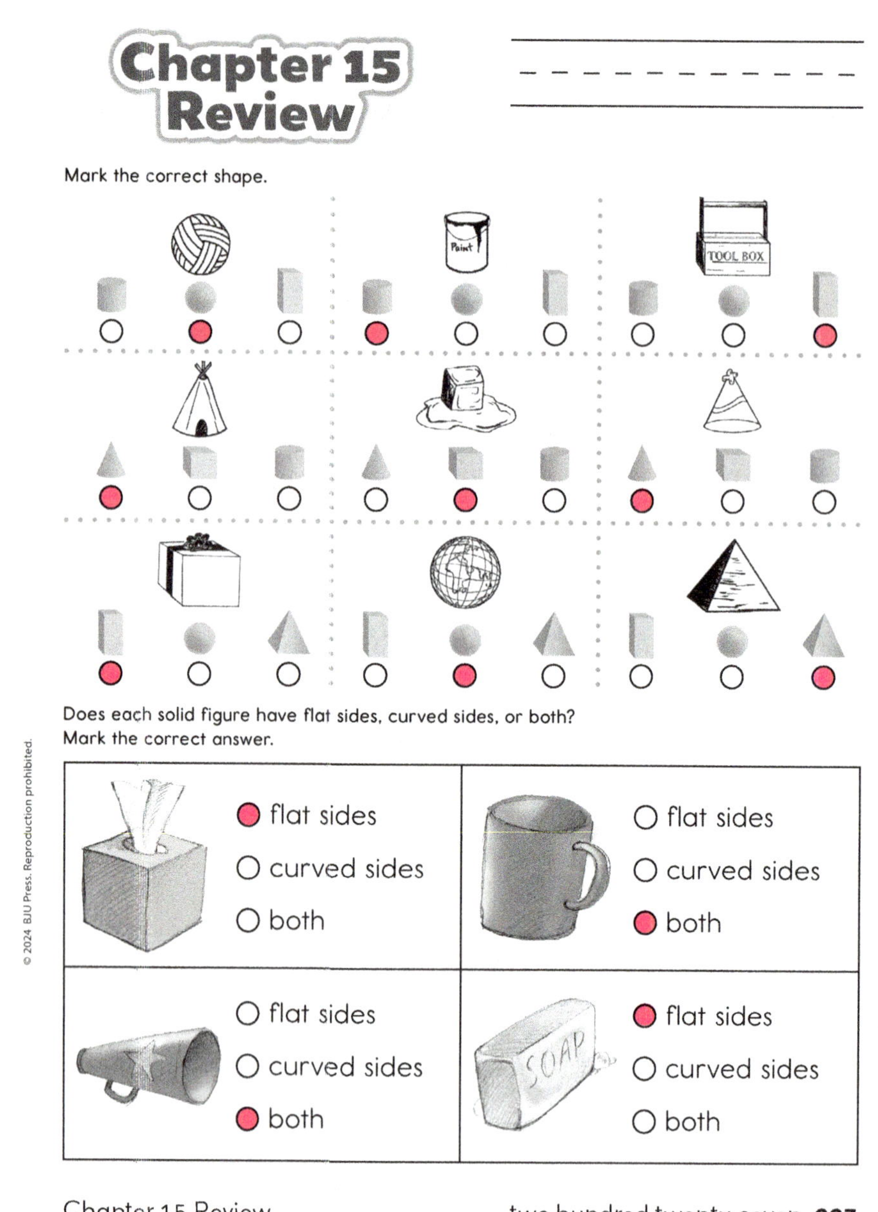

Write the number of sides and corners.

4 sides
4 corners

4 sides
4 corners

3 sides
3 corners

Mark an *X* on the figure that is the same size and shape as the first one.

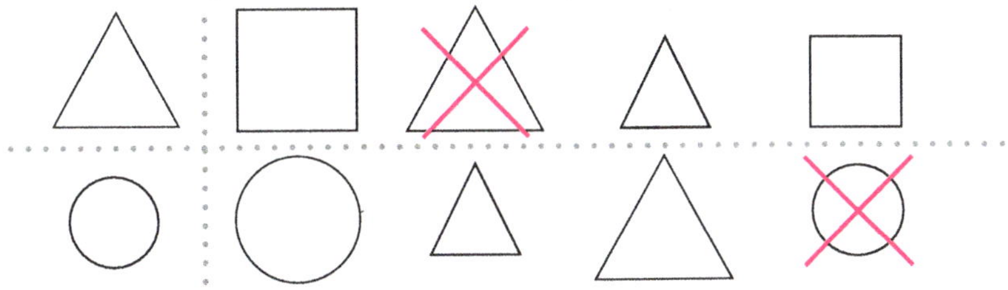

Draw the shape to complete each pattern.

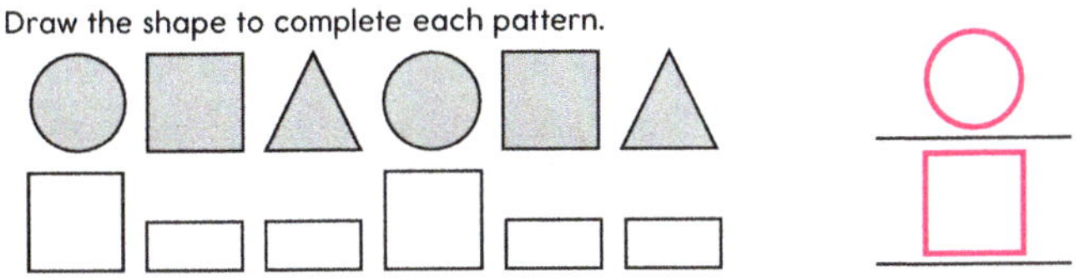

Write the letters or numbers to complete each pattern.

2 4 2 4 2 4 2 4 2 4

X Y Z X Y Z X Y Z X Y

Mark the correct answer.

Which shape are you *less likely* to pick?

○ triangle

● rectangle

Draw a line of symmetry so each shape has matching parts.

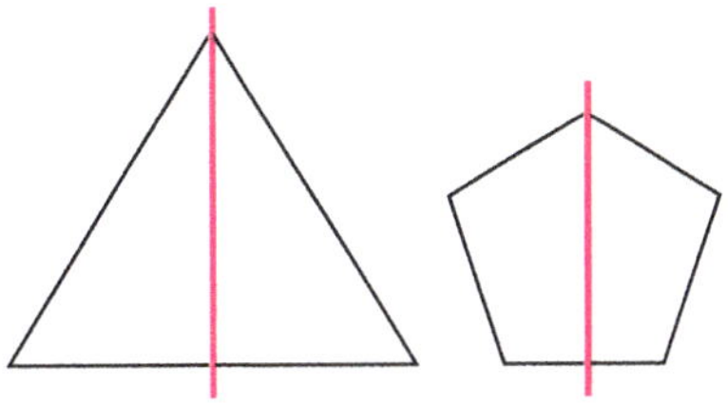

123 Worktext page 239
Reviews pages 229–30

Chapter 15 Test

Administer the Chapter 15 Test.

Cumulative Concept Review

Worktext page 239

Review the following concepts. Adapt instructions and activities and provide reteaching as needed to meet the specific needs of your students.

- Comparing 2-digit numbers by using > and < (Lesson 26)
- Determining the value of a set of coins by *counting on* (Lesson 108)
- Determining whether there is enough money to purchase an item (Lesson 108)

Retain a copy of Worktext page 240 for the discussion of the Chapter 16 essential question during Lessons 124 and 133.

Reviews pages 229–30

Use the Reviews pages to help students retain previously learned skills.

Extended Activity

Make Shape Art

Procedure

Draw a circle, a triangle, a rectangle, and a square for display. Direct the students to draw and color a house, using each shape at least once. Discuss ideas for possible uses of shapes such as triangles for the roof; squares, circles, and rectangles for windows; and rectangles for doors.

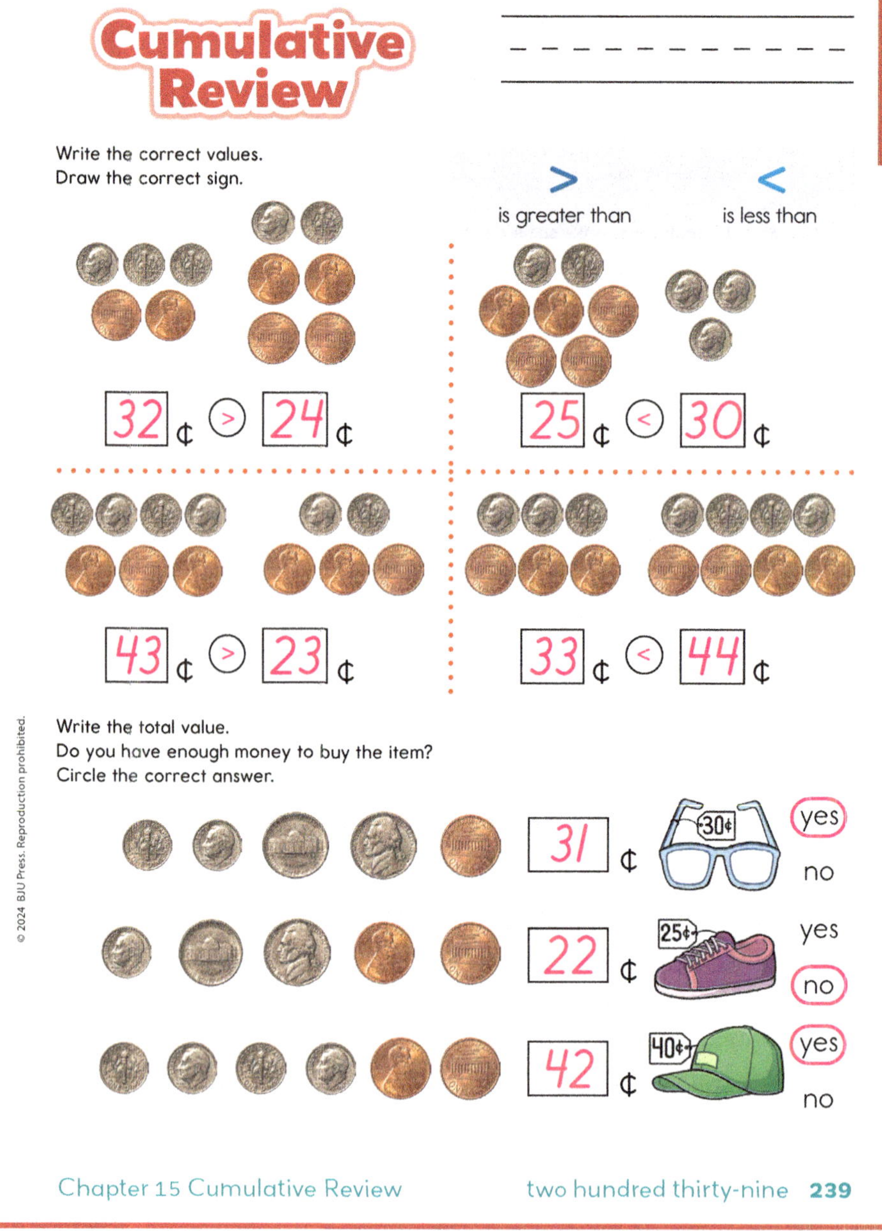

Cumulative Review

Write the correct values.
Draw the correct sign.

> is greater than
< is less than

32 ¢ > 24 ¢

25 ¢ < 30 ¢

43 ¢ > 23 ¢

33 ¢ < 44 ¢

Write the total value.
Do you have enough money to buy the item?
Circle the correct answer.

31 ¢ — 30¢ — yes / no

22 ¢ — 25¢ — yes / no

42 ¢ — 40¢ — yes / no

Add.

43	61	33	74
+ 33	+ 24	+ 16	+ 25
76	85	49	99

Subtract.

58	75	94	86
− 34	− 43	− 41	− 23
24	32	53	63

Write an equation for each word problem.
Solve.

Sid has a box with 24 crayons and a box with 12 crayons. How many crayons does Sid have in all?

work space

$$\begin{array}{r} 24 \\ +\,12 \\ \hline 36 \end{array}$$

24 + 12 = 36 crayons

Jill has a box with 48 crayons. Sue has a box with 24 crayons. How many more crayons does Jill have than Sue?

work space

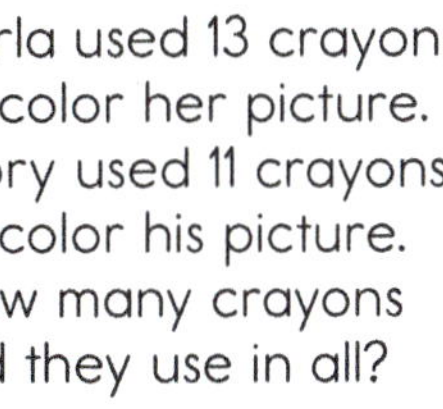

$$\begin{array}{r} 48 \\ -\,24 \\ \hline 24 \end{array}$$

48 − 24 = 24 crayons

Karla used 13 crayons to color her picture. Cory used 11 crayons to color his picture. How many crayons did they use in all?

work space

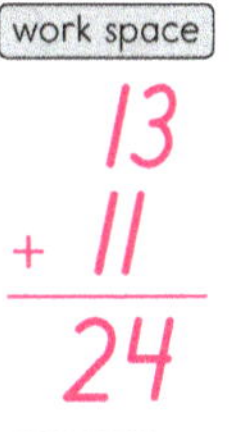

$$\begin{array}{r} 13 \\ +\,11 \\ \hline 24 \end{array}$$

 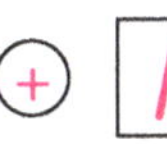  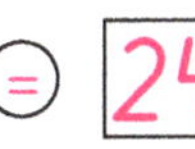

13 + 11 = 24 crayons

123

Addition Fact Review

Add.

7 + 1 8	3 + 4 7	6 + 6 12	5 + 4 9	7 + 5 12
4 + 2 6	5 + 7 12	6 + 1 7	2 + 8 10	3 + 5 8
9 + 3 12	4 + 4 8	5 + 2 7	9 + 1 10	2 + 2 4
6 + 5 11	8 + 2 10	4 + 3 7	0 + 0 0	6 + 3 9
5 + 1 6	7 + 3 10	6 + 4 10	5 + 5 10	4 + 5 9
3 + 3 6	8 + 4 12	5 + 3 8	4 + 6 10	0 + 8 8

CHAPTER 16: PLACE VALUE: THREE-DIGIT NUMBERS

PAGES	OBJECTIVES	RESOURCES & MATERIALS	ASSESSMENTS
Lesson 124	**Hundreds, Tens, Ones**		
Teacher Edition 533–37 **Worktext** 240–42	**124.1** Count by 100s to 1,000. **124.2** Identify the hundreds, tens, and ones in a 3-digit number. **124.3** Represent 3-digit numbers by using manipulatives.	**Visuals** • Place Value Kit: hundreds, tens, ones **Student Manipulatives Packet** • Hundreds/Tens/Ones Mat • Place Value Kit: hundreds, tens, ones **BJU Press Trove*** • Video: Ch 16 Intro • Web Link: Virtual Manipulatives: Base Ten Number Pieces • Games/Enrichment: Fact Reviews (Addition Facts to 12) • Games/Enrichment: Addition Flashcards • Games/Enrichment: Fact Fun Activities • PowerPoint® presentation	**Reviews** • Pages 231–32
Lesson 125	**Representing 3-Digit Numbers**		
Teacher Edition 538–41 **Worktext** 243–44	**125.1** Identify the hundreds, tens, and ones in a 3-digit number. **125.2** Represent 3-digit numbers by using manipulatives. **125.3** Rename 10 ones as 1 ten in a 3-digit number. **125.4** Complete a sequence of 3-digit numbers. **125.5** Explain how place value helps people explore God's world. BWS Exploring (explain)	**Visuals** • Place Value Kit: hundreds, tens, ones **Student Manipulatives Packet** • Hundreds/Tens/Ones Mat • Place Value Kit: hundreds, tens, ones **BJU Press Trove** • Games/Enrichment: Fact Reviews (Addition Facts to 12) • Games/Enrichment: Addition Flashcards • Games/Enrichment: Fact Fun Activities • PowerPoint® presentation	**Reviews** • Pages 233–34
Lesson 126	**Writing Numbers to 1,000**		
Teacher Edition 542–45 **Worktext** 245–46	**126.1** Identify the number of hundreds, tens, and ones in a 3-digit number. **126.2** Write 3-digit numbers by using Number Cards. **126.3** Write the number that is 1 more or 1 less. **126.4** Write the number that comes *before*, *after*, or *between*.	**Visuals** • Visual 1: *Hundred Chart* (extended) • Place Value Kit: hundreds, tens, ones **Student Manipulatives Packet** • Number Cards: 0–9 **BJU Press Trove** • Video: Place Value • Games/Enrichment: Fact Reviews (Addition Facts to 12) • Games/Enrichment: Addition Flashcards • Games/Enrichment: Fact Fun Activities • PowerPoint® presentation	**Reviews** • Pages 235–36

*Digital resources for homeschool users are available on Homeschool Hub.

PAGES	OBJECTIVES	RESOURCES & MATERIALS	ASSESSMENTS
Lesson 127 **Place Value in 3-Digit Numbers**			
Teacher Edition 546–49 **Worktext** 247–48	**127.1** Identify the number of hundreds, tens, and ones in a 3-digit number. **127.2** Write 3-digit numbers by using Number Cards. **127.3** Identify the value of the digits in a 3-digit number.	**Visuals** • Place Value Kit: hundreds, tens, ones **Student Manipulatives Packet** • Number Cards: 0–9 **BJU Press Trove** • Games/Enrichment: Fact Reviews (Addition Facts to 12) • Games/Enrichment: Addition Flashcards • Games/Enrichment: Fact Fun Activities • PowerPoint® presentation	**Reviews** • Pages 237–38
Lesson 128 **Counting More & Less**			
Teacher Edition 550–53 **Worktext** 249–50	**128.1** Identify and write the number that is 1 more or 1 less. **128.2** Identify and write the number that is 10 more or 10 less. **128.3** Count by 10s from a 3-digit number. **128.4** Identify and write the number that is 100 more or 100 less.	**Visuals** • Visual 1: *Hundred Chart* (extended) • Place Value Kit: hundreds, tens, ones **Student Manipulatives Packet** • Place Value Kit: hundreds, tens, ones • Hundreds/Tens/Ones Mat **BJU Press Trove** • Video: More and Less • Games/Enrichment: Fact Reviews (Addition Facts to 12) • Games/Enrichment: Addition Flashcards • Games/Enrichment: Fact Fun Activities • PowerPoint® presentation	**Reviews** • Pages 239–40
Lesson 129 **Comparing 3-Digit Numbers**			
Teacher Edition 554–57 **Worktext** 251–52	**129.1** Compare 3-digit numbers by using manipulatives. **129.2** Compare 3-digit numbers by using > and <. **129.3** Compare 3-digit numbers to explore God's world. BWS Exploring (apply)	**Visuals** • Place Value Kit: hundreds, tens, ones • Sign Cards: Greater than, Less than **Student Manipulatives Packet** • Sign Cards: Greater than, Less than • Number Cards: 0–9 • Dot Pattern Cards: 0–9 (backs) **BJU Press Trove** • Games/Enrichment: Addition Flashcards • Games/Enrichment: Fact Fun Activities • PowerPoint® presentation	**Reviews** • Pages 241–42

PAGES	OBJECTIVES	RESOURCES & MATERIALS	ASSESSMENTS
Lessons 130–31 Adding Large Numbers			
Teacher Edition 558–61 **Worktext** 253–54	**130.1** Join sets for 3-digit addends. **130.2** Solve a word problem by using the Problem-Solving Plan. **130.3** Write an addition equation for a word problem.	**Visuals** • Visual 21: *Problem-Solving Plan* • Place Value Kit: hundreds, tens, ones **Student Manipulatives Packet** • Hundreds/Tens/Ones Mat • Place Value Kit: hundreds, tens, ones **BJU Press Trove** • Video: Add Large Numbers • Web Link: Let's Set the Record Straight. Pizza Boxes are Recyclable. • PowerPoint® presentation	**Reviews** • Pages 243–44
Lesson 132 Practicing Adding Large Numbers			
Teacher Edition 562–65 **Worktext** 255–56	**132.1** Join sets for 3-digit numbers. **132.2** Solve a word problem by using the Problem-Solving Plan. **132.3** Write an addition equation for a word problem.	**Visuals** • Visual 21: *Problem-Solving Plan* • Place Value Kit: hundreds, tens, ones **Student Manipulatives Packet** • Hundreds/Tens/Ones Mat • Place Value Kit: hundreds, tens, ones **BJU Press Trove** • Web Link: 15 Creative Ways to Recycle Pizza Boxes • Web Link: 20 Cool Things You Can Make with a Pizza Box • PowerPoint® presentation	**Reviews** • Pages 245–46
Lesson 133 Chapter 16 Review			
Teacher Edition 566–69 **Worktext** 257–58	**133.1** Recall concepts and terms from Chapter 16.	**Visuals** • Place Value Kit: hundreds, tens, ones **Student Manipulatives Packet** • Sign Cards: Greater than, Less than • Number Cards: 0–9 • Dot Pattern Cards: 0–9 (backs) **BJU Press Trove** • Games/Enrichment: Addition Flashcards • Games/Enrichment: Fact Fun Activities • PowerPoint® presentation	**Worktext** • Chapter 16 Review **Reviews** • Pages 247–48

PAGES	OBJECTIVES	RESOURCES & MATERIALS	ASSESSMENTS
Lesson 134 Coding: Playing a Game			
Teacher Edition 570–73 **Worktext** 259–60	**134.1** Identify a pattern of repeated actions. **134.2** Translate a picture program into a clapping game. **134.3** Simplify a picture program by using loops. **134.4** Follow the Engineering Design Process to solve a problem. **134.5** Write a program for a clapping game by using loops. **134.6** Teach a friend a clapping game by using a program. BWS Exploring (apply)	**Teacher Edition** • Instructional Aid 38: *Clapping Games* • Instructional Aid 5: *STEM Engineering Design Process* **BJU Press Trove** • Web Link: Hand Clapping Games • PowerPoint® presentation	**Teacher Edition** • Instructional Aid 39: *Coding Rubric: Playing a Game*
Lesson 135 Test, Cumulative Review			
Teacher Edition 574–76 **Worktext** 261	**135.1** Demonstrate knowledge of concepts from Chapter 16 by taking the test.		**Assessments** • Chapter 16 Test **Worktext** • Cumulative Review **Reviews** • Pages 249–50

16 Place Value: Three-Digit Numbers

How do number patterns help me find someone's house?

Chapter 16

How do number patterns help me find someone's house?

Chapter Objectives

- Count by 1s, 10s, and 100s with 3-digit numbers.
- Identify the number and value of each place in a 3-digit number.
- Represent 3-digit numbers by using manipulatives.
- Write 3-digit numbers.
- Write numbers that are 1, 10, or 100 more or less than another.
- Compare 3-digit numbers.
- Use number patterns to explore God's world.
- Design a program for a game.

124 Worktext pages 240–42 Reviews pages 231–32

Objectives

- **124.1** Count by 100s to 1,000.
- **124.2** Identify the hundreds, tens, and ones in a 3-digit number.
- **124.3** Represent 3-digit numbers by using manipulatives.

Printed Resources

- Visuals: Place Value Kit (hundreds, tens, ones)
- Student Manipulatives: Hundreds/Tens/Ones Mat
- Student Manipulatives: Place Value Kit (hundreds, tens, ones)

Digital Resources

- Video: Ch 16 Intro
- Web Link: Virtual Manipulatives: Base Ten Number Pieces
- Games/Enrichment: Fact Reviews (Addition Facts to 12, page 24)
- Games/Enrichment: Addition Flash-cards (facts to 12)
- Games/Enrichment: Fact Fun Activities

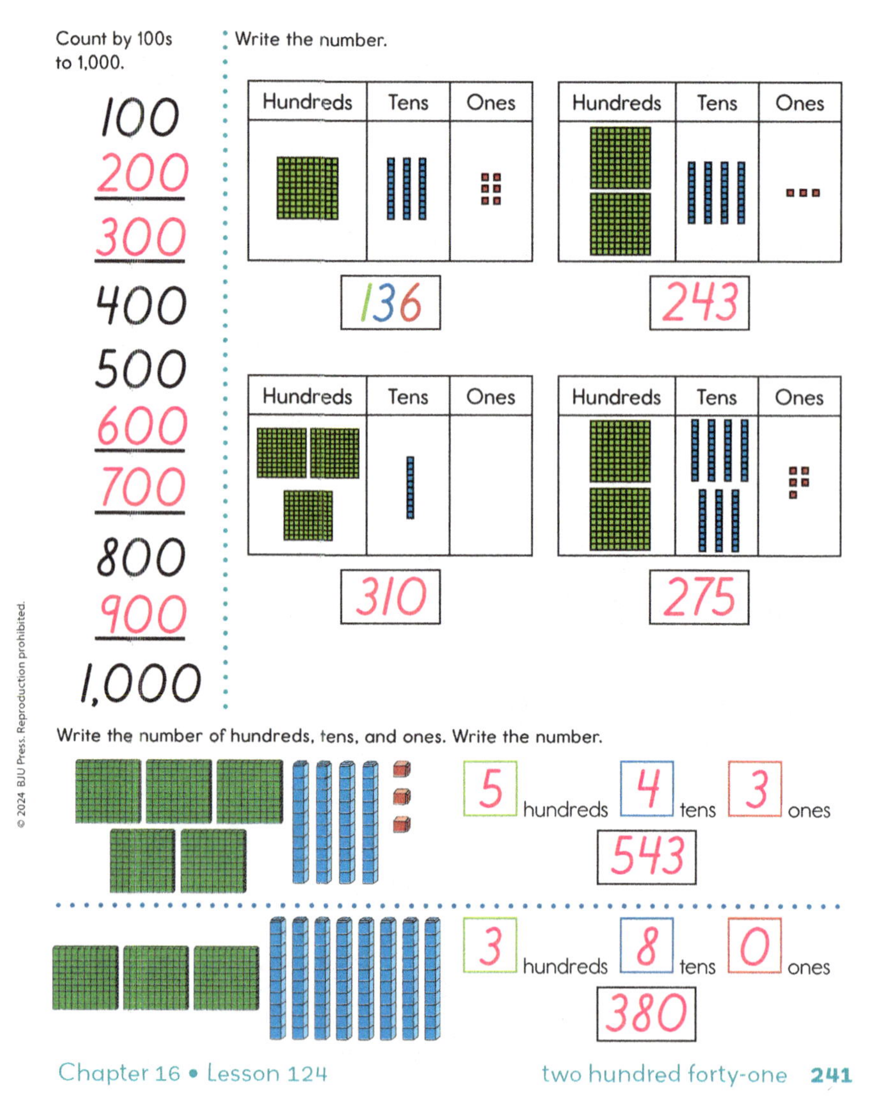

Practice & Review

Study Addition Facts to 12

Choose a game or activity from the "Fact Fun Activities" available in Trove.

Count by 10s to 100

Ask 5 students to hold up their hands with fingers extended toward the ceiling.

What is a fast way to count the students' fingers? by 10s

Guide the students as they count each pair of hands by 10s.

How many fingers are on 5 students? 50

How many students do you need to show 100 fingers? 10

Invite 5 more students to hold up their hands with fingers extended. Count together by 10s as you point to each pair of hands.

How many fingers are on 10 students? 100

How many sets of 10 make 100? 10

Instruct all the students to hold up their hands with fingers extended. Count each pair of hands together by 10s.

How many fingers are on __ students' hands?

How many sets of 10 make __?

Engage

Essential Question

Direct attention to the chapter opener on Worktext page 240 and read aloud the following scenario to introduce the chapter essential question.

DQ, Liam, and Lily squeezed into the back seat of the shiny, red car. The mouth-watering smell of pizza from the boxes in the front seat tickled their noses and made their stomachs grumble. Tate, Lily's big brother, slid into the driver's seat and checked the address of his first delivery.

"First stop is 221 Spring Street," Tate announced. He enjoyed his job as a food delivery driver and was especially happy to have Lily and her two friends Liam and DQ helping with his deliveries today. Tate carefully entered the street from the restaurant parking lot and turned right. "Spring Street crosses Main Street in a few blocks, but I'm

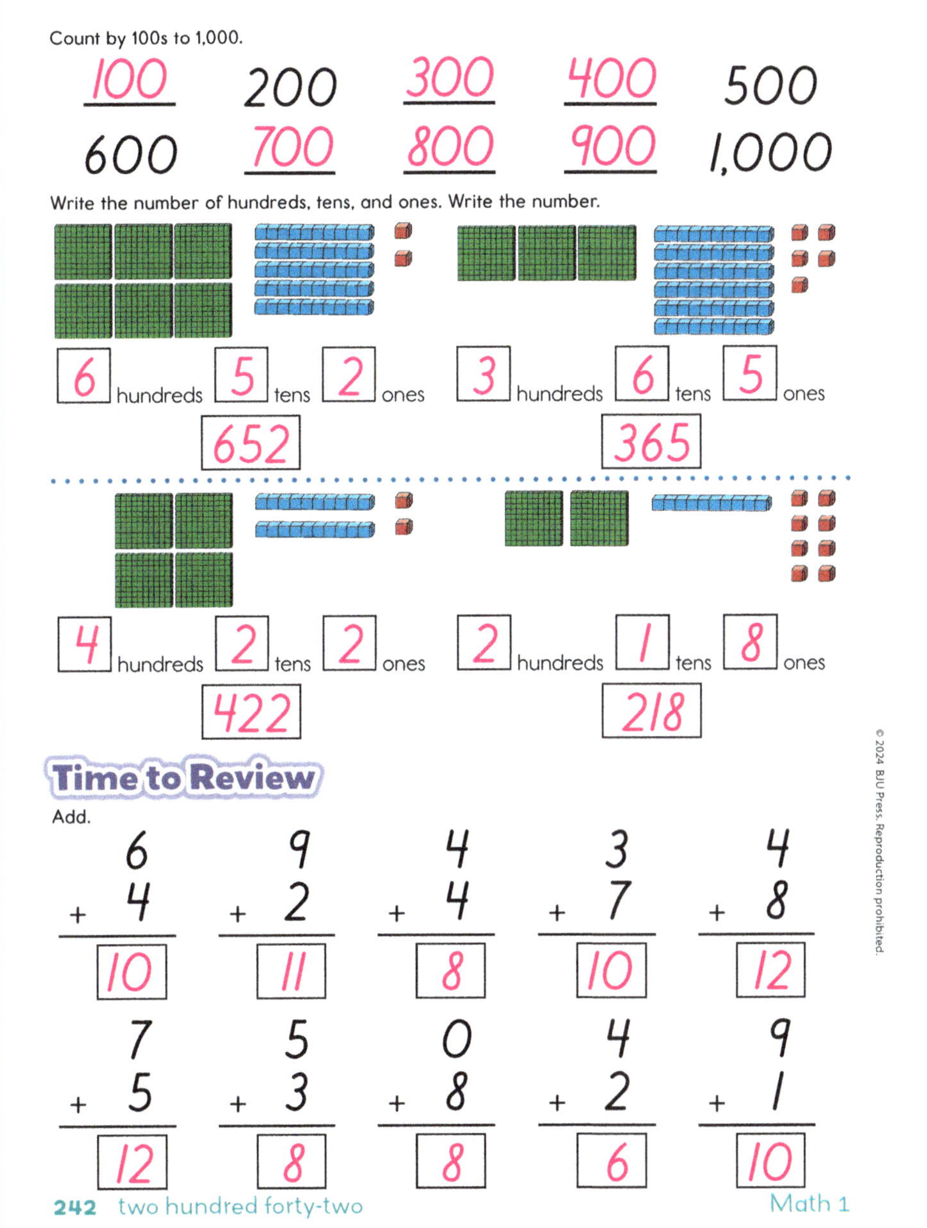

not sure where 221 Spring Street is. You'll have to help me look when we turn on the street."

The three friends sat up and looked out the car windows at the buildings as they passed. "That's my dentist's office." Liam pointed to the sign with a big grin under the name. "And that's the store where I got my new cleats for soccer." He smiled at the thought of his neon blue cleats and of the games he would soon be playing.

"We're almost at Spring Street," Tate informed his helpers. He stopped at the stop sign and turned right onto Spring Street. The mailbox at the first house on the right side of the street read *101*.

Lily read the numbers on the mailboxes on the right side of the street as they passed them. "103 . . . 105 . . . 107." She looked to the left and read "108 . . . 110 . . . 112. The numbers are in count-by-2s order!"

"That's right," explained Tate. "House numbers often follow a zigzag pattern back and forth across the street, with the even and odd numbers on opposite sides of the street. The numbers on Spring Street follow a count-by-2s pattern on each side. We still have a ways to go before we get to the 200s."

After a few minutes the car came to another stop sign and then continued along Spring Street. DQ noticed that the numbers had changed and were now in the 200s—the 200 block. *I wonder which side 221 will be on?* he thought. "Is 221 even or odd?" DQ asked his friends.

"Odd!" exclaimed Liam and Lily together.

"The odd-numbered homes are on the right side of Spring Street," observed Tate as he slowed the car to allow his helpers in the back seat to read the numbers on the mailboxes. "We'll need to keep our eyes peeled for 221 on the right side."

"There it is!" Lily pointed excitedly to the house just beyond a blossom-filled cherry tree. Tate turned into the driveway, grabbed the pizza delivery bag, and walked quickly to the door of 221 Spring Street.

After delivering the pizza, Tate checked his list for the next destination. "Two more pizzas to deliver. You three are a big help. I'm glad you learned how to read 3-digit numbers so you could help me find the right house. When we're finished, I'll treat you to your very own personal pepperoni pizzas." Tate grinned at the cheering in the back seat as he carefully backed out of the driveway and pointed his shiny, red car toward the next delivery destination.

Invite a student to read aloud the essential question. Encourage the students to consider this question as you guide them to answer it by the end of the chapter.

Instruct

Count by 100s to 1,000

Use a **visual display** to help the students count by 100s to 1,000.

Write "100" for display.

Which digit is in the Ones place? 0

Which digit is in the Tens place? 0

Direct attention to the Hundreds place in 100. Explain that just as there are places to show how many ones and tens are in a 2-digit number, there is also a special place to show how many hundreds are in a 3-digit number.

Which digit is in the Hundreds place? 1

What is this number? 100

Hundreds, Tens, Ones

Count by 100s to 1,000.

100 [200] [300] 400 [500]
600 [700] 800 [900] 1,000

Write each number.

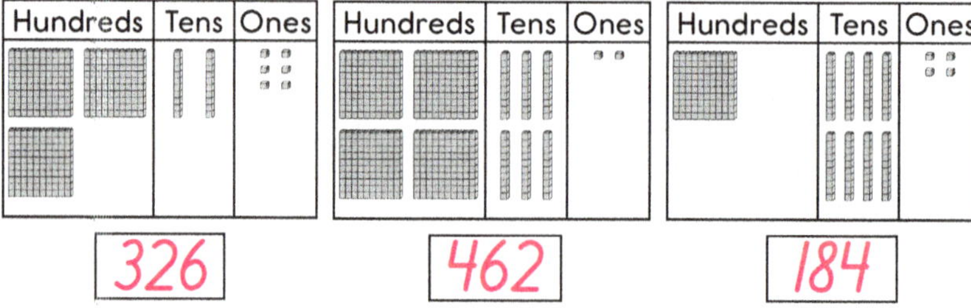

Write the number of hundreds, tens, and ones. Write the number.

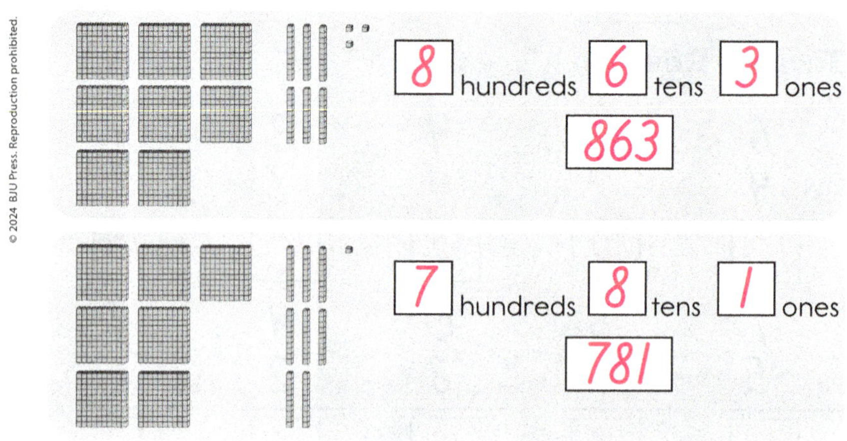

Display 1 hundred square.

How many one squares are in 1 hundred square? 100

Write "200" below 100 and display 2 hundreds.

Which digit is in the Hundreds place? 2

What is this number? 200

Write "300" below 200 and display another hundred.

Which digit is in the Hundreds place? 3

What is this number? 300

Continue the activity until 10 hundreds are displayed. Explain that 10 hundreds equal 1 thousand. Write "1,000" for display. Point out that 1,000 is a 4-digit number.

Guide the students as they count together by 100s as you direct attention to the list of numbers: 100, 200, 300 . . . 1,000.

Remind the students that they could continue counting numbers all day, every day for all year long, but they would never reach the last number. They could always say the next number in order.

Identify Hundreds, Tens & Ones

Guide a **discussion** to help the students identify place value in a 3-digit number.

Write "407" for display.

Which digit is in the Hundreds place? 4

Which digit is in the Tens place? 0

Which digit is in the Ones place? 7

Read the number together: *four hundred seven.*

Be sure to say *four hundred seven*, not *four hundred and seven*. The word *and* will be used later to denote the decimal point in a number.

Continue the activity with 342, 815, and 190.

Represent 3-Digit Numbers

Use **manipulatives** to help the students represent 3-digit numbers.

Distribute the Hundreds/Tens/Ones Mats and the Place Value Kits.

Display 6 hundreds, 3 tens, and 2 ones in a Hundreds/Tens/Ones frame.

How many hundreds are there? 6

How many tens are there? 3

How many ones are there? 2

What is this number? 632

Write "632" for display. Ask a student to read aloud the number. six hundred thirty-two

Write "314" and invite a volunteer to read the number. three hundred fourteen

Direct the students to make the number 314 on their mats. Display 3 hundreds, 1 ten, and 4 ones so that the students can check the manipulatives on their mats.

Invite a student to write a 3-digit number for display. Instruct the students to make that number on their mats.

Guide the students as they read the number.

Apply

Worktext pages 241–42

Read and guide completion of page 241.

Read and explain the directions for page 242. Assist the students as they complete the page independently.

Assess

Reviews pages 231–32

Review fact families and 2-digit subtraction on page 232.

Fact Reviews

Use "Addition Facts to 12" page 24 in Trove.

Chapter 10 Review

Write each fact family.

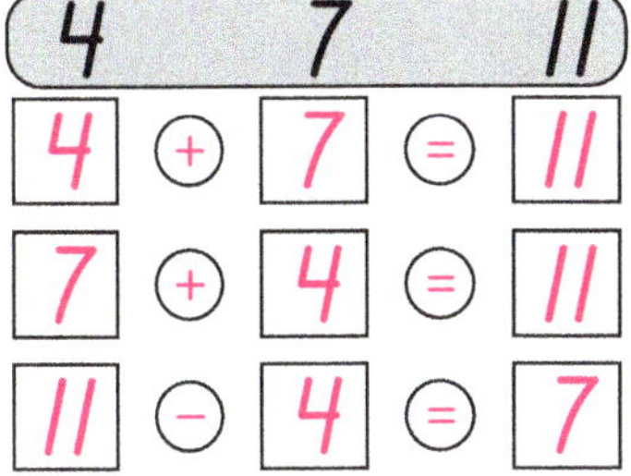

4	7	11
4 + 7 = 11		
7 + 4 = 11		
11 − 4 = 7		
11 − 7 = 4		

3	6	9
3 + 6 = 9		
6 + 3 = 9		
9 − 3 = 6		
9 − 6 = 3		

Chapter 13 Review

Subtract.

47 − 13 = 34	76 − 24 = 52	59 − 38 = 21	95 − 54 = 41
62 − 41 = 21	33 − 12 = 21	84 − 61 = 23	28 − 16 = 12

Subtraction Fact Review

Subtract.

11 − 2 = 9	10 − 5 = 5	9 − 6 = 3
10 − 3 = 7	9 − 8 = 1	11 − 7 = 4
8 − 2 = 6	12 − 4 = 8	8 − 5 = 3

232 two hundred thirty-two Math 1 Reviews

Differentiated Instruction

Identify the Number of Hundreds, Tens & Ones in a 3-Digit Number

Label a green container *Hundreds*, a blue container *Tens*, and a red container *Ones*. Place a given number of one squares (from a Student Manipulatives Packet) in the Ones container, a given number of ten strips in the Tens container, and a given number of hundred squares in the Hundreds container. Direct the students to count the number of hundreds, tens, and ones in the containers and then to say and write the number.

125 Worktext pages 243–44
Reviews pages 233–34

Objectives

- **125.1** Identify the hundreds, tens, and ones in a 3-digit number.
- **125.2** Represent 3-digit numbers by using manipulatives.
- **125.3** Rename 10 ones as 1 ten in a 3-digit number.
- **125.4** Complete a sequence of 3-digit numbers.
- **125.5** Explain how place value helps people explore God's world. BWS

Biblical Worldview Shaping

- **Exploring** (explain): Understanding place value in 3-digit numbers helps us recognize patterns in things that are numbered, like houses. Recognizing patterns is an important part of exploring God's world (125.5).

Printed Resources

- Visuals: Place Value Kit (hundreds, tens, ones)
- Student Manipulatives: Hundreds/Tens/Ones Mat
- Student Manipulatives: Place Value Kit (hundreds, tens, ones)

Digital Resources

- Games/Enrichment: Fact Reviews (Addition Facts to 12, page 25)
- Games/Enrichment: Addition Flashcards (facts to 12)
- Games/Enrichment: Fact Fun Activities

Representing 3-Digit Numbers

Write each missing number.

Write the number of hundreds, tens, and ones. Write the number.

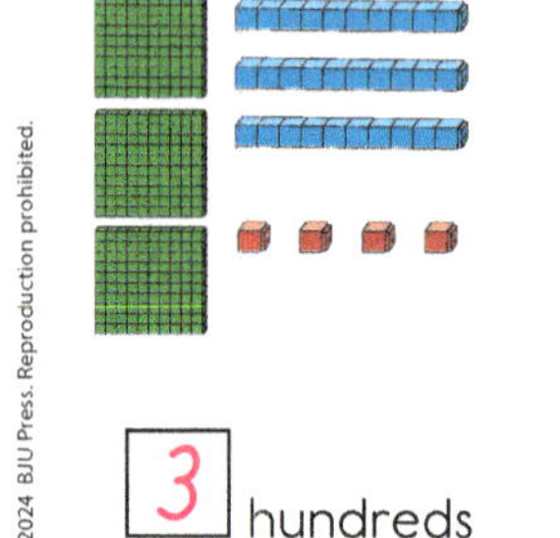

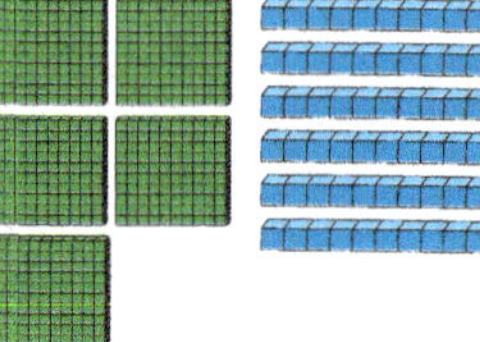

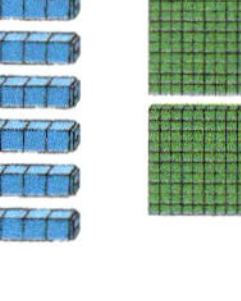

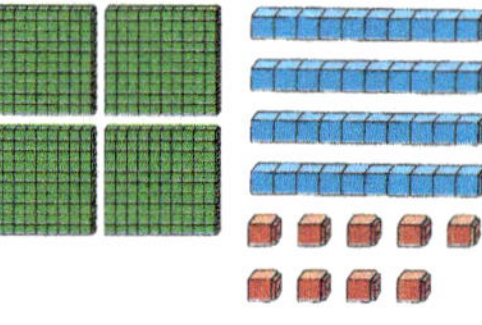

3 hundreds	5 hundreds	4 hundreds
3 tens	6 tens	4 tens
4 ones	0 ones	9 ones
334	560	449

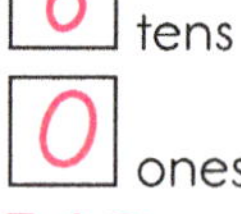

Practice & Review

Study Addition Facts to 12

Choose a game or activity from the "Fact Fun Activities" available in Trove.

Engage

Add 3 Numbers

Guide a **review activity** to help the students add 3 numbers.

Write the equation "3 + 3 + 2 = __" for display. Ask a student to read the equation aloud.

How many addends are in this equation? 3

How many addends can you add together at one time? 2

What does 3 + 3 equal? 6

Write "6" for display below 3 + 3.

What does 6 + 2 equal? 8

Display the sum to the equation. 8

Remind the students that the Grouping Principle of Addition allows them to group the addends in any order without changing the sum.

Write the original equation again, and direct attention to the 2 final addends in the equation.

What does 3 + 2 equal? 5

What does 3 + 5 equal? 8

Invite a student to complete the equation for display. Compare the sums to the equation when the addends are added in a different order to verify that the sums are the same.

Continue the activity with the following vertical equations for display. Guide the students as they find the sum of the equation by adding the addends in a different order each time.

7	2	3
1	3	3
+ 2	+ 5	+ 6
10	10	12

Write the number of hundreds, tens, and ones. Write the number.

2 hundreds 8 tens 5 ones
285

6 hundreds 0 tens 6 ones
606

Write each missing number.

639	640	641	642
99	100	101	102
309	310	311	312

Circle the word to complete the sentence.

I read 3-digit numbers to find / wash someone's house.

Time to Review

Add.

$5 + 1 + 6 = 12$

$2 + 3 + 2 = 7$

$4 + 3 + 2 = 9$

$2 + 1 + 7 = 10$

Instruct

Identify Hundreds, Tens & Ones

Guide a **discussion** to help the students identify place value in 3-digit numbers.

Write "346" for display.

Is this a 2-digit number or a 3-digit number? 3-digit number

Invite a volunteer to read aloud the number. three hundred forty-six

Which digit is in the Ones place? 6

Which digit is in the Hundreds place? 3

Which digit is in the Tens place? 4

Continue the activity for 702, 530, and 629.

Represent 3-Digit Numbers

Use **manipulatives** to help the students represent 3-digit numbers.

Distribute the Hundreds/Tens/Ones Mats and the Place Value Kits. Write "435" for display. Instruct the students to make the number on their mats, using ones, tens, and hundreds. Model the problem for the students to check their answers.

What number will you make if you add 1 more one to the Ones place? 436

Place another one in the Ones place. Write "436" after 435.

What number will you make if you add another one to the Ones place? 437

Place 1 more one in the Ones place. Write "437" after 436. Continue the activity for the numbers 438 and 439.

Direct attention to the list of numbers. Read the numbers together, beginning with the number 435.

Explain that the students will continue to use their manipulatives in the following activity.

Rename 10 Ones as 1 Ten in a 3-Digit Number

Use **manipulatives** to help the students rename 10 ones as 1 ten in a 3-digit number.

Erase the preceding list of numbers and write "348" for display. Direct the students to make the number 348 on their mats. Ask a student to say how many hundreds, tens, and ones he or she put on his or her mat. 3 hundreds, 4 tens, 8 ones

What number will you make if you add another one to the Ones place? 349

Put 1 more one in the Ones place. Write "349" after 348.

What number will you make if you add another one to the Ones place? 350

Put 1 more one in the Ones place. Write "350" after 349. Invite a volunteer to say how many hundreds, tens, and ones he or she has on his or her mat now. 3 hundreds, 4 tens, 10 ones

Guide the students as they count together the hundreds, tens, and ones (by 100s, 10s, 1s) to see whether they really have 350 on their mats: 100, 200, 300, 310, 320, 330, 340, 341, 342 . . . 350.

What can you do with the 10 ones in the Ones place? rename/trade them for 1 ten

Remove the 10 ones from the Ones place and put 1 more ten in the Tens place. Explain that 3 hundreds, 4 tens, and 10 ones is the same as 3 hundreds, 5 tens, and 0 ones since 10 ones is the same as 1 ten. Point out that whenever there are 10 ones in the Ones place, the students will rename, or trade, them for 1 ten.

What number will you make if you add another one to the Ones place? 351

Put 1 more one in the Ones place. Write "351" after 350. Ask a student to tell how many hundreds, tens, and ones he or she has on his or her mat now. 3 hundreds, 5 tens, 1 one

125

Representing 3-Digit Numbers

Write each missing number.

Write the number of hundreds, tens, and ones.
Write the number.

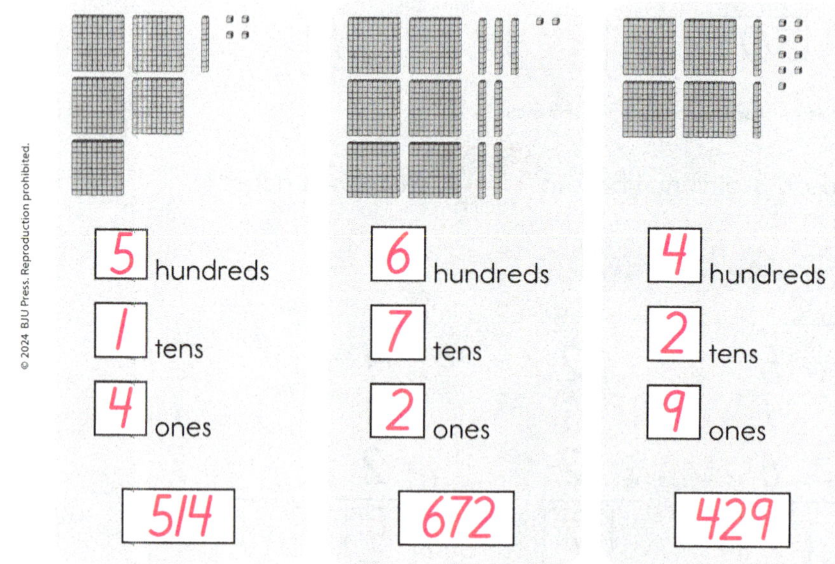

Continue the activity for the numbers 352 and 353.

Direct attention to the list of numbers. Read the numbers together, beginning with the number 348. Remind the students that after 349, the number in the Tens place changed.

For students struggling to count 3-digit numbers, cover the 3 (the digit in the Hundreds place) and ask them to give the number that comes after 49; then ask them to give the number that comes after 349.

Complete a Sequence of 3-Digit Numbers

Guide a **sequencing activity** to help the students complete a number sequence.

Write "__ 243 244 __ 246 __ 248 __ 250 __" for display. Point out that several numbers are missing.

Which number comes before 243 when you count? 242

Invite a student to write "242" in the first blank.

Which number comes after 244 when you count? 245

Invite another student to write "245" in the second blank.

Continue the activity to complete the sequence. Guide the students as they check the answers by counting the completed sequence.

Display the following sequences and guide the students as they complete them.

317 __ 319 __ 321 __ __ 324 __ 326

__ 876 877 __ 879 __ 881 __ __ 884

Exploring God's World

Guide a **discussion** to help the students explain a biblical worldview shaping truth.

Remind the students of the opener scenario in which DQ, Liam, and Lily helped Lily's older brother Tate deliver pizzas.

How did the three friends help Tate? They read the house numbers along the street to help Tate find the right address.

Read aloud the following excerpt from the opener scenario.

After delivering the pizza, Tate checked his list for the next destination. "Two more pizzas to deliver. You three are a big help. I'm glad you learned how to read 3-digit numbers so you could help me find the right house."

What skill had the 3 friends learned that made them a big help to Tate? reading 3-digit numbers

Explain that the friends' math skills helped them read the house numbers and follow the pattern of the addresses.

Point out that God can help the students use the math skills they are developing to learn more about God's world.

How can understanding place value and being able to read 3-digit numbers help you explore God's world? Reading 3-digit numbers can help me find a friend's house.

Apply

Worktext pages 243–44

Read and guide completion of page 243.

Read and explain the directions for page 244. Assist the students as they complete the page independently.

Direct attention to the biblical worldview shaping statement and guide the students as they complete the sentence.

Assess

Reviews pages 233–34

Review time and the days of the week on page 234.

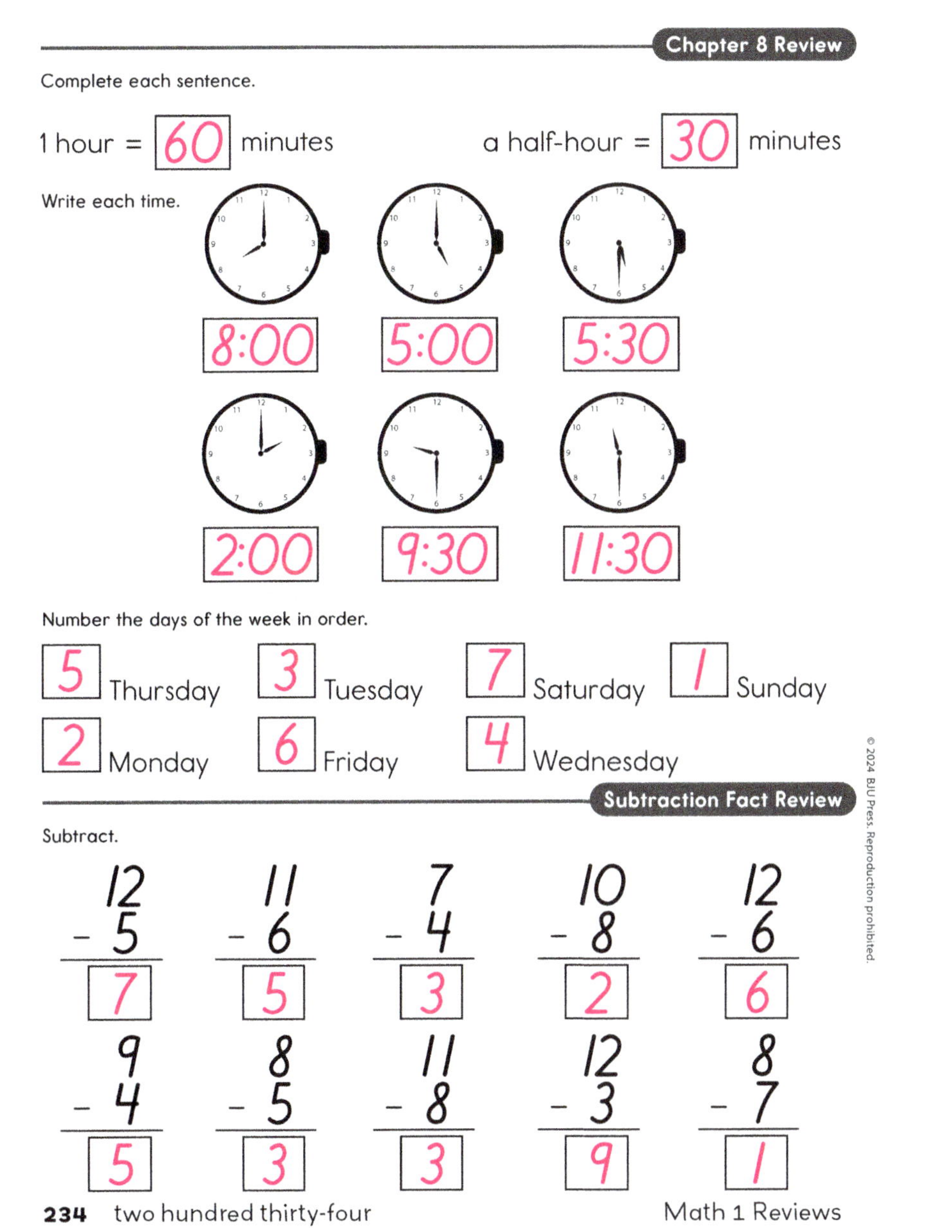

Chapter 8 Review
Complete each sentence.
1 hour = 60 minutes
a half-hour = 30 minutes
Write each time.
8:00
5:00
5:30
2:00
9:30
11:30
Number the days of the week in order.
5 Thursday
3 Tuesday
7 Saturday
1 Sunday
2 Monday
6 Friday
4 Wednesday
Subtraction Fact Review
Subtract.
12 - 5 = 7
11 - 6 = 5
7 - 4 = 3
10 - 8 = 2
12 - 6 = 6
9 - 4 = 5
8 - 5 = 3
11 - 8 = 3
12 - 3 = 9
8 - 7 = 1
© 2024 BJU Press. Reproduction prohibited.
234 two hundred thirty-four
Math 1 Reviews

Fact Reviews

Use "Addition Facts to 12" page 25 in Trove.

Extended Activity

Make 3-Digit Numbers with a Number Cube

Materials

- A number cube for each group of 3 students

Procedure

Group the students by 3s. Distribute the number cubes. Direct the students to draw 3 blanks. Instruct them to each take a turn rolling the number cube and writing the digit in a blank. After all the blanks are filled, instruct a student within the group to read the number. Continue until each student in each group has had a turn to read a number.

126

Worktext pages 245–46
Reviews pages 235–36

Objectives

- **126.1** Identify the number of hundreds, tens, and ones in a 3-digit number.
- **126.2** Write 3-digit numbers by using Number Cards.
- **126.3** Write the number that is 1 more or 1 less.
- **126.4** Write the number that comes *before, after,* or *between.*

Printed Resources

- Visual 1: *Hundred Chart* (extended)
- Visuals: Place Value Kit (hundreds, tens, ones)
- Student Manipulatives: Number Cards (0–9)

Digital Resources

- Video: Place Value
- Games/Enrichment: Fact Reviews (Addition Facts to 12, page 26)
- Games/Enrichment: Addition Flashcards (facts to 12)
- Games/Enrichment: Fact Fun Activities

Writing Numbers to 1,000

Mark the mat that shows 1 *more* than 235.

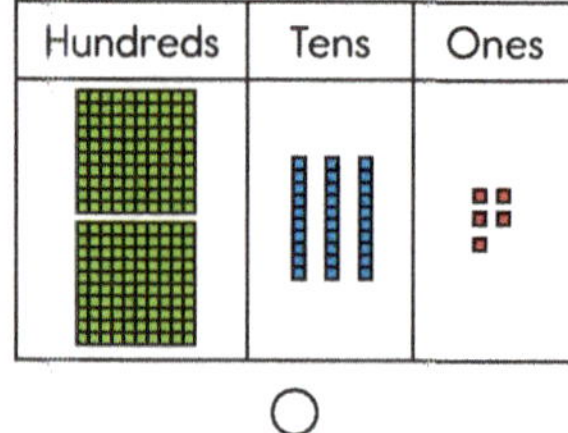

Mark the mat that shows 1 *less* than 423.

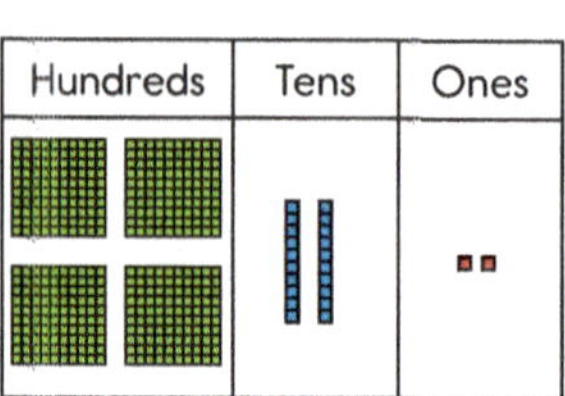

Write the number that comes just *before.*

525 526
383 384

Write the number that comes just *after.*

642 643
526 527

Write the number that comes *between.*

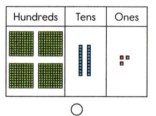

108 109 110
333 334 335
671 672 673
489 490 491
247 248 249

Practice & Review

Count by 10s from a 2-Digit Number

Direct attention to 23 on Visual 1 (extended). Guide the students as they count together by 10s by moving down the column from 23. Point out the numbers as the students count: 23, 33, 43, 53, 63, 73 . . . 143. Continue counting together by 10s to 193: 153, 163, 173, 183, 193.

Continue the activity, starting with 17.

Study Addition Facts to 12

Choose a game or activity from the "Fact Fun Activities" available in Trove.

Engage

Count 150–200

Guide a **choral counting activity** to help the students count from 150–200.

Divide the students into 2 groups. Direct the students to pretend that they are a duck by flapping their wings as they count when it is their group's turn.

Instruct Group 1 to begin counting with the number 150 and to continue counting until you clap your hands.

Instruct Group 2 to *count on* from the last number counted by Group 1 until you clap your hands. Group 1 will pick up the counting where Group 2 leaves off.

Continue to alternate groups until a group reaches 200.

Instruct

3-Digit Numbers

Use **manipulatives** to help the students represent 3-digit numbers.

Distribute the Number Cards. Draw a Hundreds/Tens/Ones frame. Display 3 hundreds, 6 tens, and 2 ones.

How many hundreds are there? 3

How many tens are there? 6

How many ones are there? 2

What is the number? three hundred sixty-two

Instruct the students to use their Number Cards to show the number. Write "362" for display. Direct the students to check their numbers to see whether they are shown correctly.

Continue the activity for 149 and 320.

Write the number that comes *between*.

200 201 202

299 300 301

529 530 531

409 410 411

616 617 618

898 899 900

Write the number that is 1 *more* or 1 *less*.

1 more	
125	126
278	279
349	350
486	487

1 less	
135	136
327	328
589	590
760	761

Time to Review

Add.

8 + 1 = 9

9 + 0 = 9

2 + 7 = 9

6 + 3 = 9

5 + 5 = 10

9 + 1 = 10

9 + 2 = 11

4 + 8 = 12

6 + 5 = 11

The Number That Is 1 More

Guide a **discussion** to help the students write the number that is 1 more.

Display 4 hundreds, 5 tens, 2 ones in a Hundreds/Tens/Ones frame.

What number is this? 452

Instruct the students to use their Number Cards to show 452.

What number will I make if I add 1 more one to 452? 453

Display 1 more one in the Ones place. Instruct the students to use their Number Cards to show the number formed when they add 1 more.

How did you change 452 to make 453? I replaced the 2 in the Ones place with a 3.

Display 3 hundreds, 4 tens, 6 ones in the frame.

What number is this? 346

Instruct the students to show the number that is 1 more than 346. 347 Invite a volunteer to add 1 more to the display.

Display 5 hundreds, 1 ten, 9 ones in the frame.

What number is this? 519

Instruct the students to show the number that is 1 more than 519. 520 Invite a volunteer to add 1 more to the display.

How many ones are in the Ones place? 10

What should I do with the 10 ones? rename or trade them for 1 ten

Remove 10 ones and add 1 ten to the Tens place.

The Number That Is 1 Less

Guide a **discussion** to help the students write the number that is 1 less.

Display 3 hundreds, 2 tens, 5 ones in a Hundreds/Tens/Ones frame.

What number is this? 325

Instruct the students to use their Number Cards to show 325.

What number will I make if I remove 1 one from 325? 324

Remove 1 one from the Ones place. Instruct the students to use their Number Cards to show the number formed when they have 1 less.

How did you change 325 to 324? I replaced the 5 in the Ones place with a 4.

Display 2 hundreds, 4 tens in the frame.

What number is this? 240

Instruct the students to show the number that is 1 less than 240. 239 Explain as you demonstrate that since there are 0 ones in the Ones place, they must first rename 1 ten as 10 ones before they can remove a one.

Display 4 hundreds, 5 tens, 1 one.

What number is this? 451

Instruct the students to show the number that is 1 less than 451. 450 Ask a student to remove 1 one from the frame.

The Number That Comes *before* or *after*

Use a **visual display** to help the students identify numbers that come *before* or *after* another number.

Display Visual 1 and direct attention to 138.

Which number comes *before* 138? 137

Is 137 1 more than or 1 less than 138? 1 less

Explain that the number that comes *before* a number is 1 *less* than that number.

Direct attention to 125 on the chart.

Which number comes *after* 125? 126

Is 126 1 more than or 1 less than 125? 1 more

Explain that the number that comes *after* a number is 1 *more* than that number.

126

Continue the activity by pointing out several numbers and asking the students which number comes *before* and which number comes *after* the numbers.

The Number That Comes *between*

Use a **visual display** to help the students identify numbers that come *between* two numbers.

Direct the students to look at numbers 114 and 116 on Visual 1.

> **Which number comes *between* 114 and 116?** 115

Explain that 114 is 1 less than 115 and that 116 is 1 more than 115; the number 115 is *between* 114 and 116.

Say 2 numbers from the chart and invite a student to tell which number comes *between* these numbers. Continue the activity with several pairs of numbers.

Apply

Worktext pages 245–46

Read and guide completion of page 245.

Read and explain the directions for page 246. Assist the students as they complete the page independently.

Assess

Reviews pages 235–36

Review fair shares and fractions of a set on page 236.

Fact Reviews

Use "Addition Facts to 12" page 26 in Trove.

Extended Activity

Write the Number That Is 1 More or 1 Less

Materials

- Number Cards: 0–9 for each student

Procedure

Distribute the Number Cards. Group the students in pairs. Instruct the first student in each pair to show a 3-digit number by using his or her Number Cards. Instruct the second student to show the number that is 1 more. Direct the students to switch roles. This time instruct the second student to show a number and the first student to show the number that is 1 less. Continue the activity until each student has made several numbers that are 1 more than and 1 less than the given number.

Writing Numbers to 1,000

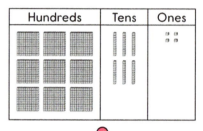

Mark the mat that shows 1 *less* than 621.

Hundreds	Tens	Ones

Hundreds	Tens	Ones

Mark the mat that shows 1 *more* than 963.

Hundreds	Tens	Ones

Hundreds	Tens	Ones

Write the number that comes just *before*.

466 467

Write the number that comes *between*.

684 685 686

322 323 324

Write the number that comes just *after*.

924 925

Chapter 9 Review

Circle the fair share for each child.
Write the number each child gets.

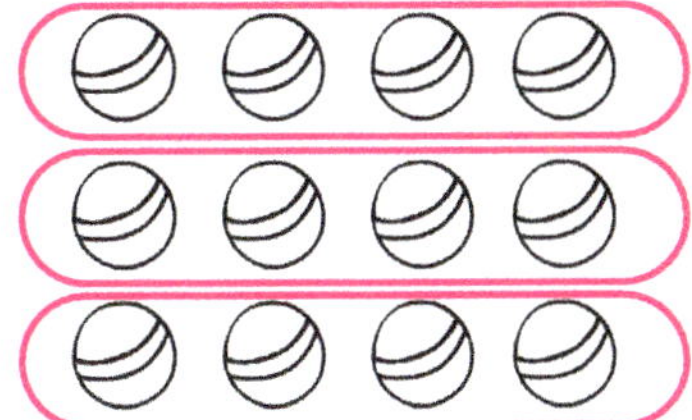

3 children get [4] each

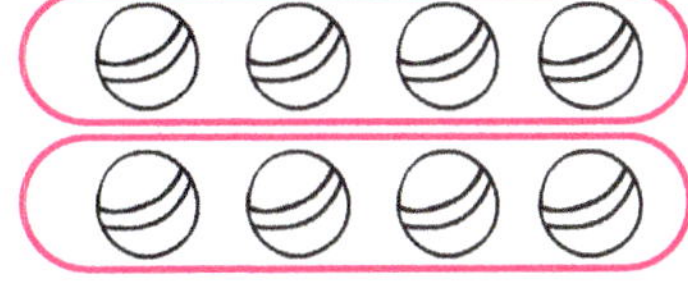

2 children get [4] each

Color to show each fraction.

$\frac{1}{2}$

$\frac{1}{3}$

$\frac{2}{4}$

Subtraction Fact Review

Subtract.

$11 - 6 = $ [5] $9 - 7 = $ [2] $8 - 6 = $ [2]

$12 - 3 = $ [9] $7 - 3 = $ [4] $9 - 6 = $ [3]

$10 - 3 = $ [7] $6 - 4 = $ [2] $11 - 4 = $ [7]

127

Worktext pages 247–48
Reviews pages 237–38

Objectives

- **127.1** Identify the number of hundreds, tens, and ones in a 3-digit number.
- **127.2** Write 3-digit numbers by using Number Cards.
- **127.3** Identify the value of the digits in a 3-digit number.

Printed Resources

- Visuals: Place Value Kit (hundreds, tens, ones)
- Student Manipulatives: Number Cards (0–9)

Digital Resources

- Games/Enrichment: Fact Reviews (Addition Facts to 12, page 27)
- Games/Enrichment: Addition Flashcards (facts to 12)
- Games/Enrichment: Fact Fun Activities

Practice & Review

Count 200–250

Explain that you will take turns counting. You will start by saying 200; then they will say the next number. Continue taking turns as you count to 250.

Study Addition Facts to 12

Choose a game or activity from the "Fact Fun Activities" available in Trove.

Place Value in 3-Digit Numbers

438

A 4 in the Hundreds place has a value of 400.
A 3 in the Tens place has a value of 30.
An 8 in the Ones place has a value of 8.

Mark the value of the underlined digit.

<u>6</u>52	● 600	○ 60	○ 6
3<u>3</u>7	○ 300	● 30	○ 3

Write the value of each underlined digit.

- 4<u>2</u>7 — 20
- 31<u>5</u> — 5
- <u>2</u>67 — 200

Read the clues and write the number.

- 7 hundreds, 6 tens, 4 ones — 764
- 4 tens, 2 ones, 3 hundreds — 342
- 5 ones, 2 hundreds, 6 tens — 265

Engage

Write 2-Digit Numbers in Expanded Form

Guide a **review** to help the students write 2-digit numbers in expanded form.

Write "30 + 6" for display. Remind the students that this is the expanded form of a number.

Which number is the same as 30 + 6? 36

Ask a student to write "36" to the right of the expanded form.

Write the following expanded forms and invite volunteers to write the standard-form numbers.

20 + 7 27
40 + 2 42
60 + 5 65

Write the following numbers and invite volunteers to write the expanded forms.

52 50 + 2
18 10 + 8
39 30 + 9

Instruct

3-Digit Numbers

Use **manipulatives** to help the students represent 3-digit numbers.

Distribute the Number Cards. Display 2 hundreds, 5 tens, and 3 ones.

How many hundreds are there? 2

How many tens are there? 5

How many ones are there? 3

What is the number? two hundred fifty-three

Ask the students to use their Number Cards to show the number. Write "253" for display. Tell the students to check their numbers to see whether they are shown correctly.

Display 4 hundreds, 0 tens, and 7 ones. Ask the students to show the number. Write "407" for display.

Which digit is in the Hundreds place? 4

Which digit is in the Tens place? 0

Which digit is in the Ones place? 7

Continue the activity with 540 and 162.

Mark the value of the orange digit.

342	300 ●	30 ○	3 ○
175	500 ○	50 ○	5 ●
287	800 ○	80 ●	8 ○

Match each clue to the correct number.

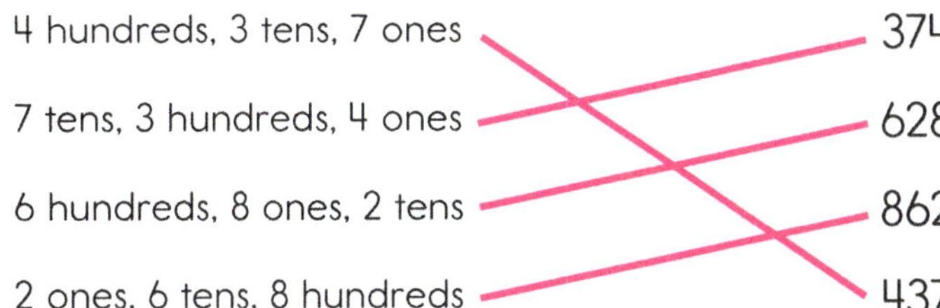

4 hundreds, 3 tens, 7 ones — 437

7 tens, 3 hundreds, 4 ones — 374

6 hundreds, 8 ones, 2 tens — 628

2 ones, 6 tens, 8 hundreds — 862

Time to Review

Write the expanded form for each number.

38 = 30 + 8

86 = 80 + 6

57 = 50 + 7

94 = 90 + 4

Write the number.

50 + 2 = 52

70 + 4 = 74

90 + 5 = 95

60 + 8 = 68

Why do the values for the two 5s differ? The 5 in the Hundreds place means there are 5 hundreds, and the 5 in the Tens place means there are 5 tens.

What is the value of the 3 in the Ones place? 3

Continue the activity for 249 and 333.

Place Value Clues

Guide a **listening activity** to help the students represent 3-digit numbers from clues they hear.

Explain to the students that they are to listen to the clues and show the mystery number by using their Number Cards.

The activity may be played as a game by dividing the students into teams and giving a point for each correct answer.

Use the following clues.

4 is in the Hundreds place, 7 is in the Tens place, 3 is in the Ones place 473

5 is in the Ones place, 2 is in the Tens place, 6 is in the Hundreds place 625

3 is in the Tens place, 7 is in the Ones place, 1 is in the Hundreds place 137

2 is in the Ones place, 5 is in the Hundreds place, 0 is in the Tens place 502

7 with the value of 70, 3 with the value of 300, 6 with the value of 6 376

2 with the value of 200, 4 with the value of 40, 0 (zero) 240

Place Value in 3-Digit Numbers

Guide a **discussion** to explain the value of the digits in 3-digit numbers.

Write the word *value* for display. Explain that they know the value of a digit by its place in a number. Write "324" for display.

Which digit is in the Hundreds place? 3

Which digit is in the Tens place? 2

Which digit is in the Ones place? 4

Write "3 hundreds, 2 tens, 4 ones" for display.

Explain that since the digit 3 is in the Hundreds place, it has a value of 300. Since the digit 2 is in the Tens place, it has a value of 20, and the digit 4 has a value of 4 since it is in the Ones place.

Write "300 + 20 + 4" in vertical form. Solve the problem and explain that if you add all the values of the digits of a number together, you will have the original number.

Write "553" for display.

Which digit is in the Hundreds place? 5

Which digit is in the Tens place? 5

Which digit is in the Ones place? 3

Write "5 hundreds, 5 tens, 3 ones" for display.

What is the value of the 5 in the Hundreds place? 500

What is the value of the 5 in the Tens place? 50

Apply

Worktext pages 247–48

Read and guide completion of page 247.

Read and explain the directions for page 248. Assist the students as they complete the page independently.

Assess

Reviews pages 237–38

Review fact families on page 238.

Fact Reviews

Use "Addition Facts to 12" page 27 in Trove.

Extended Activity

Identify the Value of the Digits

Materials

- A spinner with 10 sections labeled 0–9

Procedure

Invite 3 students to spin the spinner. Instruct the students to write for display the digits the spinner stops on. Direct the students to tell as many different 3-digit numbers as possible, using these digits.

For example, digits 4, 1, and 7 can make the following numbers: 417, 471, 741, 714, 174, and 147.

Write the numbers for display. Invite students to tell the value of each digit in the numbers.

Place Value in 3-Digit Numbers

Mark the value of each underlined digit.

775	○ 500	○ 50	● 5
234	● 200	○ 20	○ 2
186	○ 800	● 80	○ 8
923	● 900	○ 90	○ 9

Write the value of each underlined digit.

375	300	107	7
281	80	456	400
762	2	841	40

Read the clues and write the number.

3 hundreds, 7 tens, 2 ones — 372

8 hundreds, 6 ones, 4 tens — 846

5 tens, 9 hundreds, 1 one — 951

Chapter 10 Review

Write each fact family.

5 6 11

5 + 6 = 11
6 + 5 = 11
11 − 5 = 6
11 − 6 = 5

5 7 12

5 + 7 = 12
7 + 5 = 12
12 − 5 = 7
12 − 7 = 5

Mark the circle next to the fact that does *not* belong in each family.

3 9 12

- ○ 9 + 3 = 12
- ○ 3 + 9 = 12
- ○ 12 − 3 = 9
- ● 8 − 4 = 4

4 6 10

- ○ 6 + 4 = 10
- ○ 4 + 6 = 10
- ● 10 − 5 = 5
- ○ 10 − 4 = 6

3 6 9

- ○ 3 + 6 = 9
- ● 6 + 2 = 8
- ○ 9 − 3 = 6
- ○ 9 − 6 = 3

Addition Fact Review

Add.

8 + 4 = 12	9 + 3 = 12	7 + 6 = 13	8 + 2 = 10	4 + 5 = 9
7 + 3 = 10	6 + 6 = 12	5 + 7 = 12	7 + 7 = 14	8 + 3 = 11

128 Worktext pages 249–50 Reviews pages 239–40

Objectives

- **128.1** **Identify and write the number that is 1 more or 1 less.**
- **128.2** **Identify and write the number that is 10 more or 10 less.**
- **128.3** **Count by 10s from a 3-digit number.**
- **128.4** **Identify and write the number that is 100 more or 100 less.**

Printed Resources

- Visual 1: *Hundred Chart* (extended)
- Visuals: Place Value Kit (hundreds, tens, ones)
- Student Manipulatives: Place Value Kit (hundreds, tens, ones)
- Student Manipulatives: Hundreds/ Tens/Ones Mat

Digital Resources

- Video: More and Less
- Games/Enrichment: Fact Reviews (Addition Facts to 12, page 28)
- Games/Enrichment: Addition Flashcards (facts to 12)
- Games/Enrichment: Fact Fun Activities

Identifying the number that is 100 more or 100 less is included in this lesson for enrichment instruction. Present this concept based on the readiness of your students.

Counting More & Less

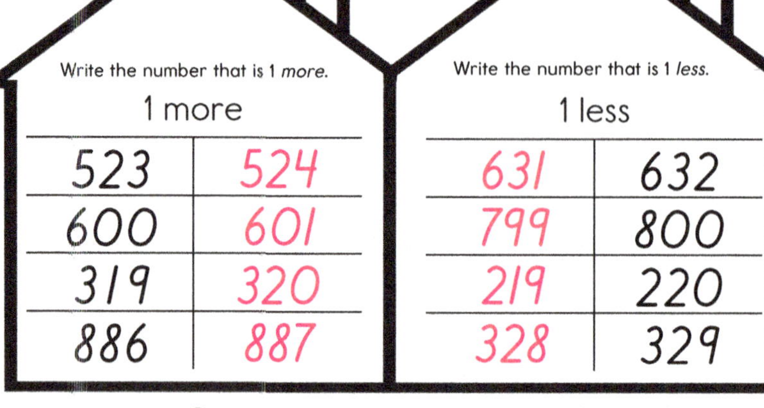

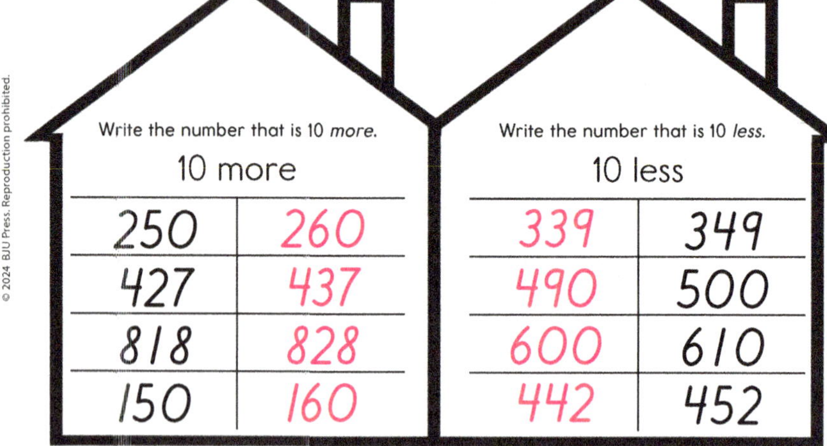

Practice & Review

Count by 10s from a 2-Digit Number

Direct attention to 26 on Visual 1. Guide the students as they count together by 10s; as you move down the rows, point out the numbers: 26, 36, 46, 56 . . . 146. Continue counting together to 196: 156, 166 . . . 196.

Repeat the activity, starting with a different number.

Add 3 Numbers

Write "2 + 4 + 3 = __" for display. Ask a student to read it aloud.

How many addends are in this equation? 3

How many addends can you add together at one time? 2

What does 2 + 4 equal? 6

Write "6" below 2 + 4.

What does 6 + 3 equal? 9

Write "9" in the answer blank.

Remind the students that the Grouping Principle of Addition allows them to group the addends in any order without changing the sum.

What does 4 + 3 equal? 7

What does 2 + 7 equal? 9

Write the following problems for display. Guide the students as they solve them.

6	2	4
1	4	4
+ 4	+ 4	+ 3
11	10	11

Study Addition Facts to 12

Choose a game or activity from the "Fact Fun Activities" available in Trove.

Engage

One More

Guide a **discussion** to help the students view a math concept from the perspective of biblical teaching.

Write the number that is 1 *more*.

529	244
530	245
801	675
802	676

Write the number that is 1 *less*.

175	781
174	780
438	356
437	355

Circle the number that is 10 *more*.

350	**649**
(360)	650
370	(659)
812	**500**
813	(510)
(822)	520

Circle the number that is 10 *less*.

786	**230**
766	240
(776)	(220)
462	**919**
(452)	(909)
442	929

Time to Review

Add.

4	5	7	4
2	1	0	4
+ 3	+ 6	+ 3	+ 4
9	12	10	12

Read aloud Luke 15:10. Explain that God rejoices when one person comes to know Him as Savior and that not only does God rejoice, but also all the angels rejoice. Point out that God is interested in each person; when a person receives Jesus as his or her Savior, God and His angels rejoice with great joy over one more Christian.

Why do you think God is interested in 1 more person? God loves each person; God sent Jesus to pay for the sins of everyone; God does not want anyone to perish.

Instruct

Identify the Number That Is 1 More or 1 Less

Guide a **hands-on activity** to help the students identify the number that is 1 more or 1 less.

Distribute the Hundreds/Tens/Ones Mats and the Place Value Kits. Write "246" for display. Direct the students to make the number on their mats.

How many hundreds, tens, and ones did you put on your mat? 2 hundreds, 4 tens, 6 ones

What number will you make if you add 1 more one to the Ones place? 247

Instruct the students to put another one in the Ones place. Write "247" after 246.

Guide the students as they remove a one from the Ones place so that 246 is on the mats again.

What number is 1 less than 246? 245

Instruct the students to remove a one from the Ones place. Write "245" before 246.

Display Visual 1 and direct attention to 124.

What number is 1 more than 124? 125

What number is 1 less than 124? 123

Continue the activity, pointing out different numbers (100–150) and allowing students to say the number that is 1 more or 1 less than a number.

Write "769" for display. Ask a student to write the number that is 1 more than 769. 770 Invite another student to write the number that is 1 less than 769. 768

Continue the activity, giving a number (100–999) and directing students to write the number that is 1 more or 1 less.

Identify the Number That Is 10 More or 10 Less

Continue the **hands-on activity** to help the students identify the number that is 10 more or 10 less.

Write "352" for display. Direct the students to make the number on their mats.

How many hundreds, tens, and ones did you put on your mat? 3 hundreds, 5 tens, 2 ones

What number will you make if you add 10 more to 352? 362

Instruct the students to put another ten in the Tens place.

Write "362" after 352.

Direct the students to remove a ten from the Tens place so that 352 is on the mats again.

What number is 10 less than 352? 342

Guide the students as they remove a ten from the Tens place.

Write "342" before 352. Ask the students to look at the digit in the Tens place. The original number had 5 tens. The number that is 10 more has 6 tens, and the number that is 10 less has 4 tens.

Continue the activity for 516. 526, 506

128

Direct attention to 138 on Visual 1. Explain that to find the number that is 10 more, they move down a row.

What is 10 more than 138? 148

Point out 138 again. Explain that to find the number that is 10 less, they move up a row.

What is 10 less than 138? 128

Continue the activity, pointing out different numbers and allowing students to tell the number that is 10 more or 10 less than a number (100–150).

Write "483" for display. Invite a student to write the number that is 10 more than 483. 493

Ask a student to write the number that is 10 less than 483. 473

Continue the activity, giving a number (100–989) and instructing the students to write the number that is 10 more or 10 less.

Count by 10s from a 3-Digit Number

Guide a **choral counting activity** to help the students count by 10s from a 3-digit number.

Direct attention to 110 on Visual 1. Count together with the students by 10s as you move down the rows, pointing out the numbers: 120, 130, 140, 150. Continue counting with the students to 200: 160, 170, 180, 190, 200.

Guide the students as they count together by 10s from 330 to 430: 330, 340, 350, 360, 370, 380, 390, 400, 410, 420, 430.

Identify the Number That Is 100 More or 100 Less

Guide a **hands-on activity** to help the students identify the number that is 100 more or 100 less.

Write "256" for display. Direct the students to make the number on their mats.

How many hundreds, tens, and ones did you put on your mat? 2 hundreds, 5 tens, 6 ones

What number will you make if you add 100 more to 256? 356

Guide the students as they put another hundred in the Hundreds place. Write "356" after 256.

Direct the students to remove a hundred from the Hundreds place so that 256 is on the mats again.

Counting More & Less

Write the number that is 1 *more.*

Write the number that is 1 *less.*

Write the number that is 10 *more.*

Write the number that is 10 *less.*

What number is 100 less than 256? 156

Instruct the students to remove a hundred from the Hundreds place.

Write "156" before 256. Ask the students to look at the digits in the Hundreds place of the 3 numbers. Point out that the original number had 2 hundreds. The number that is 100 more has 3 hundreds, and the number that is 100 less has 1 hundred.

Continue the activity with 527. 627, 427

Write "471" for display. Choose a student to write the number that is 100 more than 471 and the number that is 100 less than 471. 571, 371

Continue the activity, giving a number (100–899) and directing students to write the number that is 100 more or 100 less.

Apply

Worktext pages 249–50

Read and guide completion of page 249.

Read and explain the directions for page 250. Assist the students as they complete the page independently.

Assess

Reviews pages 239–40

Review 2-digit subtraction and measurement on page 240.

Fact Reviews

Use "Addition Facts to 12" page 28 in Trove.

Chapter 13 Review

Subtract.

65	74	87	64	38
− 32	− 21	− 22	− 30	− 16
33	53	65	34	22

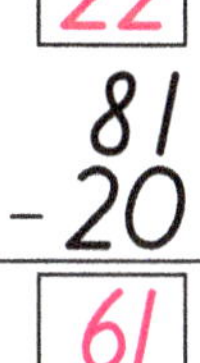

96	85	68	54	81
− 24	− 14	− 52	− 43	− 20
72	71	16	11	61

Chapter 11 Review

Guess each length. Use an inch ruler to measure.

My Guess	My Measure
___ inches	2 inches
___ inches	3 inches
___ inches	4 inches

Use an inch ruler to draw the length. Start at the star.

5 inches ★

Subtraction Fact Review

Subtract.

10 − 3 = 7	11 − 4 = 7	9 − 5 = 4
8 − 2 = 6	12 − 9 = 3	11 − 9 = 2
7 − 4 = 3	11 − 6 = 5	9 − 7 = 2

129 Worktext pages 251–52 Reviews pages 241–42

Objectives

- **129.1** Compare 3-digit numbers by using manipulatives.
- **129.2** Compare 3-digit numbers by using > and <.
- **129.3** Compare 3-digit numbers to explore God's world. BWS

Biblical Worldview Shaping

- **Exploring** (apply): Comparing numbers to determine if they fall within a certain range is an important real-world skill that helps us make wise decisions (129.3).

Printed Resources

- Visuals: Place Value Kit (hundreds, tens, ones)
- Visuals: Sign Cards (Greater than, Less than)
- Student Manipulatives: Sign Cards (Greater than, Less than)
- Student Manipulatives: Number Cards (0–9)
- Student Manipulatives: Dot Pattern Cards (0–9; backs)

Digital Resources

- Games/Enrichment: Addition Flashcards (facts to 12)
- Games/Enrichment: Fact Fun Activities

Comparing 3-Digit Numbers

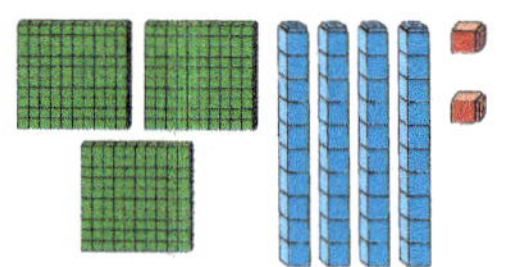

342 ? 362

- ● is less than <
- ○ is greater than >

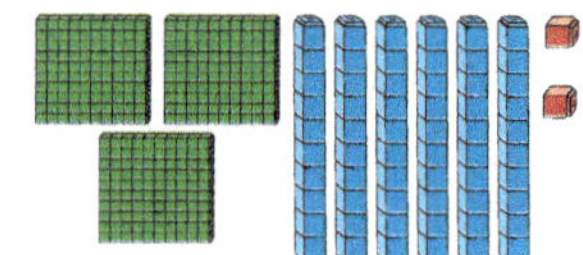

Mark the correct answer.

246 ? 146

- ○ is less than <
- ● is greater than >

632 ? 530

- ○ is less than <
- ● is greater than >

300 ? 400

- ● is less than <
- ○ is greater than >

250 ? 220

- ○ is less than <
- ● is greater than >

Draw a dot next to the smaller number.
Draw the correct sign.

200 > 100   732 < 832

460 > 420   129 > 128

Practice & Review

Count 200–300; Count by 10s to 300

Guide the students as they form a circle. Say a number between 200 and 300. Then, beginning with the student to your right, direct the students to say the next number. Continue to count around the circle until they reach 300.

Modify the activity by instructing the students to count by 10s to 300.

Study Addition Facts to 12

Choose a game or activity from the "Fact Fun Activities" available in Trove.

Engage

Distinguish between *Left & Right*

Guide the students in a **physical activity** to help them distinguish between left and right.

Instruct the students to stand by their desks and complete the following directions for left and right.

Raise your right hand.

Show me your left thumb.

Stand on your right foot.

Touch your left shoulder.

Make a fist with your right hand.

Hop on your right foot.

Touch your left elbow.

Put your left hand on your head.

Instruct

Compare 3-Digit Numbers by Using Manipulatives

Use **manipulatives** to help the students compare 3-digit numbers.

Display 7 hundreds on the left and 2 hundreds on the right.

Which set has the greater number of hundreds? the set with 7 hundreds

Which has the lesser number of hundreds? the set with 2 hundreds

Circle the correct answer.

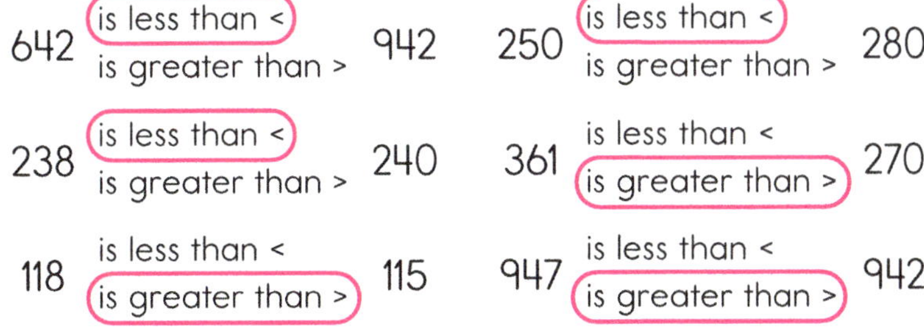

642 is less than < / is greater than > 942
250 is less than < / is greater than > 280
238 is less than < / is greater than > 240
361 is less than < / is greater than > 270
118 is less than < / is greater than > 115
947 is less than < / is greater than > 942

Draw a dot next to the smaller number.
Draw the correct sign.

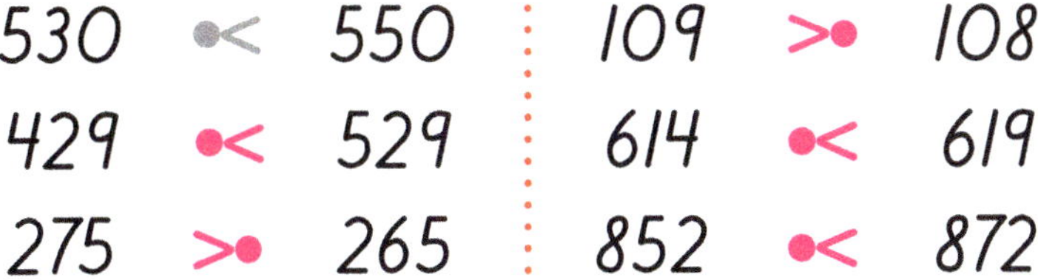

Circle the word to complete the sentence.

I can compare 3-digit numbers to help me make a foolish / wise choice.

Time to Review

Color the circle on the left red.
Color the circle on the right green.
Color the square on the left blue.
Color the square on the right yellow.
Color the triangle on the left orange.
Color the triangle on the right purple.

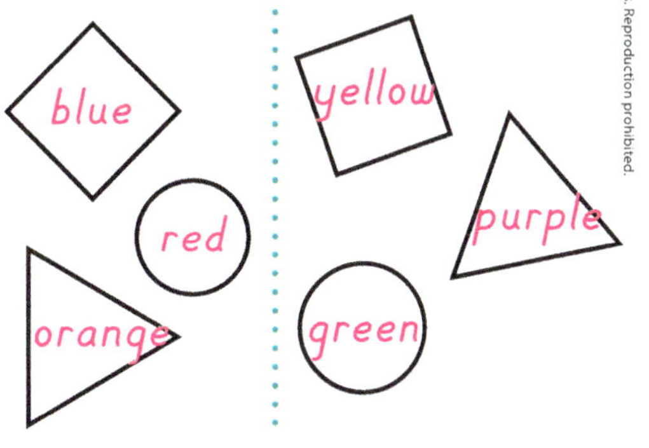

Write "700 is greater than 200" for display. Invite a volunteer to read aloud the sentence and tell whether it is correct.

Why is 700 greater than 200? 700 has more hundreds.

Remind the students that there is a shorter way to write the number sentence by replacing the words *is greater than* with a sign. Display the Greater-than Sign Card.

Rewrite the sentence "700 > 200" for display.

Display 3 hundreds on the left and 6 hundreds on the right.

Which set has the greater number of hundreds? the set with 6 hundreds

Which has the lesser number of hundreds? the set with 3 hundreds

Write "300 is less than 600" for display. Ask a student to read aloud the sentence and tell whether it is correct.

Why is 300 less than 600? 300 has fewer hundreds.

Display the Less-than Sign Card.

What words in the number sentence does this sign replace? is less than

Rewrite the sentence "300 < 600" for display.

Compare 3-Digit Numbers by Using > and <

Guide a **kinesthetic activity** to help the students write a number sentence that compares 3-digit numbers by using the greater-than and less-than signs.

Distribute the Number Cards and the Greater-than and Less-than Sign Cards. Write "162" for display on the left side and "345" on the right side. Instruct the students to use their cards to show 162 on the left side of their desks and 345 on the right side.

Which number sentence is correct: 162 is less than 345 or 162 is greater than 345? 162 is less than 345.

Guide the students as they choose the correct sign to place between their numbers. Read aloud the number sentence together with the students.

Direct the students to watch as you model drawing the less-than sign to complete your number sentence. Draw a dot between the numbers, but closer to the lesser (smaller) number. Using the dot as a point of reference, draw a sideways *V* that opens (big, wide) toward the greater number.

162 < 345

Guide the students as they use their fingers to trace over their Less-than Sign Cards by starting at the wide end of the *V* down toward the point and out to complete the *V*.

Write "526" on the left and "237" on the right for display. Instruct the students to use their cards to show 526 on the left side of their desks and 237 on the right side.

Which number sentence is correct: 526 is less than 237 or 526 is greater than 237? 526 is greater than 237.

Instruct the students to choose the correct sign to place between their numbers. Guide them as they read aloud the number sentence correctly.

Model drawing the greater-than sign to complete your number sentence. Draw a dot between the numbers, but closer to the lesser (smaller) number. Using the dot as a point of reference, draw a sideways *V* that opens (big, wide) toward the greater number.

526 > 237

Guide the students as they use their fingers to trace over their Greater-than Sign Cards by starting at the wide end of the *V* down toward the point and out to complete the *V*.

Display 126 (1 hundred, 2 tens, 6 ones) on the left and 145 (1 hundred, 4 tens, 5 ones) on the right.

Which number is greater? 145

Which number is less? 126

How do you know 145 is greater than 126? 145 has more tens.

Point out that in these numbers the tens need to be compared because the hundreds are the same—both numbers have 1 hundred.

Write "126 __ 145" for display. Direct the students to use their cards to show the numbers and to choose the correct sign to place between their numbers to complete the sentence. 126 < 145

Invite a volunteer to complete your number sentence by drawing a less-than sign. Guide the students as they read aloud the number sentence. 126 is less than 145.

Continue the activity, comparing 546 with 542 and 321 with 457. 546 > 542 and 321 < 457

Guide the students to the conclusion that they can tell whether a 3-digit number is less than or greater than another number by comparing the hundreds. If the hundreds are the same, compare the tens, and if the hundreds and the tens are the same, compare the ones.

Compare 3-Digit Numbers to Make a Choice

Guide a **discussion** to help the students apply their understanding of numbers that are greater than and less than to making a choice from the perspective of biblical teaching.

Read aloud Proverbs 13:1*a*.

What does a wise son do? He hears his father's (parents') instruction and obeys.

Read aloud the following word problem.

Liam lives at 436 Elm Lane. Liam's father says that Liam can ride his bike to visit friends on their block whose house numbers are greater than 430 but less than 450. Liam's new friend Clay lives at 467 Elm Lane. Should Liam ride his bike to Clay's house?

How can Liam make a wise choice to obey? He can compare Clay's house number with the numbers he is allowed to visit.

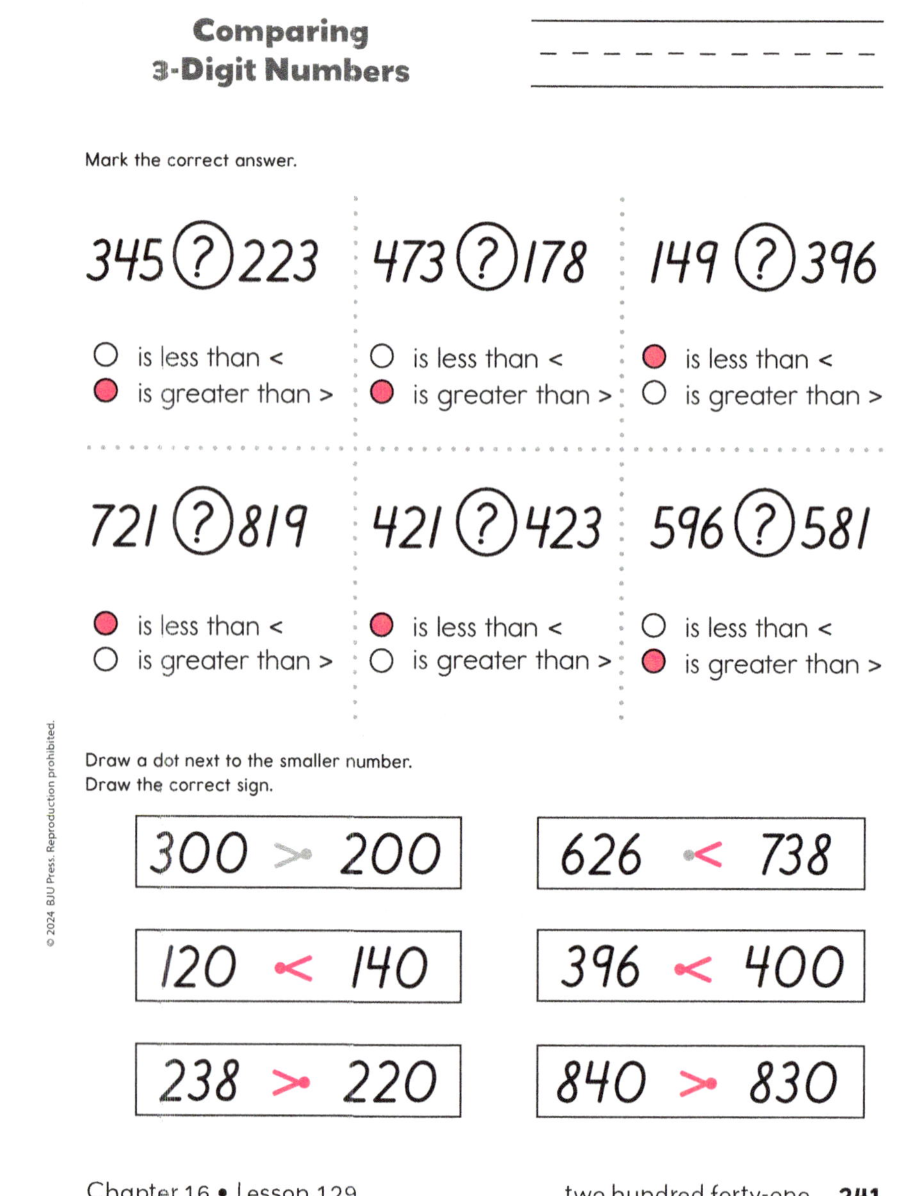

Comparing 3-Digit Numbers

Mark the correct answer.

345 ? 223 — ○ is less than < ● is greater than >

473 ? 178 — ○ is less than < ● is greater than >

149 ? 396 — ● is less than < ○ is greater than >

721 ? 819 — ● is less than < ○ is greater than >

421 ? 423 — ● is less than < ○ is greater than >

596 ? 581 — ○ is less than < ● is greater than >

Draw a dot next to the smaller number.
Draw the correct sign.

300 > 200	626 < 738
120 < 140	396 < 400
238 > 220	840 > 830

Write "467 __ 430" for display. Point out that both numbers have 4 hundreds, so the students must compare the tens. Direct the students to use their cards to show the numbers and to choose the correct sign to place between their numbers to complete the sentence. 467 > 430

Read aloud the sentence from the word problem which tells the numbers Liam is allowed to visit.

Is 467 greater than 430? yes

Explain that the students must also compare 467 with 450. Write "467 __ 450" for display. Direct the students to use their cards to show the numbers and to choose the correct sign to complete the sentence. 467 > 450

Compare the number sentence with the information from the word problem.

Is 467 less than 450? No, it is greater.

Should Liam ride his bike to Clay's house? No, that would not be a wise choice. Liam would be foolish to disobey his father.

Apply

Worktext pages 251–52

Guide a discussion about the number sentence at the top of page 251. Read the directions and guide completion of the page. Remind the students to draw a dot next to the lesser (smaller) number and then draw

a sideways *V* that opens (big, wide) toward the greater number. After the students complete the number sentences at the bottom of the page, ask volunteers to read them aloud.

Read and explain the directions for page 252. Assist the students as they complete the page independently.

Direct attention to the biblical worldview shaping statement and guide the students as they complete the sentence.

Assess

Reviews pages 241–42

Review 2-digit addition and reading a thermometer on page 242.

Extended Activity

Write a Number Sentence

Materials

- 10 index cards
- Sign Cards: Greater than, Less than

Procedure

Write a different 3-digit number on each index card. Direct two students to each choose an index card and to stand and hold it so that the number is showing. Ask a third student to choose the correct sign to complete the number sentence and to hold it up as he or she stands between the students. Guide the students as they read aloud the number sentence and decide whether it is correct.

130–31

Worktext pages 253–54
Reviews pages 243–44

Objectives

130–31.1 Join sets for 3-digit addends.

130–31.2 Solve a word problem by using the Problem-Solving Plan.

130–31.3 Write an addition equation for a word problem.

Printed Resources

- Visual 21: *Problem-Solving Plan*
- Visuals: Place Value Kit (hundreds, tens, ones)
- Student Manipulatives: Hundreds/Tens/Ones Mat
- Student Manipulatives: Place Value Kit (hundreds, tens, ones)

Digital Resources

- Video: Add Large Numbers
- Web Link: Let's Set the Record Straight. Pizza Boxes are Recyclable.

Materials

- Pizza box or picture of one

Lessons 130–32 are included as enrichment instruction. Present these lessons based on your schedule and the readiness of your students. You may spend 2 lessons covering Lessons 130–31, or you may present both lessons in 1 presentation.

All problems in this lesson should be written vertically. All word-problem equations should be written horizontally.

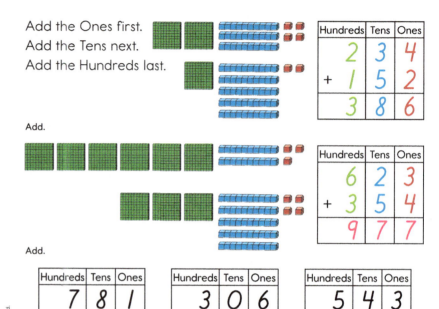

Adding Large Numbers

Add the Ones first.
Add the Tens next.
Add the Hundreds last.

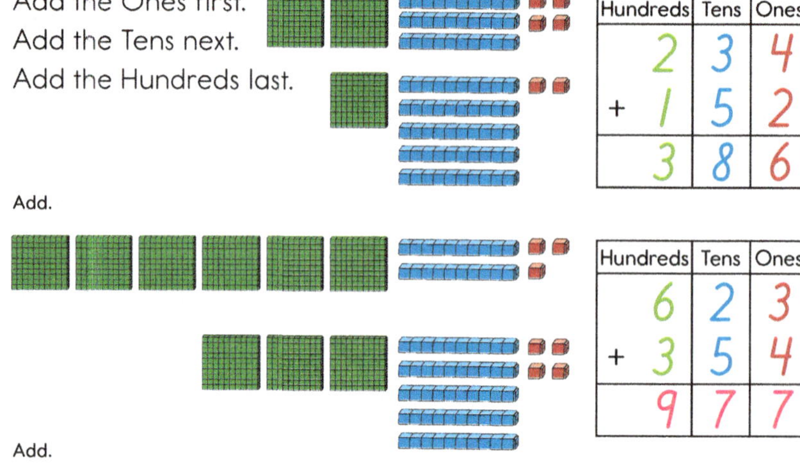

	Hundreds	Tens	Ones
	2	3	4
+	1	5	2
	3	8	6

Add.

	Hundreds	Tens	Ones
	6	2	3
+	3	5	4
	9	7	7

Add.

	Hundreds	Tens	Ones
	7	8	1
+	2	1	4
	9	9	5

	Hundreds	Tens	Ones
	3	0	6
+	1	5	3
	4	5	9

	Hundreds	Tens	Ones
	5	4	3
+	1	3	2
	6	7	5

Write an equation for the word problem.
Solve.

Tate delivered 212 pizzas last month.
He delivered 124 pizzas this month.
How many pizzas did Tate deliver in all?

212 + 124 = 336 pizzas

	Hundreds	Tens	Ones
	2	1	2
+	1	2	4
	3	3	6

Engage

Recycling

Guide a **discussion** about recycling to introduce the students to a way of exploring God's world.

Remind the students that Lily, Liam, and DQ helped Tate, Lily's big brother, deliver pizzas.

What is a pizza in when it is delivered to your house? a pizza box

Explain that pizza boxes are made from cardboard. The cardboard insulates the pizza to keep it warm until the pizza is ready to be eaten. Cardboard is a type of thick, stiff paper. It is made from trees that have been cut down and sent through a special process.

What do you do with pizza boxes when you have finished eating all the pizza? throw them away, recycle them, use them as compost in the garden, make new things from them

Explain that because pizza boxes are made from paper, they can be recycled. Point out that some recycling companies come to houses and pick up things made of cardboard and other paper that are placed in special bins; other companies have locations where people can take cardboard and paper to be recycled.

Explain that when cardboard is recycled, it goes through a special process that cleans it so that the paper fibers can be used again to make more paper.

How is recycling cardboard helpful? It cuts down on the number of trees needed to make new cardboard, cuts down on the amount of trash, and saves money.

Point out that God has put people in charge of taking care of things in His world. Recycling is a good way to make wise use of what God has given us.

Add.

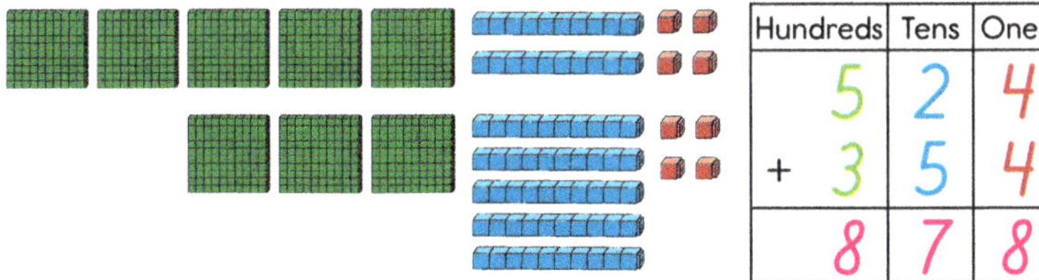

Hundreds	Tens	Ones
5	2	4
+ 3	5	4
8	7	8

Add.

Hundreds	Tens	Ones
3	2	1
+ 1	4	3
4	6	4

Hundreds	Tens	Ones
6	4	1
+ 2	3	1
8	7	2

Hundreds	Tens	Ones
7	2	6
+ 2	3	1
9	5	7

Hundreds	Tens	Ones
2	0	0
+ 3	5	6
5	5	6

Hundreds	Tens	Ones
5	0	6
+ 1	4	3
6	4	9

Hundreds	Tens	Ones
2	6	4
+ 2	3	1
4	9	5

Write an equation for the word problem.
Solve.

Best Pizza sold 352 pizzas last week.

They sold 234 pizzas this week.

How many pizzas did they sell in all?

352 + 234 = 586 pizzas

Hundreds	Tens	Ones
3	5	2
+ 2	3	4
5	8	6

Instruct

3-Digit Addends

Use **manipulatives** to help the students join sets to add 3-digit numbers.

Distribute the Hundreds/Tens/Ones Mats and the Place Value Kits. Tell the students that they will be adding 3-digit numbers on their mats, using ones, tens, and hundreds.

Write "215 + 134" for display.

Draw a Hundreds/Tens/Ones frame for display. Display 2 hundreds, 1 ten, 5 ones in the frame. Direct the students to put 2 hundreds, 1 ten, 5 ones near the top of their mats.

Display 1 hundred, 3 tens, 4 ones below the first addend. Instruct the students to put 1 hundred, 3 tens, 4 ones near the bottom of their mats. Remind the students that addition means "joining sets together."

How can these numbers be added? add (join) the ones, tens, and hundreds together

Explain that when joining sets with hundreds, tens, and ones, the squares in the Ones place should always be joined first, then the strips in the Tens place are joined, and finally the squares in the Hundreds place are joined. Direct the students to move all the squares in the Ones place together on their mats while you demonstrate.

Hundreds	Tens	Ones

How many ones are there in all? 9

Write "9" in the Ones place of the answer.

Direct the students to move all the tens in the Tens place together on their mats while you demonstrate.

How many tens are there in all? 4

Write "4" in the Tens place of the answer. Direct each student to move all the hundreds in the Hundreds place together while you demonstrate.

How many hundreds are there in all? 3

Write "3" in the Hundreds place of the answer.

What is the answer? 349

Model the following addition problems while the students do them on their mats. Emphasize adding the Ones place first, the Tens place next, and the Hundreds place last.

144	403	224
+ 223	+ 135	+ 102
367	538	326

Problem-Solving Plan

Follow the **Problem-Solving Plan** to help the students solve word problems.

Display Visual 21 and read aloud the questions. Remind the students that solving word problems is made easier by following a plan. Read aloud the following word problem several times. Explain that the students need to think about the information given in the word problem.

DQ helped Tate take pizza boxes to the recycling center. The first week they took 153 boxes to the center. The second week they took 142 boxes. How many boxes did they take to the recycling center in 2 weeks?

What is the question in this word problem? How many boxes did they take to the recycling center in 2 weeks?

What information is given? The first week they took 153 boxes, and the second week they took 142 boxes.

Do you add or subtract to find the answer? I add because I am joining sets to find the total.

What is the equation needed to solve this problem? 153 + 142 = __

Write "153 + 142 = __" for display.

Should you rewrite the equation in vertical form? Yes; it will be easier to add in vertical form since the addends are 3-digit numbers.

Invite a student to write the equation in vertical form.

What do you add first when there are 3-digit addends? the ones

Ask a student to solve the problem. 295

Does the answer make sense? yes

Explain that if they had subtracted 142 from 153, the answer would have been 11 and would not have made sense. Write "295 boxes" to complete the equation.

Apply

Worktext pages 253–54

Read and guide completion of page 253.

Read and explain the directions for page 254. Assist the students as they complete the page independently.

Assess

Reviews pages 243–44

Read the directions and guide completion of the pages.

Adding Large Numbers

Add the Ones first.
Add the Tens next.
Add the Hundreds last.

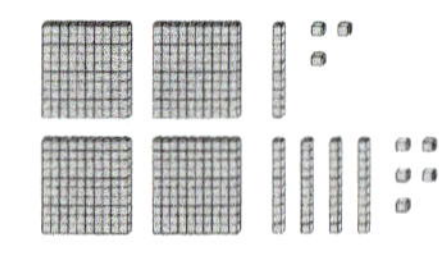

Hundreds	Tens	Ones
2	1	3
+ 2	4	5
4	5	8

Add.

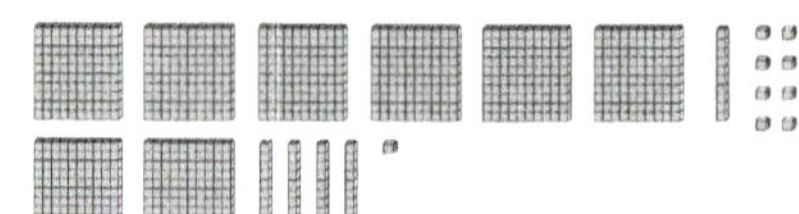

Hundreds	Tens	Ones
6	1	8
+ 2	4	1
8	5	9

Add.

Hundreds	Tens	Ones
1	3	4
+ 4	2	3
5	5	7

Hundreds	Tens	Ones
5	3	2
+ 4	1	5
9	4	7

Hundreds	Tens	Ones
1	4	2
+ 2	4	5
3	8	7

Write an equation for the word problem.
Solve by using the Hundreds/Tens/Ones frame.

Noah lived for 600 years before the Flood. He lived for 350 more years after the Flood. How many years did Noah live in all?

Hundreds	Tens	Ones
6	0	0
+ 3	5	0
9	5	0

600 + 350 = 950 years

Add.

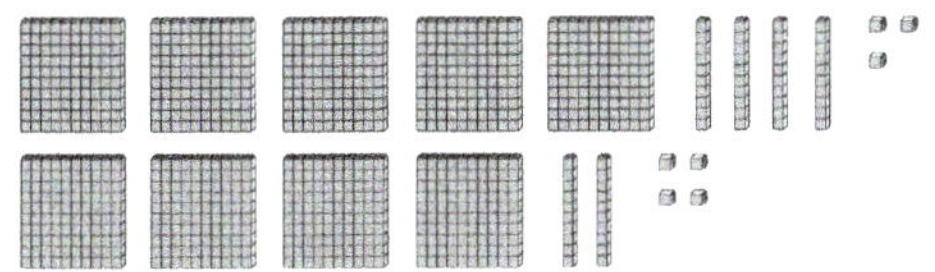

Hundreds	Tens	Ones
5	4	3
+ 4	2	4
9	6	7

Add.

Hundreds	Tens	Ones
6	8	4
+ 2	1	1
8	9	5

Hundreds	Tens	Ones
2	3	5
+ 4	3	2
6	6	7

Hundreds	Tens	Ones
5	7	4
+ 1	0	5
6	7	9

Hundreds	Tens	Ones
4	3	5
+ 1	4	2
5	7	7

Hundreds	Tens	Ones
7	3	5
+ 2	4	2
9	7	7

Hundreds	Tens	Ones
3	9	3
+ 2	0	4
5	9	7

Write an equation for the word problem.
Solve by using the Hundreds/Tens/Ones frame.

Dad has 150 nails and 125 screws in his tool kit. How many screws and nails does Dad have in all?

Hundreds	Tens	Ones
1	5	0
+ 1	2	5
2	7	5

150 + 125 = 275 screws and nails

132

Worktext pages 255–56
Reviews pages 245–46

Objectives

- **132.1** Join sets for 3-digit numbers.
- **132.2** Solve a word problem by using the Problem-Solving Plan.
- **132.3** Write an addition equation for a word problem.

Printed Resources

- Visual 21: *Problem-Solving Plan*
- Visuals: Place Value Kit (hundreds, tens, ones)
- Student Manipulatives: Hundreds/Tens/Ones Mat
- Student Manipulatives: Place Value Kit (hundreds, tens, ones)

Digital Resources

- Web Link: 15 Creative Ways to Recycle Pizza Boxes
- Web Link: 20 Cool Things You Can Make with a Pizza Box

Materials

- Pizza box or a picture of one

All problems in this lesson should be written vertically. All word-problem equations should be written horizontally.

Practicing Adding Large Numbers

Add.

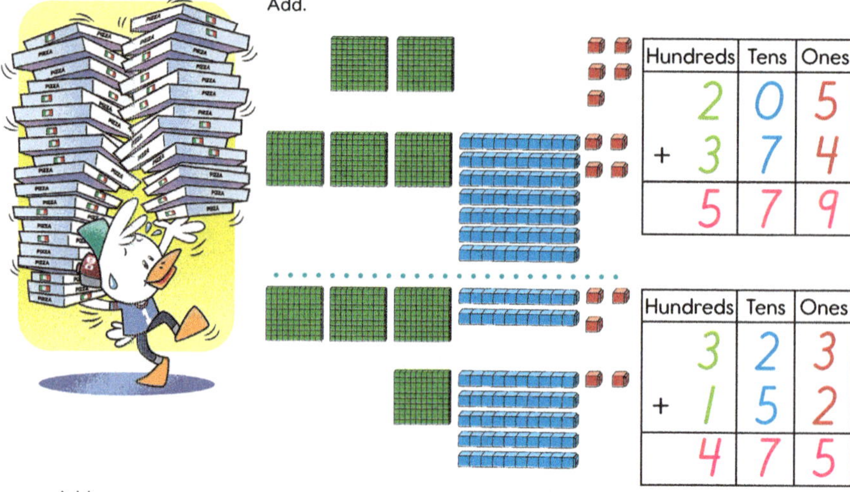

	Hundreds	Tens	Ones
	2	0	5
+	3	7	4
	5	7	9

	Hundreds	Tens	Ones
	3	2	3
+	1	5	2
	4	7	5

Add.

	Hundreds	Tens	Ones
	7	5	3
+	2	1	5
	9	6	8

	Hundreds	Tens	Ones
	6	0	3
+	2	9	4
	8	9	7

	Hundreds	Tens	Ones
	8	2	7
+	1	4	0
	9	6	7

Write an equation for the word problem.
Solve.

Best Pizza has 356 pizza boxes.
They need 213 more boxes.
How many boxes do they need in all?

356 + 213 = 569 boxes

	Hundreds	Tens	Ones
	3	5	6
+	2	1	3
	5	6	9

Chapter 16 • Lesson 132 two hundred fifty-five 255

Engage

Pizza Box DIY Ideas

Share a **visual display** of do-it-yourself ideas for using discarded pizza boxes to introduce the students to another way of exploring God's world.

Remind the students of the previous discussion of recycling pizza boxes. Point out that pizza boxes can also be reused in creative ways that save money and make wise use of what God has given us in His world.

Share some fun ideas for reusing pizza boxes from the suggested links.

Instruct

3-Digit Addends

Use **manipulatives** to help the students join sets to add 3-digit numbers.

Distribute the Hundreds/Tens/Ones Mats and the Place Value Kits. Tell the students that they will be adding 3-digit numbers on their mats.

Write "416 + 223" for display. Instruct the students to use their Place Value Kits to make the first addend near the top of their mats. Draw a Hundreds/Tens/Ones frame for display and model 416 in the frame, using hundreds, tens, and ones.

How did you make 416 on your mat? 4 hundreds, 1 ten, 6 ones

Instruct the students to make the second addend near the bottom of their mats while you model it near the bottom of your Hundreds/Tens/Ones frame.

How did you make 223 on your mat? 2 hundreds, 2 tens, 3 ones

What does addition mean? joining groups or sets together

How can these numbers be added? add (join) the ones, tens, and hundreds together

Explain that when joining sets with hundreds, tens, and ones, the Ones place should always be added together first, then the Tens place, and finally the Hundreds place.

Direct the students to move all the ones in the Ones place together on their mats while you demonstrate.

How many ones are there in all? 9

Write "9" in the Ones place of the answer.

Instruct the students to move all the tens in the Tens place together on their mats while you demonstrate.

Add.

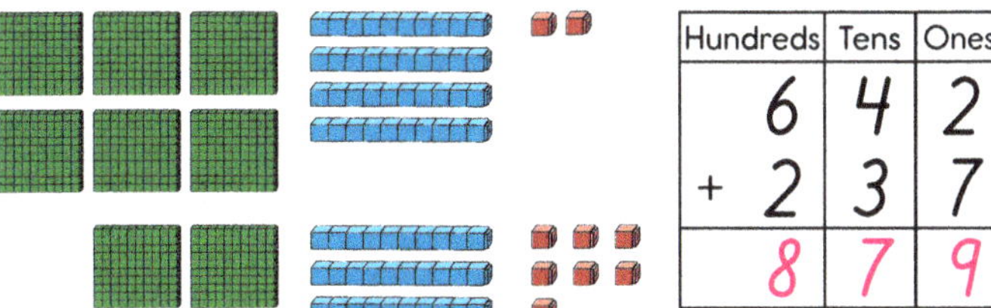

Hundreds	Tens	Ones
6	4	2
+ 2	3	7
8	7	9

Add.

Hundreds	Tens	Ones
5	6	5
+ 1	3	2
6	9	7

Hundreds	Tens	Ones
4	5	7
+ 4	2	1
8	7	8

Hundreds	Tens	Ones
5	8	4
+ 3	1	0
8	9	4

Hundreds	Tens	Ones
2	6	4
+ 2	0	3
4	6	7

Hundreds	Tens	Ones
6	2	0
+ 2	7	5
8	9	5

Hundreds	Tens	Ones
4	6	1
+ 5	3	4
9	9	5

Write an equation for the word problem.
Solve.

Tate used 513 slices of pepperoni to make pizzas on Monday.
He used 452 slices on Tuesday.
How many slices of pepperoni did Tate use in all?

Hundreds	Tens	Ones
5	1	3
+ 4	5	2
9	6	5

513 + 452 = 965 slices of pepperoni

How many tens are there in all? 3

Write 3 in the Tens place of the answer.

Direct the students to move all the hundreds in the Hundreds place together while you demonstrate.

How many hundreds are there in all? 6

Write "6" in the Hundreds place of the answer.

What is the answer? 639

Model the following addition problems while the students do them on their mats. Emphasize adding the Ones place first, the Tens place next, and the Hundreds place last.

354	314	124
+ 223	+ 104	+ 215
577	418	339

Problem-Solving Plan

Follow the **Problem-Solving Plan** to help the students solve word problems.

Display Visual 21 and read aloud the questions. Remind the students that solving word problems is made easier by following a plan. Read aloud this word problem several times. Explain that the students need to think about the information given in the word problem.

Lily and Liam's school collected pizza boxes to use for art projects. During the first week, they collected 150 boxes. During the second week, they collected 225 boxes. How many boxes did they collect in 2 weeks?

What is the question in this word problem? How many boxes did they collect in 2 weeks?

What information is given? They collected 150 boxes the first week and 225 boxes the second week.

Do you add or subtract to find the answer, and why? I add because I am joining sets to find the total.

What is the equation needed to solve this problem? 150 + 225 = __

Write "150 + 225 = __" for display.

Should you rewrite the equation in vertical form? Yes, 3-digit addends are easier to add in vertical form.

Invite a volunteer to write the equation in vertical form.

What do you add first when there are 3-digit addends? the ones

Ask a student to solve the problem. 375

Does the answer make sense? yes

Write "375 boxes" to complete the equation.

Apply

Worktext pages 255–56

Read and guide completion of page 255.

Read and explain the directions for page 256. Assist the students as they complete the page independently.

Assess

Reviews pages 245–46

Read the directions and guide completion of the pages.

Extended Activity

Add 3-Digit Numbers

Materials

- A spinner with sections labeled 1, 2, 3, and 4 for each group
- A sheet of paper for each student

Procedure

Group the students in pairs or small groups. Give each group a spinner.

Instruct a student in each group to make the first 3-digit addend by spinning the spinner 3 times; the group members should record this first addend. Direct another student in the group to make the second 3-digit addend by spinning the spinner 3 times; the group members should record this second addend.

Instruct the students to solve their problems and check them with the other students in their group. When a group is sure that they have the correct answer, check their problem.

Practicing Adding Large Numbers

Add.

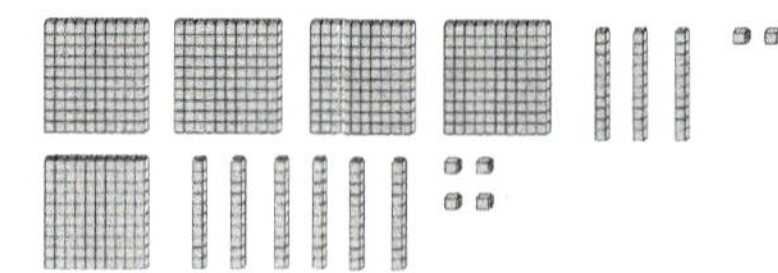

	Hundreds	Tens	Ones
	4	3	2
+	1	6	4
	5	9	6

Add.

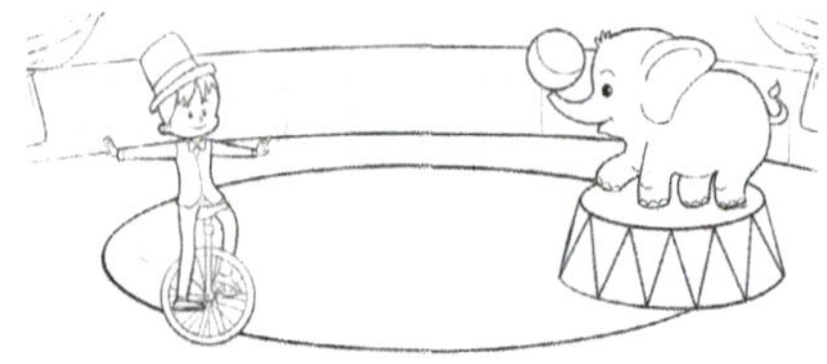

	Hundreds	Tens	Ones
	2	6	4
+	1	2	5
	3	8	9

	Hundreds	Tens	Ones
	6	7	5
+	3	2	1
	9	9	6

	Hundreds	Tens	Ones
	3	4	2
+	4	1	6
	7	5	8

	Hundreds	Tens	Ones
	7	5	4
+	2	4	3
	9	9	7

Write an equation for the word problem.
Solve by using the Hundreds/Tens/Ones frame.

On Friday 618 people came to the circus.
On Saturday 371 people came to the circus.
How many people came to see the circus?

	Hundreds	Tens	Ones
	6	1	8
+	3	7	1
	9	8	9

618  371  989 people

Add.

	Hundreds	Tens	Ones
	6	3	5
+	3	4	1
	9	7	6

	Hundreds	Tens	Ones
	4	5	3
+	2	4	2
	6	9	5

	Hundreds	Tens	Ones
	8	2	3
+	1	6	3
	9	8	6

	Hundreds	Tens	Ones
	7	0	5
+	1	9	3
	8	9	8

	Hundreds	Tens	Ones
	2	5	4
+	3	2	1
	5	7	5

	Hundreds	Tens	Ones
	3	8	2
+	4	0	7
	7	8	9

Write an equation for the word problem.
Solve by using the Hundreds/Tens/Ones frame.

The students sold 156 bags of cotton candy and 342 bags of popcorn. How many bags of food did the students sell in all?

	Hundreds	Tens	Ones
	1	5	6
+	3	4	2
	4	9	8

156 + 342 = 498 bags

133

Worktext pages 257–58
Reviews pages 247–48

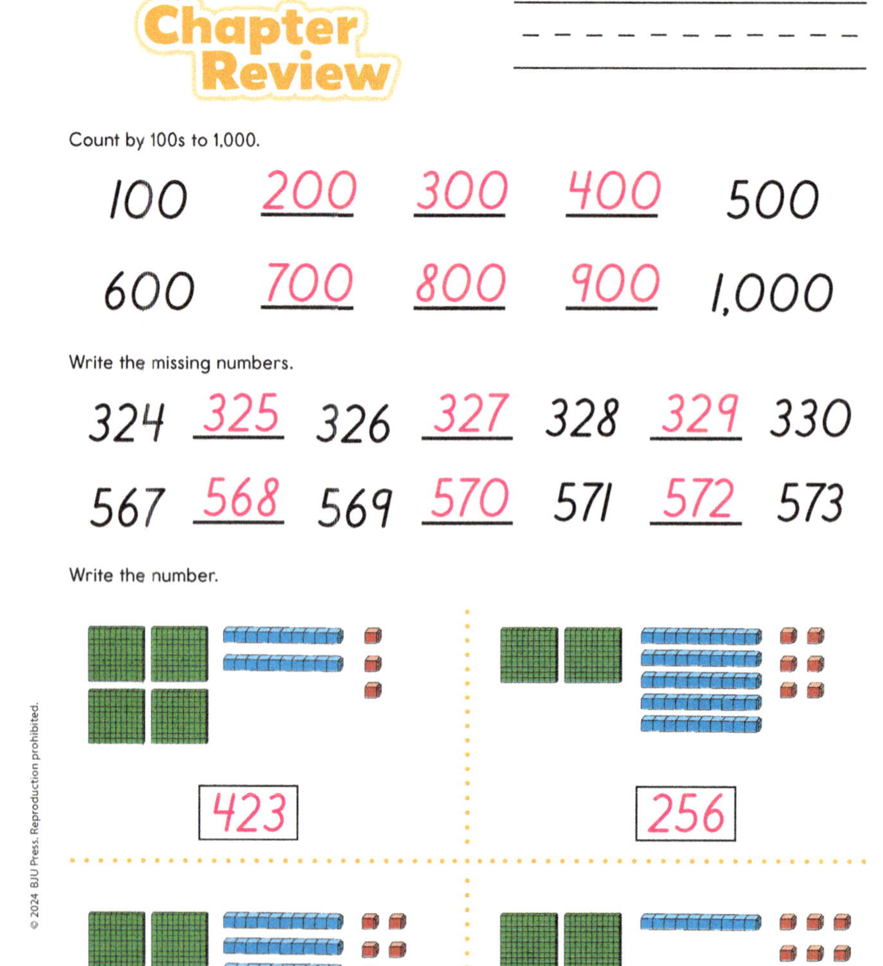
Chapter Review

Count by 100s to 1,000.

100 200 300 400 500

600 700 800 900 1,000

Write the missing numbers.

324 325 326 327 328 329 330

567 568 569 570 571 572 573

Write the number.

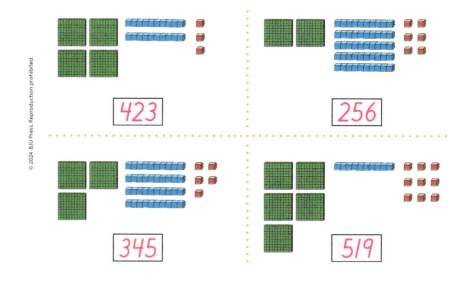

423

256

345

519

How do number patterns help me find someone's house?

Chapter Concept Review

- Practice concepts from Chapter 16 to prepare for the test.

Printed Resources

- Visuals: Place Value Kit (hundreds, tens, ones)
- Student Manipulatives: Sign Cards (Greater than, Less than)
- Student Manipulatives: Number Cards (0–9)
- Student Manipulatives: Dot Pattern Cards (0–9; backs)

Digital Resources

- Games/Enrichment: Addition Flashcards (facts to 12)
- Games/Enrichment: Fact Fun Activities

Practice & Review

Count 300–400

Instruct the students to stand in a circle. Say a number between 300 and 400. Then, beginning with the student to your right, direct each student to say the next number. Continue to count around the circle until you reach 400.

Study Addition Facts to 12

Choose a game or activity from the "Fact Fun Activities" available in Trove.

Instruct

Count by 100s

Distribute 1 hundred to each of 10 students. Instruct the students with a hundred to hold it up. Guide the students as they count the hundreds by 100s. Explain that 10 hundreds is the same as 1,000.

Identify the Number of Hundreds, Tens, Ones

Write "357" for display. Ask a student to read aloud the number. three hundred fifty-seven

What are the 3, 5, and 7 called? digits

Which digit is in the Hundreds place? 3

Which digit is in the Ones place? 7

Which digit is in the Tens place? 5

Identify the Value of the Digits in a 3-Digit Number

Write "429" for display. Ask a student to read aloud the number. four hundred twenty-nine

What is the value of the 2 in the Tens place? 20

What is the value of the 4 in the Hundreds place? 400

What is the value of the 9 in the Ones place? 9

Write "888" for display. Ask a student to read aloud the number. eight hundred eight-eight

What is the value of the 8 in the Hundreds place? 800

What is the value of the 8 in the Ones place? 8

What is the value of the 8 in the Tens place? 80

Identify the Number That Is 1 More or 1 Less, 10 More or 10 Less, 100 More or 100 Less

Distribute the Number Cards. Display 3 hundreds, 5 tens, 7 ones.

What number is displayed? 357

133

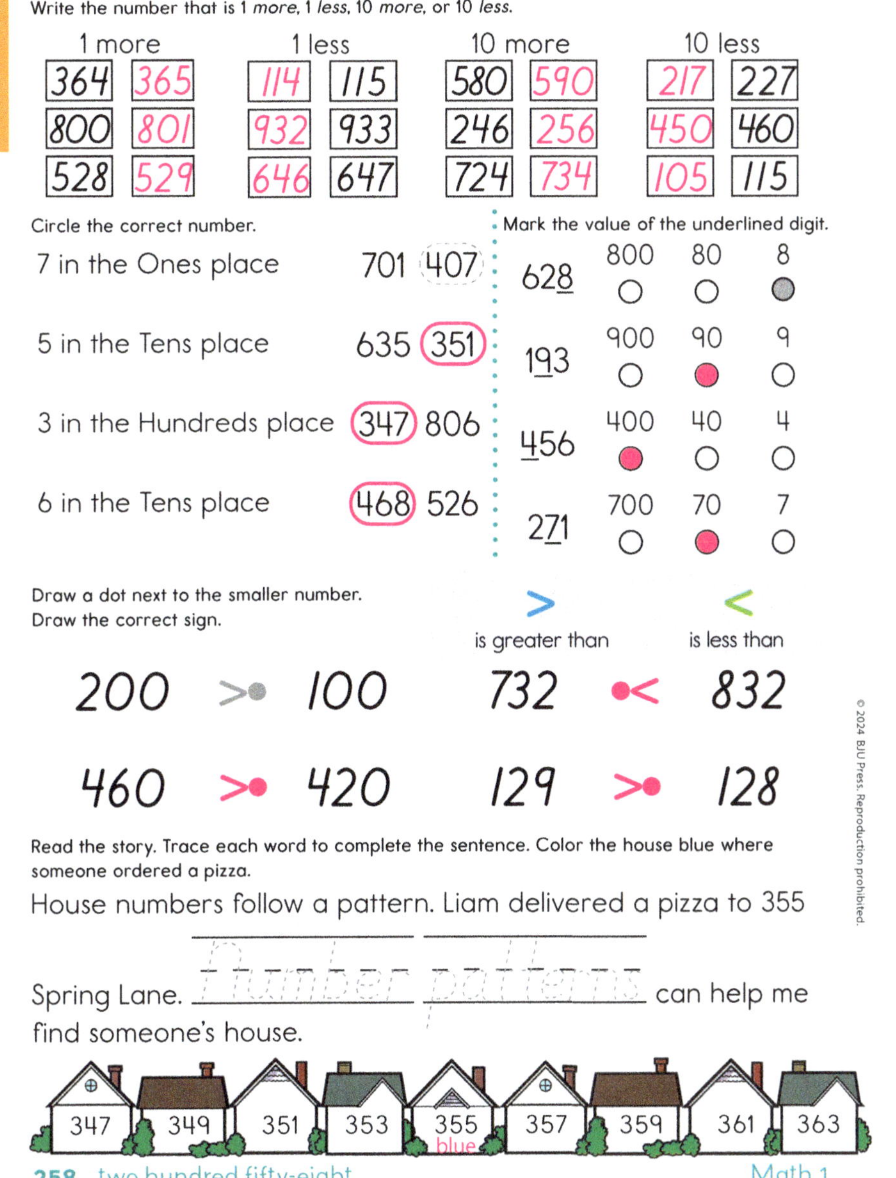
Write the number that is 1 *more*, 1 *less*, 10 *more*, or 10 *less*.

1 more		1 less		10 more		10 less	
364	365	114	115	580	590	217	227
800	801	932	933	246	256	450	460
528	529	646	647	724	734	105	115

Circle the correct number.

7 in the Ones place	701	(407)
5 in the Tens place	635	(351)
3 in the Hundreds place	(347)	806
6 in the Tens place	(468)	526

Mark the value of the underlined digit.

628	800	80	8 ●
193	900	90 ●	9
456	400 ●	40	4
271	700	70 ●	7

Draw a dot next to the smaller number.
Draw the correct sign.

> is greater than   < is less than

200 > 100   732 < 832

460 > 420   129 > 128

Read the story. Trace each word to complete the sentence. Color the house blue where someone ordered a pizza.

House numbers follow a pattern. Liam delivered a pizza to 355 Spring Lane. Number patterns can help me find someone's house.

258 two hundred fifty-eight   Math 1

Direct the students to use their Number Cards to show the number. Write "357" for display.

Add 1 one to your display of 357.

What number is 1 more than 357? 358

Direct the boys to change the 7 in the Ones place to an 8 to show 358. Write "358" after 357.

Remove a one from the Ones place so that the original number 357 is displayed.

What number is 1 less than 357? 356

Write "356" before 357. Remove 1 one from 357. Direct the girls to change the 7 in the Ones place to a 6 to show 356.

Display 2 hundreds, 4 tens, 6 ones.

What number is displayed? 246

Direct the students to use their Number Cards to show the number. Write "246" for display.

Ask the girls to change 246 so that it is 10 more. Ask the boys to change 246 so that it is 10 less.

What number is 10 more than 246? 256

What number is 10 less than 246? 236

Write "236" before 246. Write "256" after 246.

Which digit did you change to make the number 10 more or 10 less? I changed the digit 4, because it is in the Tens place.

Display 5 hundreds, 7 tens, 2 ones.

What number is displayed? 572

Direct the students to use their Number Cards to show the number. Write "572" for display.

Ask the girls to change 572 so that it is 100 more. Ask the boys to change 572 so that it is 100 less.

What number is 100 more than 572? 672

What number is 100 less than 572? 472

Write "472" before 572. Write "672" after 572.

Which digit did you change to make the number 100 more or 100 less? I changed the digit 5, because it is in the Hundreds place.

Compare 3-Digit Numbers by Using > and <

Distribute the Greater-than and Less-than Sign Cards. Write "245" on the left side of the board and "268" on the right side. Direct the students to use their Number Cards to show 245 on the left side of their desks and 268 on the right side.

Which number sentence is correct: 245 is less than 268 or 245 is greater than 268? 245 is less than 268.

Instruct the students to choose the correct sign to place between their numbers. Guide the students as they read aloud the number sentence correctly.

Write "307" for display on the left and "289" on the right. Instruct the students to use their cards to show 307 on the left side of their desks and 289 on the right side of their desks. Instruct the students to choose the correct sign to place between their numbers.

Which sign correctly completes this number sentence? a greater-than sign

Guide the students as they read aloud the number sentence. Remind the students that the point of the arrow in the greater-than and less-than signs always points to the lesser (smaller) number.

Apply

Worktext pages 257–58

Read and guide completion of page 257.

Read and explain the directions for page 258. Assist the students as they complete the page independently.

133

Direct attention to the biblical worldview shaping section on page 258. Display the picture from the chapter opener and read aloud the essential question. Remind the students that DQ, Liam, and Lily helped deliver pizzas. To read the house numbers, they had to read and compare 3-digit numbers. They learned that the numbers followed a pattern.

How did the number pattern help them find the correct house? They learned that the odd numbers were on one side of the street and the even numbers on the other side. They counted by 2s to find the correct house number.

Read aloud the directions. Guide the students as they read aloud the story (word problem) and trace the words to complete the final sentence.

How will you know which house to color blue? I must follow the number pattern to find the house where Liam delivered a pizza.

Guide the students as they complete the problem.

Assess

Reviews pages 247–48

Use the Reviews pages to provide additional preparation for the chapter test.

Chapter 16 Review

Count by 100s to 1,000.

100 200 300 400 500

600 700 800 900 1,000

Write each missing number.

455 456 457 458 459 460 461

739 740 741 742 743 744 745

Write each number.

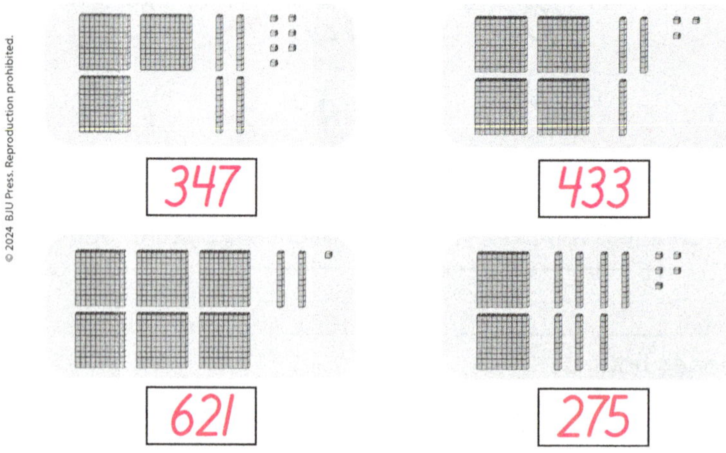

347 433

621 275

Write the number that is 1 *more*, 1 *less*, 10 *more*, or 10 *less*.

1 more		1 less		10 more		10 less	
245	246	724	725	480	490	818	828
300	301	559	560	326	336	230	240
861	862	192	193	655	665	935	945

Draw a circle around the correct number.

8 in the Ones place: 483  (608)

5 in the Hundreds place: (563) 452

2 in the Tens place: (623) 392

Mark the value of each underlined digit.

75<u>3</u>	300 ○	30 ○	3 ●
3<u>2</u>4	200 ○	20 ●	2 ○
<u>9</u>16	900 ●	90 ○	9 ○
5<u>6</u>9	600 ○	60 ●	6 ○

Draw a dot next to the smaller number.
Draw the correct sign.

> is greater than < is less than

195 < 643

356 < 456

700 > 500

903 > 890

134 Worktext pages 259–60

STEM

Objectives

- **134.1** Identify a pattern of repeated actions.
- **134.2** Translate a picture program into a clapping game.
- **134.3** Simplify a picture program by using loops.
- **134.4** Follow the Engineering Design Process to solve a problem.
- **134.5** Write a program for a clapping game by using loops.
- **134.6** Teach a friend a clapping game by using a program. BWS

Biblical Worldview Shaping

- **Exploring** (apply): The ability to recognize and repeat patterns is a gift from God. We can share games that we find enjoyable with others and help them discover a new pattern too (134.6).

Printed Resources

- Instructional Aids 38a–b: *Clapping Games*
- Instructional Aid 5: *STEM Engineering Design Process*
- Instructional Aid 39: *Coding Rubric: Playing a Game*

Digital Resources

- Web Link: Hand Clapping Games

Mark loops in the clapping game. Play the game with a partner.

double, double, this, this

double, double, that, that

double this, double that

double, double, this, that

Ask. Think. Plan.

Choose from the pictures. Plan a new clapping program for 1 line of the song.

A sail - or went to sea, sea, sea

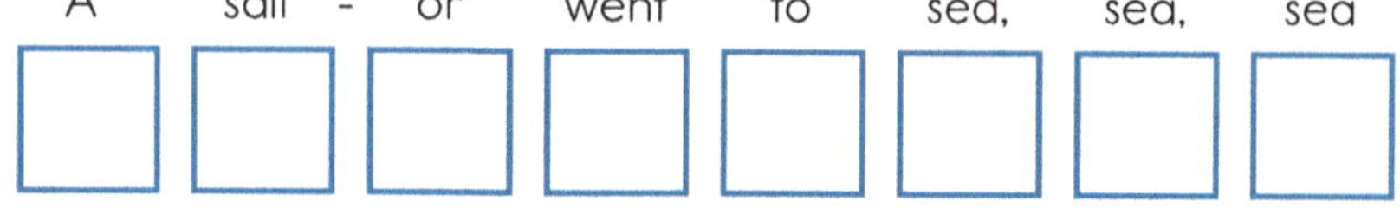

Make. Test. Make Better.

Test the program. Make it better. Glue the pictures in place.

Engage

Sing a Song

Sing a song together to introduce the STEM lesson.

Sing or recite the first verse of the song *Bingo* together.

There was a farmer had a dog,
And Bingo was his name-o.
B-I-N-G-O!
B-I-N-G-O!
B-I-N-G-O!
And Bingo was his name-o!

Explain that today the students will learn a clapping game to go with the song.

Instruct

Clapping Game

Model a clapping game to prepare the students to identify a pattern of repeated actions.

Invite the students to repeat the following actions with you.

Tap: Tap your open palms against your lap.

Clap: Clap your palms together.

Pat: Clap your open palms against someone else's palms.

Demonstrate clapping the following pattern: tap, tap, clap, clap, pat, pat, clap, clap.

Guide the students as they follow you to practice the pattern without a partner.

Once the students are comfortable with the clapping pattern, demonstrate clapping the pattern to *Bingo*.

Begin with the first tap on the word *was* in the song.

Choose from the pictures on page 259. Write your program with loops.

Write a name to complete the sentence.

I will teach ______________ [name] a clapping game by using a program.

Use your program to clap the entire song. Teach someone your program.

A Sailor Went to Sea, Sea, Sea

A sailor went to sea, sea, sea
To see what he could see, see, see.
But all that he could see, see, see
Was the bottom of the deep blue sea, sea, sea.

260 two hundred sixty Math 1

There was a farmer had a dog,

TAP! TAP! CLAP! CLAP!

And Bingo was his name-o.

PAT! PAT! CLAP! CLAP!

B - I - N - G - O !

TAP! TAP! CLAP! CLAP!

B - I - N - G - O !

PAT! PAT! CLAP! CLAP!

B - I - N - G - O !

TAP! TAP! CLAP! CLAP!

And Bingo was his name-o!

PAT! PAT! CLAP! CLAP!

Identify a Pattern

Guide a **physical activity** to help the students identify a pattern of repeated actions.

Invite a student to join you to model clapping the patterns with you as you sing or recite the first verse of *Bingo.*

Pair the students. Guide them as they follow you and your clapping partner to clap the pattern with their partners. Repeat until the students are comfortable with the game.

Remind the students that an *algorithm* is a list of steps that tells someone how to complete a task. Today their task is to play a game. Explain that they can write an algorithm that tells them the steps to follow to complete the clapping game.

134

What clapping pattern do you follow in the game? tap, tap, clap, clap, pat, pat, clap, clap

Write the algorithm in words for display as the students identify it.

Translate a Picture Program into a Game

Use a **visual** to guide the students as they translate a picture program into the actions needed to play a clapping game.

Display the *Clapping Games* 38a page. Point out the pictures at the top and ask a student to demonstrate the action that goes with each picture.

Direct attention to the actions pictured under the title *Bingo.* Invite a volunteer to say the actions in lines 1–4. tap, tap, clap, clap, pat, pat, clap, clap

Point out that the pictures match the displayed algorithm. Explain that the pictures show the algorithm for the clapping game to *Bingo* as a picture *program.*

Review the idea that an algorithm written in code (such as pictures) that someone or something (such as a machine) can follow is called a *program.*

Clap the picture program to *Bingo* together as you follow each picture.

Simplify by Using Loops

Guide a **visual analysis** to help the students identify the repeating elements in the pictured program.

What actions repeat in the program? Each action repeats in each line, and the 4-line tap-tap-clap-clap-pat-pat-clap-clap pattern repeats.

Remind the students that something they do more than once in a row in an algorithm or program is called a *loop.*

How could you show the picture program with loops? I could first show each line as a loop of a repeating action. Then I could show each pattern of 4 repeating actions as a loop (a series of repeating actions).

Mark the program with loops for display, directing attention to the loops in each step. Point out the loops within a loop.

Explain that a clapping pattern can be repeated many times in a clapping game. Direct the students to clap the pattern again as you point out the loops on the pictured program.

Introduce another **clapping game** to reinforce the use of loops.

Display the clapping action "salute" at the top of the *Clapping Games* 38b page. Demonstrate how to do a salute by touching your forehead with the side of your palm.

Salute: Touch your forehead with the side of your palm.

Invite the students to salute.

Display the words to the song "A Sailor Went to Sea, Sea, Sea" found on the *Clapping Games* 38b page. Sing or recite the words to the song. Model the following actions for the clapping game.

A Sailor Went to Sea, Sea, Sea

Was the bottom of the deep blue

sea, sea, sea.

Display the picture program for the song on *Clapping Games* 38b. Guide the students as they identify the repeated steps. Invite students to mark the repeated steps as loops on the display.

Point out the other clapping actions at the top of the page. Explain that they may use these actions on their Worktext pages. Demonstrate the actions with a volunteer.

Cross pat right: Cross with your right hand to pat your partner's right palm with your right palm.

Cross pat left: Cross with your left hand to pat your partner's left palm with your left palm.

Back pat: Pat the back of your partner's hands with the back of your hands.

Apply

Worktext pages 259–60

Read and explain the directions for the "Double This, Double That" exercise. Guide the students as they work with a partner to mark the loops in the picture program. Provide a time later for the students to play the game with a partner, if possible.

Explain that they will complete the page by using the Engineering Design Process.

Ask. Think. Plan.

Guide a **collaborative activity** to help the students follow the Engineering Design Process to solve a problem. Explain that each student will work with a partner to plan a new clapping game program to go with "A Sailor Went to Sea, Sea, Sea."

Direct attention to the steps on the *STEM Engineering Design Process* page.

What problem do you need to solve? how to write a new clapping game program to go with the song

Guide the students through the remaining steps of the Engineering Design Process as they use the Worktext pages to plan their programs. Explain that first they will think about their programs with their partners. Then they will cut out only the pictures that they need for their programs and lay them in order in the boxes below the words for the first line.

Instruct the students to cut out their pictures carefully since they will use the remaining pictures to complete Worktext page 260.

Pair the students. Direct them to complete the Ask. Think. Plan. activity on their Worktext pages. Assist the students as needed.

Make. Test. Make Better.

After the students have their plans in place but before they glue the pictures, direct them to test their programs with their partners by clapping it together. Instruct them to make it better if they need to. Explain that when they are satisfied with their programs, they may glue the pictures in place.

Write a Program with Loops

Continue the **collaborative activity** to guide the students as they write their programs with loops.

Direct attention to the picture program they wrote on Worktext page 259.

Did you repeat any actions in your program? If you did, how could you make your program shorter and simpler? I could use loops for the repeated actions.

Direct attention to the loop exercise at the top of Worktext page 260. Explain that the students will continue to work with their partners as they write their programs with loops. Explain that they may mark loops on their picture programs on page 259 as they plan.

When they have their programs with loops ready, they will then cut out pictures from the ones they have left from page 259 and use them to write their programs with loops at the top of page 260.

The students may not need all the boxes provided for their loops on page 260.

Explain that the students may write in a number before a picture to show that the action is repeated as a loop on page 260.

When the student pairs have completed their loop exercises, encourage them to practice their programs with their partners. Explain that they will use the same program for each line of the song.

Instruct the students to quietly practice clapping the entire song with their partners. Choose students to demonstrate their newly created clapping game.

Teaching a Friend

Guide a **discussion** to help the students see that God helps them use patterns in math to make their lives better and more enjoyable.

> **Who gave you the ability to do a clapping game?** God

Read aloud Psalm 23:3. Explain that God is kind to give us times when He makes us feel joyful. Point out God's goodness in giving us the ability to see patterns and use them to do fun things that refresh our soul and body.

Explain that they can share that joy by teaching a friend a clapping game. Direct attention to the biblical worldview shaping statement on Worktext page 260. Instruct them to write the name of someone that they would like to teach a clapping game.

Encourage students to pair with other students to teach and learn their original clapping games by using their picture programs.

Assess

Rubric

Use the *Coding Rubric: Playing a Game* page to assess the project. The rubric may be customized to include your chosen criteria.

Chapter 16 Test

Administer the Chapter 16 Test.

Cumulative Concept Review

Worktext page 261

Review the following concepts. Adapt instructions and activities and provide reteaching as needed to meet the specific needs of your students.

- Writing time to the hour and half-hour (Lesson 59)
- Reading a table and determining the elapsed time (Lesson 58)
- Identifying even and odd numbers (Lesson 23)

Retain a copy of Worktext page 261 for the discussion of the Chapter 17 essential question during Lessons 136 and 142.

Reviews pages 249–50

Use the Reviews pages to help students retain previously learned skills.

Extended Activity

Write Numbers That Are More or Less

Materials

- 2 different-colored markers

Procedure

Write "462" for display. Ask a student to use one color to write the number that is 10 less than 462 above the number. Then ask another student to use the other color to write the number that is 10 more than 462 below the number. 452, 472

What happens to the digit in the Tens place? 10 less—digit is one less than 6; 10 more—digit is one more than 6

Continue the activity but choose students to write the numbers that are 100 more and 100 less than 462. 562, 362

Continue the activity with other numbers.

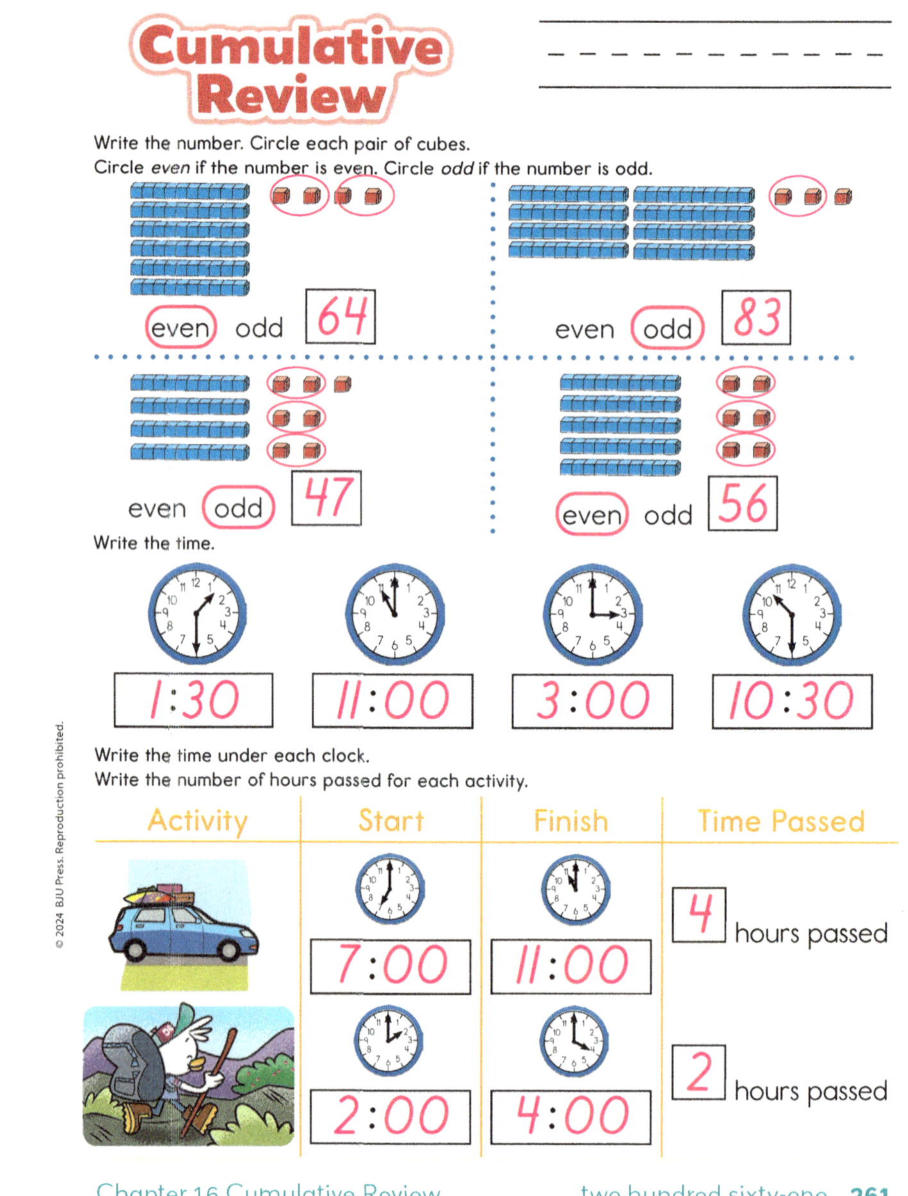

Add.

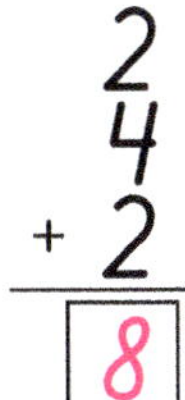

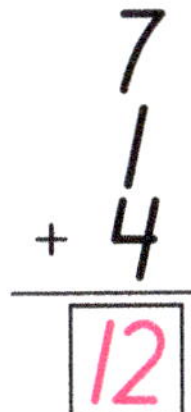

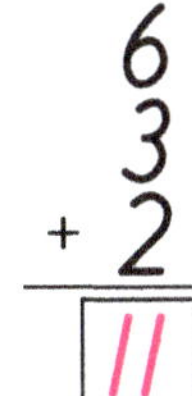

2	7	6	5
4	1	3	2
+ 2	+ 4	+ 2	+ 3
8	12	11	10

Write an equation for each word problem. Solve.

Mother used 42 apples to make applesauce. She used 23 apples to make pies. How many apples did Mother use in all?

work space

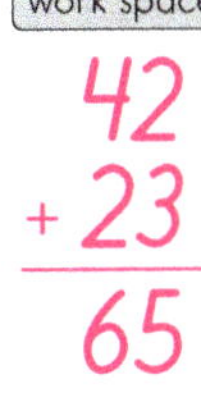

$42 + 23 = 65$

42 + 23 = 65 apples

Paul has 27 apples in his basket. He sold 16 apples. How many apples does Paul have left?

work space

$27 - 16 = 11$

27 − 16 = 11 apples

Write the total value.

42 ¢

18 ¢

56 ¢

Subtraction Fact Review

Subtract.

$5 - 1 = 4$	$9 - 9 = 0$	$6 - 2 = 4$	$10 - 8 = 2$	$11 - 2 = 9$
$7 - 5 = 2$	$10 - 6 = 4$	$11 - 3 = 8$	$9 - 7 = 2$	$6 - 4 = 2$
$9 - 4 = 5$	$9 - 5 = 4$	$7 - 4 = 3$	$5 - 2 = 3$	$8 - 5 = 3$
$5 - 3 = 2$	$7 - 6 = 1$	$5 - 4 = 1$	$10 - 2 = 8$	$8 - 3 = 5$
$4 - 4 = 0$	$11 - 9 = 2$	$8 - 0 = 8$	$6 - 1 = 5$	$11 - 7 = 4$
$12 - 7 = 5$	$6 - 5 = 1$	$6 - 3 = 3$	$8 - 4 = 4$	$12 - 8 = 4$

CHAPTER 17: METRIC MEASUREMENT

PAGES	OBJECTIVES	RESOURCES	ASSESSMENTS
Lesson 136 Measure in Centimeters			
Teacher Edition 581–85 **Worktext** 262–64	**136.1** Compare the length of objects by using the words *longer* and *shorter*. **136.2** Identify a metric measuring unit for length. **136.3** Measure the length and height of objects in centimeters. **136.4** Compose an addition word problem. **136.5** Explain how measuring length and height helps people care for others. BWS	**Teacher Edition** • Instructional Aid 35: *Symmetrical Shapes* • Instructional Aid 36: *Asymmetrical Shapes* • Instructional Aid 40: *Centimeter Measurement Worksheet* **Visuals** • Rulers: Centimeter, Inch **Student Manipulatives Packet** • Ruler: Centimeter **BJU Press Trove*** • Video: Ch 17 Intro • Games/Enrichment: Fact Reviews (Subtraction Facts to 12) • Games/Enrichment: Subtraction Flashcards • Games/Enrichment: Fact Fun Activities • PowerPoint® presentation	**Reviews** • Pages 251–52
Lesson 137 Estimate & Measure in Centimeters			
Teacher Edition 586–89 **Worktext** 265–66	**137.1** Measure the length and height of objects in centimeters. **137.2** Measure distance in centimeters. **137.3** Estimate the length of an object to the nearest centimeter. **137.4** Measure the length of an object to the nearest centimeter. **137.5** Solve a comparison subtraction problem.	**Visuals** • Visual 19: *Metric Measurement* • Rulers: Centimeter, Paper Clip • Counters: Duck Tracks • Money Kit: 5 nickels, 7 dimes **Student Manipulatives Packet** • Rulers: Centimeter, Paper Clip **BJU Press Trove** • Games/Enrichment: Fact Reviews (Subtraction Facts to 12) • PowerPoint® presentation	**Reviews** • Pages 253–54

*Digital resources for homeschool users are available on Homeschool Hub.

PAGES	OBJECTIVES	RESOURCES	ASSESSMENTS
Lesson 138 Using Liters			
Teacher Edition 590–93 **Worktext** 267–68	**138.1** Identify a metric measuring unit for capacity. **138.2** Compare the capacity of objects by using the words *more* and *less*. **138.3** Solve a comparison subtraction problem.	**Visuals** • Visual 19: *Metric Measurement* **Student Manipulatives Packet** • Number Cards: 0–9 **BJU Press Trove** • Games/Enrichment: Fact Reviews (Subtraction Facts to 12) • Games/Enrichment: Subtraction Flashcards • Games/Enrichment: Fact Fun Activities • PowerPoint® presentation	**Reviews** • Pages 255–56
Lesson 139 Using Kilograms			
Teacher Edition 594–97 **Worktext** 269–70	**139.1** Identify a metric measuring unit for mass. **139.2** Identify a tool for measuring mass. **139.3** Compare the mass of objects by using the words *more* or *less*. **139.4** Compose a word problem.	**Visuals** • Visual 19: *Metric Measurement* **BJU Press Trove** • Games/Enrichment: Fact Reviews (Subtraction Facts to 12) • PowerPoint® presentation	**Reviews** • Pages 257–58
Lesson 140 Read a Celsius Thermometer			
Teacher Edition 598–601 **Worktext** 271–72	**140.1** Identify a metric measuring unit for temperature. **140.2** Read a Celsius thermometer. **140.3** Match weather-related activities to Celsius temperature. **140.4** Identify measurement units and tools. **140.5** Explain how reading a thermometer helps people care for others. BWS	**Teacher Edition** • Instructional Aid 41: *Celsius Thermometer* **Visuals** • Visual 20: *Temperature* • Ruler: Centimeter (the red back) **Student Manipulatives Packet** • Ruler: Centimeter (the red back) **BJU Press Trove** • Games/Enrichment: Fact Reviews (Subtraction Facts to 12) • Games/Enrichment: Subtraction Flashcards • Games/Enrichment: Fact Fun Activities • PowerPoint® presentation	**Reviews** • Pages 259–60

PAGES	OBJECTIVES	RESOURCES	ASSESSMENTS
Lesson 141 **Measure Perimeter**			
Teacher Edition 602–5 **Worktext** 273–74	**141.1** Find the perimeter of a figure by using a nonstandard unit of measurement. **141.2** Find the perimeter of a figure in centimeters.	**Visuals** • Ruler: Centimeter **Student Manipulatives Packet** • Ruler: Centimeter **BJU Press Trove** • PowerPoint® presentation	**Reviews** • Pages 261–62
Lesson 142 **Chapter 17 Review**			
Teacher Edition 606–9 **Worktext** 275–76	**142.1** Recall concepts and terms from Chapter 17.	**Teacher Edition** • Instructional Aid 42: *Metric Worksheet* • Instructional Aid 41: *Celsius Thermometer* **Student Manipulatives Packet** • Ruler: Centimeter **BJU Press Trove** • Games/Enrichment: Subtraction Flashcards • Games/Enrichment: Fact Fun Activities • PowerPoint® presentation	**Worktext** • Chapter 17 Review **Reviews** • Pages 263–64
Lesson 143 **Test, Cumulative Review**			
Teacher Edition 610–12 **Worktext** 277	**143.1** Demonstrate knowledge of concepts from Chapter 17 by taking the test.		**Assessments** • Chapter 17 Test **Worktext** • Cumulative Review **Reviews** • Pages 265–66

Chapter 17

How can I use measurement to make a friend's day better?

Chapter Objectives

- Measure with units of length and capacity.
- Compare measurements in units of length, capacity, and mass.
- Read a Celsius thermometer.
- Solve word problems involving measurements.
- Compose word problems involving measurements.
- Explain how metric measurement can be used to care for others.

136

Worktext pages 262–64
Reviews pages 251–52

Objectives

- **136.1** Compare the length of objects by using the words *longer* and *shorter*.
- **136.2** Identify a metric measuring unit for length.
- **136.3** Measure the length and height of objects in centimeters.
- **136.4** Compose an addition word problem.
- **136.5** Explain how measuring length and height helps people care for others. BWS

Biblical Worldview Shaping

- **Caring** (explain): Measuring length and height accurately can help us make an attractive craft. Making a craft like a card or a picture to give to someone else is a good way to show we care about them (136.5).

Printed Resources

- Instructional Aid 35: *Symmetrical Shapes*
- Instructional Aid 36: *Asymmetrical Shapes*
- Instructional Aid 40: *Centimeter Measurement Worksheet* (for the teacher and for each student)
- Visuals: Rulers (Centimeter, Inch)
- Student Manipulatives: Ruler (Centimeter)

Digital Resources

- Video: Ch 17 Intro
- Games/Enrichment: Fact Reviews (Subtraction Facts to 12, page 49)
- Games/Enrichment: Subtraction Flashcards
- Games/Enrichment: Fact Fun Activities

Measure in Centimeters

Mark an *X* on the longer object. Circle the shorter object.

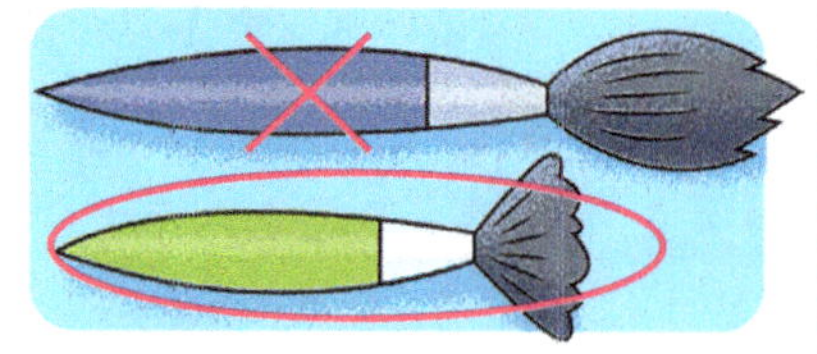

Mark an *X* on the taller object. Circle the shorter object.

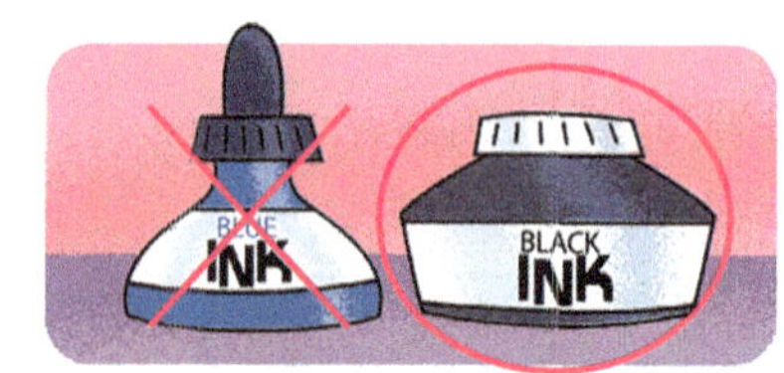

Write the number of centimeters.

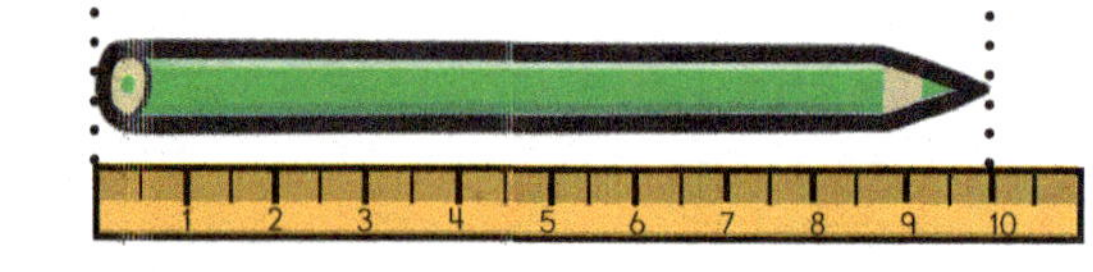

10 centimeters

Use a centimeter ruler to measure.

4 centimeters

7 centimeters

6 centimeters

Practice & Review

Draw a Line of Symmetry

Draw a square for display.

What is this shape? a square

How do you know if a figure has symmetry? if it can be divided into 2 equal parts; if I can fold it and the 2 parts match

Ask a volunteer to draw a line of symmetry on the square. The line can be vertical, horizontal, or diagonal.

Is this the only way a square can be divided to show symmetry? no

Erase the line drawn by the student and ask another volunteer to draw a different line.

Draw a symmetrical triangle.

What is this shape? a triangle

Ask a student to draw a line of symmetry.

Write a *W* for display. Ask a volunteer to draw a line of symmetry.

Identify Symmetrical & Asymmetrical Figures

Display several figures from the *Symmetrical Shapes* and *Asymmetrical Shapes* pages. As you hold up each shape, direct a student to tell whether the shape has symmetry and to explain his or her answer.

Study Subtraction Facts to 12

Select a game or activity from the "Fact Fun Activities" in Trove. Practice the previously memorized subtraction facts.

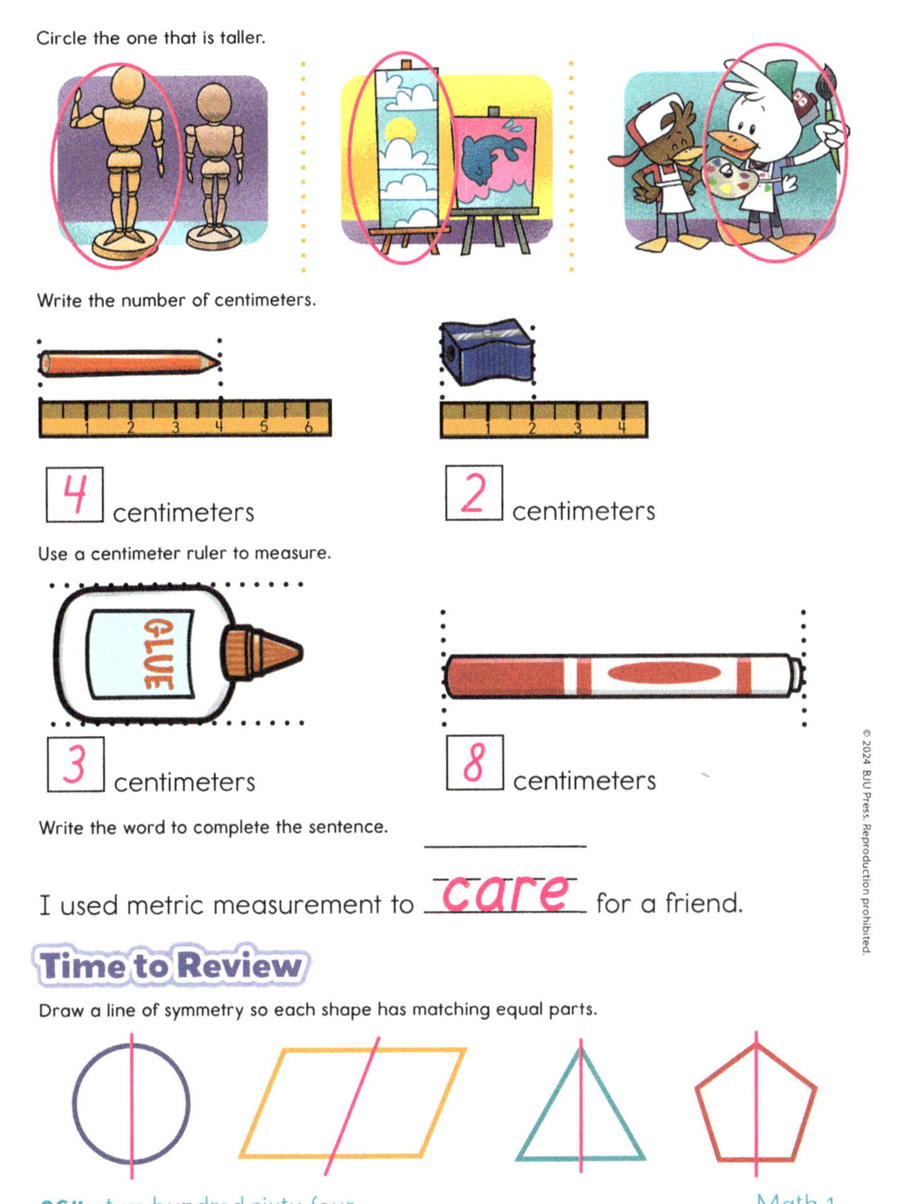

Circle the one that is taller.

Write the number of centimeters.

4 centimeters

2 centimeters

Use a centimeter ruler to measure.

3 centimeters

8 centimeters

Write the word to complete the sentence.

I used metric measurement to care for a friend.

Time to Review

Draw a line of symmetry so each shape has matching equal parts.

Engage

Essential Question

Direct attention to the chapter opener on Worktext page 262 and read aloud the following scenario to introduce the chapter essential question.

DQ bounded out of bed Saturday morning and straightened his white feathers. He remembered that he and Liam were invited to Lily's house to paint pictures. Painting was one of DQ's favorite things to do.

Lily welcomed her friends with a cheery "hello" when they arrived. She gave them each a painting apron and helped them tie it on.

"Don't want to get any of that colored paint on your slick, white feathers, DQ," she warned playfully.

Lily then showed them the supplies she had set out—white painting paper, cans of paint, paintbrushes, scissors, pencils, centimeter rulers, cups of water, and brightly colored construction paper.

"We could give our pictures to Rosa. She has been sick for a while," said DQ. "That will really show we care about her. It may cheer her up too."

"What a great idea!" Lily and Liam agreed.

The friends worked hard to complete their paintings. Liam's was a farm scene, Lily's pictured a garden of flowers, and DQ's showed ducks paddling in a pond.

"What is the colored paper for?" asked Liam, noticing the sheets of construction paper stacked nearby.

"It's for showcasing our artwork. Every picture looks better in a frame," explained Lily. "We should measure our pictures now, so we know how large to make the frames."

They used a centimeter ruler to measure their pictures. They were 10 centimeters wide by 12 centimeters long.

"We'll need to make the frame larger, so it makes a border around the picture," said Liam.

They found that cutting a rectangle 14 by 16 centimeters worked well for mounting their pictures to give them a frame. With their knowledge of measuring, they finished the frames quickly.

"I'm really excited to give these to Rosa," said Liam. "Let's see if we can go and drop these off at her house now."

"Yes, good idea, but not before we clean up this mess first!" laughed Lily.

Invite a student to read aloud the essential question. Encourage the students to consider this question as you lead them to answer it by the end of the chapter.

Instruct

Compare the Length of Objects

Use a **visual** to help the students compare the length of objects.

Distribute a copy of the *Centimeter Measurement Worksheet* to each student and display your copy. Write "L = longer" and "S = shorter" for display. Direct attention to the rectangles containing a star on the page.

Which star rectangle is longer?
rectangle 1

Direct the students to write an *L* in the rectangle.

Which star rectangle is shorter?
rectangle 2

Direct the students to write an *S* in the rectangle.

136

Repeat the procedure with the rectangles containing hearts. L rectangle 4; S rectangle 3

What are you comparing about these rectangles? their length

Direct attention to the rectangles containing triangles.

Write "T = taller" for display.

Which rectangle is taller? the one on the left

Direct the students to write a *T* in the rectangle.

Which rectangle is shorter? the one on the right

Direct the students to write an *S* in the rectangle.

What are you comparing about these rectangles? their height

Identify a Metric Measuring Unit for Length

Use a **manipulative** to identify a measuring unit for length.

Display the Inch Ruler and Centimeter Ruler. Invite a student to identify the Inch Ruler.

What unit of measurement do you use with an inch ruler? inches

Distribute a Centimeter Ruler to each student. Explain that the students will use a centimeter ruler to measure, and that, in many countries, centimeters are used for measuring length and height.

Direct the students to place their index fingers in a space on the Centimeter Rulers as you demonstrate, explaining that 1 centimeter is about the width of an index finger.

Measure the Length & Height of Objects in Centimeters

Guide a **hands-on activity** to familiarize the students with measuring.

Direct attention to the rectangles containing stars on the page. Explain that the students will measure the length of the rectangles in centimeters.

What is the correct way to place the Centimeter Ruler? line up the left end of the ruler with the edge of the object

As each rectangle is measured, ask a volunteer to tell how long each rectangle is. 6 centimeters, 4 centimeters

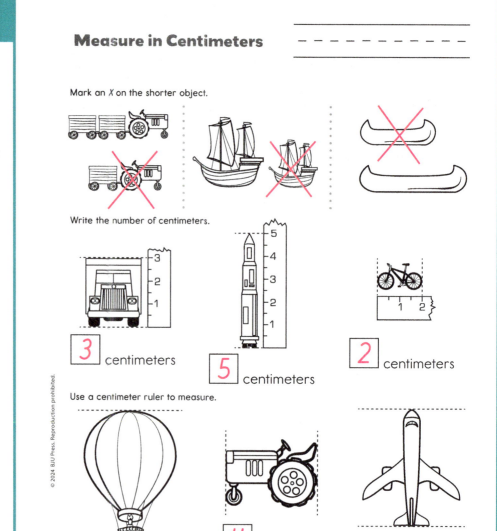

Measure in Centimeters

Mark an *X* on the shorter object.

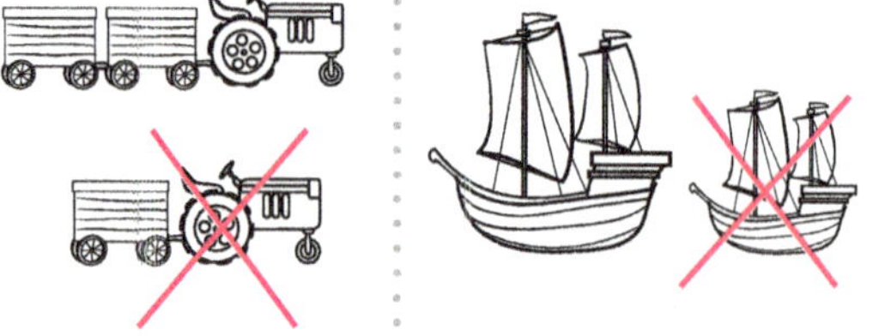

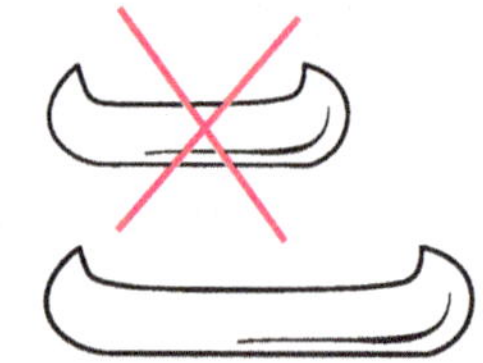

Write the number of centimeters.

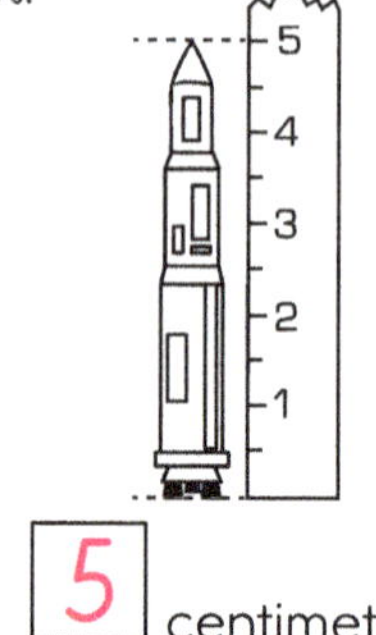

3 centimeters

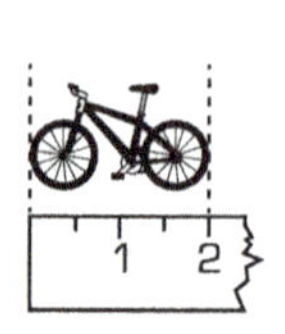

5 centimeters

2 centimeters

Use a centimeter ruler to measure.

6 centimeters

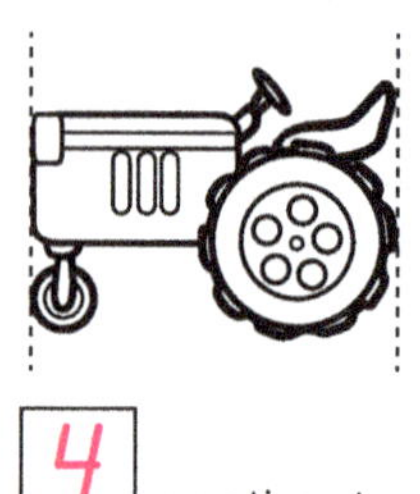

4 centimeters

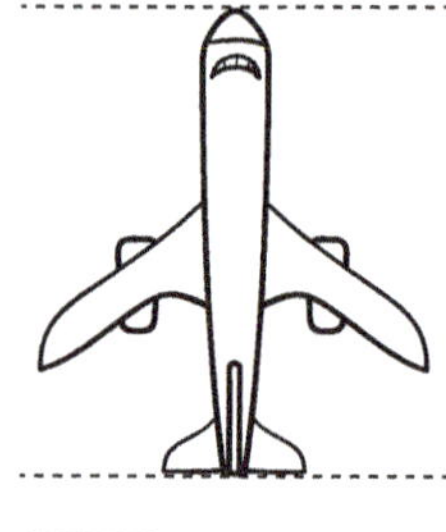

5 centimeters

Continue the activity with the rectangles containing hearts and the rectangles containing triangles.

Hearts: 5 centimeters, 7 centimeters

Triangles: 10 centimeters, 9 centimeters

Direct attention to the numbered rectangles on the page.

Which numbered rectangle is the longest? number 4

Compose an Addition Word Problem

Write an **equation** to help the students practice composing a word problem.

Write "25 + 34 = __ centimeters" for display. Ask a student to compose a word problem for the equation.

Rewrite the equation in vertical form and ask a volunteer to solve it.

Write a new equation and direct students to compose another word problem.

Measuring Helps People Care for Others

Guide a **discussion** to help the students explain a biblical worldview shaping truth.

What were DQ and his friends measuring in the beginning of the lesson? frames for pictures

Why were they painting pictures? to give to their friend

What unit were they using to measure? centimeter

Chapter 9 Review

Match the shape to the correct fraction.

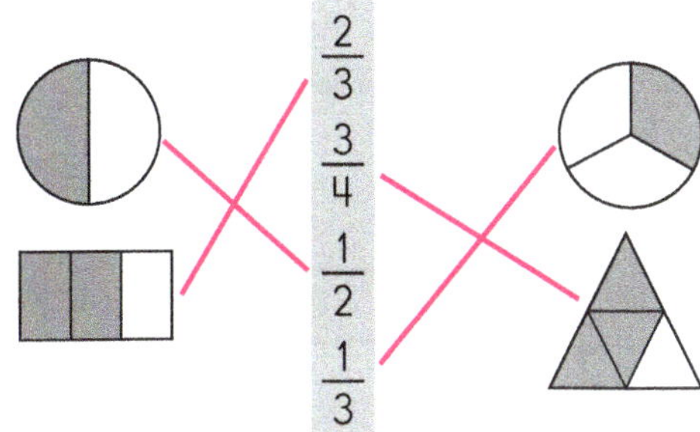

Color the shape to match the fraction.

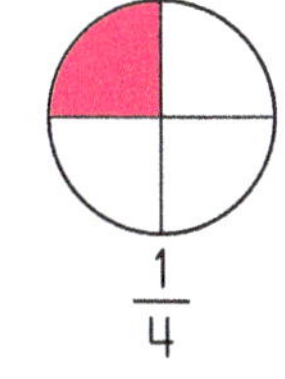
$\frac{1}{4}$

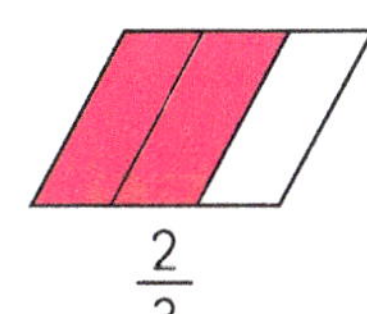
$\frac{2}{3}$

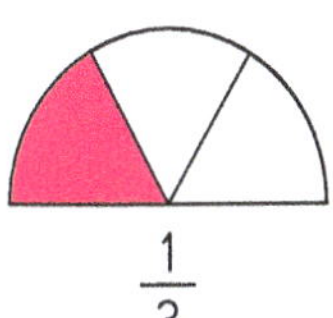
$\frac{1}{3}$

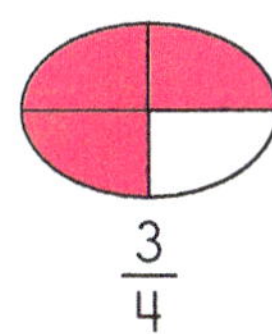
$\frac{3}{4}$

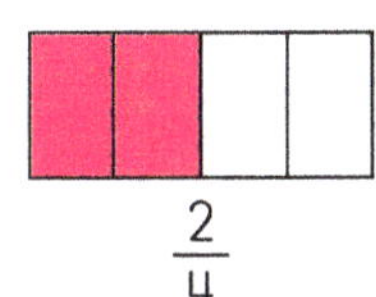
$\frac{2}{4}$

Addition Fact Review

Add.

5 + 3 = 8 8 + 4 = 12 7 + 2 = 9

4 + 2 = 6 3 + 6 = 9 0 + 9 = 9

252 two hundred fifty-two Math 1 Reviews

Explain that math is useful for solving problems. DQ and his friends measured in centimeters in order to make picture frames for their friend. In doing so they were showing that they care for her.

How can measuring length help you care for others? Measuring helps me make a gift to give to a friend.

Apply

Worktext pages 263–64

Read and guide completion of page 263. Remind the students to line up the left end of the Centimeter Ruler with the edge of the object.

Read and explain the directions for page 264. Assist the students as they complete the page independently.

Direct attention to the biblical worldview shaping statement and guide the students as they complete the sentence.

Assess

Reviews pages 251–52

Review coloring the shape to match the fraction on page 252.

Fact Reviews

Use "Subtraction Facts to 12" page 49 in Trove.

Extended Activity

Estimate & Measure Objects

Materials

- Objects to measure: tissue box, chalk-board eraser, box of staples, stapler, lunchbox, pencil, scissors
- Centimeter Ruler for each student

Procedure

Pair the students. Instruct the first partner to estimate the length of an object and the second partner to measure the object.

Differentiated Instruction

Read a Centimeter Ruler

Provide the students with a centimeter ruler. Instruct students to point to the number of centimeters you say. For example, "Point to 4 centimeters." Help the students find and point to 4 centimeters on the ruler. Continue the activity with different centimeter measurements.

137

Worktext pages 265–66
Reviews pages 253–54

Objectives

- **137.1** Measure the length and height of objects in centimeters.
- **137.2** Measure distance in centimeters.
- **137.3** Estimate the length of an object to the nearest centimeter.
- **137.4** Measure the length of an object to the nearest centimeter.
- **137.5** Solve a comparison subtraction problem.

Printed Resources

- Visual 19: *Metric Measurement*
- Visuals: Rulers (Centimeter, Paper Clip)
- Visuals: Counters (Duck Tracks)
- Visuals: Money Kit (5 nickels, 7 dimes)
- Student Manipulatives: Rulers (Centimeter, Paper Clip)

Digital Resources

- Games/Enrichment: Fact Reviews (Subtraction Facts to 12, page 50)

Materials

- A small picture frame
- A paintbrush
- A small container of paint

Estimate & Measure in Centimeters

Guess the length. Use a centimeter ruler to measure.

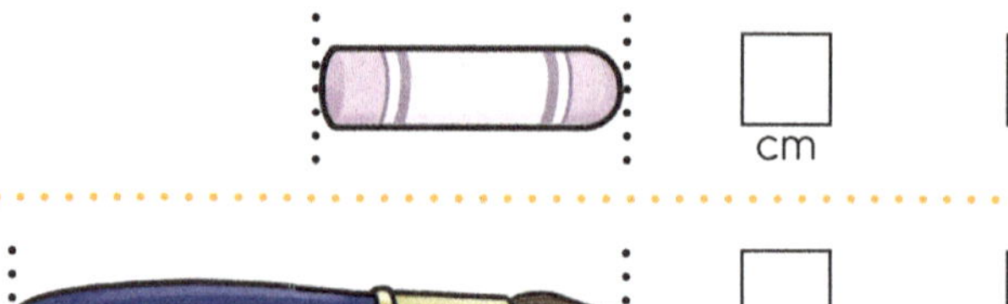

	My Guess	My Measure
eraser	☐ cm	4 cm
paintbrush	☐ cm	8 cm
crayon	☐ cm	4 cm
small eraser	☐ cm	1 cm

Use a centimeter ruler to measure. Mark the correct answer.

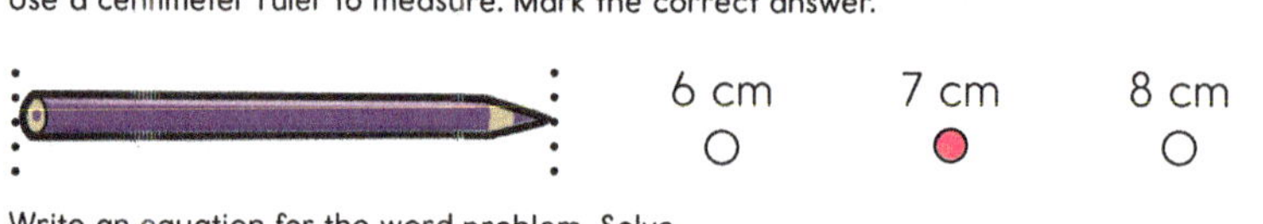

6 cm ○ 7 cm ● 8 cm ○

Write an equation for the word problem. Solve.

DQ has a paintbrush that is 29 cm long. Liam has a paintbrush that is 14 cm long. How much longer is DQ's brush than Liam's brush?

29 − 14 = 15 cm

(work space)

$$\begin{array}{r} 29 \\ -\ 14 \\ \hline 15 \end{array}$$

Practice & Review

Determine the Value of a Set of Coins by Counting on

Display 2 dimes and 3 nickels in random order.

What do you do first when counting the value of a set of coins? separate the different coins into groups

Which coins do you count first? coins with the greatest value

Ask a volunteer to separate the coins into nickels and dimes. Count the coins together: 10, 20, (deep breath), 25, 30, 35¢.

Continue the activity with 3 dimes and 5 nickels: 10, 20, 30, (deep breath), 35, 40, 45, 50, 55¢.

Compare Numbers by Using > and <

Display 5 dimes and 5 nickels on the left. Count the money as demonstrated previously. Write "75¢" for display.

Display 7 dimes and 3 nickels on the right. Count the money together. Write "85¢" to the right of 75¢, leaving a space.

Which amount is smaller? 75¢

Place a dot next to the smaller amount. Draw the < between the 2 amounts. Guide the students as they read the number sentence together. 75¢ is less than 85¢.

Continue the activity for 5 dimes, 3 nickels compared to 6 dimes. 65¢ > 60¢

Write "25 __ 41."

What place do you compare first? Tens

Which number has fewer tens? 25

Draw the <.

Write "57 __ 53."

What place do you compare first? Tens

Which number has fewer tens? Neither; they are the same.

What place do you compare next? Ones

Which number has more ones? 57

Draw the >.

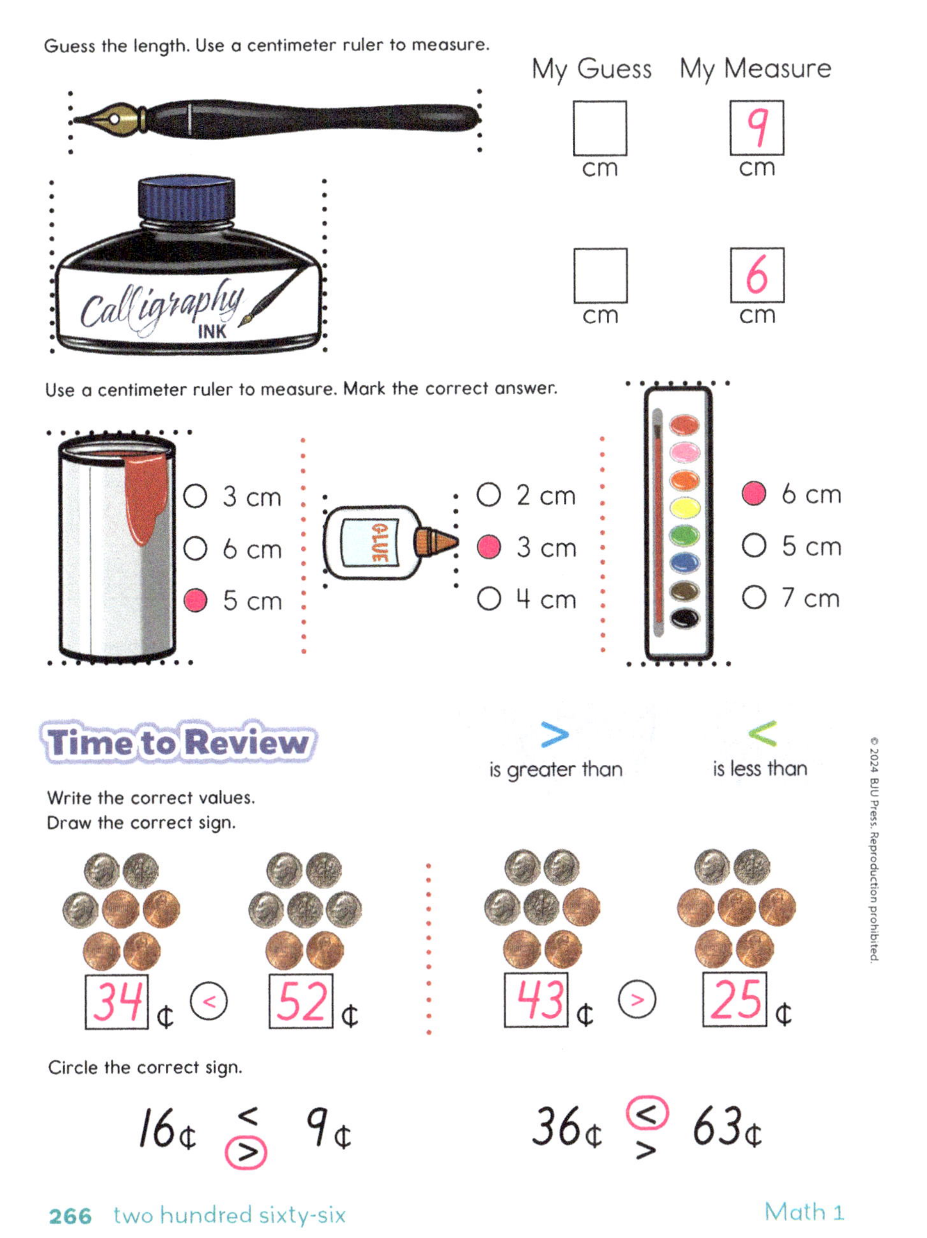

Guess the length. Use a centimeter ruler to measure.

	My Guess	My Measure
pen	___ cm	9 cm
ink bottle	___ cm	6 cm

Use a centimeter ruler to measure. Mark the correct answer.

- ○ 3 cm ○ 6 cm ● 5 cm
- ○ 2 cm ● 3 cm ○ 4 cm
- ● 6 cm ○ 5 cm ○ 7 cm

Time to Review

\> is greater than < is less than

Write the correct values.
Draw the correct sign.

34¢ < 52¢ 43¢ > 25¢

Circle the correct sign.

16¢ > 9¢ 36¢ < 63¢

137

Place the Centimeter Ruler correctly on the paintbrush.

How long is the paintbrush?

Write the measurement for display, using the symbol *cm*. Explain that *cm* is the symbol for *centimeters*.

Metric symbols do not have a period.

Continue the activity by measuring the length of the picture frame and the height of the paint container.

Measure Distance in Centimeters

Use **counters** to help the students measure distance.

Explain that DQ walked to the school. Display a duck track. Place another duck track 15 centimeters from the first duck track.

Measure the distance between the 2 tracks. Count together as you point to each centimeter on the Centimeter Ruler.

What is the distance between the 2 duck tracks? 15 centimeters

Continue the activity by placing the tracks 10 centimeters apart.

Estimate & Measure the Length of an Object to the Nearest Centimeter

Use **Guess-and-Check** to give students practice in measuring length.

Distribute the Paper Clip Ruler and the Centimeter Ruler. Write "My Guess" and "My Measure" for display.

Ask the students how many centimeters long they think the Paper Clip Ruler is.

Write the estimate under "My Guess." Direct the students to measure the length of their Paper Clip Rulers.

About how long is the Paper Clip Ruler? 23 cm

Write "23 cm" under "My Measure." Compare the estimate and the actual length.

Continue the activity, estimating and measuring the following lengths.

4 paper clips on the Paper Clip Ruler 12 cm

5 paper clips on the Paper Clip Ruler 15 cm

1 paper clip on the Paper Clip Ruler 3 cm

Engage

Review a Metric Measuring Unit

Review the term *centimeter* to prepare the students for the content of the lesson.

Direct attention to the centimeter ruler in the length section on Visual 19.

How long is each space on the centimeter ruler? 1 centimeter

How long is the width of your index finger? 1 centimeter

How long is the comb? 10 centimeters

How do you place the ruler when you are measuring an object? place the left end of the centimeter ruler even with the left edge of the object

Instruct

Measure the Length & Height of Objects by Using Centimeters

Use **objects** to help the students measure correctly.

Display the small picture frame, container of paint, and paintbrush. Explain that the students are going to help DQ measure these art supplies.

Instruct a student to hold the paintbrush horizontally as you place the left end of the Centimeter Ruler on the center of the paintbrush.

Is this the correct way to measure? No; the end of the ruler must line up with the end of the paintbrush.

137

Encourage the students to measure other items. Encourage them to estimate before measuring.

Solve a Comparison Subtraction Problem

Use the **Problem-Solving Plan** to solve a subtraction problem.

Read the following word problem several times, asking the students to think about the information so that they can develop a plan to find the answer.

A museum has a gold picture frame and a silver picture frame. The gold frame is 45 cm long. The silver frame is 30 cm long. How much longer is the gold frame than the silver frame?

What is the question? How much longer is the gold frame than the silver frame?

What information is given? The gold frame is 45 cm long, and the silver frame is 30 cm long.

Do you add or subtract to find the answer? subtract

What equation would you use to solve the problem? 45 – 30 = __

Write "45 – 30 = __" for display.

Should you rewrite the equation in vertical form? Yes; it will be easier to subtract in vertical form since the problem has 2-digit numbers.

Ask a volunteer to write the equation in vertical form.

What is the answer to this problem? 15 cm

Does the answer make sense? yes

Continue the activity with the following word problem.

An artist bought 2 paintbrushes. He bought a sable brush that is 25 cm long and a fan brush that is 21 cm long. How much longer is the sable brush than the fan brush?
25 – 21 = 4 cm

Apply

Worktext pages 265–66

Read and guide completion of page 265. Remind the students to line up the left end of the Centimeter Ruler with the edge of the object.

Estimate & Measure in Centimeters

Guess each length. Use a centimeter ruler to measure.

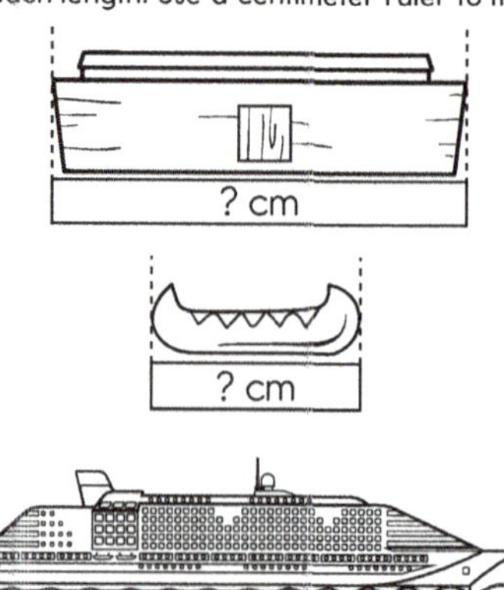

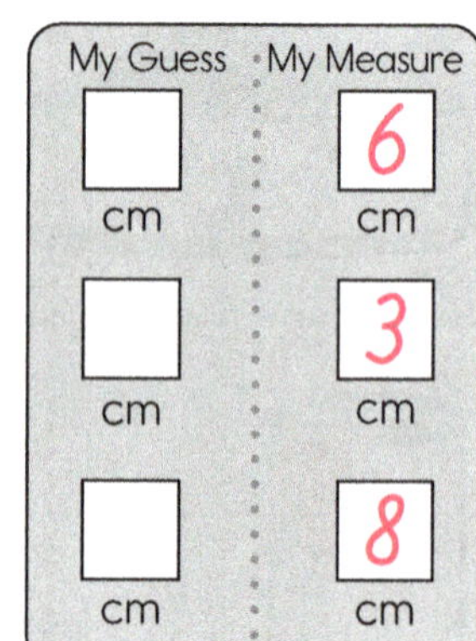

My Guess
My Measure
cm
6 cm
cm
3 cm
cm
8 cm

Use a centimeter ruler to measure.

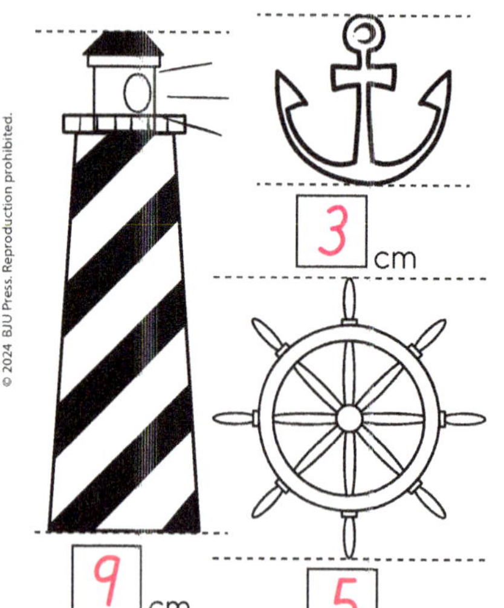

Write an equation for the word problem. Solve.

John's boat is 33 centimeters long. Sam's boat is 21 centimeters long. How much longer is John's boat than Sam's boat?

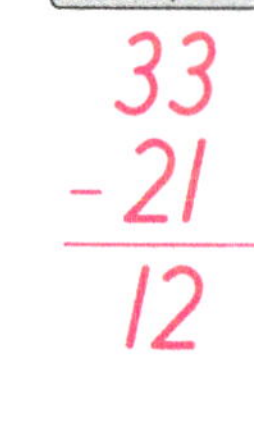

33 − 21 = 12 cm

Read and explain the directions for page 266. Assist the students as they complete the page independently.

Assess

Reviews pages 253–54

Review determining the month that comes next on page 254.

Fact Reviews

Use "Subtraction Facts to 12" page 50 in Trove.

Chapter 8 Review

Color the paw print with the month that comes next.

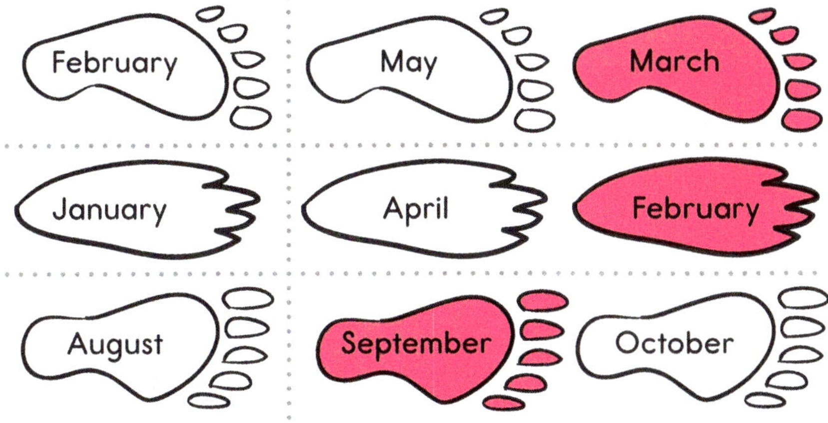

Subtraction Fact Review

Subtract.

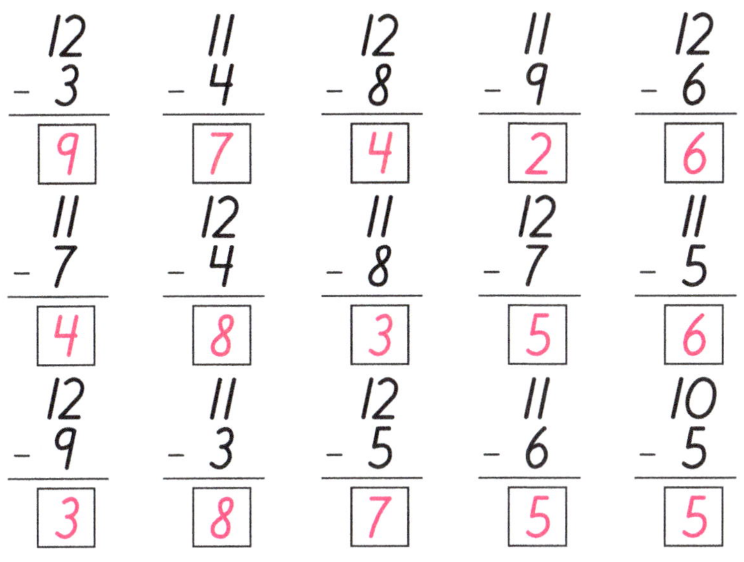

$$\begin{array}{r} 12 \\ -\ 3 \\ \hline \boxed{9} \end{array} \quad \begin{array}{r} 11 \\ -\ 4 \\ \hline \boxed{7} \end{array} \quad \begin{array}{r} 12 \\ -\ 8 \\ \hline \boxed{4} \end{array} \quad \begin{array}{r} 11 \\ -\ 9 \\ \hline \boxed{2} \end{array} \quad \begin{array}{r} 12 \\ -\ 6 \\ \hline \boxed{6} \end{array}$$

$$\begin{array}{r} 11 \\ -\ 7 \\ \hline \boxed{4} \end{array} \quad \begin{array}{r} 12 \\ -\ 4 \\ \hline \boxed{8} \end{array} \quad \begin{array}{r} 11 \\ -\ 8 \\ \hline \boxed{3} \end{array} \quad \begin{array}{r} 12 \\ -\ 7 \\ \hline \boxed{5} \end{array} \quad \begin{array}{r} 11 \\ -\ 5 \\ \hline \boxed{6} \end{array}$$

$$\begin{array}{r} 12 \\ -\ 9 \\ \hline \boxed{3} \end{array} \quad \begin{array}{r} 11 \\ -\ 3 \\ \hline \boxed{8} \end{array} \quad \begin{array}{r} 12 \\ -\ 5 \\ \hline \boxed{7} \end{array} \quad \begin{array}{r} 11 \\ -\ 6 \\ \hline \boxed{5} \end{array} \quad \begin{array}{r} 10 \\ -\ 5 \\ \hline \boxed{5} \end{array}$$

254 two hundred fifty-four

Math 1 Reviews

138

Worktext pages 267–68
Reviews pages 255–56

Objectives

- **138.1** Identify a metric measuring unit for capacity.
- **138.2** Compare the capacity of objects by using the words *more* and *less*.
- **138.3** Solve a comparison subtraction problem.

Printed Resources

- Visual 19: *Metric Measurement*
- Student Manipulatives: Number Cards (0–9)

Digital Resources

- Games/Enrichment: Fact Reviews (Subtraction Facts to 12, page 51)
- Games/Enrichment: Subtraction Flashcards
- Games/Enrichment: Fact Fun Activities

Materials

- A 1-gallon jug
- A 1-liter container of colored water
- A cereal bowl
- 5 eight-ounce glasses
- A felt-tip pen
- A funnel (optional)

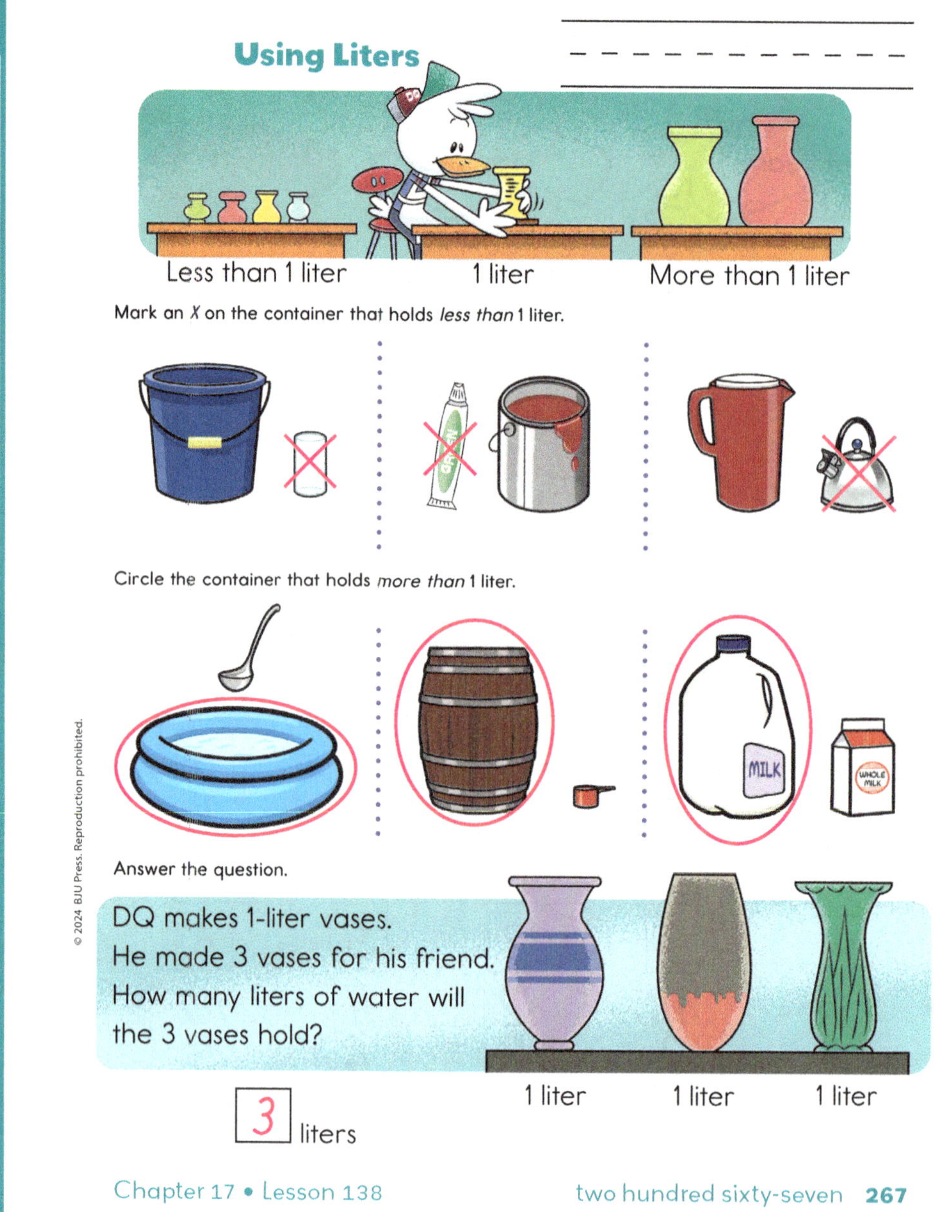

Practice & Review

Write Numbers 200–250

Distribute Number Cards 0–9. Direct the students to use their Number Cards to show 214. Ask a volunteer to write "214" for display for the students to check their cards.

Continue the activity for numbers 237, 250, 249, and 208.

Write Numbers That Come *before, after,* or *between*

Distribute Number Cards 0–9. Instruct the students to use their Number Cards to write the number that matches each clue. Invite a student to write the correct number for display for the students to check their cards.

Which number comes *before* 220? 219

Which number comes *after* 238? 239

Which number comes *between* 214 and 216? 215

Continue the activity with similar clues.

Study Subtraction Facts to 12

Select a game or activity from the "Fact Fun Activities" in Trove. Practice the previously memorized subtraction facts.

Engage

Units of Capacity

Guide a **review** of units of capacity to activate prior knowledge.

Display the 1-gallon jug.

How much does this container hold? 1 gallon

What is usually sold in gallons? milk, juice, paint, gasoline

Explain that gallons, quarts, and cups are customary units of capacity that students have learned.

Point out that there is also a metric unit of capacity that students will learn more about in this lesson.

Instruct

Identify a Metric Measuring Unit for Capacity

Use a **demonstration** to identify a metric measuring unit.

What containers do you see in this room? wastebasket, tissue box, lunchbox, plastic box, pencil box

Mark an X on each container that holds *less than* 1 liter.
Circle each container that holds *more than* 1 liter.

Write an equation for each word problem. Solve.

Lily bought 8 liters of orange drink. She bought 4 liters of grape drink. How many more liters of orange drink did she buy?

8 − 4 = 4 liters

Liam's pot holds 6 liters of water. Lily's pot holds 3 liters. How many more liters of water does Liam's pot hold?

6 − 3 = 3 liters

Time to Review

Subtract.

12 − 9 = 3 10 − 7 = 3 8 − 2 = 6

11 − 6 = 5 9 − 5 = 4 7 − 3 = 4

Direct attention to the capacity section on Visual 19. Point to the 1-liter container as you explain that the liter is a unit of measurement used to tell how much something holds. Explain that in many countries liters are used to measure how much a container holds. Direct attention to the 2-liter container.

Where have you seen a 2-liter container before? at home or in a store

Display 5 glasses. Hold up the 1-liter container of water as you explain that it contains 1 liter of water.

Do you think that 1 glass will hold all the water from the 1-liter container? no

Ask the students how many glasses they think can be filled with the 1 liter of water.

Pour the water into the glasses.

How many glasses did I fill? 4

How many glasses have just a small amount? 1

If all the water amounts in the glasses were added together, how much water would it be? 1 liter

Pour the water back into the 1-liter container.

Compare the Capacity of Objects

Use a measuring **activity** to help the students compare the capacity of objects.

Display the 1-gallon jug.

Does this hold more or less than 1 liter of water? more

Ask the students to guess how much of the jug will be filled when the 1 liter of water is poured into it.

Direct several students to use a felt-tip pen to mark where they think the water will be when you pour 1 liter of water into the jug.

Pour the water into the jug. Compare the marks to the water level on the jug.

Ask the students to determine whose guess was the closest.

Pour the water back into the 1-liter container.

Use the same procedure with a cereal bowl.

Ask several students to stand by containers in the classroom that would hold more than 1 liter. wastebasket, gallon jug, plastic box

Ask several students to show items that would hold less than 1 liter. glue bottle, water bottle, crayon box

If a student chooses an item that is close to a liter, explain that it would be hard to tell whether it was a liter by just looking at it. The capacity of the item would need to be measured to know for sure.

Solve a Comparison Subtraction Problem

Use the **Problem-Solving Plan** to solve a comparison problem.

Read aloud the following word problem.

Liam used 10 liters of yellow paint on his porch swing and 2 liters of blue paint on the trim. How many more liters of yellow paint than blue paint did Liam use?

What is the question? How many more liters of yellow paint than blue paint did Liam use?

What information is given? Liam used 10 liters of yellow paint and 2 liters of blue paint.

Do you add or subtract to find the answer? subtract

What subtraction equation can you write to show the difference between the liters of yellow paint and the liters of blue paint? 10 − 2 = ___

Write "10 − 2 = ___ liters" for display. Ask a volunteer to solve the equation. 8

138

Does the answer make sense? Yes; I am finding how much more yellow paint he used than blue paint; I am comparing to find the difference.

Continue the activity with the following word problem.

DQ and Liam helped their neighbor paint his workshop. They used 12 liters of tan paint for the walls. They used 4 liters of brown paint for the door and window frames. How many more liters of tan paint than brown paint did DQ and Liam use? 12 – 4 = 8 liters

Apply

Worktext pages 267–68

Read and guide completion of page 267.

Read and explain the directions for page 268. Assist the students as they complete the page independently.

Assess

Reviews pages 255–56

Review writing fact families and determining the value of a digit on page 256.

Fact Reviews

Use "Subtraction Facts to 12" page 51 in Trove.

Extended Activities

Determine the Capacity of a Container

Materials

- Different sizes of cups, boxes, jars, glasses, and bottles
- A 1-liter container filled with rice
- A funnel
- A dishpan

Procedure

Display a funnel, a dishpan, and a 1-liter container of rice. Instruct each student to choose one container and to estimate whether it would hold more than 1 liter, less than 1 liter, or 1 liter. Direct them to place the containers in the dishpan and to use the funnel to pour the rice into the containers to find out whether the estimates are correct.

Using Liters

Mark an *X* on each container that holds *less than* 1 liter.
Circle each container that holds *more than* 1 liter.

Write an equation for each word problem. Solve.

Uncle Kyle has 9 liters of blue paint and 6 liters of green paint. How many more liters of blue paint does Uncle Kyle have than green paint?

9 − 6 = 3 liters

Zoe is making punch for a party. She needs 5 liters of orange drink and 3 liters of ginger ale. How many more liters of orange drink than ginger ale does she need?

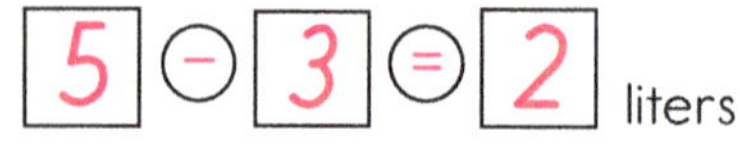
5 − 3 = 2 liters

Chapter 17 • Lesson 138 two hundred fifty-five 255

Determine Whether a Container Has a Capacity of *More Than* or *Less Than* 1 Liter

Procedure

Explain that you will name an object. Instruct the students to stand if the object holds *more than* 1 liter and to raise their hands if it holds *less than* 1 liter.

kitchen sink more	barrel more
soup ladle less	flower vase less
medicine bottle less	bathtub more
swimming pool more	drinking cup less

Chapter 10 Review

Write each fact family.

3	9	12

3 + 9 = 12
9 + 3 = 12
12 − 3 = 9
12 − 9 = 3

4	8	12

4 + 8 = 12
8 + 4 = 12
12 − 4 = 8
12 − 8 = 4

Chapter 16 Review

Mark the value of each underlined digit.

6_2_3	200 ○	20 ●	2 ○
_5_40	500 ●	50 ○	5 ○
97_2_	200 ○	20 ○	2 ●
_7_56	700 ●	70 ○	7 ○
2_3_5	300 ○	30 ●	3 ○
19_9_	900 ○	90 ○	9 ●
_5_63	500 ●	50 ○	5 ○
27_1_	100 ○	10 ○	1 ●

139

Worktext pages 269–70
Reviews pages 257–58

Objectives

- **139.1** Identify a metric measuring unit for mass.
- **139.2** Identify a tool for measuring mass.
- **139.3** Compare the mass of objects by using the words *more* or *less*.
- **139.4** Compose a word problem.

Printed Resources

- Visual 19: *Metric Measurement*

Digital Resources

- Games/Enrichment: Fact Reviews (Subtraction Facts to 12, page 52)

Materials

- 1 object to represent a cube (for example, wooden block, Rubik's Cube, photo cube, game die)
- 1 object to represent a rectangular prism (for example, tissue box, book, wooden block)
- 1 object to represent a cylinder (for example, oatmeal container, soft drink can, roll of paper towels)
- 1 object to represent a cone (for example, funnel, birthday hat, wooden block)
- 1 object to represent a sphere (for example, globe, orange, marble)
- 1 object weighing about 1 pound (for example, melon, pair of shoes, bag or box of sugar)
- A 1-liter container of colored water (from Lesson 138)
- A paper cup
- A paper clip
- A 5-pound bag of flour or other object, weighing more than 1 kilogram (2.2 pounds)
- A balance scale

The concept of mass has been simplified in this lesson and in the Worktext to allow the students to build on their foundational understanding of weight. However, you may focus more on the difference between weight and mass in your instruction.

Using Kilograms

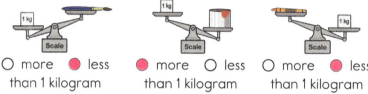

Mark the correct answer.

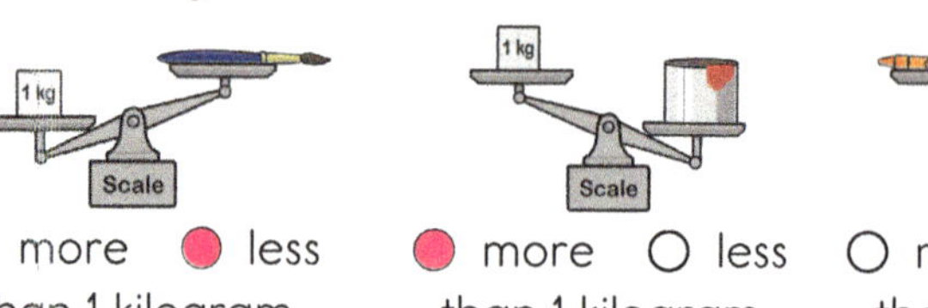

○ more ● less
than 1 kilogram

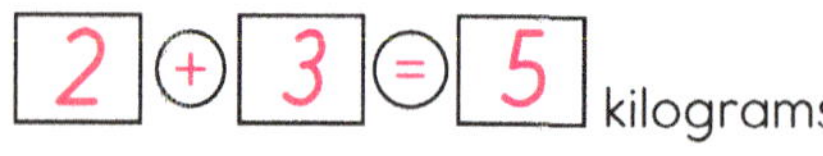

○ more ● less
than 1 kilogram

● more ○ less
than 1 kilogram

○ more ● less
than 1 kilogram

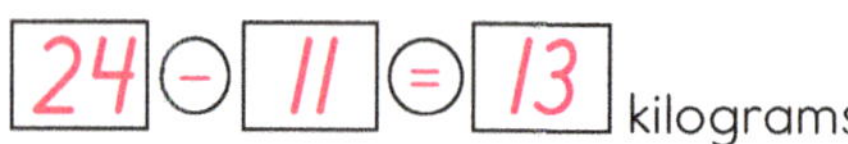

Write an equation for each word problem. Solve.

Lily is making treats for the art show. She needs 2 kilograms of flour for cookies. The cake needs 3 kilograms of flour. How many kilograms of flour will Lily need?

2 + 3 = 5 kilograms

(work space)

$$\begin{array}{r} 2 \\ +\ 3 \\ \hline 5 \end{array}$$

Jason's pet dog weighs 24 kilograms. Jordan's pet dog weighs 11 kilograms. How many more kilograms does Jason's dog weigh than Jordan's dog?

24 − 11 = 13 kilograms

(work space)

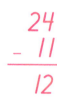

$$\begin{array}{r} 24 \\ -\ 11 \\ \hline 12 \end{array}$$

If a balance scale is not available, you may make one with a clothes hanger, string, and 2 equal-sized berry baskets. (See Lesson 88 for the illustration.)

Practice & Review

Identify Curves, Faces & Corners

Display the rectangular prism, cylinder, cone, sphere, and cube.

How do you identify a solid figure? by its faces (flat sides), curves (curved sides), and corners

Invite a student to find the sphere in the set of objects.

Does a sphere have any corners? no, because there are no points and no flat sides to make corners

Does a sphere have any faces? no, because there are no flat sides

Does a sphere have any curves? yes

Explain that if a solid figure rolls, it has a curved side. Direct a student to roll the sphere.

Does a sphere roll? yes

Continue the activity with each solid figure.

Engage

Introduction to Mass

Guide a **review** of weight to activate prior knowledge.

Display the object weighing about 1 pound.

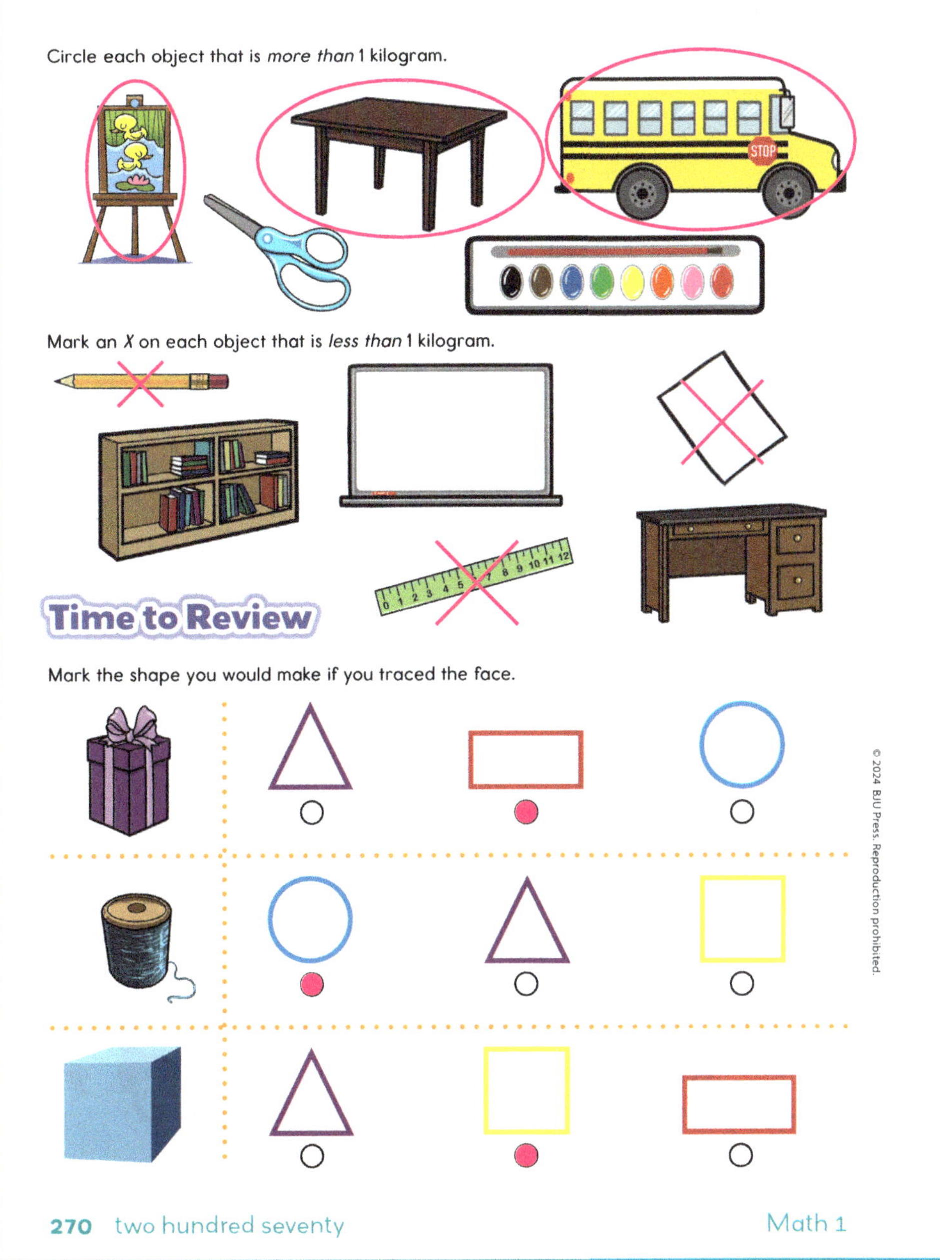

Circle each object that is *more than* 1 kilogram.
STOP
Mark an *X* on each object that is *less than* 1 kilogram.
Time to Review
Mark the shape you would make if you traced the face.
© 2024 BJU Press. Reproduction prohibited.
270 two hundred seventy
Math 1

Explain that you will hold up an object and ask the students if it weighs more than or less than 1 pound.

(Select items from the materials list)

paper cup	less than a pound
marble	less than a pound
melon	more than a pound
birthday hat	less than a pound
liter of water	more than a pound

Remind students that the pound is a customary unit of weight.

Point out that students will be learning about a new measuring unit in this lesson.

Instruct

Identify a Metric Measuring Unit for Mass

Use a **visual aid** to help the students learn the concept of mass.

What have you used to measure the length of an object? centimeter ruler

What have you used to measure how much something holds? a 1-liter container

Display the 1-liter container of water.

Ask the students whether they can use a liter to tell how heavy something is.

Explain that the students will be learning about the amount of matter in an object, or its *mass*. The mass of a liter of water is 1 kilogram. Give each student the opportunity to hold the 1-liter container to become familiar with its mass.

Direct attention to the mass section on Visual 19.

Which measurement tool is used to measure mass? a balance scale

What is the mass of the pair of shoes? 1 kilogram

What is the mass of the bag of oranges? 5 kilograms

Identify a Tool for Measuring Mass

Use a **demonstration** to help the students understand the balance scale.

Display the balance scale, explaining that this tool is one way to measure the mass of an object.

Direct attention to the pans on the balance scale. Explain that the scale works by placing an object in each pan. The object that is heavier or has more mass will go down, and the lighter side with less mass will rise.

What will happen if the objects have the same mass? Conclude together that the pans on the scale will be level with each other and will be balanced.

Direct the students to stand beside their desks with their arms outstretched. Explain that they are going to practice pretending to be balance scales. Demonstrate using your arms after each question.

How does your balance scale show that the object in your right hand has more mass than the object in your left hand? My right hand is down, and my left hand is up.

How does your balance scale show that the object in your left hand has more mass than the object in your right hand? My left hand is down, and my right hand is up.

What happens if the two objects have the same mass? Both arms will be straight out from the shoulders at the same height.

139

Compare the Mass of Objects

Guide a **movement activity** to show the students how to compare mass.

Place the 1 liter of water on the balance scale and hold up a paper cup.

What is the mass of 1 liter of water? 1 kilogram

Ask the students whether they think the mass of the paper cup is more than or less than 1 kilogram.

Place the paper cup on the other side of the balance scale.

Is the mass of the paper cup more than or less than 1 kilogram? less than

Is the side of the balance scale with the paper cup up or down? up; because the mass of the paper cup is less than 1 kilogram

Continue the activity with a paper clip and then a bag of flour. less than 1 kilogram; more than 1 kilogram

Instruct the students to stand and pretend to be balance scales with 1 liter of water in their right hands.

What is the mass of 1 liter of water? 1 kilogram

Explain that you will suggest different objects that the students should pretend to hold in their left hands. They should tip their scales up or down, depending on whether the mass of the object is more than or less than 1 kilogram.

feather right hand down, left hand up

dog right hand up, left hand down

bicycle right hand up, left hand down

hat right hand down, left hand up

toothbrush right hand down, left hand up

stove right hand up, left hand down

Compose a Word Problem

Write an **equation** to help the students practice composing a word problem.

Write "17 + 12 = __kilograms" for display. Ask a student to compose a word problem for this equation. Rewrite the equation in vertical form and ask a volunteer to solve it. 29

Continue the activity with 23 − 11 = __ kilograms. 12

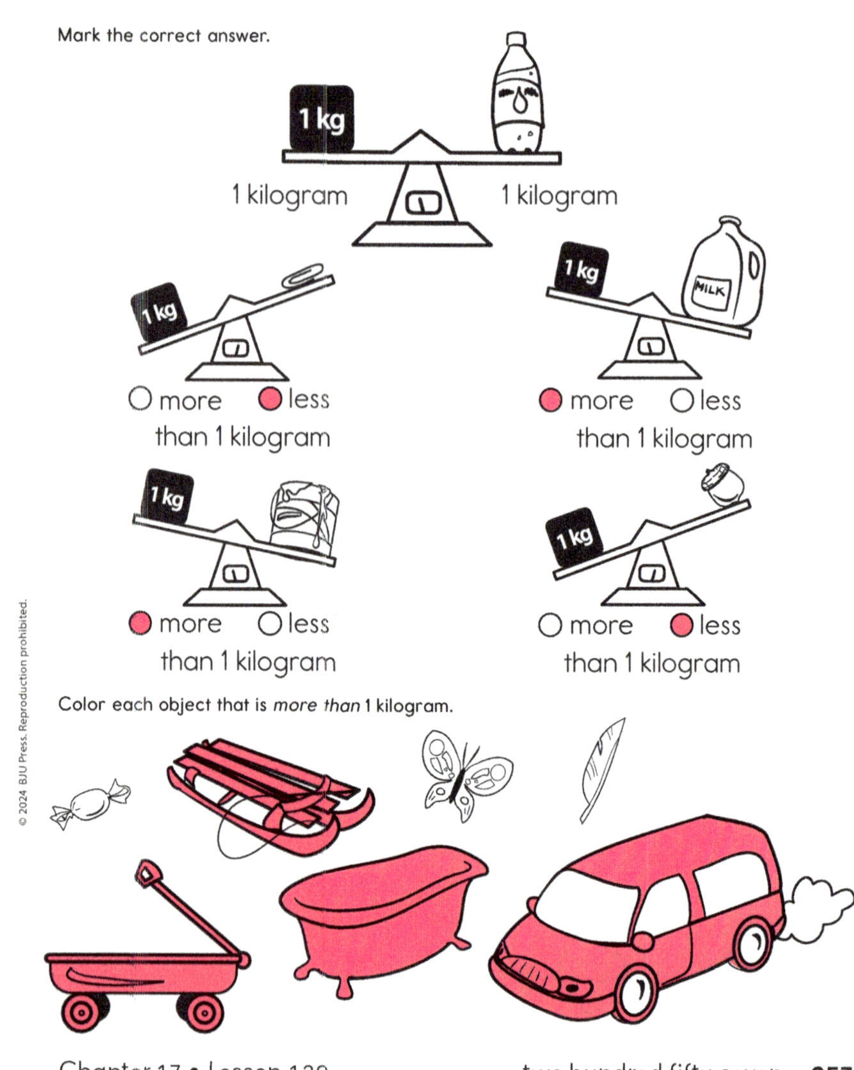

Apply

Worktext pages 269–70

Read and guide completion of page 269.

Read and explain the directions for page 270. Assist the students as they complete the page independently.

Assess

Reviews pages 257–58

Review writing the total value of a set of coins on page 258.

Fact Reviews

Use "Subtraction Facts to 12" page 52 in Trove.

Extended Activity

Determine Whether an Object Is *More Than* or *Less Than* 1 Kilogram

Materials

- Drawing paper for each student

Procedure

Instruct each student to fold a piece of drawing paper in half. Direct the students to draw an object on one side of the paper that would be *less than* 1 kilogram and another object on the other side that would be *more than* 1 kilogram. Instruct the students to exchange papers. Direct each student to circle the object that is *more than* 1 kilogram and to mark an *X* on the object that is *less than* 1 kilogram.

Chapter 14 Review

Write the total value.

37¢

12¢

Subtraction Fact Review

Subtract.

8 − 4 = 4 7 − 5 = 2 10 − 8 = 2

7 − 3 = 4 8 − 3 = 5 11 − 7 = 4

9 − 8 = 1 10 − 6 = 4 9 − 7 = 2

10 − 2 = 8 9 − 6 = 3 10 − 4 = 6

140

Worktext pages 271–72
Reviews pages 259–60

Objectives

- **140.1** Identify a metric measuring unit for temperature.
- **140.2** Read a Celsius thermometer.
- **140.3** Match weather-related activities to Celsius temperature.
- **140.4** Identify measurement units and tools.
- **140.5** Explain how reading a thermometer helps people care for others. BWS

Biblical Worldview Shaping

- **Caring** (explain): Reading a thermometer and understanding how to interpret the temperature helps us know what outdoor events are appropriate and how to dress for them. We can care for others by helping them prepare too (140.5).

Printed Resources

- Instructional Aid 41: *Celsius Thermometer* (for the teacher and for each student)
- Visual 20: *Temperature*
- Visuals: Ruler (Centimeter; the red back)
- Student Manipulatives: Ruler (Centimeter; the red back)

Digital Resources

- Games/Enrichment: Fact Reviews (Subtraction Facts to 12, page 53)
- Games/Enrichment: Subtraction Flashcards
- Games/Enrichment: Fact Fun Activities

Materials

- A small can of paint
- A paintbrush

Practice & Review

Study Subtraction Facts to 12

Select a game or activity from the "Fact Fun Activities" in Trove. Practice the previously memorized subtraction facts.

Read a Celsius Thermometer

Write the temperature.
Draw a line to match each thermometer to the correct picture.

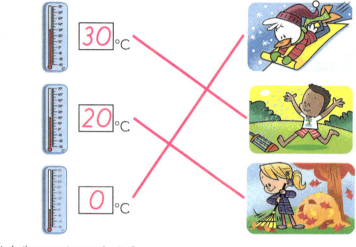

30 °C
20 °C
0 °C

Circle the correct measuring tool.

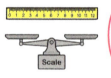

Engage

Review

Guide the students in a **review** to reinforce the Problem-Solving Plan.

Read aloud the following word problem.

DQ is shipping two picture frames to his uncle. The mass of one frame is 12 kilograms, and the mass of the other frame is 10 kilograms. What is the mass of both frames together?

What is the question? What is the mass of both frames together?

What information is given? The mass of one frame is 12 kilograms, and the mass of the other frame is 10 kilograms.

Should you add or subtract to find the answer? add

What equation should you use to solve this word problem? 12 + 10 = __

Write "12 + 10 = __ kilograms" and write the problem in vertical form.

Ask a volunteer to solve the problem. 22

Does the answer make sense? yes, because I am joining the 2 sets to get a larger set

Continue the activity with the following word problem.

DQ has 86 cm of blue ribbon. He cut off 54 cm to put on the package. How much ribbon does he have left? 86 – 54 = 32 cm

Write the temperature.
Draw a line to match each thermometer to the correct picture.

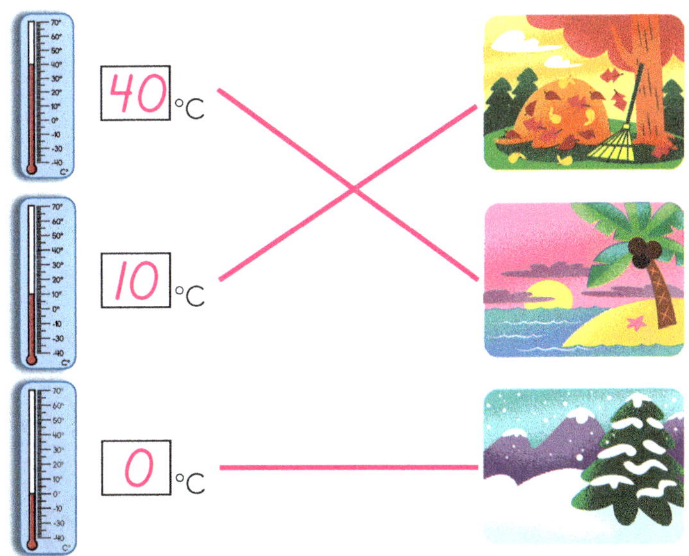

Circle the word(s) to complete the sentence.

I can help keep other people warm by reading the temperature / food labels.

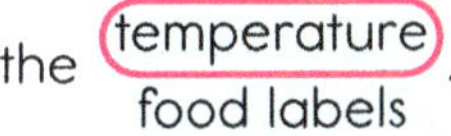

Subtract.

10 − 3 = 7   12 − 8 = 4   7 − 4 = 3

11 − 7 = 4   9 − 5 = 4   8 − 0 = 8

Instruct

Identify a Metric Measuring Unit for Temperature

Guide a **discussion** to identify a measuring unit for temperature.

Display Visual 20.

How do you measure temperature? with a thermometer

Which unit is used to measure temperature? degrees

Explain that students will measure temperature by using *degrees Celsius*. Point to the word *Celsius* and the symbol °C.

Direct students to look at the first picture.

Is it winter or spring? spring

What is on the tree? flowers

Direct attention to the Celsius thermometer under the spring picture. Remind the students that the number at the top of the red-shaded area of the thermometer tells the temperature.

What is the temperature for this spring picture? 20° Celsius

Say the temperature together. 20° Celsius

What happens to the thermometer when the temperature begins to get warmer? The red liquid on the thermometer goes up to indicate warmer temperature.

Direct attention to the 30° Celsius thermometer.

Is this temperature colder or warmer than the first one? warmer

What is the temperature? 30° Celsius

Direct students to look at the picture.

What time of year is it? summer

Direct attention to the third Celsius thermometer.

What happens to the thermometer when the temperature begins to get colder? The red liquid goes down to indicate colder temperature.

What is the temperature on this thermometer? 0° Celsius

Direct students to look at the picture.

What time of year is it? winter

Read a Celsius Thermometer

Use a **manipulative** to help the students read a Celsius thermometer.

Distribute the Centimeter Rulers (red back) and a *Celsius Thermometer* page to each student. Display Visual 20 and use the red back of the Centimeter Ruler to show 20° Celsius.

What is the temperature? 20° Celsius

Write "20° Celsius" for display. Direct the students to show this temperature on their thermometers. Explain that this is about what the temperature is in the room.

If the temperature outside is 5° Celsius, should you wear a jacket at recess? yes

Why should you move the red line down on the thermometer to show 5° Celsius? It is colder than 20° Celsius.

Display the temperature at 5° Celsius and write "5° Celsius."

Allow time for the students to show 5° Celsius on their thermometers.

Direct the students to show 0° Celsius on their thermometers.

How cold do you think it is?

Explain that it is so cold that water freezes.

Show 37° Celsius. Explain that 37° Celsius is normal body temperature for one who is not sick and does not have a fever.

Direct the students to show 100° Celsius on their thermometers.

Is this hot or cold? hot

How hot do you think this is?

Explain that it is so hot that water boils.

Match Weather-Related Activities to Celsius Temperature

Use a **manipulative** to help the students show a Celsius temperature.

Write "0°C," "20°C," and "40°C" for display. Write "cold" below 0°C, "warm" below 20°C, and "hot" below 40°C.

Instruct the students to use their thermometers to show the degrees Celsius for the weather when they are swimming in an outdoor pool.

What temperature did you show? 40° Celsius

What other activities might you do when the temperature is 40° Celsius? go to the beach, play indoor games, eat ice cream

Direct the students to show the temperature for the weather when they are ice-skating on a pond.

What temperature did you show? 0° Celsius

What other activities might you do when the temperature is 0° Celsius? build a snowman, ski, drink hot cocoa

Direct the students to show the temperature for the weather when they are having a picnic.

What temperature did you show? 20° Celsius

What other activities might you do when the temperature is 20° Celsius? ride a bike, go hiking, play sports

Identify Measurement Units & Tools

Use **visual aids** to help the students identify measurement units.

Explain that you have some items from DQ's painting activity. Hold up the paintbrush.

Which measuring tool should I use to tell how long this paintbrush is? a ruler

What is the metric unit of measurement for length? centimeter

Hold up the can of paint.

Read a Celsius Thermometer

Write each temperature.

10 °C | 0 °C | 50 °C | 40 °C

Draw a line from each thermometer to the correct picture.

Which tool should I use to measure the mass of the can of paint? a balance scale

What is the metric unit of measurement for mass? kilogram

Which metric unit of measurement is used to tell how much paint is in the can? liter

Reading a Thermometer Helps People Care for Others

Guide a **discussion** to help the students explain a biblical worldview shaping truth.

Read aloud the following scenario.

DQ and his friends were planning a picnic on Saturday. Lily was packing some tasty food in a basket, and Liam was looking for his kickball. He thought about bringing it along for a game after lunch. Sam, Ken, and Jen were coming too.

DQ checked the thermometer outside. It was 15° Celsius.

"We'll need to take jackets today," he said. "That way we'll be more comfortable while the weather is cool this morning."

The friends all agreed. Soon they were walking out the door, happy and prepared for a great time together.

How did DQ show that he cared about his friends? He made sure they were dressed warmly enough.

Point out that skills the students learn in math, such as reading a thermometer, can be used to care for the needs of others.

Chapter 16 Review
Write the number that is 1 more on each flower.
329 684 270 433 118
328 683 269 432 117
Write the number that is 1 less on each flowerpot.
415 273 446 652 898
414 272 445 651 897
Write the number that is 10 more on each flower.
340 636 441 758 560
330 626 431 748 550
Addition Fact Review
Add.
8 + 2 = 10
7 + 4 = 11
3 + 3 = 6
5 + 2 = 7
9 + 0 = 9
3 + 6 = 9
9 + 2 = 11
6 + 1 = 7
4 + 3 = 7
8 + 4 = 12
260 two hundred sixty
Math 1 Reviews
© 2024 BJU Press. Reproduction prohibited.

How can reading a thermometer help you care for others? It can help me keep others warm and comfortable.

Apply

Worktext pages 271–72

Read and guide completion of page 271.

Read and explain the directions for page 272. Assist the students as they complete the page independently.

Direct attention to the biblical worldview shaping statement and guide the students as they complete the sentence.

Assess

Reviews pages 259–60

Review writing the number that is 1 more, 1 less, or 10 more on page 260.

Fact Reviews

Use "Subtraction Facts to 12" page 53 in Trove.

141

Worktext pages 273–74
Reviews pages 261–62

Objectives

- **141.1** Find the perimeter of a figure by using a nonstandard unit of measurement.
- **141.2** Find the perimeter of a figure in centimeters.

Printed Resources

- Visuals: Ruler (Centimeter)
- Student Manipulatives: Ruler (Centimeter)

Materials

- A calculator (optional)
- A copy of Worktext page 273 (for the teacher)

Guide the students as they work with 4 addends. The students may use calculators if they have already been introduced.

Practice & Review

Identify the Value of the Digits in a 3-Digit Number

Write the number 439 for display.

Which digit is in the Hundreds place? 4

Which digit is in the Ones place? 9

Which digit is in the Tens place? 3

What is the value of the 4 in the Hundreds place? 400

What is the value of the 3 in the Tens place? 30

What is the value of the 9 in the Ones place? 9

Repeat this activity for the numbers 572 and 916.

Compare 3-Digit Numbers by Using > and <

Write "346 __ 264" for display.

To compare these numbers, which place do you look at first? Hundreds place

Which digit is in the Hundreds place in 346? 3

Which digit is in the Hundreds place in 264? 2

Which number—346 or 264—has more hundreds? 346

Measure Perimeter

Listen to your teacher.

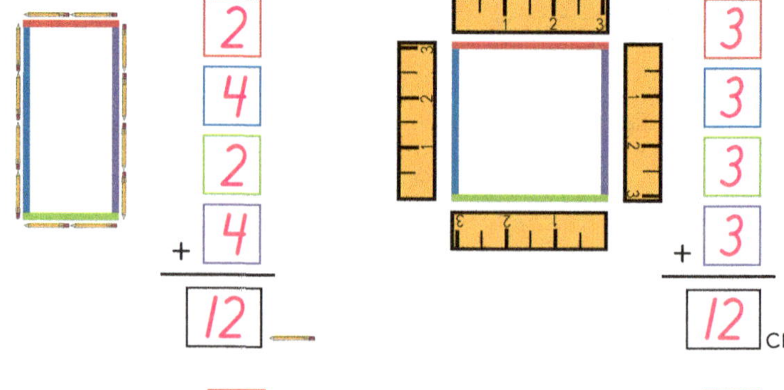

2 + 4 + 2 + 4 = 12

3 + 3 + 3 + 3 = 12 cm

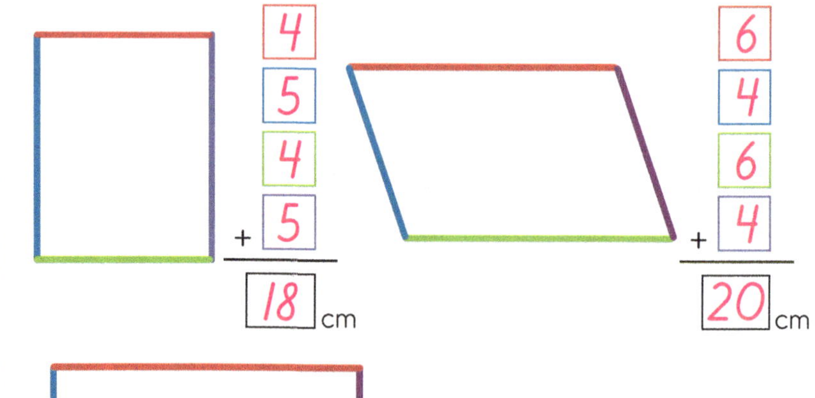

4 + 5 + 4 + 5 = 18 cm

6 + 4 + 6 + 4 = 20 cm

7 + 3 + 7 + 3 = 20 cm

Chapter 17 • Lesson 141 two hundred seventy-three 273

Do you draw a greater-than or less-than sign between the numbers? greater-than sign

Draw the > between the numbers, reminding the students to place a dot next to the smaller number to help them draw the sign.

Write "452 __ 491."

Which digit is in the Hundreds place in the number 452? 4

Which digit is in the Hundreds place in the number 491? 4

Which number has more hundreds? Both have the same.

Which place do you look at next? the Tens place

Which digit is in the Tens place in 452? 5

Which digit is in the Tens place in 491? 9

Which number—452 or 491—has more tens? 491

Do you draw a greater-than or less-than sign between the numbers? less-than sign

Draw the < between the numbers.

Continue the activity with 262 < 264.

Engage

Introduction to Perimeter

Direct a **movement activity** to determine perimeter.

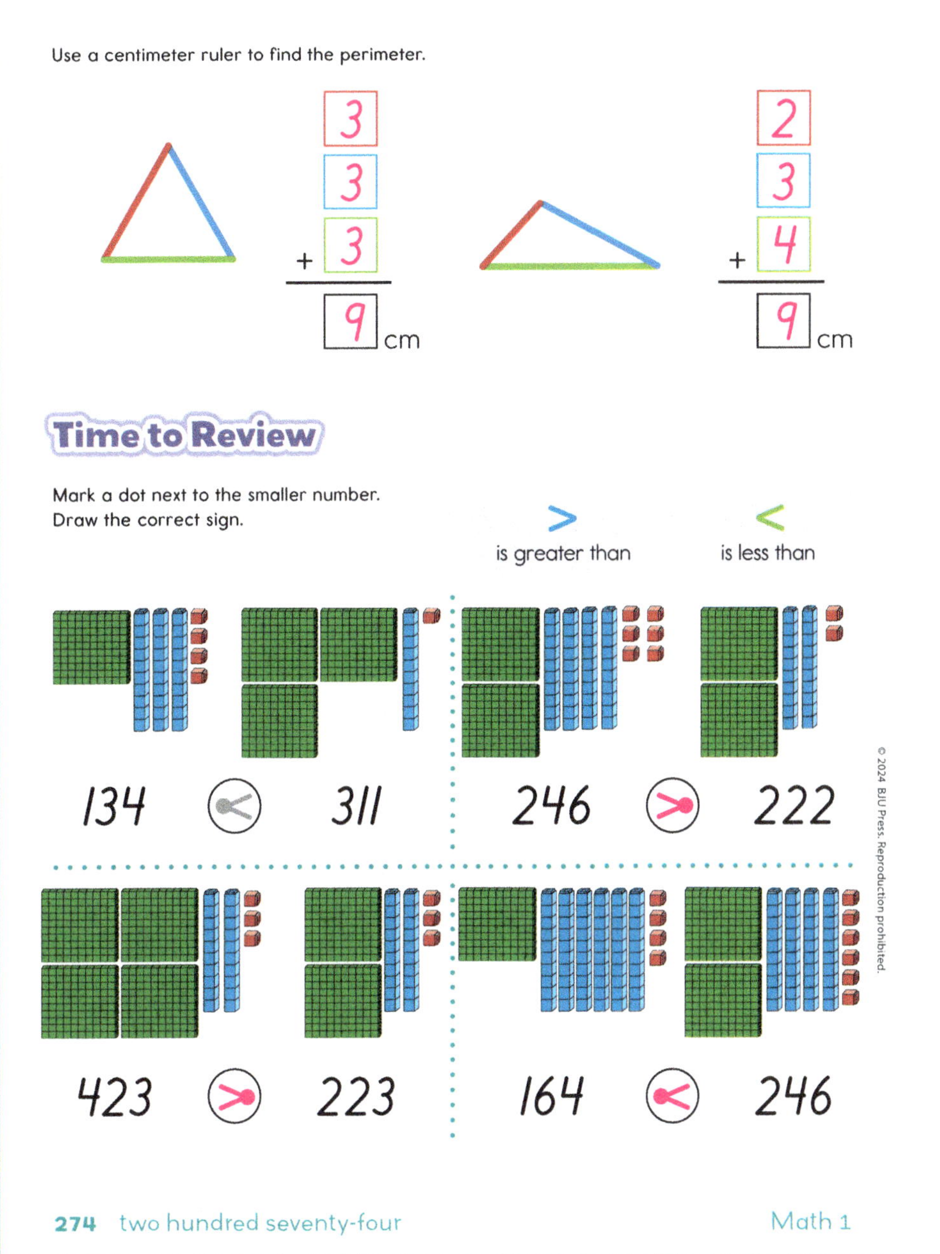

Use a centimeter ruler to find the perimeter.
3
3
+ 3
9 cm
2
3
+ 4
9 cm
Time to Review
Mark a dot next to the smaller number.
Draw the correct sign.
> is greater than
< is less than
134 < 311
246 > 222
423 > 223
164 < 246
© 2024 BJU Press. Reproduction prohibited.
274 two hundred seventy-four
Math 1

Instruct the students to walk around the outside edge of the room in a line.

Walking around the room may not be an exact perimeter because of cabinets and bookshelves, but it will help the students understand the concept that perimeter is the "distance around an object."

What did we do? walked around the room

Explain that you want the students to measure the distance around the room, using footsteps. Ask a volunteer to walk around the outside edge of the room, placing one foot in front of the other. Count together the number of footsteps the student takes as he or she walks.

What is the distance around the room, using footsteps as the measuring unit?

Write *perimeter* for display. Explain that the distance around an object is called the *perimeter*. Encourage several other students to measure the room, using their footsteps.

Instruct

Find the Perimeter of a Figure by Using a Nonstandard Unit of Measurement

Demonstrate finding the perimeter, using a nonstandard unit.

Distribute Worktext page 273 and a Centimeter Ruler to each student.

Direct attention to the first rectangle. Write the problem for display as the students work on their pages.

What is used to measure the perimeter of the rectangle? pencils

How many pencils are across the top or the red side? 2

Direct the students to write the number 2 in the red box.

How many pencils long is the blue side? 4

Direct the students to write the number 4 in the blue box.

Repeat the procedure with the green and purple sides.

Add the 4 measurements while the students watch. $2 + 4 + 2 + 4 = 12$ Direct the students to complete the problem on their pages.

What is the perimeter of the rectangle? 12 pencils

How do you find the perimeter of an object? by adding the lengths of all the sides together

Find the Perimeter of a Figure in Centimeters

Use a **centimeter ruler** to determine perimeter.

Direct attention to the square on the Worktext page. Write the problem for display as the students work on their pages.

What is used to measure the perimeter of the square? centimeter rulers

How many centimeters long is the top or red side? 3 cm

Direct the students to write the number 3 in the red box.

How many centimeters long is the blue side? 3 cm

Direct the students to write the number 3 in the blue box.

Repeat the procedure with the green and purple sides.

Add the 4 measurements while the students watch. $3 + 3 + 3 + 3 = 12$

Direct the students to complete the problem on their pages.

What is the perimeter of the square? 12 centimeters

How did you find the perimeter of the square? by adding the lengths of all the sides together

Direct attention to the first rectangle on the second row. Explain that the students will use their Centimeter Rulers to find the perimeter of this rectangle. Write the problem for display as the students work on their pages.

Direct them to place their Centimeter Rulers along the top or red side.

How should your Centimeter Ruler be lined up? with the left edge lined up with the left end of the red line

How long is the red side? 4 cm

Direct the students to write the number 4 in the red box.

Instruct the students to measure the blue side.

How long is the blue side? 5 cm

Direct the students to write the number 5 in the blue box.

Repeat the procedure with the green and purple sides.

Add the 4 measurements while the students watch. 4 + 5 + 4 + 5 = 18 cm Direct the students to complete the written problem.

What is the perimeter of the rectangle? 18 cm

How did you find the perimeter of the figure? by adding the lengths of all the sides together

Repeat the procedure with the 2 remaining figures on Worktext page 273. 6 + 4 + 6 + 4 = 20 cm, 7 + 3 + 7 + 3 = 20 cm

Apply

Worktext pages 273–74

Page 273 was completed during the lesson.

Read and explain the directions for page 274. Assist the students as they complete the page independently.

Assess

Reviews pages 261–62

Review writing the number of hundreds, tens, and ones on page 262.

Measure Perimeter

Use a centimeter ruler to find each perimeter. Mark an X on each side after you measure it.

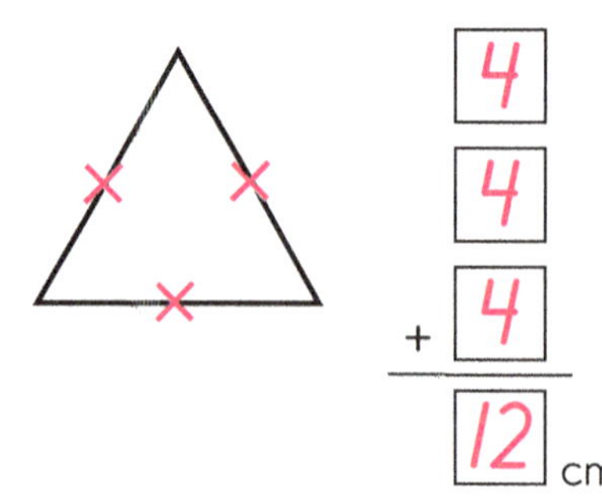

4 + 4 + 4 = 12 cm

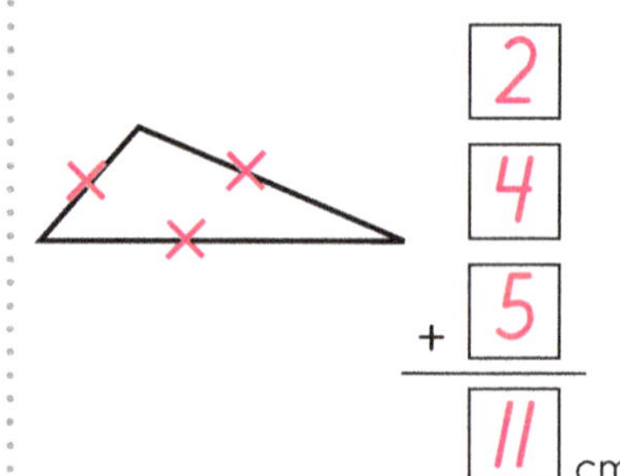

2 + 4 + 5 = 11 cm

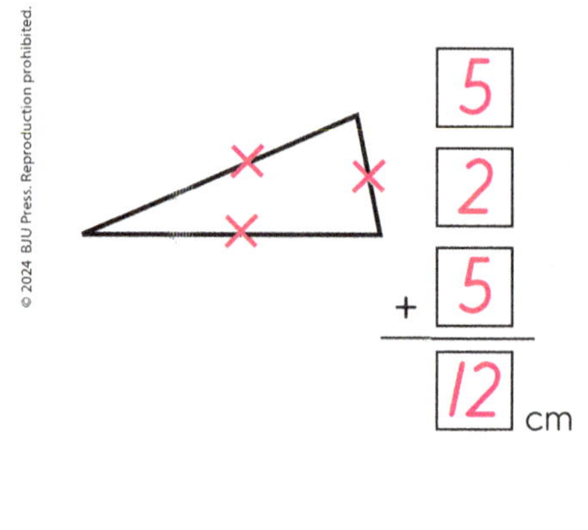

5 + 2 + 5 = 12 cm

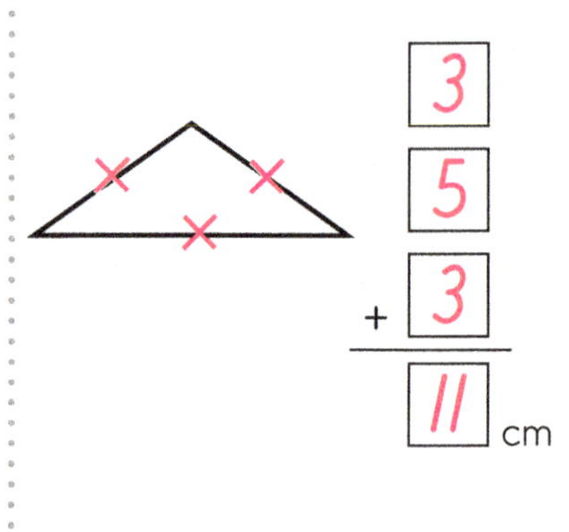

3 + 5 + 3 = 11 cm

Chapter 17 • Lesson 141 two hundred sixty-one **261**

Extended Activity

Estimate & Measure Height

Materials

- 15 UNIFIX® Cubes for each student
- Centimeter Ruler for each student

Procedure

Distribute 15 UNIFIX Cubes to the students; instruct each student to build a tower that he or she thinks would be 10 centimeters high. about 5 or 6 cubes

Direct the students to measure their towers and compare their measurements with the estimate. Repeat the activity, using the following heights for the towers.

15 centimeters high about 8 cubes

6 centimeters high about 3 cubes

24 centimeters high about 13 cubes

Chapter 16 Review

Write the hundreds, tens, and ones.
Write the number.

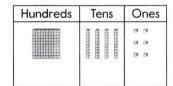

Hundreds	Tens	Ones
2	1	4

214

Hundreds	Tens	Ones
1	4	6

146

Hundreds	Tens	Ones
4	2	3

423

Hundreds	Tens	Ones
3	3	8

338

Subtraction Fact Review

Subtract.

$7 - 3 = 4$ $10 - 2 = 8$ $9 - 8 = 1$

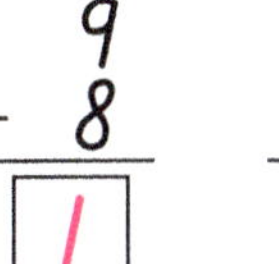

$8 - 8 = 0$ $7 - 0 = 7$

$8 - 5 = 3$

$10 - 9 = 1$ $8 - 6 = 2$

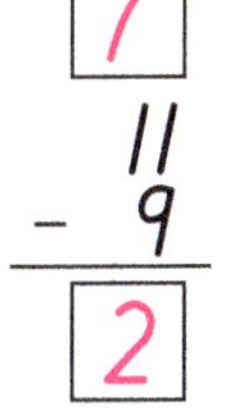

$9 - 5 = 4$ $11 - 9 = 2$

142

Worktext pages 275–76
Reviews pages 263–64

How can I use measurement to make a friend's day better?

Chapter Concept Review

- Practice concepts from Chapter 17 to prepare for the test.

Printed Resources

- Instructional Aid 42: *Metric Worksheet* (for each student)
- Instructional Aid 41: *Celsius Thermometer* (for each student)
- Student Manipulatives: Ruler (Centimeter)

Digital Resources

- Games/Enrichment: Subtraction Flashcards
- Games/Enrichment: Fact Fun Activities

Materials

- A cup
- A jug with a capacity of more than 1 liter
- A 1-liter container of colored water (from Lesson 138)
- A wastebasket

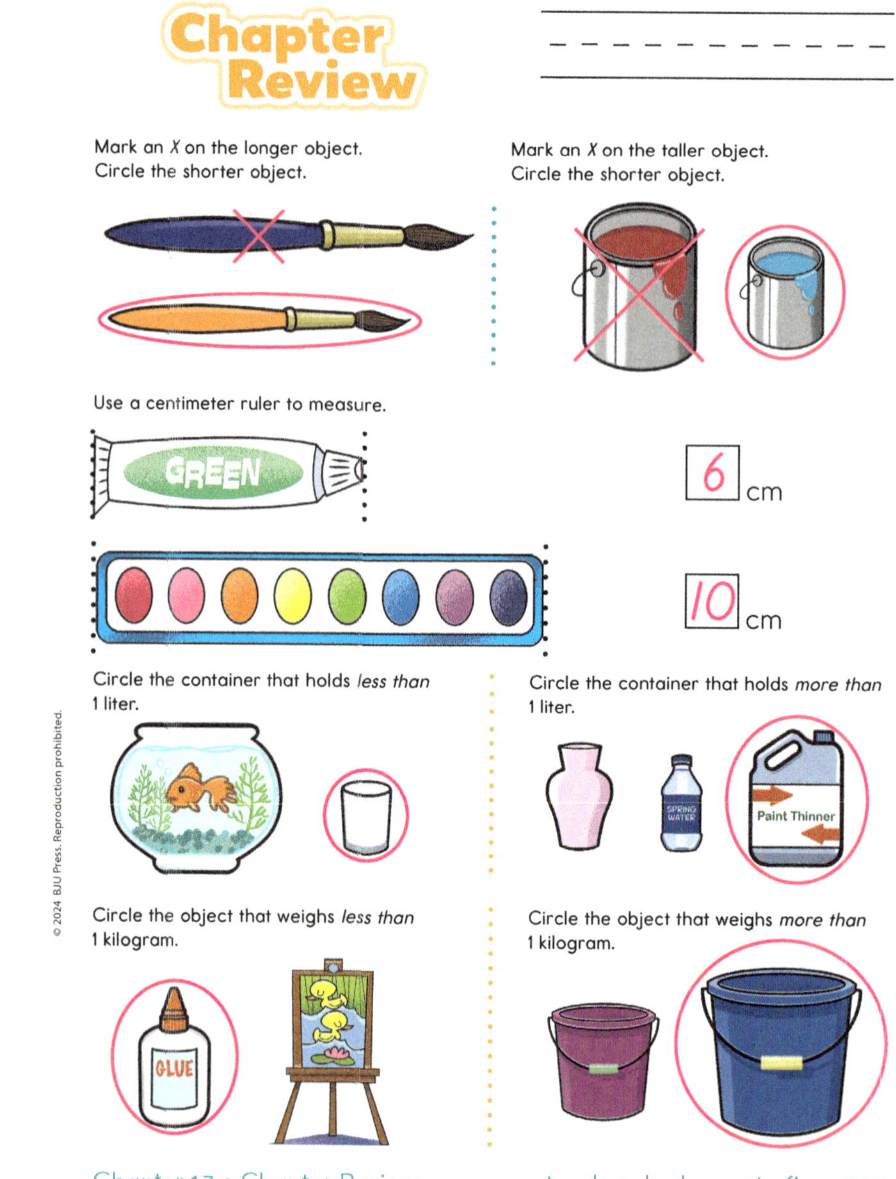

Chapter Review

Mark an *X* on the longer object.
Circle the shorter object.

Mark an *X* on the taller object.
Circle the shorter object.

Use a centimeter ruler to measure.

6 cm

10 cm

Circle the container that holds *less than* 1 liter.

Circle the container that holds *more than* 1 liter.

Circle the object that weighs *less than* 1 kilogram.

Circle the object that weighs *more than* 1 kilogram.

Practice & Review

Study Subtraction Facts to 12

Select a game or activity from the "Fact Fun Activities" in Trove. Practice the previously memorized subtraction facts.

Instruct

Identify the Correct Measuring Tool

Display the 1-liter container of water.

Which measurement tool do you need to find how much water my bottle holds? a 1-liter (or metric measurement) container

Which tool do you need to find how many kilograms the water bottle holds? a scale

Which tool do you need to find the height of my water bottle? a centimeter ruler

If you are going on a hike today, which tool will you need to tell you what to wear? a thermometer

Compare Lengths

Distribute a copy of the *Metric Worksheet* page and a Centimeter Ruler to each student.

What do you see on the bottom of the page? paintbrushes

Which paintbrush is longer? B

Direct the students to circle the answer.

Direct attention to the picture frames displayed on the page.

Direct the students to mark an *X* on the picture frame that is longer.

Which picture frame is longer? the one with the picture of the giraffe

Estimate & Measure the Length/Height of an Object

Direct attention to the sign at the top of the *Metric Worksheet* page.

What does the sign say? Measurement Gallery

Direct several students to estimate the length of the sign. Direct the students to measure the sign by using their Centimeter Rulers.

Where should the ruler be placed? The left edge of the ruler should be lined up with the left edge of the sign.

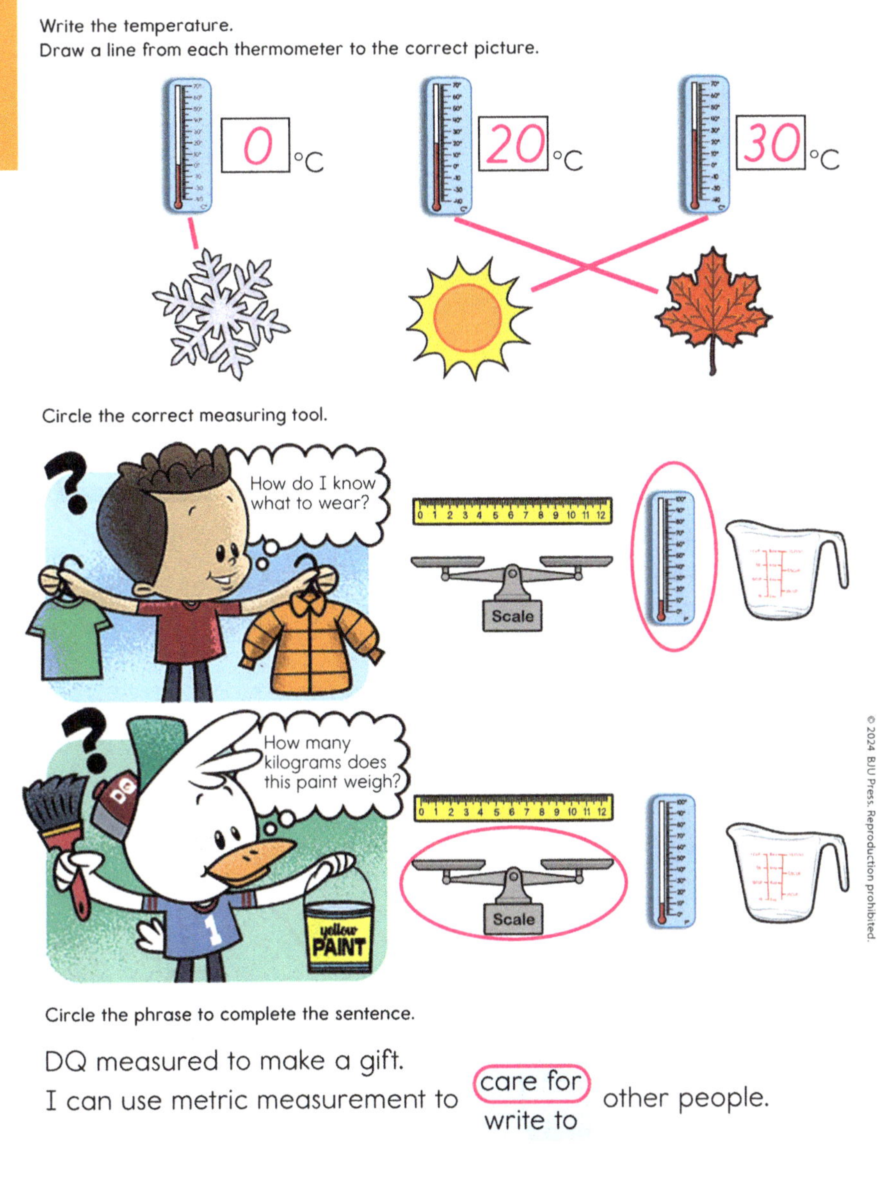

Write the temperature.
Draw a line from each thermometer to the correct picture.
0 °C
20 °C
30 °C
Circle the correct measuring tool.
How do I know what to wear?
Scale
How many kilograms does this paint weigh?
DQ
yellow PAINT
Scale
© 2024 BJU Press. Reproduction prohibited.
Circle the phrase to complete the sentence.
DQ measured to make a gift.
I can use metric measurement to care for / write to other people.
276 two hundred seventy-six
Math 1

How long is the sign? 11 cm

Direct the students to write the measurement on their copies of the page.

Repeat the procedure for the width of the picture frames 4 cm, 3 cm and for the height of the painting and easel. 7 cm

Compare a Container's Capacity to That of a Liter

Display the 1-liter container of water, a wastebasket, a cup, and the jug.

Which tool measures how much something holds? a liter container

Hold up the liter container of water and the wastebasket.

Does this wastebasket hold more than or less than 1 liter? more than

Continue the activity with the cup and the jug.

Direct attention to the number 1 shelf on the *Metric Worksheet* page.

Which container holds less than 1 liter? the cup

Direct the students to circle it.

Direct the students to look at the number 2 shelf on the page.

Which container holds more than 1 liter? the green paint can

Direct the students to mark an *X* on it.

Compare an Object's Mass to That of a Kilogram

Instruct each student to pretend that he or she is a balance scale.

How does your balance scale show that the object in your right hand is heavier than the object in your left hand? The right hand is down, and the left hand is up.

What does your balance scale look like if the object in your left hand is heavier than the object in your right hand? The left hand is down, and the right hand is up.

What will happen if the 2 objects have the same mass? Both arms will be straight out from the shoulders.

Display the 1-liter container of water.

What is the mass of a 1-liter container of water? 1 kilogram

Is a car more than or less than 1 kilogram? more than

Is a pencil more than or less than 1 kilogram? less than

Direct attention to the balance scale on the number 3 shelf on the *Metric Worksheet*.

Is the bowl on the scale more than or less than 1 kilogram? less than

Which object on the number 4 shelf is more than 1 kilogram? the paint thinner

Direct the students to circle it.

Read a Celsius Thermometer

Direct attention to the thermometer on the *Metric Worksheet* page.

What is the temperature? 0°C

Is 0° Celsius cold or hot? cold

What would happen to a snowman if the temperature got warmer? It would melt.

Distribute a copy of the *Celsius Thermometer* page to each student. Direct the students to show the temperature at which water freezes, using the red back of the Centimeter Ruler.

At which temperature does water freeze? 0° Celsius

Do you remember at which temperature water boils? 100° Celsius

Show the temperature on your thermometer.

Instruct the students to show 30° Celsius on their thermometers.

What activity could you do if the temperature was 30° Celsius? any summer activity

Instruct the students to show 10° Celsius.

Which season of the year might it be? autumn or fall

Apply

Worktext pages 275–76

Read the directions and guide completion of the pages.

Direct attention to the biblical worldview shaping section on page 276. Display the picture from the chapter opener and remind the students that DQ measured to make picture frames for his friends.

Remind the students that DQ also read a thermometer to help make sure his friends were dressed comfortably for a picnic.

How can I measure to make my friend's day better? I can measure to make a gift or to help my friend be comfortable.

Provide prompts as necessary to guide the students to the answer on the Worktext page.

Assess

Reviews pages 263–64

Use the Reviews pages to provide additional preparation for the chapter test.

Extended Activity

Measure Items from a Scavenger Hunt

Materials

- Centimeter Ruler for each student

Procedure

Take the students to the playground; pair the students. Instruct each pair to find the following items.

a leaf longer than 3 centimeters

a stone less than 1 kilogram

a stick shorter than 20 centimeters

Return to the room; direct the students to measure the objects to check the accuracy of their findings. Encourage the students to share their results.

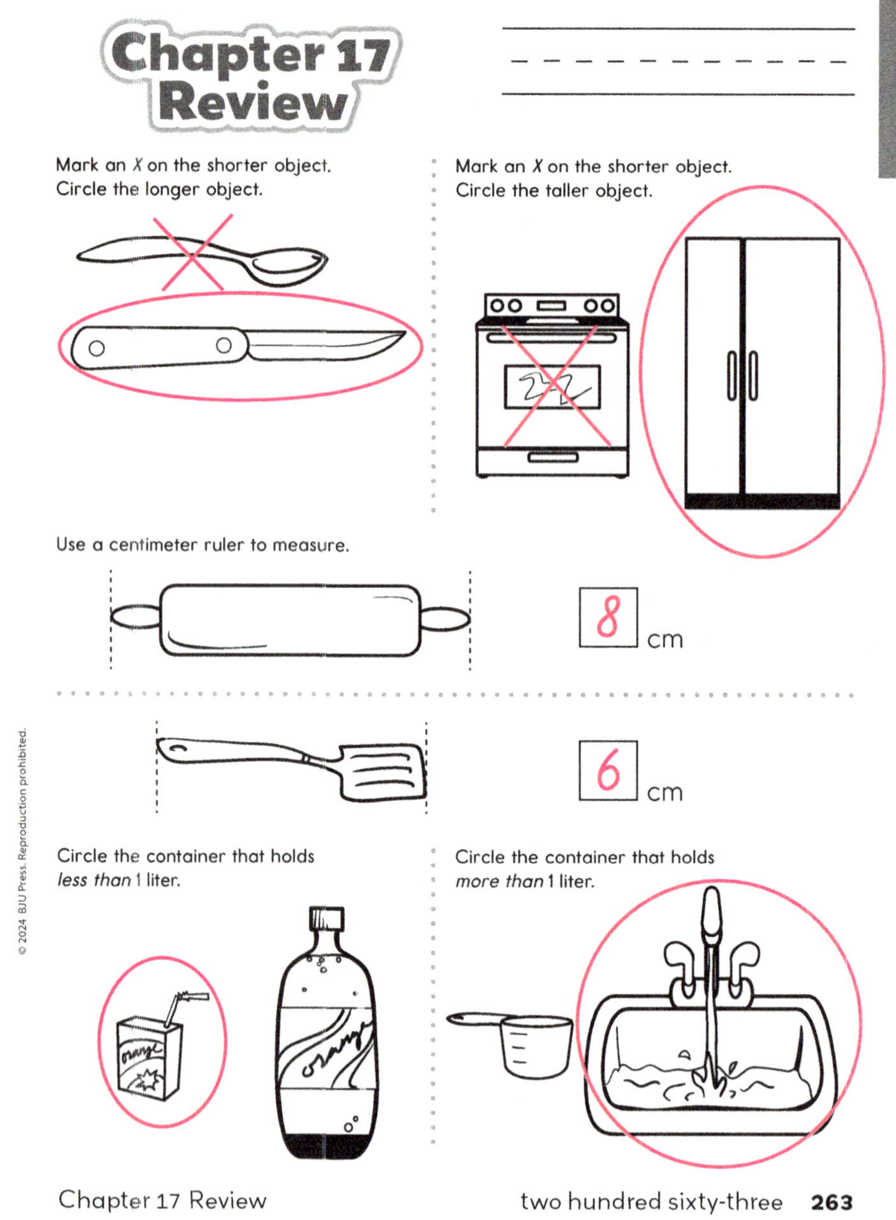

Chapter 17 Review
Mark an X on the shorter object.
Circle the longer object.
Mark an X on the shorter object.
Circle the taller object.
Use a centimeter ruler to measure.
8 cm
6 cm
Circle the container that holds less than 1 liter.
Circle the container that holds more than 1 liter.
© 2024 BJU Press. Reproduction prohibited.
Chapter 17 Review
two hundred sixty-three 263

142

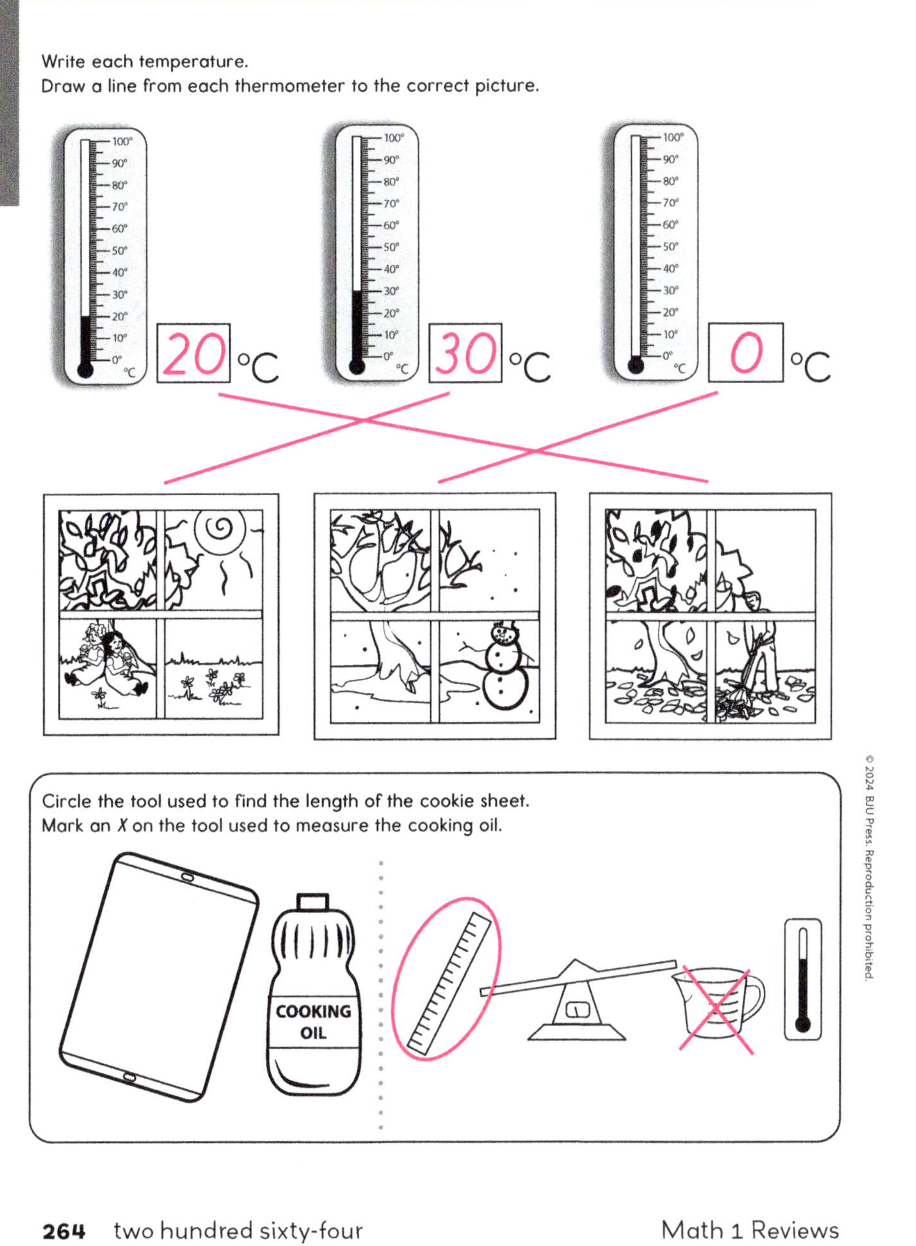
Write each temperature.
Draw a line from each thermometer to the correct picture.
20 °C
30 °C
0 °C
Circle the tool used to find the length of the cookie sheet.
Mark an X on the tool used to measure the cooking oil.
COOKING OIL
© 2024 BJU Press. Reproduction prohibited.
264 two hundred sixty-four
Math 1 Reviews

143 Worktext page 277 Reviews pages 265–66

Chapter 17 Test

Administer the Chapter 17 Test.

Cumulative Concept Review

Worktext page 277

Review the following concepts. Adapt instructions and activities and provide reteaching as needed to meet the specific needs of your students.

- Copying and extending patterns of shapes (Lesson 121)
- Matching part of a set to the correct fraction (Lesson 70)
- Predicting the probability of an activity (Lesson 72)

Retain a copy of Worktext page 278 for the discussion of the Chapter 18 essential question during Lessons 144 and 153.

Reviews pages 265–66

Use the Reviews pages to help students retain previously learned skills.

Extended Activity

Match Weather-Related Activities to Celsius Temperature

Materials

- Prepare word cards for display: *beach towel, mittens, ice skates, roller skates, sled, jump rope, winter hat, summer hat, bathing suit, snow skis, umbrella, snow boots*
- 2 gym bags: 1 labeled *summer bag,* 1 labeled *winter bag*

Procedure

Explain that DQ is going on a trip, but what he packs depends on the weather. Explain that the temperature is 35° Celsius.

Should DQ take his mittens or a beach towel? beach towel

Invite a student to place the correct word card in the correct bag. Continue the activity.

If the temperature was 0° Celsius, should he pack ice skates or roller skates? ice skates

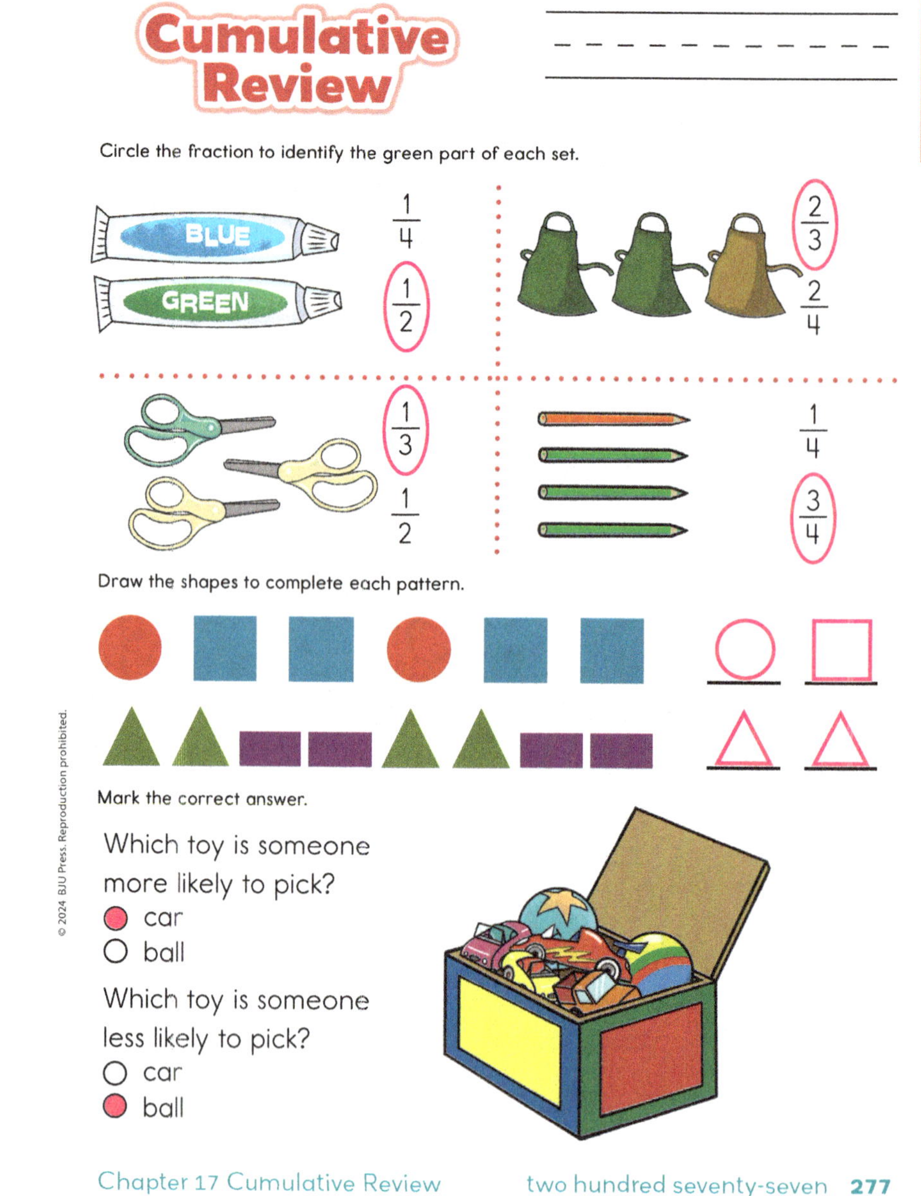

Cumulative Review

Circle the fraction to identify the green part of each set.

$\frac{1}{4}$ $\frac{1}{2}$ (circled)

$\frac{2}{3}$ (circled) $\frac{2}{4}$

$\frac{1}{3}$ (circled) $\frac{1}{2}$

$\frac{1}{4}$ $\frac{3}{4}$ (circled)

Draw the shapes to complete each pattern.

Mark the correct answer.

Which toy is someone more likely to pick?
- car (marked)
- ball

Which toy is someone less likely to pick?
- car
- ball (marked)

If the temperature was 20° Celsius, should he pack a sled or a jump rope? jump rope

If the temperature was 5° Celsius, should he pack a winter hat or a summer hat? winter hat

143

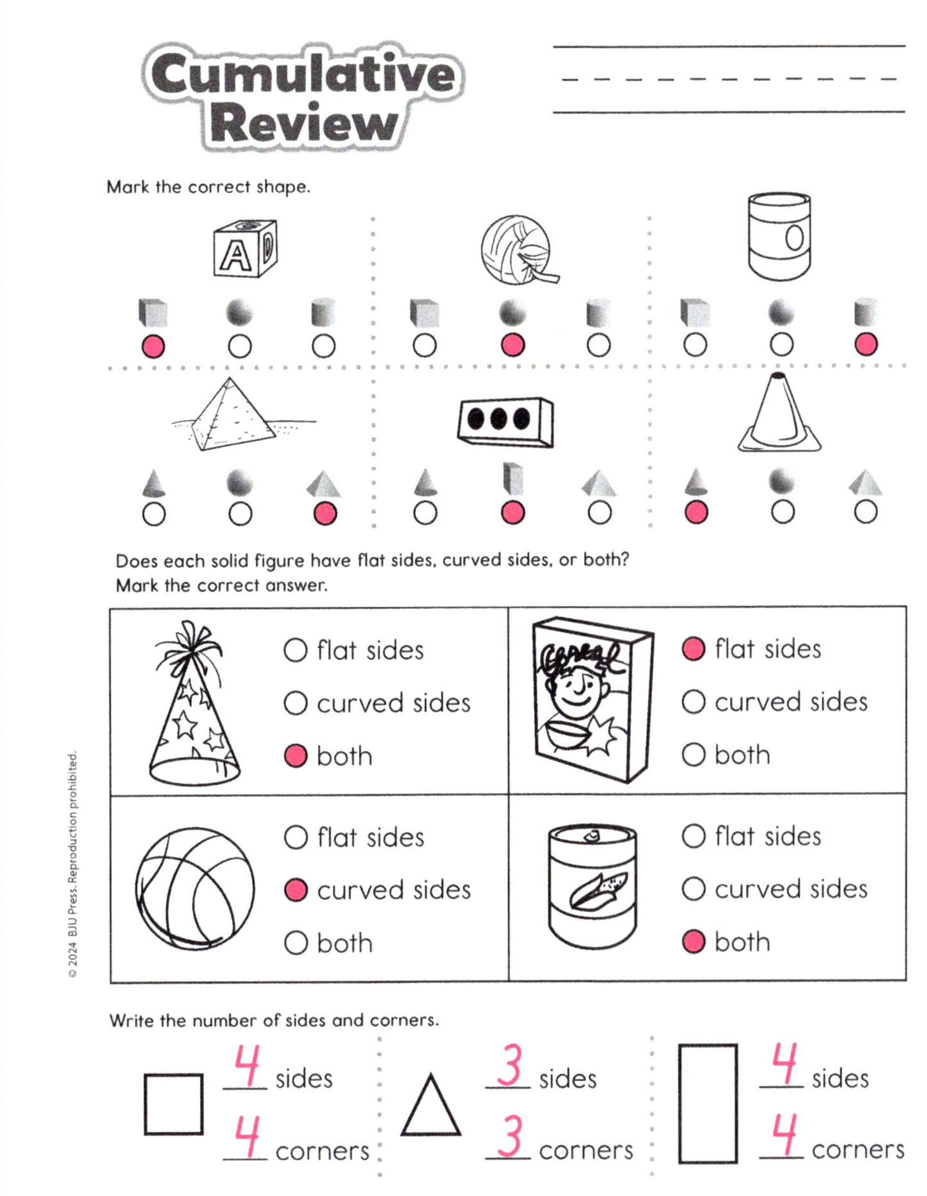
Cumulative Review
Mark the correct shape.
Does each solid figure have flat sides, curved sides, or both?
Mark the correct answer.
flat sides
curved sides
both
flat sides
curved sides
both
flat sides
curved sides
both
flat sides
curved sides
both
Write the number of sides and corners.
4 sides
4 corners
3 sides
3 corners
4 sides
4 corners
Chapter 17 Cumulative Review
two hundred sixty-five 265
© 2024 BJU Press. Reproduction prohibited.

Subtraction Fact Review

Subtract.

7 − 7 = 0	12 − 9 = 3	10 − 2 = 8	6 − 4 = 2	7 − 2 = 5
8 − 6 = 2	10 − 1 = 9	7 − 5 = 2	12 − 8 = 4	10 − 5 = 5
11 − 4 = 7	8 − 3 = 5	12 − 6 = 6	6 − 5 = 1	9 − 7 = 2
12 − 7 = 5	9 − 3 = 6	8 − 7 = 1	5 − 1 = 4	11 − 8 = 3
8 − 2 = 6	12 − 3 = 9	10 − 9 = 1	7 − 4 = 3	11 − 7 = 4
9 − 8 = 1	11 − 6 = 5	8 − 4 = 4	9 − 9 = 0	12 − 4 = 8

CHAPTER 18: ADDITION & SUBTRACTION FACTS TO 20

PAGES	OBJECTIVES	RESOURCES & MATERIALS	ASSESSMENTS
Lesson 144 Add Doubles			
Teacher Edition 617–21 **Worktext** 278–80	**144.1** Add double facts for addends 1–10. **144.3** Complete missing addend equations by using doubles. **144.3** Solve a word problem by adding doubles. **144.4** Explain how adding double facts helps people do work. BWS Working (explain)	**Visuals** • Place Value Kit: tens, ones **BJU Press Trove*** • Video: Ch 18 Intro • Web Link: Virtual Manipulatives: Base Ten Number Pieces • Games/Enrichment: Fact Reviews (Subtraction Facts to 12) • PowerPoint® presentation	**Reviews** • Pages 267–68
Lesson 145 Add Near Doubles			
Teacher Edition 622–25 **Worktext** 281–82	**145.1** Write an addition equation for a near double fact (double + 1). **145.2** Add near double facts. **145.3** Solve word problems by using near doubles.	**Visuals** • Counters: 20 ducks **Student Manipulatives Packet** • Ruler: Centimeter • Counters: 20 ducks **BJU Press Trove** • Games/Enrichment: Fact Reviews (Subtraction Facts to 12) • PowerPoint® presentation	**Reviews** • Pages 269–70
Lesson 146 Add 9 or 10			
Teacher Edition 626–29 **Worktext** 283–84	**146.1** Add 10 by using the Ten Bar. **146.2** Add 9 by making 10 (using the Ten Bar). **146.3** Solve addition word problems.	**Teacher Edition** • Instructional Aid 44: *Ten Bar* **Student Manipulatives Packet** • Ten Bar Mat **BJU Press Trove** • Web Link: Virtual Manipulatives: Number Frames • Games/Enrichment: Fact Reviews (Subtraction Facts to 12) • Games/Enrichment: Subtraction Flashcards • Games/Enrichment: Fact Fun Activities • PowerPoint® presentation	**Reviews** • Pages 271–72

*Digital resources for homeschool users are available on Homeschool Hub.

PAGES	OBJECTIVES	RESOURCES & MATERIALS	ASSESSMENTS
Lesson 147	**Add 6, 7, 8**		
Teacher Edition 630–33 **Worktext** 285–86	**147.1** Add 9 or 10 by using the Ten Bar. **147.2** Add 6, 7, or 8 by making 10 (using the Ten Bar). **147.3** Make 10 mentally while adding. **147.4** Solve a word problem by using the Problem-Solving Plan.	**Teacher Edition** • Instructional Aid 44: *Ten Bar* **Visuals** • Dot Pattern Cards: 0–10 **Student Manipulatives Packet** • Dot Pattern Cards: 1–9 • Ten Bar Mat **BJU Press Trove** • Web Link: Virtual Manipulatives: Number Frames • Games/Enrichment: Fact Reviews (Subtraction Facts to 12) • PowerPoint® presentation	**Reviews** • Pages 273–74
Lesson 148	**3 Addends**		
Teacher Edition 634–37 **Worktext** 287–88	**148.1** Solve a 3-addend equation for a word problem. **148.2** Solve 3-addend equations by using manipulatives. **148.3** Add 3 addends by using the Grouping Principle. **148.4** Solve 3-addend vertical problems. **148.5** Propose a reason to persevere in solving a 3-addend equation for a word problem. BWS Working (formulate)	**Visuals** • Counters: 15 ducks **BJU Press Trove** • Games/Enrichment: Fact Reviews (Subtraction Facts to 12) • PowerPoint® presentation	**Reviews** • Pages 275–76
Lesson 149	**Double Fact Families**		
Teacher Edition 638–41 **Worktext** 289–90	**149.1** Relate subtraction to addition. **149.2** Add double facts for 1–6. **149.3** Identify the fact families for double facts for 7–10. **149.4** Add double facts for 7–10. **149.5** Solve a word problem involving double facts.	**Visuals** • Shapes Kit: 20 circles **BJU Press Trove** • Web Link: Virtual Manipulatives: Number Frames • PowerPoint® presentation	**Reviews** • Pages 277–78
Lesson 150	**Fact Families for 13, 14; Missing Addend**		
Teacher Edition 642–45 **Worktext** 291–92	**150.1** Separate a set to demonstrate subtraction from 13 and 14. **150.2** Write addition and subtraction equations for fact families for 13 and 14. **150.3** Complete missing addend equations for word problems.	**Student Manipulatives Packet** • Ten Bar Mat **BJU Press Trove** • PowerPoint® presentation	**Reviews** • Pages 279–80

PAGES	OBJECTIVES	RESOURCES & MATERIALS	ASSESSMENTS
Lesson 151 Fact Families for 15, 16			
Teacher Edition 646–49 **Worktext** 293–94	**151.1** Separate a set to demonstrate subtraction from 15. **151.2** Write addition and subtraction equations for fact families for 15 and 16. **151.3** Solve a word problem by using the Problem-Solving Plan.	**Student Manipulatives Packet** • Counters: 4 ducks • Ten Bar Mat **BJU Press Trove** • PowerPoint® presentation	**Reviews** • Pages 281–82
Lesson 152 Fact Families for 17, 18; Multistep Problems			
Teacher Edition 650–53 **Worktext** 295–96	**152.1** Write addition and subtraction equations for fact families for 17 and 18. **152.2** Solve a multistep word problem.	**Visuals** • Counters: 8 ducks **Student Manipulatives Packet** • Number Cards: 0–9 • Red Mat • Counters: 8 ducks **BJU Press Trove** • Games/Enrichment: Addition Flashcards (facts to 12) • Games/Enrichment: Fact Fun Activities • PowerPoint® presentation	**Reviews** • Pages 283–84
Lesson 153 Chapter 18 Review			
Teacher Edition 654–57 **Worktext** 297–98	**153.1** Recall concepts and terms from Chapter 18.	**Teacher Edition** • Instructional Aid 44: *Ten Bar* **Student Manipulatives Packet** • Ten Bar Mat **BJU Press Trove** • Web Link: Virtual Manipulatives: Number Frames • Web Link: Virtual Manipulatives: UNIFIX® Cubes • Web Link: Virtual Manipulatives: Base Ten Number Pieces • PowerPoint® presentation	**Worktext** • Chapter 18 Review **Reviews** • Pages 285–86
Lesson 154 Test, Cumulative Review			
Teacher Edition 658–60 **Worktext** 299	**154.1** Demonstrate knowledge of concepts from Chapter 18 by taking the test.		**Assessments** • Chapter 18 Test **Worktext** • Cumulative Review **Reviews** • Pages 287–88

18 Addition & Subtraction Facts to 20

How can math help me take care of animals?

Help the class find the monkey house.

START

FINISH

ZOO

PEANUTS

Chapter 18

How can math help me take care of animals?

Chapter Objectives

- Add by using strategies including make 10, add doubles, and add near doubles.
- Solve 3-addend problems by using the Grouping Principle.
- Write addition and subtraction facts for fact families.
- Subtract by separating a set.
- Solve addition and subtraction word problems, including multistep problems.
- Formulate a reason for solving a word problem.

144

Worktext pages 278–80
Reviews pages 267–68

Objectives

- **144.1** Add double facts for addends 1–10.
- **144.2** Complete missing addend equations by using doubles.
- **144.3** Solve a word problem by adding doubles.
- **144.4** Explain how adding double facts helps people do work. BWS

Biblical Worldview Shaping

- **Working** (explain): Recognizing patterns in God's world often helps us work more efficiently. Since many real-world problems involve adding doubles, knowing double facts helps us complete our work (144.4).

Printed Resources

- Visuals: Place Value Kit (tens, ones)

Digital Resources

- Video: Ch 18 Intro
- Web Link: Virtual Manipulatives: Base Ten Number Pieces
- Games/Enrichment: Fact Reviews (Subtraction Facts to 12, page 54)

Materials

- 20 UNIFIX® Cubes
- A month page from an old calendar (for each student; varied pages)

Add Doubles

0 1 2 3 4 5 6 7 8 9 10 11 12 13 14 15 16 17 18 19 20

Add.

$8 + 8 = 16$ $\quad$ $10 + 10 = 20$

$9 + 9 = 18$ $\quad$ $7 + 7 = 14$

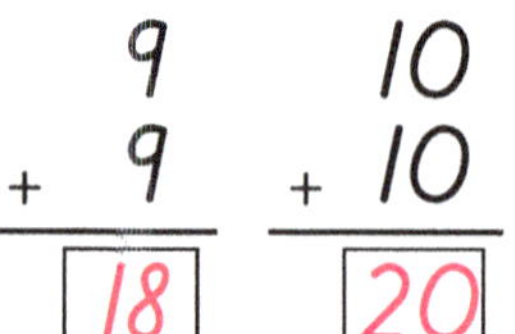
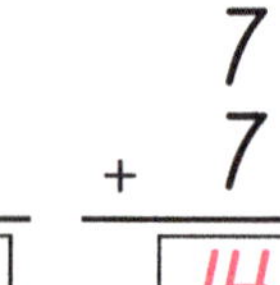
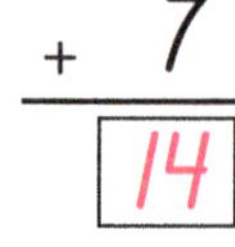

9	10	8	7
+ 9	+ 10	+ 8	+ 7
18	20	16	14

Complete each missing addend equation.

$9 + 9 = 18$

$6 + 6 = 12$

$10 + 10 = 20$

$8 + 8 = 16$

Although 10 + 10 = 20 is not considered a fact since it is not included in the 100 basic facts that the students memorize, it is included in the teaching of this lesson.

Practice & Review

Identify the Position of Days on a Calendar

Distribute the monthly calendar pages and UNIFIX Cubes. Explain that as you say various days and dates, the students should cover that day or date on their calendars with a cube. The first student to fill an entire week with cubes is the winner. Call out dates such as the following.

last day of the month
second Friday
14th
first Tuesday

Join Sets to Add 2-Digit Numbers

Write "47 + 12" vertically. Use tens and ones to demonstrate each step of the problem.

How do I show 47? 4 tens, 7 ones

How do I show 12? 1 ten, 2 ones

What do you add first? the ones

How many ones are there in all? 9

How many tens are there in all? 5

What is the sum of the addition problem? 59

Ask a student to complete the problem. 59

Continue the activity for the following problems.

36	68
+ 21	+ 20
57	88

Engage

Essential Question

Direct attention to the chapter opener on Worktext page 278 and **read aloud** the following scenario to introduce the chapter essential question.

"I want to see the giraffes!"

"Can we go to the tiger area first?"

"Will we get to touch the snakes?"

0 1 2 3 4 5 6 7 8 9 10 11 12 13 14 15 16 17 18 19 20

Add.

$7 + 7 = 14$

$8 + 8 = 16$

$10 + 10 = 20$

$6 + 6 = 12$

Complete each missing addend equation.

$8 + 8 = 16$

$7 + 7 = 14$

$9 + 9 = 18$

$10 + 10 = 20$

Write an equation for each word problem. Solve.

The snake has 8 red stripes and 8 black stripes. How many stripes does the snake have?

8 + 8 = 16 stripes

Cora saw 7 large tigers and 7 small tigers. How many tigers did she see?

7 + 7 = 14 tigers

Trace the word to complete the sentence.

I can solve a feeding problem by adding doubles.

Time to Review

Add.

34 + 43 = 77	46 + 32 = 78	73 + 13 = 86	65 + 13 = 78
14 + 52 = 66	68 + 31 = 99	54 + 45 = 99	50 + 26 = 76

DQ, Lily, and Liam gathered excitedly around their teacher at the entrance to the zoo. They could hardly wait to see their favorite animals!

Miss Markle laughed. She was happy to see her students excited about the zoo. "We'll visit as many of your favorite exhibits as we can." She scanned the group. "Hold up your water bottles so I can see that everyone has one." The sunny day promised to be perfect for the first-grade zoo field trip.

"Hi, little guy." Lily touched the glass on the huge aquarium as the sea lion swam close. The sea lion's big eyes and whiskers reminded Lily of a puppy. "He's so cute. I wish I could go swimming with him!"

"I wish I could ride that elephant." Liam pointed toward the broad fenced-in elephant habitat. He raised his eyebrows in surprise as he read the information on the sign. "I didn't know elephants were good swimmers. I'd like to see one of them do a cannonball!" The other children laughed at the thought of an elephant in goggles holding his trunk and sending a tidal wave of water out of the pool with his dive.

"Let's head to the monkey exhibit next." Miss Markle's eyes twinkled with excitement. "I have a surprise for you."

"Hello, class! My name is Jess, and my job is to feed the monkeys. Would you like to help?" The class cheered their approval. DQ quacked his agreement.

"I have the carrots, cucumbers, green beans, apples, eggs, and monkey chow ready for them. You get to add the peanuts, one of their favorite treats," Jess said. "Each bucket of food will feed 2 monkeys, and each monkey gets 7 peanuts. How many peanuts should you put in each bucket?"

DQ closed his eyes and pictured 7 peanuts in his left hand and 7 in his right. He was so happy that the class had been studying its double addition facts. He opened his eyes and raised his hand excitedly. "I know—it's 14! 7 + 7 = 14. We need 14 peanuts in each bucket."

"Excellent!" Jess said approvingly. "You can add 14 peanuts to this bucket." Together the students counted out the peanuts for each bucket and watched as the monkeys eagerly ate the treat the students had helped prepare for them.

Tired but happy, the students walked toward the exit at the end of the day. "I may not get to ride an elephant—or swim with one," Liam chuckled as they passed the elephants again. "But maybe I can become a zookeeper some day and feed one. Elephants eat a lot. I'd better get busy learning my math facts for that!"

Invite a student to read aloud the essential question. Encourage the students to consider this question as you guide them to answer it by the end of the chapter.

Instruct

Double Facts for Addends 1–6

Model double addition facts by using UNIFIX Cubes.

Hold 1 UNIFIX Cube in your right hand.

> **How many cubes are in this set?** 1
>
> **How many cubes are needed to make another set with the same number of cubes?** 1

Hold 1 more cube in your left hand. Ask a student to write an equation that shows the joining of the 2 sets. 1 + 1 = 2

> **What do you call facts that have the same addends?** double facts or doubles

Invite volunteers to write the double facts for 2, 3, 4, 5, and 6. 2 + 2 = 4, 3 + 3 = 6, 4 + 4 = 8, 5 + 5 = 10, 6 + 6 = 12

144

Double Facts for Addends 7–10

Use **UNIFIX Cubes** to help the students write double facts for addends 7–10.

Distribute the UNIFIX Cubes. Instruct the students to join 7 cubes.

How many cubes are needed to make another set with the same number of cubes? 7

Instruct the students to make another set of 7 cubes.

What equation shows the joining of these 2 sets? 7 + 7 = __

Join the 2 sets. How many cubes do you have in all? 14

Write "7 + 7 = 14" for display.

Is this equation a double? Yes; the addends are the same.

Continue the activity to write the double facts for 8, 9, and 10. 8 + 8 = 16, 9 + 9 = 18, 10 + 10 = 20

Missing Addend Equations That Use Doubles

Write "7 + __ = 14" for display. Explain that this is a missing addend equation. The sum is known, but 1 addend is missing.

How can you find out how many more cubes are needed? *count on* from 7, add cubes, use a number line, use fact families

Instruct the students to place 7 cubes on their desks to represent the first addend. Guide the students as they *count on* as they add cubes to make 14.

How many cubes did you add to make 14? 7

Complete the equation.

Write "8 + __ = 16" for display. Direct the students to place 8 cubes on their desks and to determine the missing addend. Invite a volunteer to complete the equation. 8

Continue the activity with 9 + __ = 18 9 and 10 + __ = 20 10.

What do you notice about these missing addend equations? They are doubles.

Problem-Solving Plan

Use the **Problem-Solving Plan** to help the students solve word problems.

Read aloud the following word problem.

Add Doubles

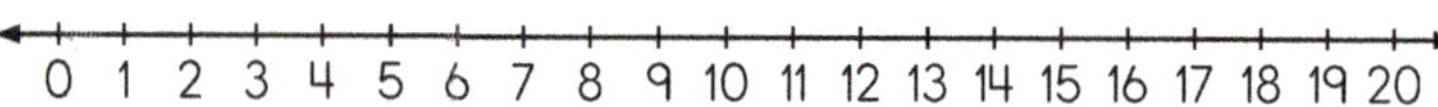

Add. Use the number line if needed.

8 + 8 = 16

9 + 9 = 18

7 + 7 = 14

$$\begin{array}{r} 9 \\ +\ 9 \\ \hline 18 \end{array} \qquad \begin{array}{r} 10 \\ +\ 10 \\ \hline 20 \end{array} \qquad \begin{array}{r} 8 \\ +\ 8 \\ \hline 16 \end{array} \qquad \begin{array}{r} 7 \\ +\ 7 \\ \hline 14 \end{array}$$

Complete each missing addend equation.

9 + 9 = 18

10 + 10 = 20

7 + 7 = 14

8 + 8 = 16

Sam fed 8 ducks at the zoo. Meg fed 8 more ducks. How many ducks did they feed in all?

What is the question? How many ducks did they feed in all?

What information is given? Sam fed 8 ducks, and Meg fed 8 ducks.

Do you add or subtract to find the answer? add

What is the equation that you can use to solve this word problem? 8 + 8 = __

Write "8 + 8 = __" for display.

Ask a student to solve the problem. 16 Write *ducks* after the answer.

Does the answer make sense? yes

Continue the activity with the following word problem.

Jen bought 6 red balloons and 6 blue balloons at the zoo. How many balloons did she buy at the zoo? 6 + 6 = 12 balloons

Feeding Problem

Guide a **discussion** to help the students explain a biblical worldview shaping truth.

Remind the students of the monkey feeding problem that DQ solved in the opener scenario.

What problem did DQ solve? how many peanuts to put in each food bucket for the monkeys

What math skill did DQ use to solve the problem? adding double facts

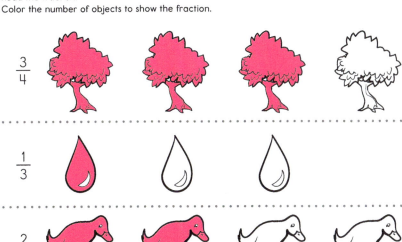

Chapter 9 Review

Read the fraction.
Color the number of objects to show the fraction.

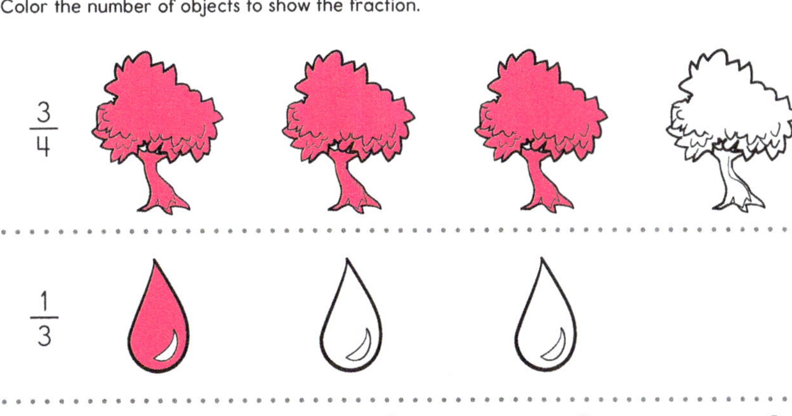

$\frac{3}{4}$

$\frac{1}{3}$

$\frac{2}{4}$

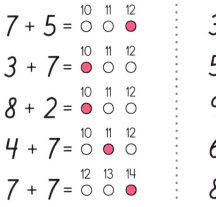

Addition Fact Review

Add. Mark the correct circle.

	10	11	12
7 + 5 =	○	○	●
3 + 7 =	●	○	○
8 + 2 =	●	○	○
4 + 7 =	○	●	○

	12	13	14
7 + 7 =	○	○	●

	10	11	12
3 + 8 =	○	●	○
5 + 6 =	○	●	○
9 + 3 =	○	○	●
6 + 6 =	○	○	●
8 + 4 =	○	○	●

What task were the students able to complete after DQ solved the problem? They were able to prepare the food buckets for the monkeys.

How did adding double facts help the students complete their work? Adding double facts helped them know how many peanuts to put in each bucket so they could prepare the monkeys' food properly.

Point out that God uses the math skills the students are learning to help them solve problems and do the work He has for them.

Apply

Worktext pages 279–80

Read and guide completion of page 279.

Read and explain the directions for page 280. Assist the students as they complete the page independently.

Direct attention to the biblical worldview shaping statement and guide the students as they complete the sentence.

Assess

Reviews pages 267–68

Review fractions on page 268.

Fact Reviews

Use "Subtraction Facts to 12" page 54 in Trove.

Extended Activity

Add Doubles with Squares

Materials

- Instructional Aid 43: *Square Grid* for each student

Procedure

Distribute the *Square Grid* pages. Write "9 + 9 = __" for display. Direct each student to color 9 squares across the top row with one crayon and then 9 more squares on the next row with a different crayon.

What kind of addition equation is this? a double

In the third row, instruct each student to write the equation and to solve it by adding together the colored squares. Allow the students to choose other double facts for addends 1–10 to illustrate in the same manner.

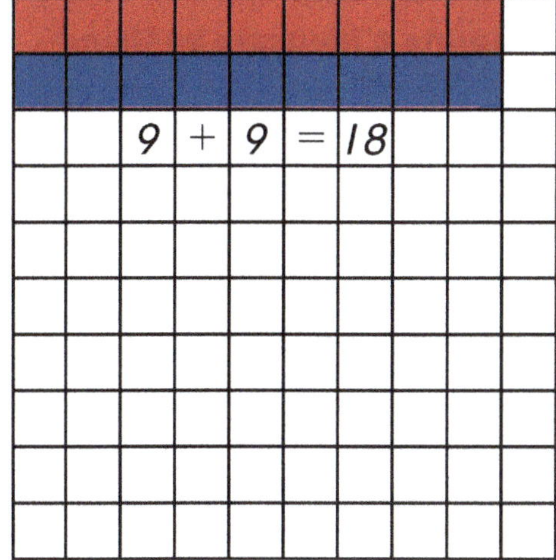

Differentiated Instruction

Recognize Facts for Fact Families

Write the following fact families on cards: 5-6-11 and 2-9-11. Write the following corresponding facts on cards: 6 + 5 = 11, 5 + 6 = 11, 11 – 5 = 6, 11 – 6 = 5, 2 + 9 = 11, 9 + 2 = 11, 11 – 2 = 9, and 11 – 9 = 2. Display the two fact family cards. Mix up the fact cards and give the students one card at a time to match to the correct fact family card. When they are finished sorting all the facts, direct them to read aloud all the facts for each family.

145

Worktext pages 281–82
Reviews pages 269–70

Objectives

- **145.1** Write an addition equation for a near double fact (double + 1).
- **145.2** Add near double facts.
- **145.3** Solve word problems by using near doubles.

Printed Resources

- Visuals: Counters (20 ducks)
- Student Manipulatives: Ruler (Centimeter)
- Student Manipulatives: Counters (20 ducks)

Digital Resources

- Games/Enrichment: Fact Reviews (Subtraction Facts to 12, page 55)

Practice & Review

Estimate & Measure by Using Centimeters

Distribute the Centimeter Rulers. Direct the students to put their Math Worktexts on their desks.

Any book that the student has available may be used as long as none of the measurements are more than 25 cm.

What is each space on the ruler called? a centimeter

Pair the students. Explain that one student is to estimate the length of the book and the other student is to measure the length of the book. Continue the activity for the width of the book.

Direct each pair to stack their 2 Worktexts. Explain that one student is to estimate the height of the stack and the other student is to measure the height of the stack of Worktexts.

Count 450–500

Guide the students as they take turns counting together with you by ones.

Explain that you will start by saying 450; then they will say the next number. Continue to take turns as you count to 500.

Add Near Doubles

Near Doubles Strategy
Think of the double fact for the smaller addend. Add 1 to the double fact sum.

Think 4 + 4 = 8
(8 + 1) 4 + 5 = 9

Add.

5 + 5 = 10
5 + 6 = 11

7 + 7 = 14
7 + 8 = 15

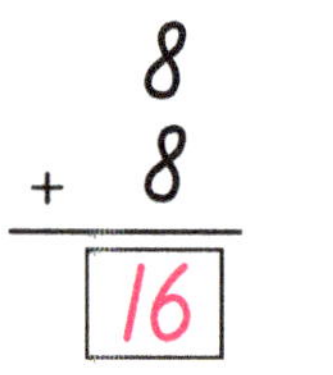

8 + 8 = 16

8 + 9 = 17

6 + 6 = 12

6 + 7 = 13

Write an equation for each word problem. Solve.

There are 8 monkeys climbing in the trees and 7 monkeys playing on the ground. How many monkeys are there in all?

8 + 7 = 15 monkeys

At the zoo 6 bears are rolling in the grass. The other 7 bears are asleep on the rocks. How many bears are at the zoo?

6 + 7 = 13 bears

Engage

Add Double Facts

Guide an **activity** to help the students review double addition facts.

Pair the students. In each pair, the first student will say the double fact equation and the second student will say the answer. Proceed around the room, one pair at a time, reviewing the double facts for addends 1–10. When all the facts have been reviewed, instruct the students to switch tasks and review the facts again.

Instruct

Equations for Near Double Facts (Double + 1)

Involve the students in a **demonstration** to help them write an equation for a near double fact.

Ask 2 students to stand on the right and 2 students to stand on the left in the front of the room.

What addition equation do these students show? 2 + 2 = __

What is the sum of 2 + 2? 4

Write "2 + 2 = 4" for display.

What do you notice about 2 + 2 = 4? It is a double fact.

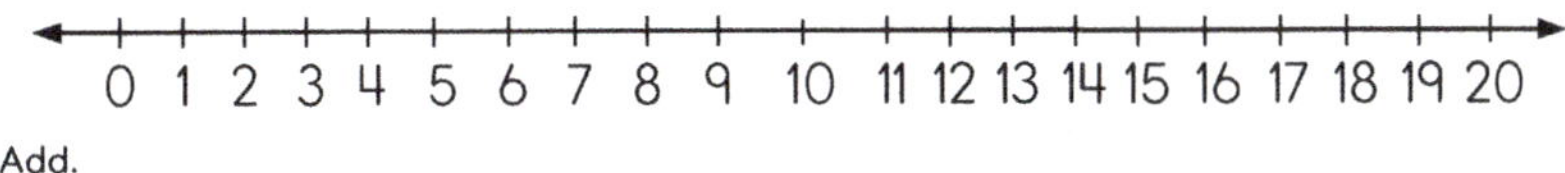

Add.

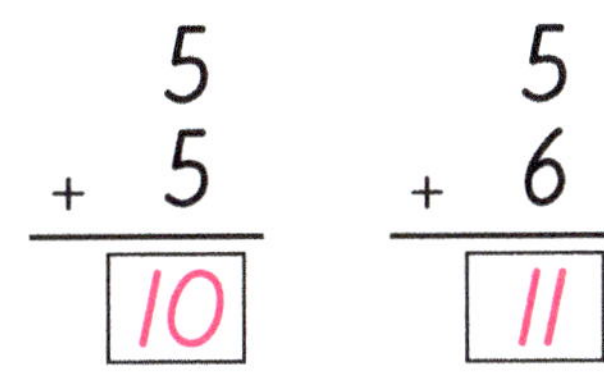

$$\begin{array}{r} 5 \\ +\ 5 \\ \hline 10 \end{array} \qquad \begin{array}{r} 5 \\ +\ 6 \\ \hline 11 \end{array} \qquad \begin{array}{r} 6 \\ +\ 6 \\ \hline 12 \end{array} \qquad \begin{array}{r} 6 \\ +\ 7 \\ \hline 13 \end{array}$$

8 + 8 = 16

8 + 9 = 17

7 + 7 = 14

7 + 8 = 15

Write an equation for each word problem. Solve.

Hayden counted 6 black roaches in the cage. He counted 7 brown ones. How many roaches did he count?

   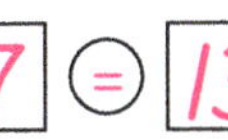

6 + 7 = 13 roaches

Kris saw 5 turtles sitting on a log. He also saw 6 turtles swimming. How many turtles did Kris see?

5 + 6 = 11 turtles

Guess the length. Use a centimeter ruler to measure.

	My Guess	My Measure
red panda	cm	8 cm
bird	cm	3 cm

Invite another student to stand with the students on the left.

What is the new addition equation? 2 + 3 = __

What is the sum of 2 + 3? 5

Write "2 + 3 = 5."

Direct attention to the equation 2 + 3 = 5. Explain that the addend 3 in the equation makes the sum 1 more than the double fact 2 + 2 = 4.

Continue the activity, using students to illustrate 3 + 3 = 6 and 3 + 4 = 7.

Add Near Double Facts

Use **manipulatives** to help the students add near double facts.

Display 4 duck counters in a row.

How many ducks are there? 4

Align 4 ducks in a second row directly under the first row of ducks.

How many ducks are in the second row? 4

Point out that the ducks show a double fact.

Which double fact do the ducks show? 4 + 4 = 8

Write "4 + 4 = 8" for display.

Place a duck at the end of the second row.

Which equation is represented now? 4 + 5 = __

Write "4 + 5 = __"; then draw a vertical line as shown as you explain that they know that 4 + 4 = 8.

Point out that for the equation 4 + 5 = __, they are adding 1 more to the double fact 4 + 4.

What is the sum of 4 + 5? 9

Ask a student to complete the near double fact equation.

Distribute the duck counters to the students. Demonstrate the following equations as the students show their work.

Write "5 + 5 = __" for display. Direct the students to show the double fact equation with their ducks. Remind them to make 2 rows showing the ducks directly under each other.

What is the answer to this double fact? 10

Place 1 more duck in the second row.

How many more have you added to the double fact 5 + 5? 1 more

What equation do the ducks show now? 5 + 6 = 11

Instruct the students to use their ducks to show the double fact for 6. 2 rows of 6 ducks

Write "6 + 6 = __."

What is the answer to this double fact? 12

Instruct the students to place 1 more duck in the top row this time.

What equation is shown now? 7 + 6 = __

What is 1 more than the double fact of 6 + 6? 13

Continue the activity for the following equations.

8 + 8 = 16 and 8 + 9 = 17

9 + 9 = 18 and 9 + 10 = 19

Say the following facts for oral practice without using the ducks. Guide the students as they mentally think of the double fact as they answer each near double fact.

$$\begin{array}{r} 5 \\ +\ 6 \\ \hline 11 \end{array} \qquad \begin{array}{r} 7 \\ +\ 8 \\ \hline 15 \end{array} \qquad \begin{array}{r} 6 \\ +\ 7 \\ \hline 13 \end{array} \qquad \begin{array}{r} 8 \\ +\ 9 \\ \hline 17 \end{array}$$

145

Solve Word Problems by Using Near Doubles

Follow the **Problem-Solving Plan** to help the students solve word problems by using near doubles.

The zoo has 6 spider monkeys and 7 howler monkeys. How many monkeys does the zoo have in all?

What is the question? How many monkeys does the zoo have in all?

What information is given? The zoo has 6 spider monkeys and 7 howler monkeys.

Do you add or subtract to find the answer? add

What is the equation that you can use to solve this word problem? 6 + 7 = __

Write "6 + 7 = __" for display.

Which double fact can help you solve this problem? 6 + 6 = 12

Ask a student to solve the equation. 13 Write *monkeys* after the answer.

Does the answer make sense? yes

Continue the activity with the following word problem.

The zookeeper scattered 8 bales of hay in the giraffe pen and 9 bales of hay in the bison pen. How many bales of hay did the zookeeper use in all? 8 + 9 = 17 bales

Apply

Worktext pages 281–82

Review the steps to mentally work a near doubles equation, using the top portion of page 281. Read and guide completion of the page.

Read and explain the directions for page 282. Assist the students as they complete the page independently.

Assess

Reviews pages 269–70

Review measurement on page 270.

Fact Reviews

Use "Subtraction Facts to 12" page 55 in Trove.

Add Near Doubles

Think 3 + 3 = 6.

6 + 1 → 3 + 4 = 7

Near Doubles Strategy
Think of the double fact for the smaller addend. Add 1 to the double fact sum.

Add.

8 + 8 = 16   8 + 9 = 17   9 + 9 = 18   9 + 10 = 19

7 + 7 = 14   7 + 8 = 15   5 + 5 = 10   5 + 6 = 11

Write an equation for each word problem. Draw pictures to show each equation. Solve.

Mason counted 8 black dogs at the pet store. He counted 9 white dogs. How many dogs did he count in all?

8 + 9 = 17 dogs

There are 5 dogs sleeping in the pen. There are 6 dogs eating. How many dogs are there in all?

5 + 6 = 11 dogs

Extended Activity

Add Near Doubles

Procedure

Play "Animal Charades." Ask 2 groups of students to represent an animal (for example, a group of 5 lions and a group of 6 lions). Instruct the other students to write an equation to show how many animals there are in all. Invite students to tell which double fact helps to solve the near double fact. Continue the activity and allow each student to act out an animal. Use the following equations.

1 + 2   2 + 3   3 + 4   4 + 5
5 + 6   6 + 7   7 + 8   8 + 9

Chapter 11 Review

Use an inch ruler to measure.

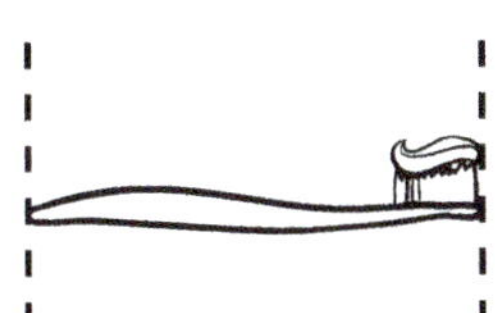

2 inches

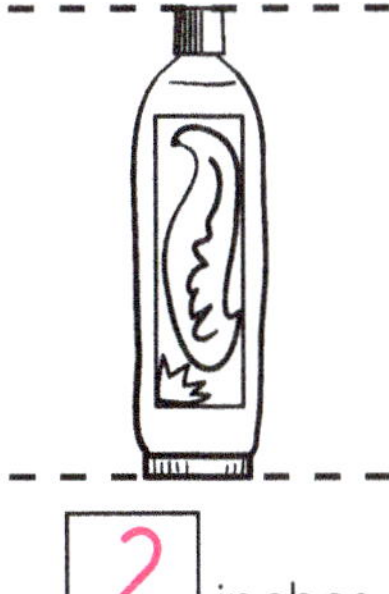

2 inches

Circle the container that holds *more*.

Circle the correct measuring tool.

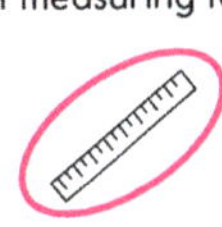

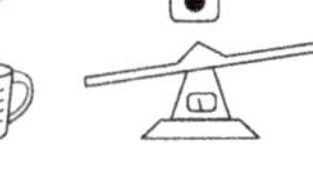

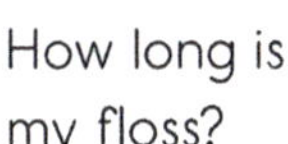

How long is my floss?

How hot is my cocoa?

Subtraction Fact Review

Subtract. Mark the correct circle.

Problem	Choices	Marked
$9 - 4 =$	3, 4, 5	5
$11 - 6 =$	3, 4, 5	5
$3 - 1 =$	1, 2, 3	2
$11 - 8 =$	3, 4, 5	3
$6 - 3 =$	3, 4, 5	3
$12 - 6 =$	6, 7, 8	6

146 Worktext pages 283–84 Reviews pages 271–72

Objectives

- **146.1** Add 10 by using the Ten Bar.
- **146.2** Add 9 by making 10 (using the Ten Bar).
- **146.3** Solve addition word problems.

Printed Resources

- Instructional Aid 44: *Ten Bar*
- Student Manipulatives: Ten Bar Mat

Digital Resources

- Web Link: Virtual Manipulatives: Number Frames
- Games/Enrichment: Fact Reviews (Subtraction Facts to 12, page 56)
- Games/Enrichment: Subtraction Flashcards (facts to 12)
- Games/Enrichment: Fact Fun Activities

Materials

- 19 UNIFIX® Cubes (for the teacher and for each student)
- A sheet of paper with "10 + __ = __" written at the bottom horizontally

Although 10 is not part of the 100 basic facts that the students memorize, adding 10 is used in this lesson to help in adding 9.

Add 9 or 10

Color the first addend. Complete the Ten Bar with the second addend. Add.

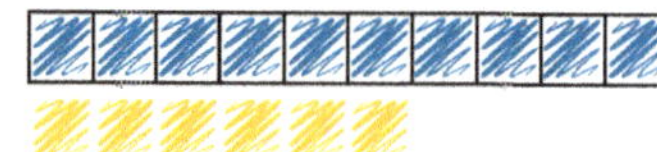

10 + 6 = 16

10 + 8 = 18

10 + 4 = 14

10 + 2 = 12

Color the first addend. Complete the Ten Bar with the second addend. Add.

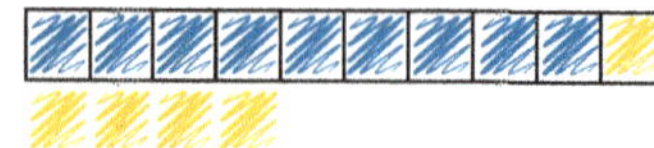

9 + 5 = 14

9 + 4 = 13

9 + 7 = 16

9 + 6 = 15

Write an equation for the word problem. Solve.

Jeff fed 9 crackers to the goats. He fed 3 crackers to the sheep. How many crackers did Jeff feed the animals?

9 + 3 = 12 crackers

Practice & Review

Count 500–600

Instruct the students to stand in a circle. Say a number between 500 and 600. Then, beginning with the student on your right, ask the students to say the next number. Go around the circle until you reach 600.

Study Subtraction Facts to 12

Choose a game or activity from the "Fact Fun Activities" available in Trove.

Engage

10 Fingers

Guide the preparation of a **visual representation** of 10 fingers to prepare the students to add 10.

Distribute the "10 + __ = __" sheets of paper. Instruct the students to trace both of their hands on the paper. Write the equation "10 + __ = __" for display and trace your hands above it.

What are some things that you do with your hands? write, pick up things, eat, dress myself

Point out that many things we do with our hands involve using our fingers together as a group. Explain that making groups of 10 can make it easier to add large numbers.

Continue to use the drawings of the students' hands in the activity that follows.

Instruct

Add 10 by Using Fingers

Guide a **drawing activity** to help the students add 10.

Direct the students' attention to the drawings of their hands.

How many fingers are in your picture? 10

What is something you might find on someone's fingers? rings

Number the students from 1–9, repeating the numbering until each student has been assigned a number 1–9.

Color the first addend. Complete the Ten Bar with the second addend. Add.

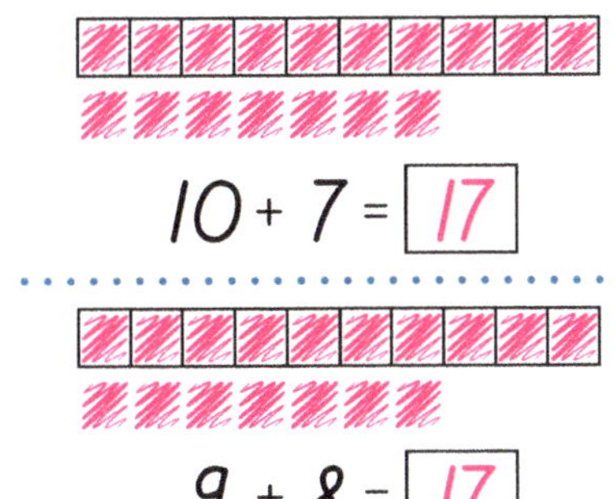

$10 + 7 = 17$

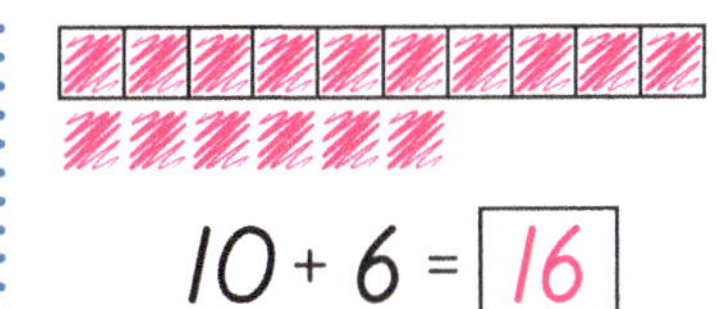

$10 + 6 = 16$

$9 + 8 = 17$

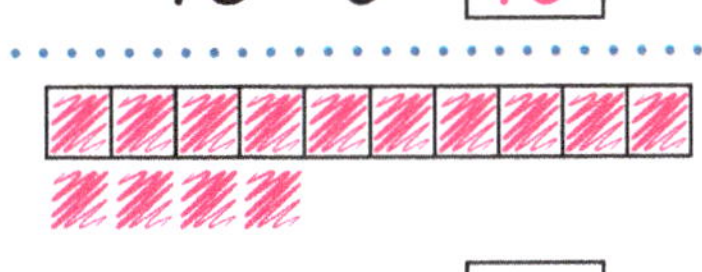

$9 + 5 = 14$

Add.

9	9	10	10	10
+ 3	+ 2	+ 5	+ 3	+ 8
12	11	15	13	18

Time to Review

Subtract.

$10 - 8 = 2$

$12 - 8 = 4$

$11 - 6 = 5$

$8 - 5 = 3$

$10 - 6 = 4$

$9 - 5 = 4$

$11 - 3 = 8$

Instruct the students to draw a ring on the assigned number of fingers in their pictures. Draw a ring on 5 fingers in your picture.

Direct attention to the displayed equation 10 + __ = __. Explain that the first addend represents their 10 fingers, and the second addend will represent the number of rings on the fingers in their drawings. Instruct the students to write in the first blank the number of rings that they drew. Write "5" in your first blank.

Explain that it is not necessary to count the fingers again because they know that there are 10. Explain that they can *count on* from 10 to find the answer. Use your picture to demonstrate adding the 5 rings by *counting on* from 10: 10, 11, 12, 13, 14, 15.

What is the answer to my equation 10 + 5? 15

Ask the students to solve their equations by *counting on* from 10 and writing the answer in the blank after the equals sign. Direct the students who drew 1 ring to hold up their hands.

What is your equation? 10 + 1 = 11

Continue the activity with the following equations.

10	10	10	10
+ 2	+ 3	+ 4	+ 5
12	13	14	15
10	10	10	10
+ 6	+ 7	+ 8	+ 9
16	17	18	19

Add 10 by Using the Ten Bar

Use the **Ten Bar** to help the students add 10.

Distribute the Ten Bar Mats and UNIFIX Cubes. Guide the students as they count the number of spaces on the Ten Bar. 10 Write "10 + 7 = __" for display.

What is the first addend in this equation? 10

Direct the students to place 10 cubes on their Ten Bars.

What is the second addend in the equation? 7

Direct the students to place 7 cubes below their Ten Bars.

Do you need to count the cubes on the Ten Bar to find the sum? No; I know that there are 10.

Explain that they can *count on* from 10 to find the sum. Guide the students as they *count on* from 10: 10, 11, 12, 13, 14, 15, 16, 17.

What is the sum of 10 + 7? 17

Write "10 + 7" in vertical form. Point out the placement of the 7 in the Ones place. Explain that the number 7 has no tens, so it is important that it is placed in the Ones place. Complete the equation.

Continue the activity with 10 + 3 = 13 and 10 + 9 = 19.

Add 9 by Making 10

Use the **Ten Bar** to help the students add 9.

Explain to the students that knowing how to add 10 with a Ten Bar will help them to add 9. Write "9 + 3 = __" for display. Demonstrate each step on the Ten Bar for display.

What is the first addend? 9

Instruct the students to place 9 cubes on the Ten Bar.

What is the second addend? 3

Instruct the students to place 3 cubes below the Ten Bar.

How many more cubes do you need in the Ten Bar to make a group of 10? 1

Guide the students as they move 1 of the 3 cubes to the Ten Bar.

Do you need to count the cubes on

the bar to solve the equation? No; I can *count on* from 10.

Guide the students as they *count on* from 10: 10, 11, 12.

What is the answer to this problem? 12

Complete the equation.

Continue the activity with 9 + 5 = 14 and 9 + 7 = 16.

Say the following facts for oral practice without the Ten Bar. Guide the students as they mentally add 10 before they answer each fact.

$$\begin{array}{r}9\\+2\\\hline 11\end{array}\quad\begin{array}{r}9\\+4\\\hline 13\end{array}\quad\begin{array}{r}9\\+6\\\hline 15\end{array}\quad\begin{array}{r}9\\+9\\\hline 18\end{array}$$

Word Problems

Guide the students as they use the **Problem-Solving Plan** to solve word problems.

Emmanuel counted 10 canaries and 6 finches in the zoo display. How many birds did Emmanuel count?

What is the question? How many birds did Emmanuel count?

What information is given? He counted 10 canaries and 6 finches.

Do you add or subtract to find the answer? add

What is the equation that you can use to solve this word problem? 10 + 6 = __

Write "10 + 6 = __" for display. Guide the students as they use the Ten Bar Mats and UNIFIX Cubes to find the sum.

Invite a volunteer to complete the equation. 16 Write *birds* after the answer.

Does the answer make sense? yes

Continue the activity with the following word problem.

There are 9 adults and 8 children watching the penguins being fed. How many people in all are watching the feeding of the penguins? 9 + 8 = 17 people

Apply

Worktext pages 283–84

Explain that instead of placing UNIFIX Cubes on the Ten Bar, the students will color the addends. Read and guide completion of page 283.

Add 9 or 10

Color each addend. Add.
Write the answer.

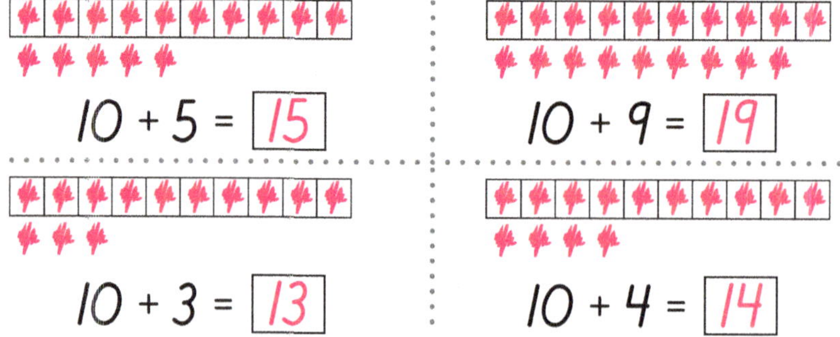

Color each addend. Complete the Ten Bar with the second addend.
Add.

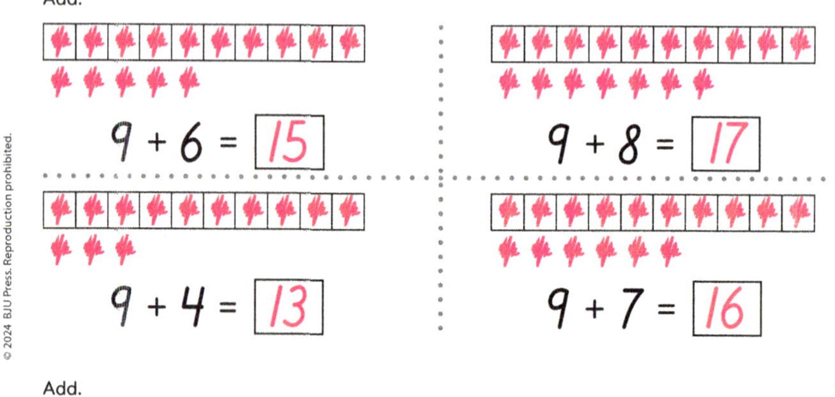

Add.

$$\begin{array}{r}10\\+7\\\hline 17\end{array}\quad\begin{array}{r}9\\+2\\\hline 11\end{array}\quad\begin{array}{r}10\\+8\\\hline 18\end{array}\quad\begin{array}{r}10\\+2\\\hline 12\end{array}\quad\begin{array}{r}10\\+6\\\hline 16\end{array}\quad\begin{array}{r}9\\+3\\\hline 12\end{array}$$

Chapter 18 • Lesson 146 two hundred seventy-one **271**

Read and explain the directions for page 284. Assist the students as they complete the page independently.

Assess

Reviews pages 271–72

Review writing the number of hundreds, tens, and ones on page 272.

Fact Reviews

Use "Subtraction Facts to 12" page 56 in Trove.

Chapter 16 Review

Write the number of hundreds, tens, and ones.

278 2 hundreds 7 tens 8 ones

461 4 hundreds 6 tens 1 ones

589 5 hundreds 8 tens 9 ones

Match the clues to the correct number.

Clues	Number
7 ones, 4 tens, 3 hundreds	864
6 tens, 8 hundreds, 4 ones	728
7 hundreds, 8 ones, 2 tens	347

(Matches: 7 ones, 4 tens, 3 hundreds — 347; 6 tens, 8 hundreds, 4 ones — 864; 7 hundreds, 8 ones, 2 tens — 728)

Addition Fact Review

Add.

$5 + 7 = 12$ $3 + 8 = 11$ $6 + 5 = 11$ $6 + 4 = 10$ $3 + 2 = 5$

$2 + 7 = 9$ $9 + 1 = 10$ $5 + 4 = 9$ $3 + 9 = 12$ $8 + 4 = 12$

147

Worktext pages 285–86
Reviews pages 273–74

Objectives

- **147.1** Add 9 or 10 by using the Ten Bar.
- **147.2** Add 6, 7, or 8 by making 10 (using the Ten Bar).
- **147.3** Make 10 mentally while adding.
- **147.4** Solve a word problem by using the Problem-Solving Plan.

Printed Resources

- Instructional Aid 44: *Ten Bar*
- Visuals: Dot Pattern Cards (0–10)
- Student Manipulatives: Dot Pattern Cards (1–9)
- Student Manipulatives: Ten Bar Mat

Digital Resources

- Web Link: Virtual Manipulatives: Number Frames
- Games/Enrichment: Fact Reviews (Subtraction Facts to 12, page 57)

Materials

- 16 UNIFIX® Cubes (for the teacher and for each student)

Add 6, 7, 8

Color the first addend.
Complete the Ten Bar with the second addend.
Add.

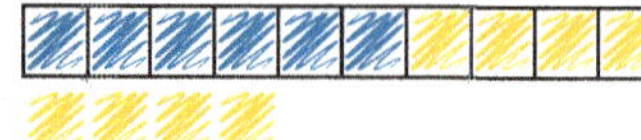

6 + 8 = 14

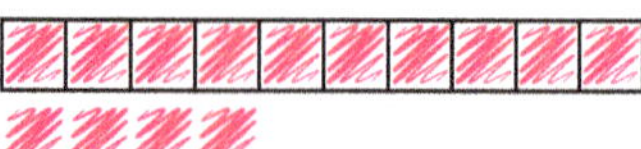

8 + 6 = 14

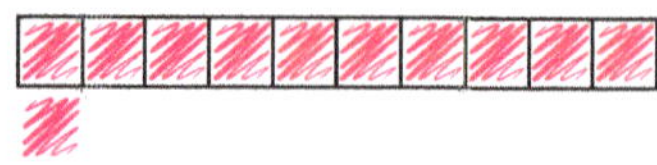

7 + 4 = 11

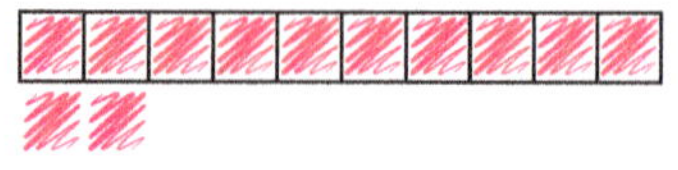

8 + 4 = 12

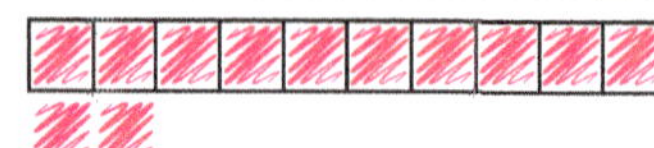

7 + 5 = 12

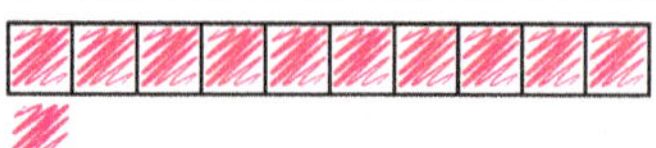

8 + 3 = 11

Write an equation for the word problem. Solve.

Jada saw 8 rams on the mountain.
She saw 5 rams eating grass.
How many rams did Jada see?

 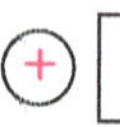   

8 + 5 = 13 rams

Practice & Review

Solve 3-Addend Problems

Write "2 + 8 + 2 = __" for display. Invite a volunteer to read the equation aloud.

How many addends are in this equation? 3

Ask a student to find the sum. 12

Continue the activity with the following problems.

7	4
3	4
+ 7	+ 3
17	11

Engage

Identify Dot Patterns 0–10

Guide a **review activity** to help the students identify the dot patterns for numbers 0–10.

Use Dot Pattern Cards as flashcards. Display the cards in random order and ask a student to identify each pattern.

Instruct

Add 9 or 10 by Using the Ten Bar

Use the **Ten Bar** to help students add 9 or 10.

Distribute the Ten Bar Mats and the UNIFIX Cubes. Write the equation "10 + 3 = __" for display. Demonstrate each step on the Ten Bar for display.

What is the first step in finding the sum of 10 + 3? place the first addend (10 cubes) on the Ten Bar

Instruct the students to place 10 cubes on the bar.

What is the second step? place the second addend below the bar (3 cubes)

Are all the spaces on the Ten Bar filled? yes

Do you need to count each of the cubes in the Ten Bar? No; I can *count on* from 10.

Guide the students as they *count on* from 10: 10, 11, 12, 13.

What is the sum of 10 + 3? 13

Complete the equation.

Write "9 + 7 = __" for display. Demonstrate each step.

What do you do first? place 9 cubes on the Ten Bar

What do you do second? place 7 cubes below the bar

Are the spaces on the Ten Bar filled? no

Color the first addend. Complete the Ten Bar with the second addend. Add.

8 + 4 = 12

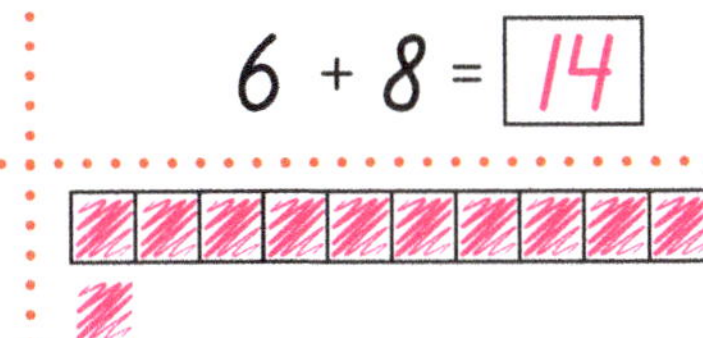

6 + 8 = 14

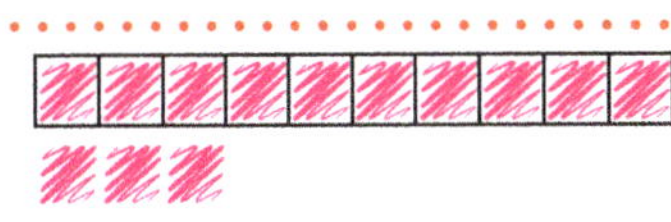

8 + 5 = 13

7 + 4 = 11

Write an equation for the word problem. Solve.

There are 8 fish swimming at the top of the tank.
There are 6 fish swimming near the sand.
How many fish are in the tank?

 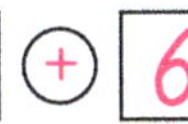   

8 + 6 = 14 fish

Time to Review

Add.

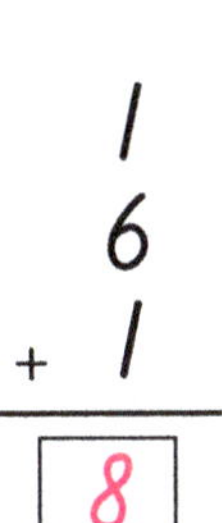 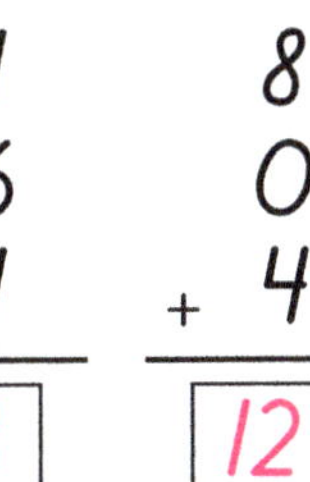 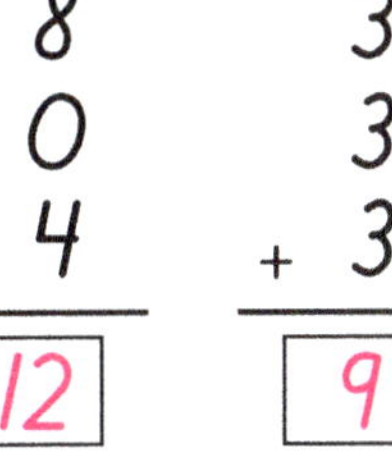 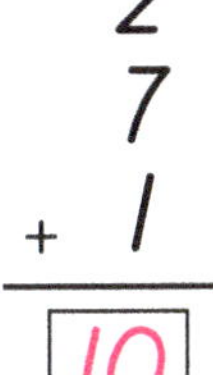

4	1	8	3	2
3	6	0	3	7
+ 2	+ 1	+ 4	+ 3	+ 1
9	8	12	9	10

What should you do? move 1 cube up to fill the Ten Bar

Direct the students to move up 1 cube.

How do you count the cubes? Because the Ten Bar makes a group of 10, I begin *counting on* from 10.

Guide the students as they *count on* from 10: 10, 11, 12, 13, 14, 15, 16.

What is the sum? 16

Complete the equation.

Add 6, 7, or 8 by Making 10

Use the **Ten Bar** to help the students add other addends by making 10.

Explain that the Ten Bar can be used to add 6, 7, or 8. Write "6 + 9 = __" for display.

Demonstrate each step on the Ten Bar for display.

What do you do first to solve this problem? place 6 cubes in the spaces on the Ten Bar

What is the next step? place 9 cubes below the bar

What do you do next? move 4 cubes into the spaces on the Ten Bar to make a group of 10

Where can you begin counting? *count on* from 10

Guide the students as they *count on* from 10: 10, 11, 12, 13, 14, 15.

What is the answer? 15

Complete the equation.

Continue the activity with the following equations.

6 + 8 = 14	7 + 5 = 12
7 + 8 = 15	8 + 6 = 14

Write "8 + 5" in vertical form. Follow the same steps to solve the problem with the Ten Bar.

Make 10 Mentally

What number do you add to 1 to make 10? 9

What number do you add to 2 to make 10? 8

Continue the activity as you guide the students to identify the addends for the equations which equal 10.

3 + 7 = 10	6 + 4 = 10	9 + 1 = 10
4 + 6 = 10	7 + 3 = 10	
5 + 5 = 10	8 + 2 = 10	

Make 10 Mentally While Adding

Use **Dot Pattern Cards** to guide the students as they make 10 mentally.

Distribute the Dot Pattern Cards. Write "8 + 3 = __" for display.

Direct the students to show the equation, using the 8 and 3 Dot Pattern Cards. Explain that making 10 can make it easier to add.

Direct the students' attention to the dot pattern for 8.

What can you add to 8 to make 10? 2

Guide the students to think "8 + 2 = 10."

Explain that the students can look at the 3 dot pattern and think *2 dots moved with the 8 dots will make 10.*

How many dots are left to add? 1

Explain that the students can now think "10 + 1 = 11."

Continue the activity with 7 + 5 = 12 and 8 + 6 = 14.

Some first graders find the making-10 strategy confusing because of the mental skills involved. This strategy will be reintroduced in second grade.

Problem-Solving Plan

Guide the students as they use the **Problem-Solving Plan** to solve a word problem.

Read aloud the following word problem.

147

Charity saw 8 prairie dogs on the dirt hills. Then 4 more prairie dogs popped their heads up from underground. How many prairie dogs did Charity see in all?

What is the question? How many prairie dogs did Charity see in all?

What information is given? Charity saw 8 prairie dogs, and 4 more popped their heads up.

Do you add or subtract to find the answer? add

What equation can you use to solve this word problem? 8 + 4 = __

Write "8 + 4 = __" for display. Guide the students as they use their UNIFIX Cubes and the Ten Bar Mat to solve the problem, if needed.

Invite a volunteer to write the answer. 12 Write *prairie dogs* after the answer.

Does the answer make sense? yes

Continue the activity with the following word problem.

Joel saw 6 flamingos standing in the water. He saw 5 more flamingos nesting on the bank. How many flamingos did Joel see in all? 6 + 5 = 11 flamingos

Apply

Worktext pages 285–86

Read and guide completion of page 285.

Read and explain the directions for page 286. Assist the students as they complete the page independently.

Assess

Reviews pages 273–74

Review liter capacity on page 274.

Fact Reviews

Use "Subtraction Facts to 12" page 57 in Trove.

Add 6, 7, 8

Color each addend. Complete the Ten Bar with the second addend. Add.

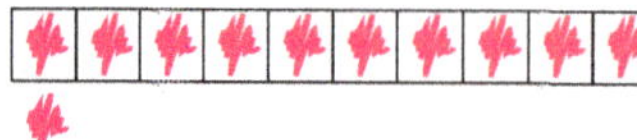

8 + 3 = 11

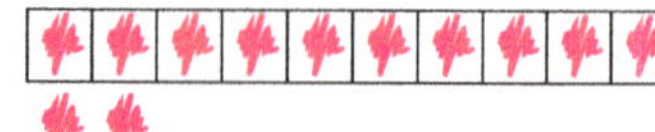

7 + 5 = 12

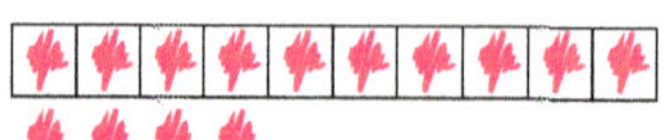

6 + 8 = 14

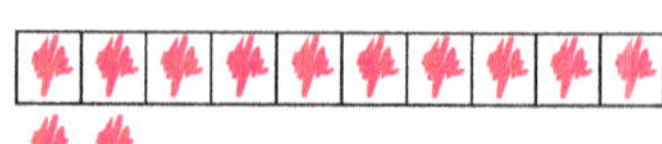

8 + 4 = 12

7 + 4 = 11

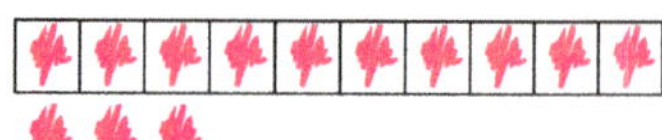

8 + 5 = 13

Write an equation for the word problem. Solve.

Ava caught 8 bugs.
She put them in a jar.
She caught 6 more bugs.
She put them in a different jar.
How many bugs did Ava catch?

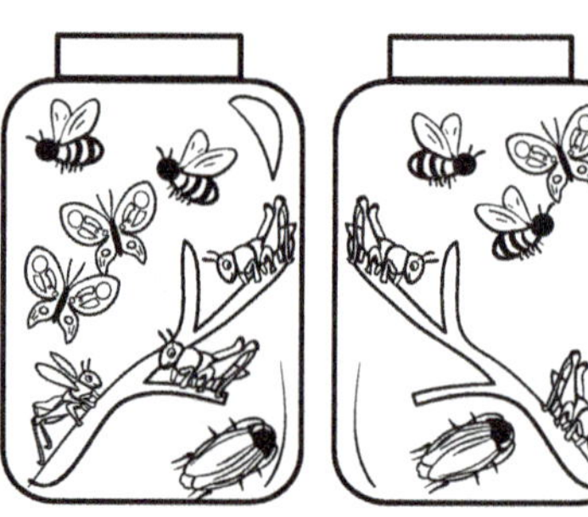

8 + 6 = 14 bugs

Extended Activity

Make 10 Mentally

Materials

- Number Cards: 1–9

Procedure

Display the Number Card for 2.

What number do you add to 2 to make 10? 8

Continue the activity by using different Number Cards.

Chapter 17 Review

Circle each container that holds *more than* 1 liter.
Mark an *X* on each container that holds *less than* 1 liter.

Subtraction Fact Review

Subtract.

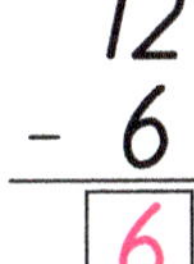
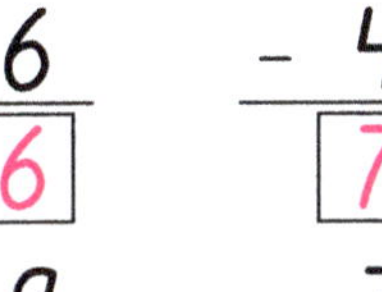

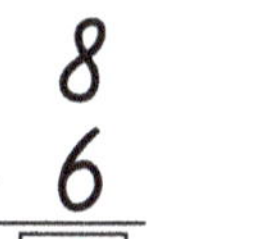
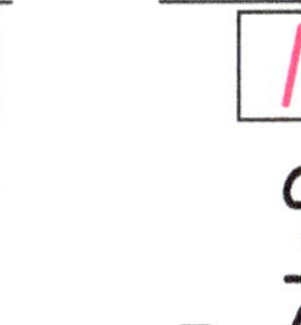

$9 - 3 = 6$	$12 - 6 = 6$	$11 - 4 = 7$	$8 - 6 = 2$	$10 - 9 = 1$
$8 - 7 = 1$	$9 - 4 = 5$	$7 - 6 = 1$	$12 - 8 = 4$	$9 - 7 = 2$

148 Worktext pages 287–88 Reviews pages 275–76

Objectives

- **148.1** Solve a 3-addend equation for a word problem.
- **148.2** Solve 3-addend equations by using manipulatives.
- **148.3** Add 3 addends by using the Grouping Principle.
- **148.4** Solve 3-addend vertical problems.
- **148.5** Propose a reason to persevere in solving a 3-addend equation for a word problem. BWS

Biblical Worldview Shaping

- **Working** (formulate): Sometimes real-world problem solving requires a lot of work. The Bible teaches us that we should do our work heartily, even if it is not easy, because that will please God (148.5).

Printed Resources

- Visuals: Counters (15 ducks)

Digital Resources

- Games/Enrichment: Fact Reviews (Subtraction Facts to 12, page 58)

Materials

- 20 UNIFIX® Cubes (for each student)

3 Addends

There are 3 eagles flying.
There are 3 eagles eating.
There are 7 eagles in the tree.

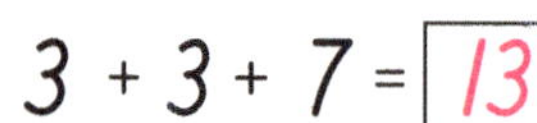

3 + 3 + 7 = 13

Use counters to add.
Write the sum.
Use the code to find a promise of God.

8	9	10	11	12	13	14
a	c	e	f	m	o	r

God promises to

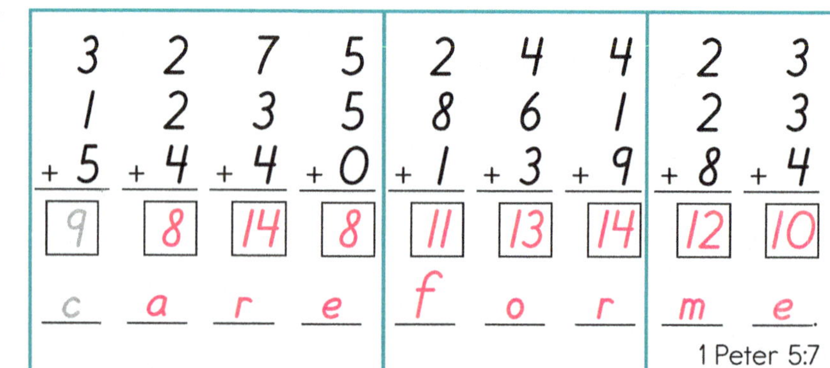

3 + 1 + 5	2 + 2 + 4	7 + 3 + 4	5 + 5 + 0	2 + 8 + 1	4 + 6 + 3	4 + 1 + 9	2 + 2 + 8	3 + 3 + 4
9	8	14	8	11	13	14	12	10
c	a	r	e	f	o	r	m	e

1 Peter 5:7

Practice & Review

Grouping Principle of Addition

Distribute the UNIFIX Cubes. Write "2 + 3 + 3" for display.

How many addends are in this problem? 3

Instruct the students to group the cubes into 3 sets on their desks, putting 2 cubes at the top, 3 in the middle, and 3 at the bottom. Direct the boys to add up to get the sum and the girls to add down.

What is the sum? 8

Which principle are you showing by the boys adding up and the girls adding down to find the sum? Grouping Principle of Addition

What does the Grouping Principle tell you? When the grouping of the addends is changed, the sum stays the same.

Continue the activity with the following problems.

$$\begin{array}{r} 4 \\ 6 \\ +\,1 \\ \hline 11 \end{array} \qquad \begin{array}{r} 2 \\ 2 \\ +\,8 \\ \hline 12 \end{array}$$

Engage

Diligence with Math

Guide a **discussion** to prepare the students to formulate a reason for persevering with math to solve a problem.

Read aloud Colossians 3:23.

What does the Bible teach us about work, even if the work is hard? Christians should work heartily or joyfully with a good attitude.

Why should we work heartily? because we are doing our work to please God

Instruct

Solve a 3-Addend Equation for a Word Problem

Use **manipulatives** to help the students solve word problems involving 3 addends.

Display the 3 sets of ducks mentioned as you read the following word problem.

There are 5 mallard ducks, 5 wood ducks, and 1 Pekin duck. How many ducks are there in all?

What equation tells about this problem? 5 + 5 + 1 = __

Write "5 + 5 + 1 = __" for display.

Use counters to add. Write the sum. Use the code to find another promise of God.

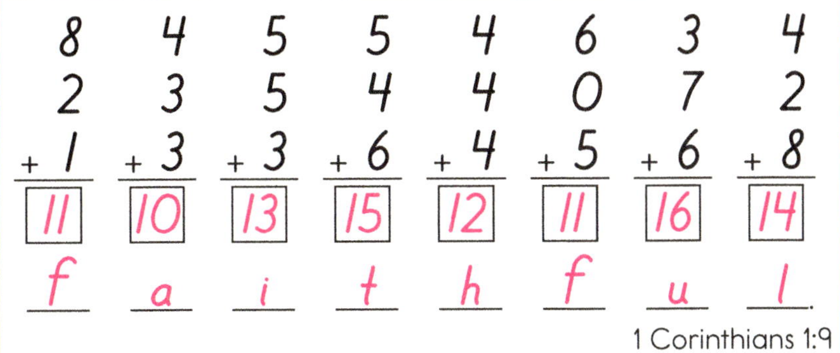

10	11	12	13	14	15	16
a	f	h	i	l	t	u

God promises to be

8	4	5	5	4	6	3	4
2	3	5	4	4	0	7	2
+ 1	+ 3	+ 3	+ 6	+ 4	+ 5	+ 6	+ 8
11	10	13	15	12	11	16	14
f	a	i	t	h	f	u	l.

1 Corinthians 1:9

Write an equation for the word problem. Solve.

There are 7 sheep on the hillside.
There are 4 sheep under the tree.
There are 3 sheep drinking at the stream.
How many sheep are there?

7 + 4 + 3 = 14 sheep

Circle the phrase that completes the sentence.

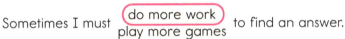

Sometimes I must (do more work / play more games) to find an answer.

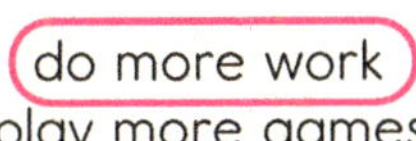

Time to Review

Trace the shape. Draw the shape.

What do you notice about the addends in this equation? There are 3.

Join the sets of ducks.

How many ducks are there in all? 11

Complete the equation with the label *ducks*.

Continue the activity for the following word problem.

There are 8 ducks swimming, 2 ducks playing with a ball, and 5 ducks sunning themselves on the rocks. How many ducks are there in all? 8 + 2 + 5 = 15 ducks

Grouping Addends

Guide the students as they use **UNIFIX Cubes** to group addends to help them solve a 3-addend equation.

Distribute the UNIFIX Cubes. Read aloud the following word problem.

Anya collects stuffed animals. She has 7 teddy bears, 7 rabbits, and 3 giraffes. How many stuffed animals does Anya have?

Guide the students as they make a bar of 7 cubes on the left of their desks, a bar of 7 cubes in the middle, and a bar of 3 cubes on the right.

Instruct the students to join the sets of cubes together.

How many stuffed animals does Anya have? 17 stuffed animals

Instruct the students to use their cubes to show the 3 sets of stuffed animals again.

Write "7 + 7 + 3 = __" for display. Direct the students to put the 7 animals and the 7 animals together.

How many animals are there in these 2 sets? 14

$$7 + 7 + 3 =$$
$$14 + 3 = 17 \text{ animals}$$

Write "14" below the 7 + 7. Ask the students to put Anya's 3 animals with the set of 14 animals. Write "+ 3 =" next to 14.

How many animals are there in all? 17 animals

Write "17 animals." Leave the equations displayed.

Instruct the students to separate their cubes again into 7 animals, 7 animals, and 3 animals. Write "7 + 7 + 3 = __" for display. Direct the students to join a set of 7 animals and the set of 3 animals.

How many animals are in the 2 sets? 10

Write the 10 under the 7 + 3. Instruct the students to join the set of 10 animals to the set of 7 animals. Write "7 +" in front of 10.

How many animals are there in all? 17 animals

Complete the equation. Leave the equations displayed.

$$7 + 7 + 3 =$$
$$7 + 10 = 17 \text{ animals}$$

Grouping Principle of Addition

Guide a **discussion** to help the students apply the Grouping Principle of Addition to solve 3-addend equations.

Direct attention to the first set of displayed equations.

Which addends did you add together the first time? 7, 7

Point out the second set of equations.

Which addends did you add together the second time? 7, 3

What was the sum whether you joined the sets of 7 and 7 together first or the sets of 7 and 3 together first? 17 animals

What is this math principle called? Grouping Principle of Addition

148

Remind the students that the Grouping Principle of Addition says that when the grouping of the addends is changed the sum stays the same.

Write "8 + 8 = 16" for display.

What kind of fact is this? a double

Write "8 + 8 + 2 = __" for display. Instruct the students to show the equation with UNIFIX Cubes. Explain that they can use the facts or strategies that they already know to solve this problem, because the Grouping Principle tells them that the grouping of the addends can change without changing the sum.

Which 2 addends could you group together since you know the double fact? 8, 8

Write "16" below 8 + 8.

What is the sum of 16 and 2? 18

Write "7 + 3 + 5 = __" for display. Instruct the students to show the equation with cubes.

Do you see a strategy that will help you solve this equation? making 10 (numbers that equal 10)

Which 2 addends should you group together? 7, 3

Show the grouping. Ask the students to find the sum. 15

Continue the activity with 3 + 8 + 2 = __. 13

Vertical Problems with 3 Addends

Guide a **discussion** to help the students solve 3-addend vertical problems.

Write "1 + 9 + 9" vertically. Remind the students that to solve the problem they can begin with the 1 and add down or they can begin with the 9 and add up.

Direct the students to copy the problem. Ask the boys to begin at 1, add down, and then write their answer. Instruct the girls to begin at 9, add up, and then write their answer. Students may use their UNIFIX Cubes if needed.

What is the sum? 19

Which addends did the boys group together when adding down? 1, 9

Which addends did the girls group together when adding up? 9, 9

3 Addends

There are 4 children praying. 

There are 3 children reading. 

There are 5 children singing. 

4 + 3 + 5 = 12 children

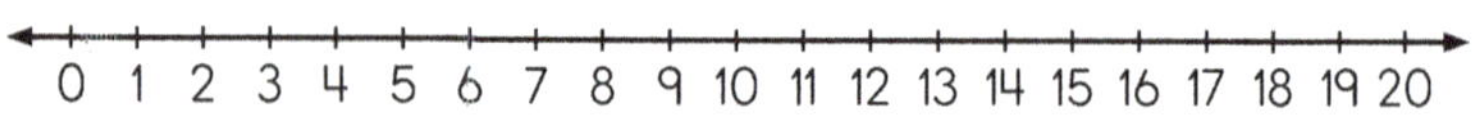

Use the number line to add.
Write the sum.
Use the code to find the answer.

10	11	12	13	14	15	16	17
d	f	g	i	l	o	r	y

We should

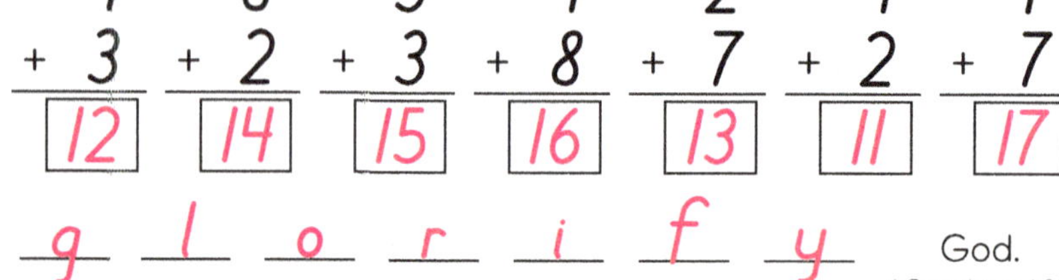

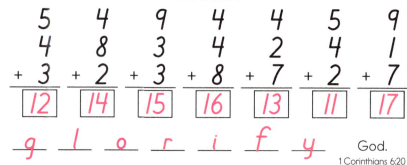

5	4	9	4	4	5	9
4	8	3	4	2	4	1
+ 3	+ 2	+ 3	+ 8	+ 7	+ 2	+ 7
12	14	15	16	13	11	17
g	l	o	r	i	f	y

God.

1 Corinthians 6:20

Which principle did you use when some of you were adding up and some were adding down? Grouping Principle

How might this principle help when working addition problems? I can use strategies or facts that I know to solve a problem more easily.

Explain that the students may also work the problem one way and then work it the other way to check their answers.

Continue the activity with 6 + 4 + 4 = __. 14

Persevere with Math

Guide a **discussion** to help the students propose a reason to persevere in solving a 3-addend equation.

Invite the students to pretend that they have been given the job of helping the zookeeper prepare food for the monkeys at the zoo today. Explain that the zookeeper chooses the monkeys' food carefully to help them stay healthy. The monkeys eat monkey chow, like a dog eats dry dog food, to make sure they get enough of the right things in their diet each day. It is the students' job to figure out how much monkey chow to give their monkeys today.

Read aloud the following word problem.

Each monkey needs 4 scoops of monkey chow in the morning, 4 scoops at lunch, and 5 scoops at supper. How many scoops of chow in all should you feed your monkey today?

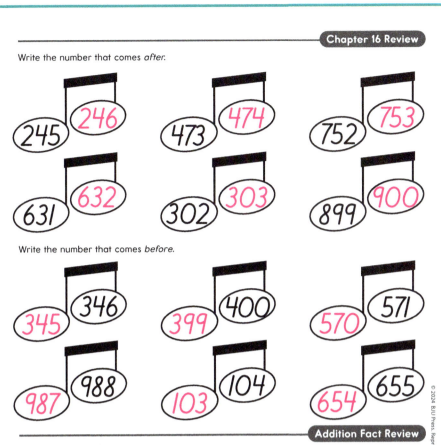

Chapter 16 Review

Write the number that comes *after*.

631 632

302 303

899 900

Write the number that comes *before*.

987 988

103 104

654 655

Addition Fact Review

Mark the correct circle.

$8 + 1 =$ 9 ● 10 ○ 11 ○

$7 + 4 =$ 9 ○ 10 ○ 11 ●

$9 + 1 =$ 9 ○ 10 ● 11 ○

$6 + 5 =$ 9 ○ 10 ○ 11 ●

$5 + 5 =$ 8 ○ 9 ○ 10 ●

$8 + 3 =$ 9 ○ 10 ○ 11 ●

$7 + 3 =$ 9 ○ 10 ● 11 ○

$6 + 2 =$ 8 ● 9 ○ 10 ○

What is the question in this word problem? How many scoops of chow in all should you feed your monkey today?

What information is given? Each monkey needs 4 scoops in the morning, 4 at lunch, and 5 at supper.

Do you add or subtract to find the answer? I add because I am joining sets to find the total.

What is the equation needed to solve this problem? $4 + 4 + 5 =$ __

Write "$4 + 4 + 5 =$ __" for display.

Guide the students as they use the Grouping Principle to solve the equation. Invite a student to complete the equation. 13

Add *scoops* after the answer.

Does the answer make sense? yes

Why would it be important for you to solve this math if you were helping the zookeeper? My job was to feed the monkey the healthy amount of monkey chow. Math helps me do my work the right way.

Why should we work hard to solve a math problem even when it is not easy? God tells us to do our work joyfully to please Him. (Colossians 3:23)

Apply

Worktext pages 287–88

Read and guide completion of page 287.

Read and explain the directions for page 288. Assist the students as they complete the page independently.

Direct attention to the biblical worldview shaping statement and guide the students as they complete the sentence.

Assess

Reviews pages 275–76

Review numbers that come *before* and *after* on page 276.

Fact Reviews

Use "Subtraction Facts to 12" page 58 in Trove.

149

Worktext pages 289–90
Reviews pages 277–78

Objectives

- **149.1** Relate subtraction to addition.
- **149.2** Add double facts for 1–6.
- **149.3** Identify the fact families for double facts for 7–10.
- **149.4** Add double facts for 7–10.
- **149.5** Solve a word problem involving double facts.

Printed Resources

- Visual: Shapes Kit (20 circles)

Digital Resources

- Web Link: Virtual Manipulatives: Number Frames

Materials

- A 2-liter beverage bottle

Double Fact Families

Write an addition and subtraction equation for each dot pattern.

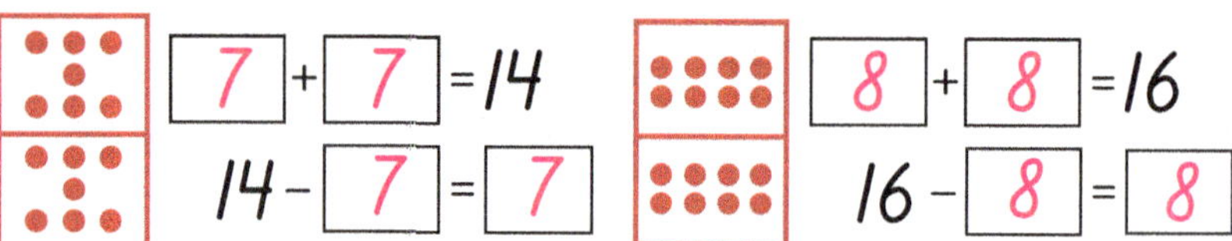

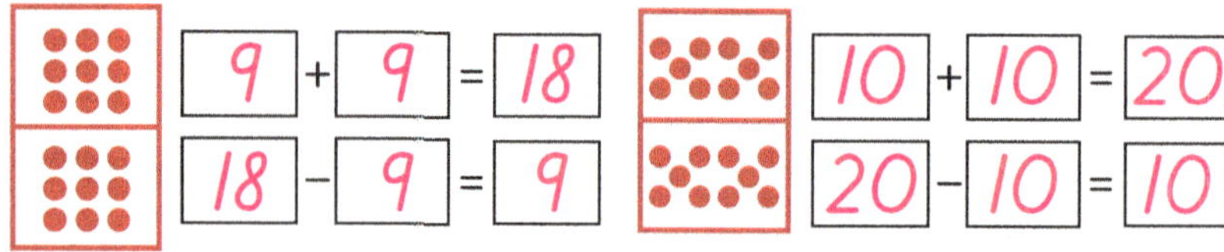

Write an equation for each word problem. Solve.

DQ has 20 peanuts.
He feeds 10 peanuts to the elephant.
How many peanuts does DQ have left?

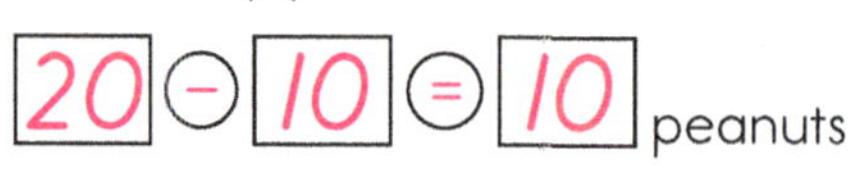

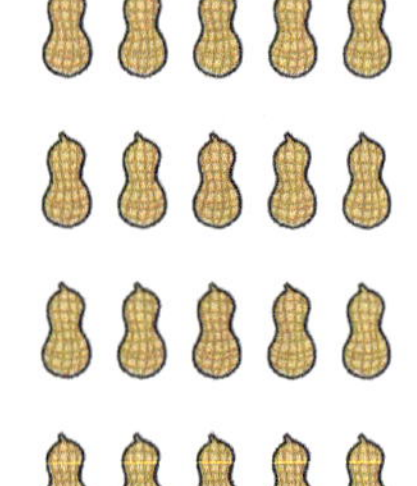

The zookeeper has 18 fish to feed to the seals. He feeds them 9 fish.
How many fish does he have left?

Practice & Review

Count 500–600

Divide the students into 2 groups. Instruct the students in Group 1 to pretend that they are lions, counting in deep voices, and the students in Group 2 to pretend that they are monkeys, counting in high voices.

Instruct Group 1 to begin counting with 500 and to continue counting until you clap your hands. Instruct Group 2 to *count on* from the last number counted by Group 1 until you clap your hands. Continue to alternate groups until they reach 600.

Engage

Identify the Correct Metric Measuring Tool

Guide a **review** of metric measuring tools.

Display the 2-liter bottle.

Which measurement tool do you need to find how much water my bottle holds? a 1-liter (metric measuring) container

Which tool do you need to find how many kilograms the bottle weighs? a scale

Which tool do you need to find the height of my bottle? a centimeter ruler

When you go outside today, which tool do you need to check to know what to wear? a thermometer

Instruct

Relate Subtraction to Addition

Use **dot patterns** to help the students relate subtraction to addition.

Display the dot pattern for 4 by arranging 4 circles.

What number does this dot pattern show? 4

What dot pattern for another number do you see within the pattern for 4? 2

Remove 2 circles.

Which dot pattern is shown now? the dot pattern for 2

Explain that removing 2 dots from the dot pattern for 4 makes the dot pattern for 2. Write "4 – 2 = 2" for display.

Replace the 2 circles and point out that they have the dot pattern for 4 again.

What addition equation describes this action? 2 + 2 = 4

Write the addition equation below the subtraction equation.

How are these addition and subtraction equations alike? The numbers are the same. They both have equals signs.

How are these equations different? The numbers are in a different order. The first equation has a minus sign since it is

Add or subtract.
Write the numbers for each fact family.

$7 + 7 = 14$
$14 - 7 = 7$

$8 + 8 = 16$
$16 - 8 = 8$

Write an addition and subtraction equation for each double fact family.

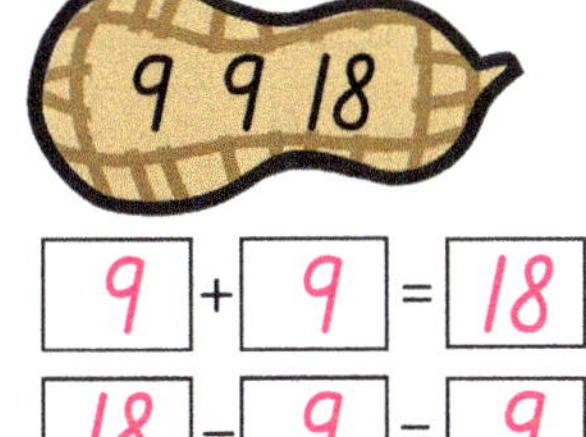

$9 + 9 = 18$
$18 - 9 = 9$

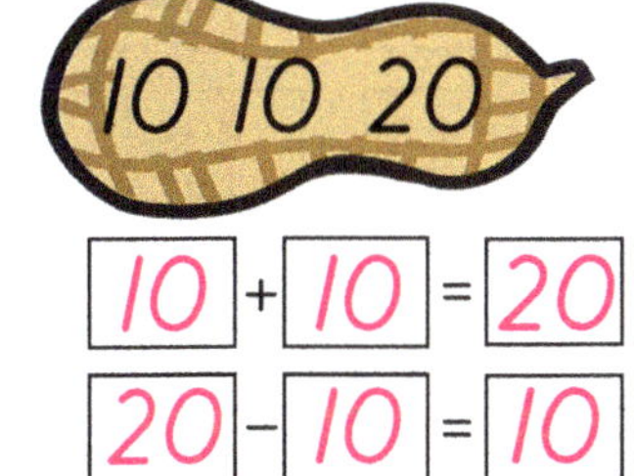

$10 + 10 = 20$
$20 - 10 = 10$

Circle the correct measuring tool.

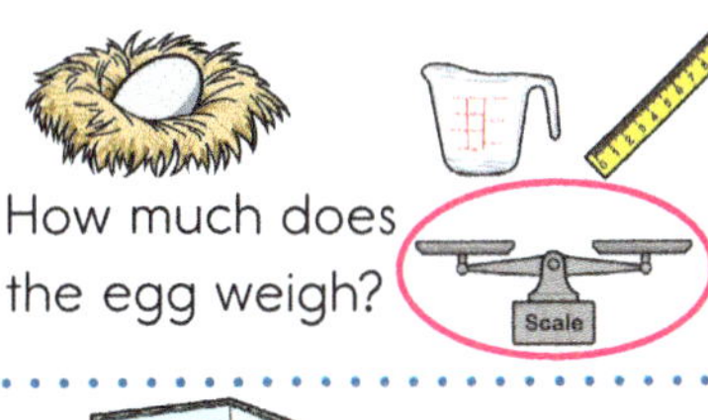

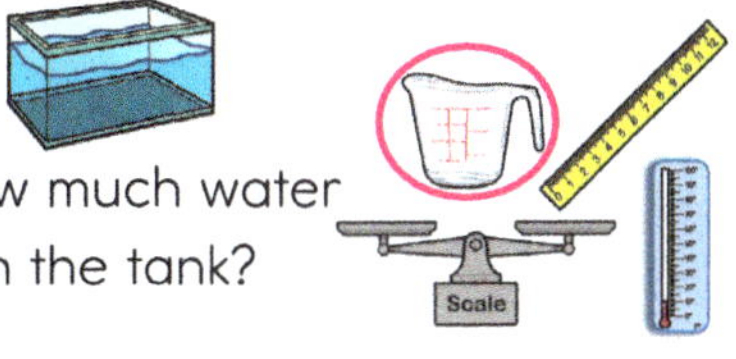

a subtraction equation. The second equation has a plus sign since it is an addition equation.

What do you call addition facts that have addends that are the same number? doubles

Remind the students that equations that have the same numbers are related since the equations are in the same fact family.

Continue the activity, making dot patterns for the following facts.

$6 - 3 = 3$ $8 - 4 = 4$ $10 - 5 = 5$ $12 - 6 = 6$

$3 + 3 = 6$ $4 + 4 = 8$ $5 + 5 = 10$ $6 + 6 = 12$

Double Facts for 1–6

Use ***if/then* statements** to help the students identify related double facts for 1–6.

Guide the students as they repeat the following *if/then* statements. After repeating the first two statements, give the *if* statement and let the students complete the statement by saying the *then* statement.

If 1 + 1 = 2, then 2 – 1 = 1.

If 2 + 2 = 4, then 4 – 2 = 2.

If 3 + 3 = 6, then 6 – 3 = 3.

If 4 + 4 = 8, then 8 – 4 = 4.

If 5 + 5 = 10, then 10 – 5 = 5.

If 6 + 6 = 12, then 12 – 6 = 6.

Fact Families

Use **dot patterns** to help the students identify fact families for double facts for 7–10.

Display the dot pattern for 14 by displaying 14 circles.

Make the dot pattern for 14 by making 2 dot patterns for 7.

What number does this dot pattern represent? 14

What dot pattern for another number do you see within the pattern for 14? 7

Remove 7 circles.

Which dot pattern is represented now? the dot pattern for 7

What equation did I make by removing 7 dots from the dot pattern for 14? 14 – 7 = 7

Write "14 – 7 = 7" for display.

Replace the 7 circles and point out that you have the dot pattern for 14 again.

What addition equation describes this action? 7 + 7 = 14

Write the addition equation below the subtraction equation.

What do you call addition facts that have addends that are the same number? doubles

Continue the activity, making dot patterns for the following facts:

$16 - 8 = 8$ $18 - 9 = 9$ $20 - 10 = 10$

$8 + 8 = 16$ $9 + 9 = 18$ $10 + 10 = 20$

What do you call equations that have the same numbers and are related? a fact family

Draw a box to the left of each pair of equations. Ask the students to look at the numbers in each pair and to tell the family that they belong to. Write the numbers in each box.

7 7 14 | 8 8 16 | 9 9 18 | 10 10 20

Point out that these subtraction facts are often easier to remember since they are related to the addition facts that are doubles.

Add Double Facts for 7–10

Use ***if/then* statements** to help the students identify related double facts for 7–10.

Guide the students as they repeat the following *if/then* statements

If 7 + 7 = 14, then 14 – 7 = 7.

If 8 + 8 = 16, then 16 – 8 = 8.

If 9 + 9 = 18, then 18 – 9 = 9.

If 10 + 10 = 20, then 20 – 10 = 10.

149

Continue the activity, giving the *if* statement and letting the students complete the statement by saying the *then* statement.

Problem-Solving Plan

Use the **Problem-Solving Plan** to help the students solve a word problem.

Read aloud the following word problem.

The zookeeper gave 16 monkeys a piece of carrot each. Immediately 8 monkeys ate their carrots. How many monkeys have not eaten their carrots?

What is the question? How many monkeys have not eaten their carrots?

What information is given? 16 monkeys got a piece of carrot, and 8 monkeys ate theirs immediately.

Do you add or subtract to find the answer? subtract

What is the equation that you can use to solve this word problem? 16 – 8 = __

Write "16 – 8 = __" for display. Use circles to make the dot pattern for 16 (2 dot patterns for 8).

Invite a volunteer to solve the problem. 8 Write *monkeys* after the answer.

Does the answer make sense? yes

Continue the activity with the following word problem.

There were 14 penguins eating on the bank. After they finished, 7 penguins dived into the water. How many penguins were left on the bank? 14 – 7 = 7 penguins

Double Fact Families

Write an addition and subtraction equation for each dot pattern.

10 + 10 = 20
20 – 10 = 10

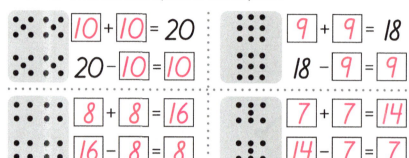

9 + 9 = 18
18 – 9 = 9

8 + 8 = 16
16 – 8 = 8

7 + 7 = 14
14 – 7 = 7

Add or subtract.
Write the numbers for each fact family.

7 + 7 = 14
14 – 7 = 8

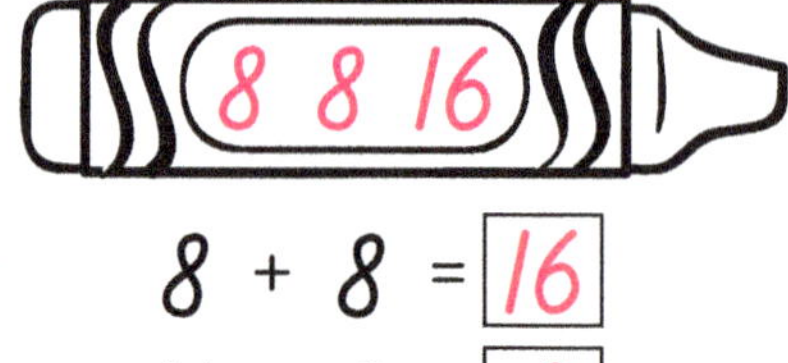

8 + 8 = 16
16 – 8 = 8

Write an addition and subtraction equation for each double fact family.

10 + 10 = 20
20 – 10 = 10

9 + 9 = 18
18 – 9 = 9

Chapter 18 • Lesson 149 two hundred seventy-seven **277**

Apply

Worktext pages 289–90

Read and guide completion of page 289.

Read and explain the directions for page 290. Assist the students as they complete the page independently.

Assess

Reviews pages 277–78

Review matching weather-related activities to Celsius temperature on page 278.

Extended Activity

Add Doubles

Materials

- A white unlined 3 × 5 card for each student
- Zoo postcards or pictures of zoo animals

Procedure

Display the zoo postcards. Distribute the 3 × 5 cards and guide the students as they make their own postcards.

Invite 5 students to come to the front of the room to show their postcards. Invite 5 more students to come to the front of the room to show their cards.

What equation can be written to show the number of students with postcards at the front of the room? 5 + 5 = 10 students

Continue the activity; vary the number of students in the groups to represent other doubles.

Chapter 17 Review

Draw a line to match each thermometer With the correct picture.

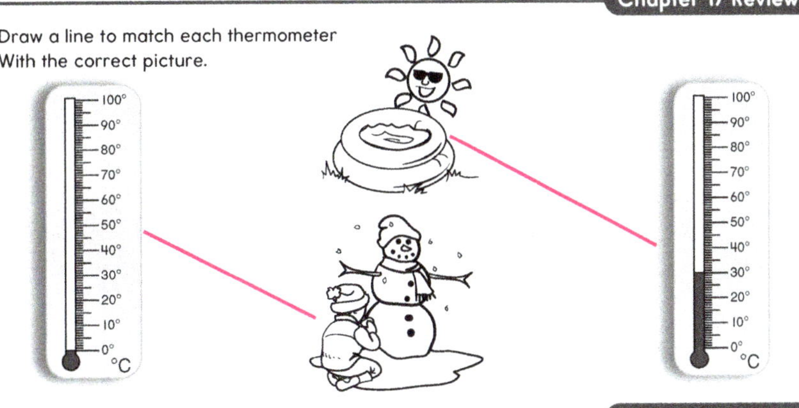

Chapter 12 Review

Add.

32	25	43	81	74
+ 15	+ 10	+ 16	+ 17	+ 21
47	35	59	98	95

64	53	21	34	27
+ 33	+ 40	+ 43	+ 14	+ 62
97	93	64	48	89

Subtraction Fact Review

Mark the correct circle.

10 − 1 = 9 (marked) 10 11

8 − 6 = 1 2 (marked) 3

11 − 4 = 7 (marked) 8 9

12 − 6 = 4 5 6 (marked)

9 − 7 = 1 2 (marked) 3

11 − 8 = 1 2 3 (marked)

278 two hundred seventy-eight Math 1 Reviews

Chapter 18 • Addition & Subtraction Facts to 20 Lesson 149 • 641

150

Worktext pages 291–92
Reviews pages 279–80

Objectives

- **150.1** Separate a set to demonstrate subtraction from 13 and 14.
- **150.2** Write addition and subtraction equations for fact families for 13 and 14.
- **150.3** Complete missing addend equations for word problems.

Printed Resources

- Student Manipulatives: Ten Bar Mat

Materials

- 14 teddy bear crackers (or UNIFIX® Cubes)

Fact Families for 13, 14; Missing Addend

Add or subtract.
Write the numbers for each fact family.

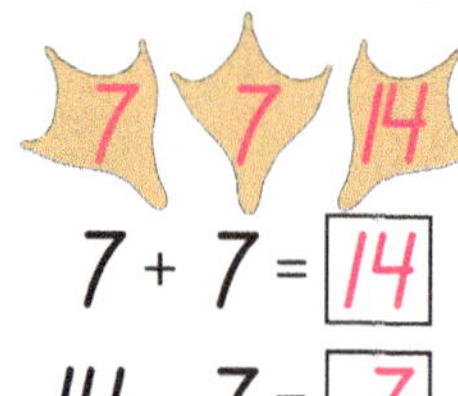

7 + 7 = 14
14 − 7 = 7

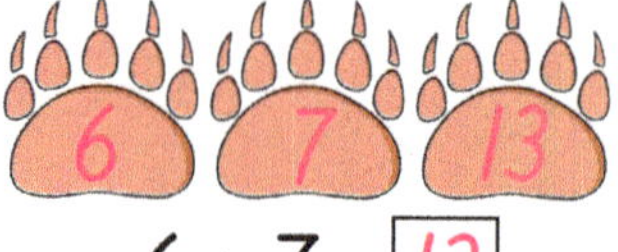

6 + 7 = 13
7 + 6 = 13
13 − 6 = 7
13 − 7 = 6

Write each fact family.

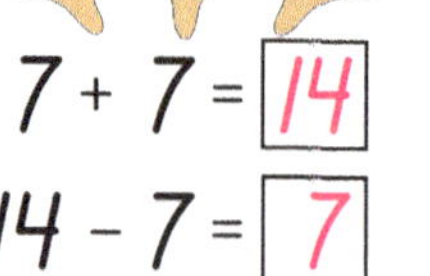

4 + 9 = 13
9 + 4 = 13
13 − 4 = 9
13 − 9 = 4

6 8 14

6 + 8 = 14
8 + 6 = 14
14 − 6 = 8
14 − 8 = 6

Complete the equation for the word problem.
Ruth is making 13 DQ bookmarks for friends. She has made 9. How many more bookmarks does Ruth need to make?

9 + 4 = 13
bookmarks

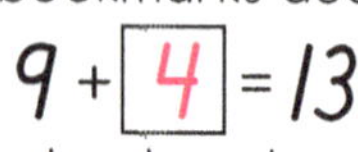

Practice & Review

Complete a Sequence of 3-Digit Numbers

Write "__ 353 354 __ 356 __ 358 __ 360 __" for display.

Point out that several numbers are missing.

Which number comes before 353 when you count? 352

Invite a volunteer to write "352" in the first blank.

Which number comes after 354 when you count? 355

Ask a student to write "355" in the second blank.

Continue the activity to complete the sequence. Check the answers by counting (reading) the completed sequence together with the students.

Continue the activity, using the following sequences.

717 __ 719 __ 721 __ __ 724 __ 726

__ 536 537 __ 539 __ 541 __ __ 544

Engage

Value of the Digits in a 3-Digit Number

Guide a **review** of place value in 3-digit numbers.

Write "857" for display. Ask a student to read aloud the number. eight hundred fifty-seven

What is the value of 5 in this number? 50

What is the value of 8 in this number? 800

What is the value of 7 in this number? 7

Write "777" for display. Invite a volunteer to read the number. seven hundred seventy-seven

What is the value of 7 in the Hundreds place? 700

What is the value of 7 in the Ones place? 7

What is the value of 7 in the Tens place? 70

Instruct

Separate a Set to Demonstrate Subtraction from 13

Use **manipulatives** to help the students subtract from 13.

Distribute a Ten Bar Mat and 14 teddy bear crackers to each student. Direct the students to place 13 bears on their Ten Bar Mats. Remind them that 13 is 10 and 3 more.

How many bears are on your bar? 10

How many bears are below your bar? 3

Ask the girls to remove 4 bears and the boys to remove 9 bears. Instruct all the students to put the bears they removed along the bottom of their mats.

Add or subtract.
Write the numbers for the fact family.

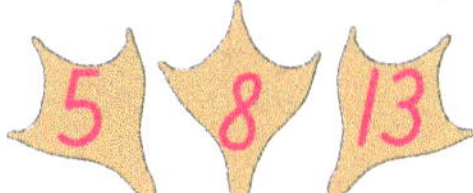

$5 + 8 = \boxed{13}$

$8 + 5 = \boxed{13}$

$13 - 5 = \boxed{8}$

$13 - 8 = \boxed{5}$

Write the fact family.

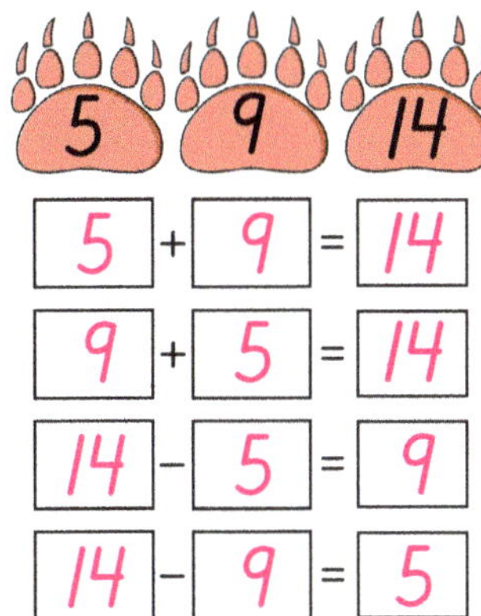

$\boxed{5} + \boxed{9} = \boxed{14}$

$\boxed{9} + \boxed{5} = \boxed{14}$

$\boxed{14} - \boxed{5} = \boxed{9}$

$\boxed{14} - \boxed{9} = \boxed{5}$

Complete the equation for each word problem.

Amber had 14 paw prints to put in front of the zoo gate. She has laid down 9 paw prints. How many more does Amber need to lay down?

$9 + \boxed{5} = 14$ paw prints

Mother made a duck cake. She had 13 birthday candles. She put 5 candles on the cake. How many more candles does she have to put on the cake?

$5 + \boxed{8} = 13$ candles

Time to Review

Write the value of the underlined digit.

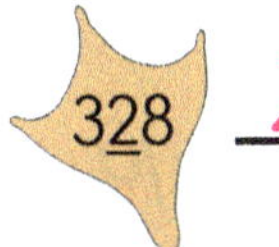

20

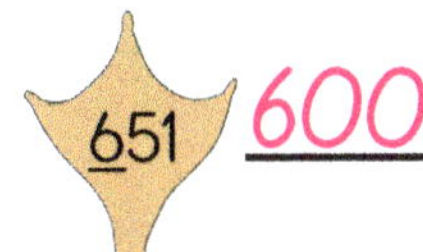

600

849 9

How many bears remain on the Ten Bar? girls: 9; boys: 4

Why do the girls and boys have different answers? Not everyone removed the same number of bears.

Invite volunteers to give the subtraction equations for the bears they removed. Write the related equations in pairs for display.

13 – 9 = 4   13 – 4 = 9

Continue the activity, removing the following pairs.

5 bears and 8 bears: 13 – 8 = 5, 13 – 5 = 8

6 bears and 7 bears: 13 – 6 = 7, 13 – 7 = 6

Continue to display the equations.

Fact Families for 13

Guide a **discussion** to help the students write related equations for fact families for 13.

Direct attention to the first pair of displayed equations.

How are the subtraction equations alike? The numbers are the same. The signs are the same.

How are the subtraction equations different? The numbers are in a different order.

Remind the students that each pair of equations that have the same numbers is related since the equations are in the same family.

Draw a box to the left of each pair of equations.

Look at the numbers in each pair. Which fact family do they belong to?

Write the numbers in each box.

4 9 13	5 8 13	6 7 13

Point out that the addition equations are missing from the families. Invite volunteers to write the addition equations that are a part of the families.

9 + 4 = 13   4 + 9 = 13

8 + 5 = 13   5 + 8 = 13

7 + 6 = 13   6 + 7 = 13

Erase all the addition and subtraction equations but leave the list of 3 fact families displayed in boxes to use for the final activity in the lesson.

Separate a Set to Demonstrate Subtraction from 14

Continue to use **manipulatives** to help the students subtract from 14.

Direct the students to place 14 bears on their Ten Bar Mats.

How many bears are on your bar? 10

How many bears are below your bar? 4

Instruct the girls to remove 5 bears, instruct the boys to remove 9 bears, and instruct all the students to put the bears they removed along the bottom of their mats.

Invite volunteers to give the subtraction equations for the bears that they removed. Write the related equations in pairs for display.

14 – 9 = 5   14 – 5 = 9

Continue the activity, removing the following pairs.

6 bears and 8 bears: 14 – 8 = 6, 14 – 6 = 8

7 bears: 14 – 7 = 7

Continue to display the equations.

Fact Families for 14

Guide a **discussion** to help the students write related equations for fact families for 14.

Direct attention to the first pair of displayed equations for 14.

How is each pair of equations related? They have the same numbers. They are part of the same family.

150

Draw a box to the left of each pair of equations and the single equation.

Ask the students to look at the numbers in each pair and in the last equation. Guide them as they identify the fact family to which they belong. Write the numbers in each box.

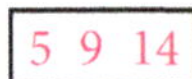

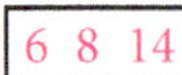

Why is there no related subtraction equation for 14 – 7 = 7? It would be the same; it is a double; there are two 7s.

What is missing from each family? addition equations

Ask students to write the addition equations.

9 + 5 = 14   5 + 9 = 14

8 + 6 = 14   6 + 8 = 14

7 + 7 = 14

Why is there no related addition equation for the 7-7-14 family? They would be the same; it is a double; there are two 7s.

Erase all the addition and subtraction equations but leave the list of 3 fact families displayed in boxes to use for the following activity. Collect the counters so that the students can no longer use them to count.

Missing Addend Equations

Use **fact families** to help the students complete missing addend equations.

Direct the students to use the list of families for 13 and 14 to help them solve the following word problems.

The zookeeper needs 14 pounds of meat to feed the bears. She has only 9 pounds. How many more pounds of meat does the zookeeper need to feed the bears?

Write "9 + __ = 14" for display. Ask a student to write the addend that will complete the equation. 5

Continue the activity for the following word problems.

The zoo trainer wants each duck to have a ball. She has 13 ducks but only 7 balls. How many more balls does the zoo trainer need? 7 + 6 = 13 balls

Rosa needs 14 dollars to buy the stuffed bear in the zoo gift shop. She has only 8 dollars. How many more dollars does Rosa need to buy the stuffed bear? 8 + 6 = 14 dollars

Fact Families for 13, 14; Missing Addend

Add or subtract.
Write the number for each fact family.

6 8 14	5 9 14
6 + 8 = 14	5 + 9 = 14
8 + 6 = 14	9 + 5 = 14
14 − 6 = 8	14 − 5 = 9
14 − 8 = 6	14 − 9 = 5

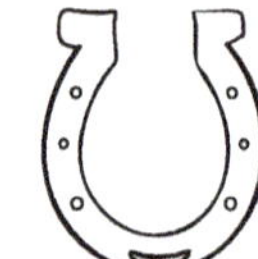

Write each fact family.

6 7 13	5 8 13
6 + 7 = 13	5 + 8 = 13
7 + 6 = 13	8 + 5 = 13
13 − 6 = 7	13 − 5 = 8
13 − 7 = 6	13 − 8 = 5

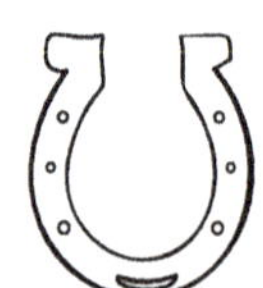

Complete the equation for the word problem.

The cowboy needs to catch 14 horses.
He has caught 7 horses.
How many more horses does the cowboy need to catch?

7 + 7 = 14
horses

Chapter 18 • Lesson 150   two hundred seventy-nine **279**

Apply

Worktext pages 291–92

Read and guide completion of page 291.

Read and explain the directions for page 292. Assist the students as they complete the page independently.

Assess

Reviews pages 279–80

Review comparing the capacity of objects to 1 liter by using the words *more* and *less* on page 280.

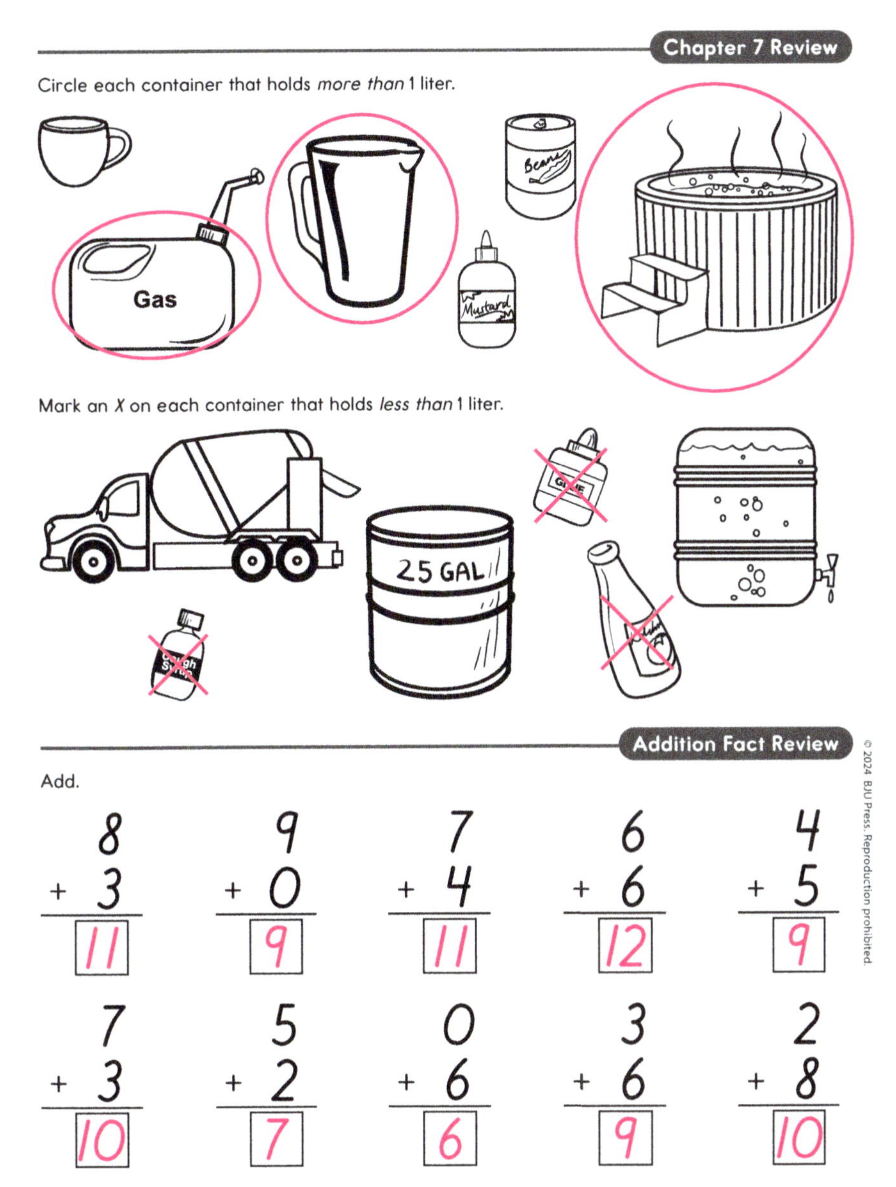

Chapter 7 Review

Circle each container that holds *more than* 1 liter.

Mark an *X* on each container that holds *less than* 1 liter.

Addition Fact Review

Add.

$8 + 3 = 11$	$9 + 0 = 9$	$7 + 4 = 11$	$6 + 6 = 12$	$4 + 5 = 9$
$7 + 3 = 10$	$5 + 2 = 7$	$0 + 6 = 6$	$3 + 6 = 9$	$2 + 8 = 10$

151 Worktext pages 293–94 Reviews pages 281–82

Objectives

- **151.1** Separate a set to demonstrate subtraction from 15.
- **151.2** Write addition and subtraction equations for fact families for 15 and 16.
- **151.3** Solve a word problem by using the Problem-Solving Plan.

Printed Resources

- Student Manipulatives: Counters (4 ducks)
- Student Manipulatives: Ten Bar Mat

Materials

- 16 fish crackers (or UNIFIX® Cubes)

Fact Families for 15, 16

Add or subtract.
Write the numbers for the fact family.

7 + 8 = 15
8 + 7 = 15
15 − 7 = 8
15 − 8 = 7

Write the fact family.

7 + 9 = 16
9 + 7 = 16
16 − 7 = 9
16 − 9 = 7

Write an addition or subtraction equation for each word problem. Solve.

There are 9 orange fish and 7 blue fish. How many fish are there in all?

9 + 7 = 16 fish

There are 15 fish in the tank. The man catches 8 fish in a net. How many fish are left in the tank?

15 − 8 = 7 fish

Practice & Review

Count 600–650

Guide the students as you take turns counting together. Explain that you will start by saying 600, then they will say the next number. Continue to take turns as you count to 650.

Engage

Determine a Fair Share

Guide a **review activity** to help the students determine a fair share.

Distribute the fish crackers and duck counters. Instruct the students to place 12 fish in the middle of their desks as you read the following word problem.

The zookeeper has 12 fish. She wants to share the fish among the 4 ducks. How many fish make a fair share?

How many ducks will share the fish? 4 ducks

How many fish are they sharing? 12 fish

Instruct the students to move 1 fish at a time to each of the 4 ducks, repeating the process until all the fish have been divided into fair shares.

How many fish make a fair share for the 4 ducks? 3 fish

Continue the activity to find the following fair shares.

6 fish shared with 3 ducks fair share = 2 fish

6 fish shared with 2 ducks fair share = 3 fish

Instruct

Separate a Set to Demonstrate Subtraction from 15

Guide a **hands-on activity** to help the students demonstrate subtraction from 15.

Distribute a Ten Bar Mat and 16 fish crackers to each student. Direct the students to place 15 fish on their Ten Bar Mats.

How many fish are on your bar? 10

How many fish are below your bar? 5

Ask the girls to remove 6 fish, ask the boys to remove 9 fish, and instruct all the students to put the fish they removed along the bottom of their mats.

Invite volunteers to give the subtraction equations for the fish they removed. Write the related equations in pairs for display.

15 − 6 = 9 15 − 9 = 6

Continue the activity, removing 7 fish and 8 fish.

15 − 7 = 8 15 − 8 = 7

Leave the equations displayed.

Fact Family Equations for 15

Guide a **discussion** to help the students write fact family equations for 15.

Direct attention to the first pair of displayed equations.

How is each pair of subtraction equations related? They have the same numbers. They are part of the same family.

Add or subtract.
Write the numbers for the fact family.

8 + 8 = 16

16 − 8 = 8

Write the fact family.

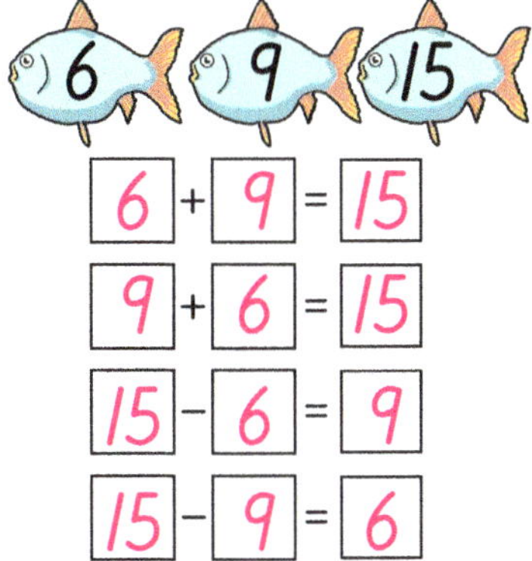

6 + 9 = 15

9 + 6 = 15

15 − 6 = 9

15 − 9 = 6

Write an equation for each word problem. Solve.

The pet store has 15 guppies for sale.
Erin buys 9 guppies.
How many guppies does the pet store have left to sell?

   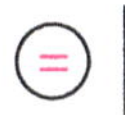 

15 − 9 = 6 guppies

The pet store has a tank with 8 catfish and 8 glassfish.
How many fish are in the tank?

8 + 8 = 16 fish

Time to Review

Circle the fair share for each child.
Write the number each child gets.

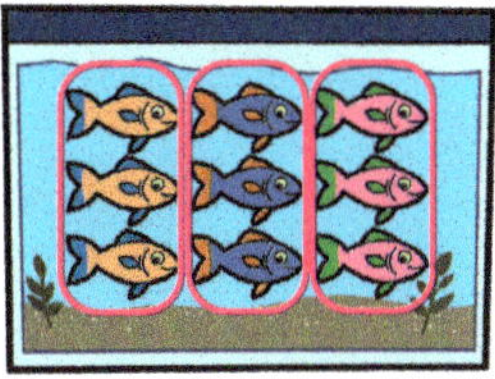

3 children get 3 each.

4 children get 2 each.

Draw a box to the left of each pair of equations.

Instruct the students to look at the numbers in each pair. Guide them as they determine the fact family to which the equations belong. Write the numbers in each box for display.

6 9 15 7 8 15

What is missing from each family? addition equations

Ask volunteers to write the addition equations that are a part of each family.

9 + 6 = 15 6 + 9 = 15

8 + 7 = 15 7 + 8 = 15

Fact Family Equations for 16

Use **manipulatives** to help the students write fact family equations for 16.

Write the numbers 7, 9, and 16 in a box. Invite the students to help you write the facts in this fact family.

7 9 16

7 + 9 = 16 16 − 7 = 9

9 + 7 = 16 16 − 9 = 7

How many addition equations are in the family? 2

How many subtraction equations are in the family? 2

Guide the students as they use 16 fish crackers and their Ten Bar Mats to check the equations.

Instruct the students to listen as you read the following facts so they can identify the 3 numbers in the fact family:
8 + 8 = 16, 16 − 8 = 8. 8-8-16

Why is there only 1 addition equation for this fact family? It is a double; there are two 8s.

Why is there no related subtraction equation for 16 − 8 = 8? It would be the same; it is a double; there are two 8s.

Problem-Solving Plan

Guide the students as they use the **Problem-Solving Plan** to solve a word problem.

Read the following word problem.

DQ saw 9 channel catfish and 6 flathead catfish. How many catfish did DQ see?

What is the question? How many catfish did DQ see?

What information is given? DQ saw 9 channel catfish and 6 flathead catfish.

Do you add or subtract to find the answer? add

What is the equation that you can use to solve this word problem? 9 + 6 = __

Write "9 + 6 = __" for display. Instruct the students to use the fish crackers and their Ten Bar Mats to solve the problem.

Ask a volunteer to write the answer for display. 15

Write *catfish* after the answer.

Does the answer make sense? yes

Continue the activity with the following word problems.

DQ saw a diver come down into the large tank at the aquarium. There were 16 fish swimming around him. When the fish were not fed, 9 fish swam away. How many fish were left swimming around the diver? 16 − 9 = 7 fish

Meg counted 7 seahorses in the tank. In the next tank, she counted 8 seahorses. How many seahorses did Meg count? 7 + 8 = 15 seahorses

Apply

Worktext pages 293–94

Read and guide completion of page 293.

Read and explain the directions for page 294. Assist the students as they complete the page independently.

Assess

Reviews pages 281–82

Review shapes on page 282.

Extended Activity

Complete a Fact Family

Materials

- A sheet of drawing paper for each student

Procedure

Distribute the drawing paper. Instruct each student to draw a picture of his or her favorite zoo animal. Ask the students to draw their animals large enough to write fact families on them. Instruct the students to write a fact family on their animals.

Fact Families for 15, 16

Add or subtract.
Write the numbers for the fact family.

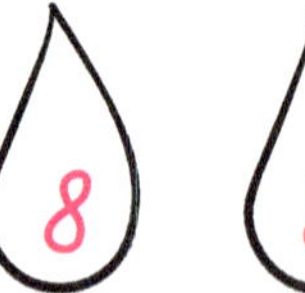

8 + 8 = 16

16 − 8 = 8

Write the fact family.

6 + 9 = 15

9 + 6 = 15

15 − 6 = 9

15 − 9 = 6

Add or subtract. Use the key to color.

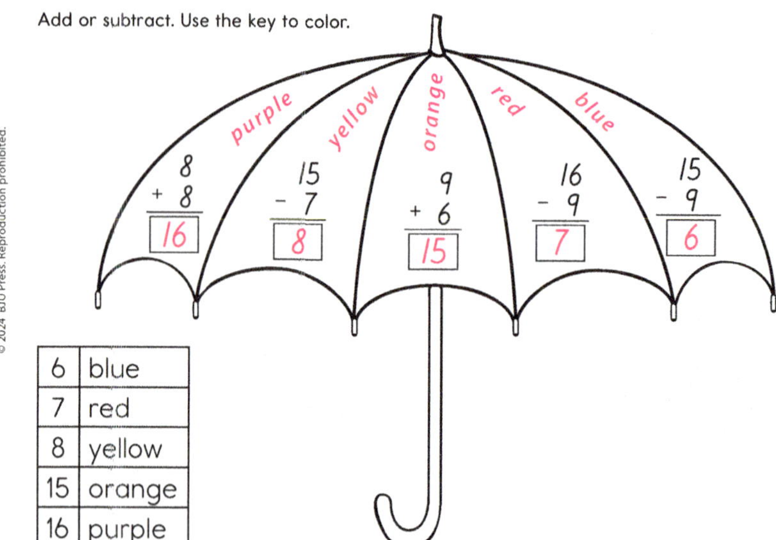

6	blue
7	red
8	yellow
15	orange
16	purple

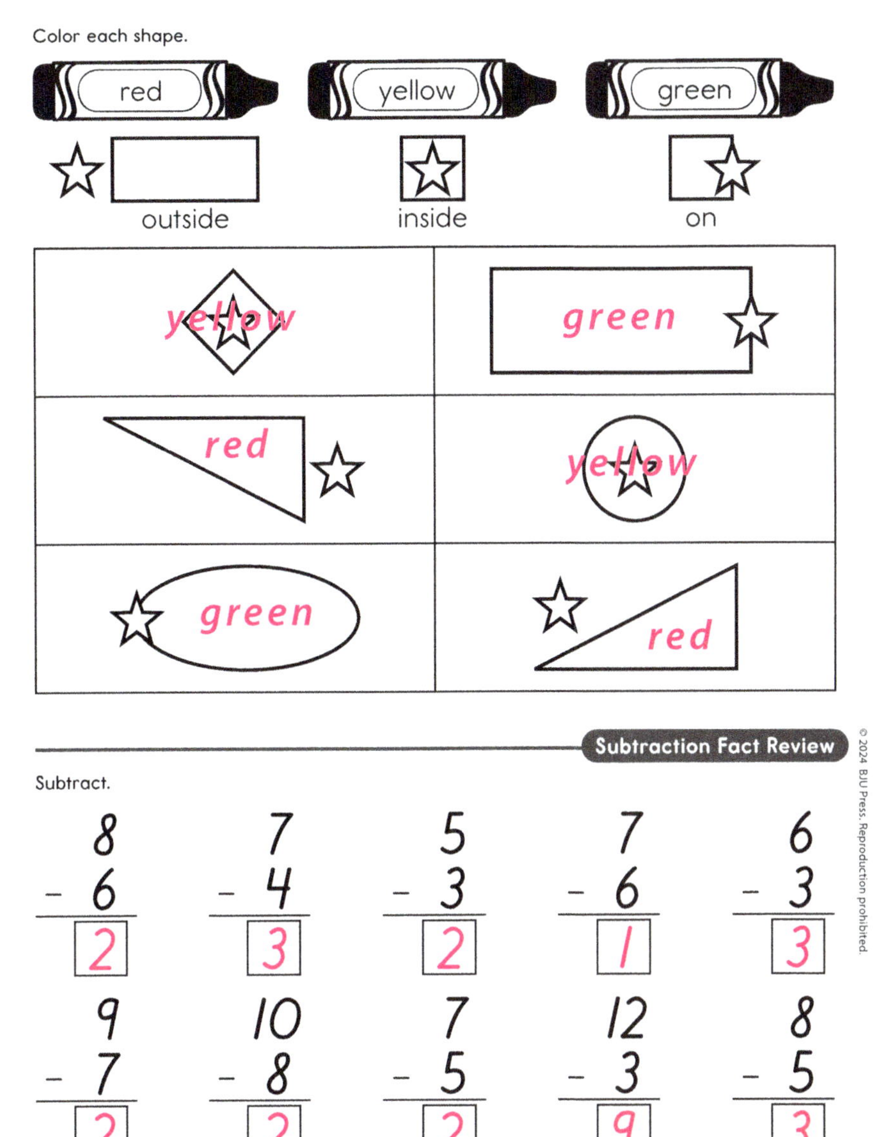

Color each shape.

red — outside
yellow — inside
green — on

yellow	green
red	yellow
green	red

Subtraction Fact Review

Subtract.

$8 - 6 = 2$ | $7 - 4 = 3$ | $5 - 3 = 2$ | $7 - 6 = 1$ | $6 - 3 = 3$

$9 - 7 = 2$ | $10 - 8 = 2$ | $7 - 5 = 2$ | $12 - 3 = 9$ | $8 - 5 = 3$

152 Worktext pages 295–96
Reviews pages 283–84

Objectives

- **152.1** Write addition and subtraction equations for fact families for 17 and 18.
- **152.2** Solve a multistep word problem.

Printed Resources

- Visuals: Counters (8 ducks)
- Student Manipulatives: Number Cards (0–9)
- Student Manipulatives: Red Mat
- Student Manipulatives: Counters (8 ducks)

Digital Resources

- Games/Enrichment: Addition Flashcards (facts to 12)
- Games/Enrichment: Fact Fun Activities

Practice & Review

Study Addition Facts to 12

Choose a game or activity from the "Fact Fun Activities" available in Trove.

Engage

Numbers That Come *before*, *after*, or *between*

Guide a **review** of sequencing 3-digit numbers.

Distribute Number Cards 0–9. Ask the students to use their cards to show the number that matches each clue.

Ask a volunteer to write the correct number for display for the students to check their cards.

Which number comes *before* 368? 367

Which number comes *after* 452? 453

Which number comes *between* 311 and 313? 312

Continue the activity with similar clues.

Instruct

Fact Family Equations for 17 & 18

Guide a **discussion** to help the students write fact family equations for 17 and 18.

Fact Families for 17, 18; Multistep Problems

Write each fact family.

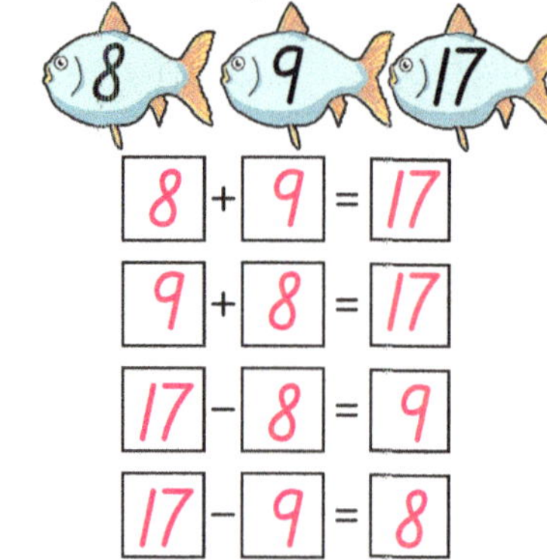

Solve the multistep problem.

Kirsten saw 10 birds drinking water. After drinking, 5 birds hopped away. Then 3 more birds came to drink. How many birds were left drinking water now?

Subtract.

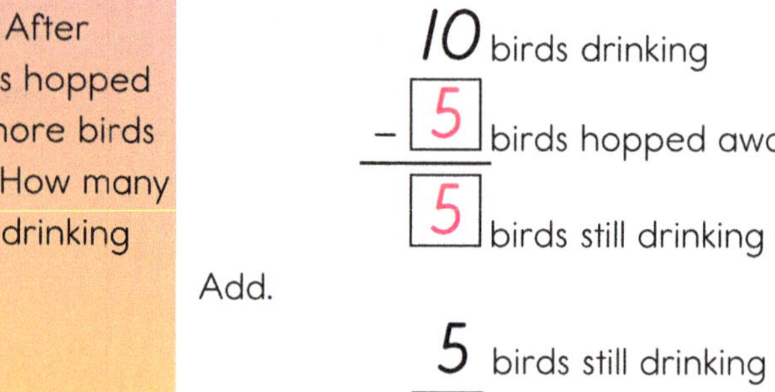

Add.

5 birds still drinking
+ 3 more birds came
8 birds in all

Write the numbers 8, 9, and 17 in a box. Guide the students as they help you write the facts in the fact family.

8 9 17

8 + 9 = 17   17 − 8 = 9
9 + 8 = 17   17 − 9 = 8

How many addition equations are in the family? 2

How many subtraction equations are in the family? 2

Instruct the students to listen for the 3 numbers in this fact family: 9 + 9 = 18, 18 − 9 = 9. 9-9-18

Why are there not 2 addition equations for this family? They would be the same; it is a double; there are two 9s.

Why is there not a related subtraction equation for 18 − 9 = 9? It would be the same; it is a double; there are two 9s.

Multistep Problems

Use **manipulatives** to help the students solve a multistep problem.

Distribute the Red Mats and ducks. Read the following word problem.

Ken saw 3 ducks in the pool at the zoo. Then 5 ducks joined them in the water. While Ken watched the ducks, 2 ducks got out of the water. How many ducks were left in the pool?

Explain that to find the answer to this word problem the students will need to answer several questions.

Solve the multistep problem.

Dan saw 3 birds sitting on a branch. Then 5 more birds came to sit on the branch. The 2 littlest birds flew away. How many birds were left?

Add.

3 birds sitting
+ 5 more birds came
8 birds in all

Subtract.

8 birds in all
− 2 birds flew away
6 birds are left

Add or subtract.
Write the numbers for each fact family.

9 + 9 = 18
18 − 9 = 9

8 + 9 = 17
9 + 8 = 17
17 − 9 = 8
17 − 8 = 9

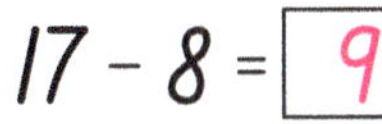

Time to Review

Add.

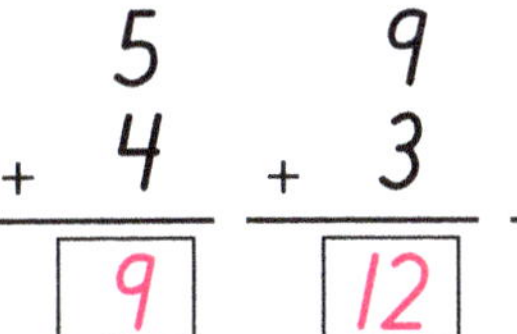

5	9	7	6	2	5
+ 4	+ 3	+ 5	+ 4	+ 8	+ 6
9	12	12	10	10	11

Demonstrate with ducks as you explain each step. Reread the word problem.

How many ducks were there in the beginning? 3 ducks

Instruct the students to place 3 ducks on the left side of their mats.

Write "3" under the 3 ducks you displayed.

How many ducks joined them? 5 ducks

Instruct the students to place 5 more ducks in the middle of their mats.

Write "5" under your ducks. Reread the word problem, stressing that 5 ducks joined the 3 ducks.

Do you add or subtract to find the answer? add

Add "+" and "=" to the equation: 3 + 5 =.

Direct the students to join the 2 groups of ducks together on their mats.

How many ducks are there in all? 8 ducks

Write "8" to complete the equation.

Ask the students to move their 8 ducks to the left side of their mats. Erase the equation, moving your ducks to the left and writing "8" beneath them.

Read the word problem again.

How many ducks got out of the water? 2 ducks

Do you add or subtract to find the answer? subtract

Instruct the students to move 2 ducks to the right side of their mats.

How many ducks are left? 6 ducks

Complete the equation: 8 − 2 = 6 ducks. Read the word problem again.

Does the answer make sense? yes

Instruct the students to remove all the ducks from their mats. Read aloud the following word problem.

Meg saw 6 ducks outside. Later 4 ducks went inside the duck house; then 3 ducks came out. How many ducks were outside then?

How many ducks did Meg see first? 6 ducks

Instruct the students to place 6 ducks on the left side of their mats.

How many ducks went inside the duck house? 4 ducks

Direct the students to move 4 ducks to the right side of their mats.

Do you add or subtract to find the answer? subtract

Invite a student to state the equation. 6 − 4 = 2 ducks

Write the equation for display. Read aloud the word problem again.

How many ducks came out of the house? 3 ducks

Instruct the students to place 3 ducks back on the left side of their mats.

Do you add or subtract to find out how many ducks were outside at the end of the problem? add

Invite a volunteer to state the equation. 2 + 3 = 5 ducks

Write the equation for display. Read aloud the word problem again.

Does the answer make sense? yes

Apply

Worktext pages 295–96

Read and guide completion of page 295.

Read and explain the directions for page 296. Assist the students as they complete the page independently.

152

Assess

Reviews pages 283–84

Review comparing numbers by using > and < on page 284.

Extended Activity

Recognize Fact Families

Materials

- Addition and subtraction equation cards for fact families for 13–18 (a total of 42 cards)

4-9-13	5-8-13	6-7-13
5-9-14	6-8-14	7-7-14
6-9-15	7-8-15	
7-9-16	8-8-16	
8-9-17		
9-9-18		

Procedure

To play "Go to the Zoo!" shuffle the fact family cards. Distribute all the cards to the players. Ask the players to put their cards together in fact families. Remind the students that some fact families have 4 equations and some have only 2 equations. The first player pulls a card from the hand of the player to his or her right and tries to match it to a fact family in his or her hand. Play continues around the circle. When a player has all the cards for a fact family, he or she places them face-up on the table. Play continues until all the fact families have been collected. The winner is the player with the most fact families.

Adjust the number of fact families included so that the players do not have too many cards to hold. A card holder for each player can be made by taking 2 plastic lids (for example, butter tub lids), placing them back-to-back, and stapling them in the middle. The cards can be put between the lids to hold them.

Fact Families for 17, 18; Multistep Problems

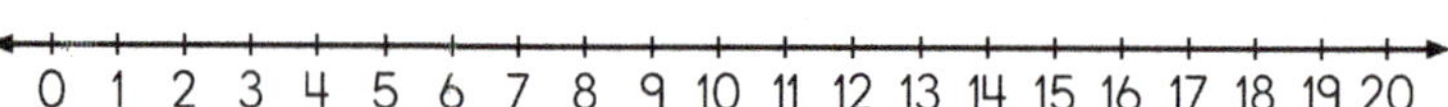

Add or subtract. Use the number line if needed.

$8 + 9 = 17$ $9 + 9 = 18$ $17 - 8 = 9$

$18 - 9 = 9$ $17 - 9 = 8$ $9 + 8 = 17$

Solve the multistep problem.

Shelby stopped to feed 1 duck.
Then 3 more ducks came to be fed.
After eating, 2 ducks waddled off.
How many ducks stayed with Shelby?

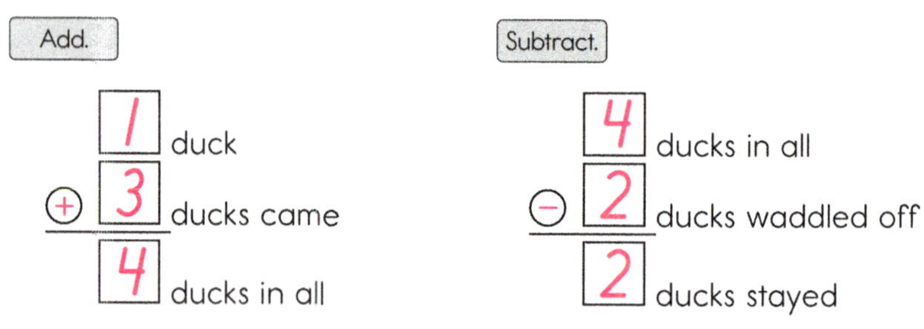

Chapter 16 Review

Draw a dot next to the smaller number. Draw the correct sign.

$>$	
is greater than	is less than

$254 < 361$	$596 < 782$
$849 > 633$	$415 < 420$
$300 > 200$	$721 < 785$

Subtraction Fact Review

Mark the correct circle.

Problem	Choices	Marked
$8 + 4 =$	10, 11, 12	12
$7 + 3 =$	10, 11, 12	10
$6 + 2 =$	8, 9, 10	8
$9 + 1 =$	8, 9, 10	10
$5 + 5 =$	10, 11, 12	10
$6 + 5 =$	9, 10, 11	11
$7 + 2 =$	9, 10, 11	9
$9 + 3 =$	10, 11, 12	12
$6 + 3 =$	9, 10, 11	9
$5 + 7 =$	10, 11, 12	12

153

Worktext pages 297–98
Reviews pages 285–86

How can math help me take care of animals?

Chapter Concept Review

- Practice concepts from Chapter 18 to prepare for the test.

Printed Resources

- Instructional Aid 44: *Ten Bar*
- Student Manipulatives: Ten Bar Mat

Digital Resources

- Web Link: Virtual Manipulatives: Number Frames
- Web Link: Virtual Manipulatives: UNIFIX® Cubes
- Web Link: Virtual Manipulatives: Base Ten Number Pieces

Materials

- 20 UNIFIX® Cubes (10 each of 2 different colors; for the teacher and for each student)

Color the first addend. Complete the Ten Bar with the second addend. Add.

10 + 4 = 14

7 + 5 = 12

6 + 8 = 14

9 + 6 = 15

Add.

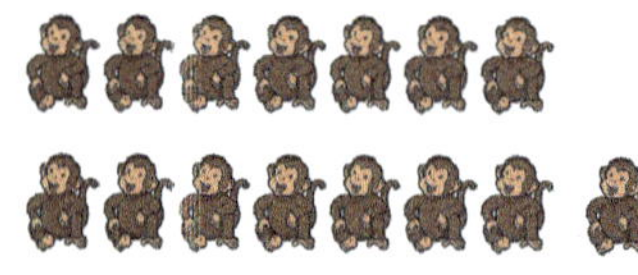

7 + 7 = 14

7 + 8 = 15

9 + 9 = 18

9 + 10 = 19

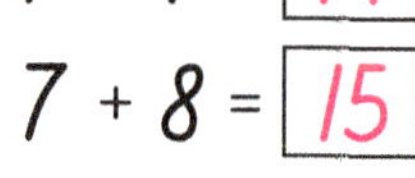

Add.

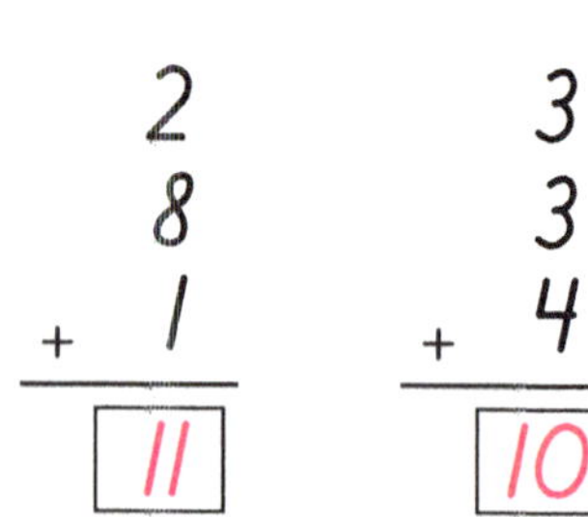

2	3	6	5
8	3	6	2
+ 1	+ 4	+ 0	+ 2
11	10	12	9

Practice & Review

Count 650–700

Group the students in 2 groups. Instruct the students in Group 1 to pretend that they are lions counting in a deep voice and Group 2 to pretend they are monkeys counting in a high voice. Ask Group 1 to begin counting with the number 650 and to continue counting until you clap your hands. Ask Group 2 to *count on* from the last number counted by Group 1 until you clap your hands. Continue to alternate groups until a group reaches 700.

Instruct

Add 10 by Using the Ten Bar; Add 6, 7, 8, or 9 by Making 10

Distribute the Ten Bar Mats and the UNIFIX Cubes. Display the Ten Bar. Write "10 + 4 = __" for display near the Ten Bar. Demonstrate each step.

What do you do first to solve this problem? place 10 cubes in the spaces on the Ten Bar

What is the next step? place 4 cubes below the bar

Where can you begin counting? *count on* from 10

Guide the students as they *count on* from 10: 10, 11, 12, 13, 14.

What is the sum of 10 + 4? 14

Complete the equation.

Write "6 + 7 = __" for display.

What do you do first to solve this problem? place 6 cubes in the spaces on the Ten Bar

What is the next step? place 7 cubes below the bar

What do you do next? move 4 cubes into the spaces on the Ten Bar to make a group of 10

Where can you begin counting? *count on* from 10

Guide the students as they *count on* from 10: 10, 11, 12, 13.

What is the answer? 13

Complete the equation.

Continue the activity with the following equations.

9 + 7 = 16   8 + 6 = 14

Add Doubles & Near Doubles

Direct the students to turn over their Ten Bar Mats to the Red Mat side. Instruct them to put 2 groups of 8 UNIFIX Cubes on their mats.

What double fact do the cubes show? 8 + 8

Write each fact family.

7 7 14
7 + 7 = 14
14 − 7 = 7
6 8 14
6 + 8 = 14
8 + 6 = 14
14 − 6 = 8
14 − 8 = 6

9 9 18

9 + 9 = 18

18 − 9 = 9

7 9 16

7 + 9 = 16

9 + 7 = 16

16 − 7 = 9

16 − 9 = 7

Solve the multistep problem.

Joshua saw 3 ducks waddling. He saw 4 ducks quacking. 2 ducks left to swim in the pond. How many ducks were left?

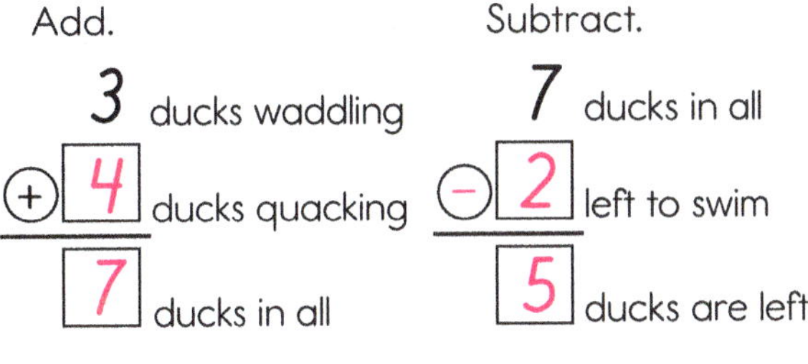

Circle the phrase that completes the sentence.

The students helped feed the monkeys. Math helped them plan the treats for the animals. Sometimes I must play more / work more to finish a job.

How do you know this is a double fact? Both addends are the same.

What is the sum of 8 + 8? 16

Write "8 + 8 = 16" for display.

Instruct the students to place a cube at the end of the second row.

Which equation is represented now? 8 + 9

Write "8 + 9 = __" for display. Remind the students that they are adding 1 more to the double fact 8 + 8.

What is the sum of 8 + 9? 17

Continue the activity with 7 + 7 = 14 and 7 + 8 = 15.

Solve a Multistep Problem

Read aloud the following word problem.

DQ saw 4 monkeys swinging from the branches. Then 7 monkeys joined them. Later 3 monkeys went to eat. How many monkeys were left swinging in the branches?

Demonstrate with UNIFIX Cubes as you explain each step. Reread the word problem.

How many monkeys were there in the beginning? 4 monkeys

Direct the students to place 4 UNIFIX Cubes on the left side of their mats.

How many monkeys joined them? 7 monkeys

Instruct the students to place 7 more cubes in the middle of their mats. Reread the word problem again, stressing that the 7 monkeys joined the 4 monkeys.

Do you add or subtract to find the answer? add

Instruct the students to join the 2 groups of cubes together on their mats.

How many cubes are there in all? 11

What is the equation? 4 + 7 = 11

Direct the students to move 11 cubes to the left of their mats.

Read the word problem aloud again.

How many monkeys went to eat? 3 monkeys

Do you add or subtract to find the answer? subtract

Guide the students as they move 3 cubes to the right side of their mats.

How many monkeys are left? 8 monkeys

Write "11 − 3 = 8 monkeys" for display. Read the word problem aloud again.

Does the answer make sense? yes

Continue the activity with the following word problem.

Stephen saw 10 baboons playing on the hill; then 6 baboons disappeared. While Stephen was watching, 4 baboons came back to play. How many baboons are playing on the hill now? 10 − 6 = 4; 4 + 4 = 8 baboons

Write Equations for Fact Families

Write the numbers 5, 8, and 13 in a box. Explain that you want the students to help you write the facts for this fact family.

5 8 13

5 + 8 = 13 13 − 5 = 8

8 + 5 = 13 13 − 8 = 5

How many addition equations are in the family? 2

How many subtraction equations are in the family? 2

Continue the activity with the following fact family.

9 6 15

6 + 9 = 15 15 − 6 = 9

9 + 6 = 15 15 − 9 = 6

Instruct the students to listen as you read the following facts so that they can provide the 3 numbers in the fact family:

7 + 7 = 14, 14 − 7 = 7. 7-7-14

153

Why are there not 2 addition equations? They would be the same; it is a double; there are two 7s.

Why is there not a related subtraction equation for 14 – 7 = 7? It would be the same; it is a double; there are two 7s.

Solve 3-Addend Equations; Apply the Grouping Principle of Addition

Write "3 + 7 + 7" vertically for display.

Instruct the students to copy the problem. Instruct the boys to begin at 3, add down, and then write the answer. Instruct the girls to begin at 7, add up, and then write the answer.

What is the sum? 17

Which addends did the boys group together when adding down? 3, 7

Which addends did the girls group together? 7, 7

Which principle did you use when some of you were adding up and some were adding down? Grouping Principle

How might this principle help when working addition problems? I can use strategies or facts to solve the problem more easily. It is also a way to check answers.

Continue the activity with 2 + 2 + 8 = __. 12

Apply

Worktext pages 297–98

Read and explain the directions for the pages. Assist the students as they complete the pages independently.

Direct attention to the biblical worldview shaping section on page 298. Display the picture from the chapter opener and read aloud the chapter essential question. Review the math work of adding doubles that the students did to help put the correct number of peanuts in the monkeys' buckets at the zoo. Review the 3-digit addend word problem they solved in another lesson to find out how much monkey chow to feed the monkeys.

Remind the students that the Bible encourages us to do our work joyfully to please God, even when it may take more work to complete a task. Guide the students as they complete the response on page 298.

Color each addend. Add. Write the answer.

10 + 6 = 16

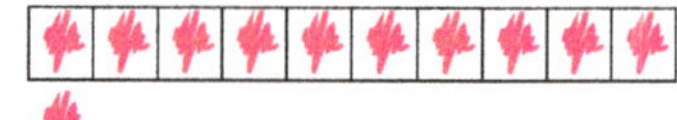

7 + 4 = 11

9 + 1 = 10

9 + 7 = 16

Add.

8 + 8 = 16

9 + 8 = 17

7 + 7 = 14

7 + 8 = 15

9 + 9 = 18

9 + 10 = 19

6 + 6 = 12

6 + 7 = 13

Assess

Reviews pages 285–86

Use the Reviews pages to provide additional preparation for the chapter test.

Write each fact family.

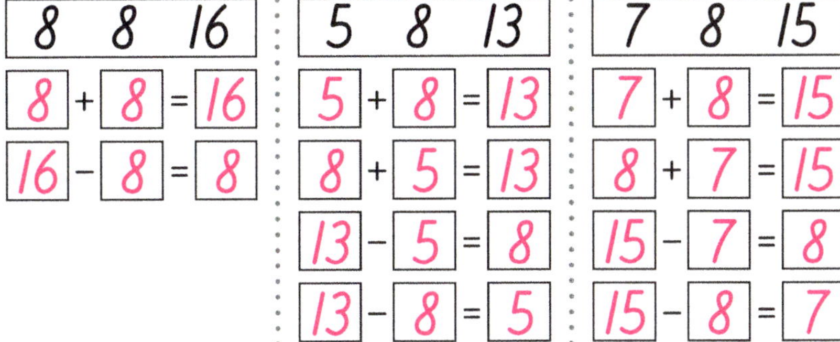

8 8 16	5 8 13	7 8 15
8 + 8 = 16	5 + 8 = 13	7 + 8 = 15
16 − 8 = 8	8 + 5 = 13	8 + 7 = 15
	13 − 5 = 8	15 − 7 = 8
	13 − 8 = 5	15 − 8 = 7

Solve the multistep problem.

The team had 8 footballs.
They lost 4 footballs.
The coach found 2 footballs in a bag.
How many footballs does the team have now?

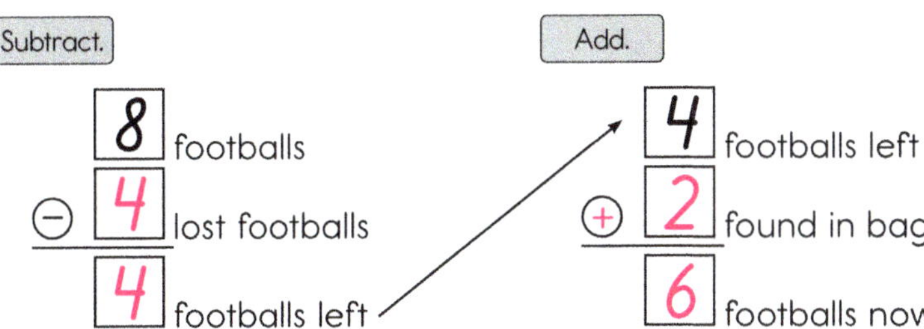

Subtract.

8 footballs
⊖ 4 lost footballs
4 footballs left

Add.

4 footballs left
⊕ 2 found in bag
6 footballs now

Add.

6	8	9	2
3	2	1	7
+ 3	+ 3	+ 1	+ 3
12	13	11	12

Chapter 18 Test

Administer the Chapter 18 Test.

Cumulative Concept Review

Worktext page 299

Review the following concepts. Adapt instructions and activities and provide reteaching as needed to meet the specific needs of your students.

- Identifying sets of pennies, nickels, and dimes equivalent to one quarter (Lesson 111)
- Writing the number that is 10 more or 10 less (Lesson 128)
- Measuring the length or height of an object in centimeters (Lesson 136)

Retain a copy of Worktext page 299 for the discussion of the Chapter 19 essential question during Lessons 155 and 163.

Reviews pages 287–88

Use the Reviews pages to help students retain previously learned skills.

Extended Activity

Add by Making 10

Materials

- 18 plastic eggs or UNIFIX® Cubes
- 1 egg carton with 2 sections cut off, leaving 2 rows of 5 sections each
- 14 index cards labeled with an equation on the front side and the answer on the back

$9 + 1 = 10$ $9 + 5 = 14$ $7 + 4 = 11$ $8 + 5 = 13$
$9 + 2 = 11$ $9 + 6 = 15$ $7 + 5 = 12$ $8 + 6 = 14$
$9 + 3 = 12$ $9 + 7 = 16$ $8 + 3 = 11$
$9 + 4 = 13$ $9 + 8 = 17$ $8 + 4 = 12$

Procedure

Instruct a student to select an index card. Ask him or her to place eggs in the egg carton to represent the first addend and eggs below the egg carton to represent the second addend. Remind the student to *count on* from 10 to find the answer. Instruct him or her to turn the index card over to check the answer.

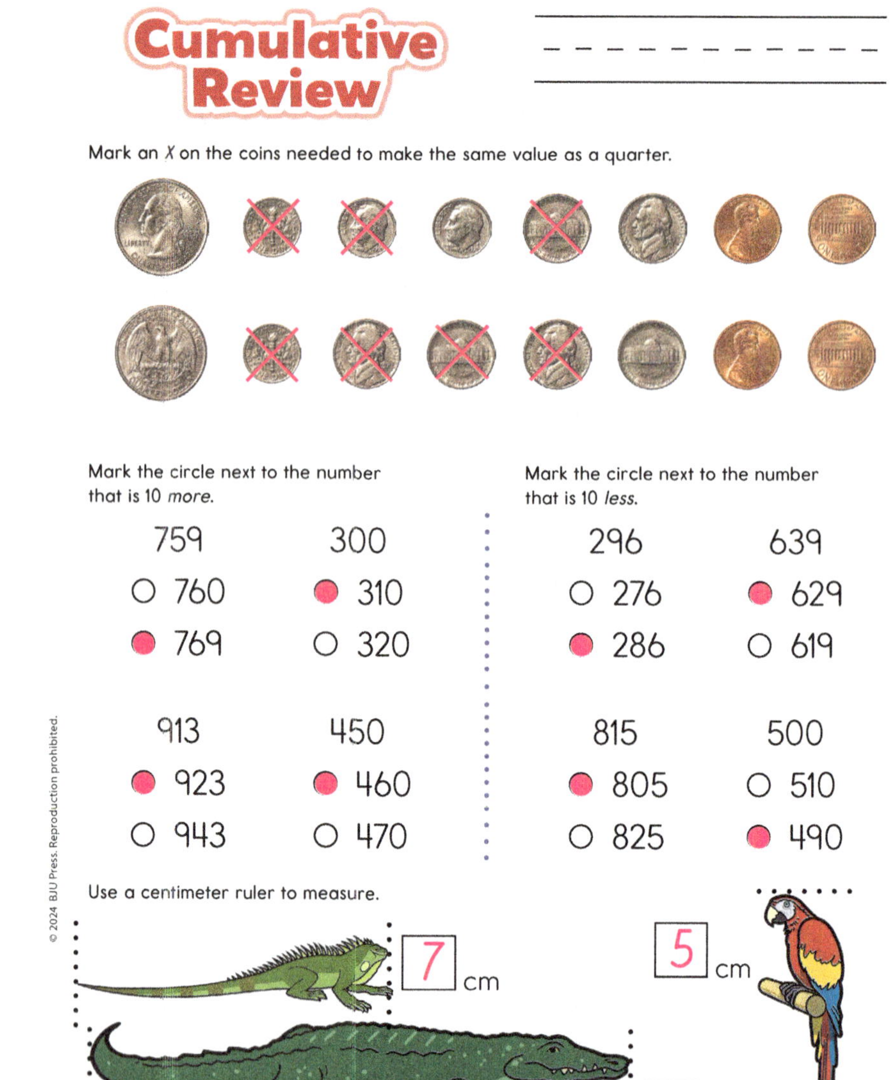

Cumulative Review

Mark an *X* on the coins needed to make the same value as a quarter.

Mark the circle next to the number that is 10 *more.*

759
○ 760
● 769

300
● 310
○ 320

913
● 923
○ 943

450
● 460
○ 470

Mark the circle next to the number that is 10 *less.*

296
○ 276
● 286

639
● 629
○ 619

815
● 805
○ 825

500
○ 510
● 490

Use a centimeter ruler to measure.

7 cm

12 cm

5 cm

Write the total value.

      42¢

     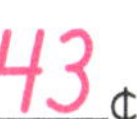 43¢

      46¢

Write the value of each underlined digit.

 10  700  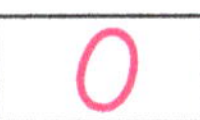 0

Color the correct clock.

4:00 

12:00   

5:30 

1:30  

Addition Fact Review

Add.

2 + 4 = 6	8 + 3 = 11	7 + 2 = 9	9 + 1 = 10	6 + 6 = 12
4 + 5 = 9	5 + 6 = 11	4 + 4 = 8	7 + 5 = 12	8 + 4 = 12
7 + 3 = 10	4 + 6 = 10	6 + 5 = 11	3 + 2 = 5	9 + 3 = 12
8 + 2 = 10	3 + 6 = 9	3 + 5 = 8	3 + 4 = 7	4 + 2 = 6
8 + 1 = 9	5 + 5 = 10	6 + 2 = 8	7 + 4 = 11	9 + 0 = 9

CHAPTER 19: WRITE EQUATIONS WITH COINS

PAGES	OBJECTIVES	RESOURCES	ASSESSMENTS
Lesson 155	**Counting Pennies with a Quarter**		
Teacher Edition 665–69 **Worktext** 300–302	**155.1** Determine the value of a set of dimes, nickels, and pennies by *counting on*. **155.2** Determine the value of a set of 1 quarter and pennies by *counting on*. **155.3** Determine whether there is enough money to purchase an item. **155.4** Identify the coins needed to purchase an item. **155.5** Explain that counting pennies with a quarter helps us explore the world by counting more quickly. BWS Exploring (explain)	**Teacher Edition** • Instructional Aid 9: *Price Tags* • Instructional Aid 45: *Coin Purse Mat* **Visuals** • Visual 14: *Money* • Money Kit: 9 pennies, 5 nickels, 5 dimes, 1 quarter **Student Manipulatives Packet** • Number Cards: 0–20 • Money Kit: 1 quarter, 3 dimes, 5 nickels, 7 pennies **BJU Press Trove*** • Video: Ch 19 Intro • Web Link: Virtual Manipulatives: Analog Clock • Web Link: Learning Coins • Web Link: Virtual Manipulatives: Money Pieces • Games/Enrichment: Addition Flashcards • Games/Enrichment: Subtraction Flashcards • PowerPoint® presentation	**Reviews** • Pages 289–90
Lesson 156	**Counting Nickels with a Quarter**		
Teacher Edition 670–73 **Worktext** 303–4	**156.1** Determine the value of a set of dimes, nickels, and pennies. **156.2** Determine the value of a set of 1 quarter and nickels. **156.3** Identify the coins needed to purchase an item. **156.4** Solve money word problems.	**Teacher Edition** • Instructional Aid 45: *Coin Purse Mat* **Visuals** • Visual 1: *Hundred Chart* • Visual 14: *Money* • Money Kit: 2 quarters, 8 dimes, 5 nickels, 6 pennies **Student Manipulatives Packet** • Number Cards: 0–20 • Money Kit: 2 quarters, 8 dimes, 5 nickels, 6 pennies **BJU Press Trove** • Video: Nickels with Quarter • Web Link: Learning Coins • Web Link: Virtual Manipulatives: Money Pieces • Games/Enrichment: Addition Flashcards • Games/Enrichment: Subtraction Flashcards • PowerPoint® presentation	**Reviews** • Pages 291–92

*Digital resources for homeschool users are available on Homeschool Hub.

PAGES	OBJECTIVES	RESOURCES	ASSESSMENTS
Lesson 157	**Counting Dimes with a Quarter**		
Teacher Edition 674–77 **Worktext** 305–6	**157.1** Determine the value of a set of dimes, nickels, and pennies up to 99¢. **157.2** Determine the value of a set of 1 quarter and dimes. **157.3** Identify the coins needed to purchase an item. **157.4** Solve money word problems.	**Teacher Edition** • Instructional Aid 45: *Coin Purse Mat* **Visuals** • Visual 1: *Hundred Chart* • Money Kit: 1 quarter, 6 dimes, 4 nickels, 7 pennies **Student Manipulatives Packet** • Number Cards: 0–20 • Money Kit: 1 quarter, 4 dimes, 4 nickels, 5 pennies **BJU Press Trove** • Video: Dimes with Quarter • Web Link: Learning Coins • Web Link: Virtual Manipulatives: Money Pieces • Games/Enrichment: Addition Flashcards • Games/Enrichment: Subtraction Flashcards • PowerPoint® presentation	**Reviews** • Pages 293–94
Lesson 158	**Counting with 2 Quarters**		
Teacher Edition 678–81 **Worktext** 307–8	**158.1** Determine the value of a set of 1 quarter and pennies and a set of 2 quarters and pennies. **158.2** Determine the value of a set of 1 quarter and nickels and a set of 2 quarters and nickels. **158.3** Determine the value of a set of 1 quarter and dimes and a set of 2 quarters and dimes. **158.4** Identify the coins needed to purchase an item. **158.5** Solve money word problems. **158.6** Express praise to God for a world where math works so well. BWS Exploring (formulate)	**Teacher Edition** • Instructional Aid 45: *Coin Purse Mat* **Visuals** • Money Kit: 2 quarters, 4 dimes, 6 nickels, 10 pennies **Student Manipulatives Packet** • DQ Clock • Number Cards: 0–20 • Money Kit: 2 quarters, 4 dimes, 4 nickels, 10 pennies **BJU Press Trove** • Web Link: Learning Coins • Web Link: Virtual Manipulatives: Money Pieces • Games/Enrichment: Addition Flashcards • Games/Enrichment: Subtraction Flashcards • PowerPoint® presentation	**Reviews** • Pages 295–96
Lesson 159	**Equal Sets**		
Teacher Edition 682–85 **Worktext** 309–10	**159.1** Identify the number of sets and the number of objects in each set. **159.2** Count by 2s, 3s, or 5s to find the total. **159.3** Make equal sets.	**Teacher Edition** • Instructional Aid 46: *Repeated Addition Mat* **Visuals** • Money Kit: 20 pennies **Student Manipulatives Packet** • Money Kit: 20 pennies **BJU Press Trove** • PowerPoint® presentation	**Reviews** • Pages 297–98

PAGES	OBJECTIVES	RESOURCES	ASSESSMENTS
Lesson 160 Repeated Addition			
Teacher Edition 686–89 **Worktext** 311–12	**160.1** Identify the number of sets and the number of objects in each set. **160.2** Count by 2s, 3s, and 5s to find the total. **160.3** Make equal sets. **160.4** Complete a repeated addition problem. **160.5** Write a repeated addition equation for a word problem.	**Teacher Edition** • Instructional Aid 46: *Repeated Addition Mat* **Visuals** • Counters: 20 cars **Student Manipulatives Packet** • Counters: 20 cars **BJU Press Trove** • PowerPoint® presentation	**Reviews** • Pages 299–300
Lessons 161–62 Writing Repeated Addition Equations			
Teacher Edition 690–93 **Worktext** 313–14	**161–62.1** Identify the number of sets and the number of objects in each set. **161–62.2** Count by 2s, 3s, and 5s to find the total. **161–62.3** Write a repeated addition equation for sets of objects. **161–62.4** Write a repeated addition equation for a word problem.	**Visuals** • Counters: 20 candies **BJU Press Trove** • PowerPoint® presentation	**Reviews** • Pages 301–2
Lesson 163 Chapter 19 Review			
Teacher Edition 694–697 **Worktext** 315–16	**163.1** Recall concepts and terms from Chapter 19.	**Teacher Edition** • Instructional Aid 45: *Coin Purse Mat* **Visuals** • Money Kit: 2 quarters, 7 dimes, 7 nickels, 5 pennies **Student Manipulatives Packet** • Money Kit: 2 quarters, 7 dimes, 7 nickels, 5 pennies **BJU Press Trove** • Web Link: Virtual Manipulatives: Money Pieces • Games/Enrichment: Addition Flashcards • Games/Enrichment: Subtraction Flashcards • PowerPoint® presentation	**Worktext** • Chapter 19 Review **Reviews** • Pages 303–4
Lesson 164 Test, Cumulative Review			
Teacher Edition 698–700 **Worktext** 317	**164.1** Demonstrate knowledge of concepts from Chapter 19 by taking the test.		**Assessments** • Chapter 19 Test **Worktext** • Cumulative Review **Reviews** • Pages 305–6

Chapter 19

19 Write Equations with Coins

What should I tell God for giving me a world where math works well?

Find the red and blue cars and figure out the total. 7 + 8 = 15

What should I tell God for giving me a world where math works well?

Chapter Objectives

- Determine the value of a set of pennies, nickels, dimes, and quarters by *counting on.*
- Compare the value of a set of coins to a purchase price.
- Identify the coins needed for a purchase.
- Solve money word problems by writing equations.
- Praise God for making a world where math works well.

155

Worktext pages 300–302
Reviews pages 289–90

Objectives

- **155.1** Determine the value of a set of dimes, nickels, and pennies by *counting on.*
- **155.2** Determine the value of a set of 1 quarter and pennies by *counting on.*
- **155.3** Determine whether there is enough money to purchase an item.
- **155.4** Identify the coins needed to purchase an item.
- **155.5** Explain that counting pennies with a quarter helps us explore the world by counting more quickly. BWS

Biblical Worldview Shaping

- **Exploring** (explain): As we explore God's world and learn more about things in it, we are able to work more efficiently. Counting pennies with a quarter is one way to count quickly and save time to explore other things (155.5).

Printed Resources

- Instructional Aid 9: *Price Tags*
- Instructional Aid 45: *Coin Purse Mat* (for each student)
- Visual 14: *Money*
- Visuals: Money Kit (9 pennies, 5 nickels, 5 dimes, 1 quarter)
- Student Manipulatives: Number Cards (0–20)
- Student Manipulatives: Money Kit (1 quarter, 3 dimes, 5 nickels, 7 pennies)

Digital Resources

- Video: Ch 19 Intro
- Web Link: Virtual Manipulatives: Analog Clock
- Web Link: Learning Coins
- Web Link: Virtual Manipulatives: Money Pieces
- Games/Enrichment: Addition Flashcards
- Games/Enrichment: Subtraction Flashcards

Materials

- Addition flashcards
- Subtraction flashcards
- Judy® Clock (or Visual 15: *Clock*)
- 3 general store items with price tags of 25¢, 30¢, and 35¢

Counting Pennies with a Quarter

Write the value of each coin.

25 ¢
quarter

1 ¢
penny

Write the value as you *count on*. Write the total.

25 ¢ 26 ¢ 27 ¢ 28 ¢ 29 ¢ — 29 ¢

25 ¢ 26 ¢ 27 ¢ — 27 ¢

25 ¢ 26 ¢ 27 ¢ 28 ¢ — 28 ¢

Mark an X on the coins needed to buy each item.

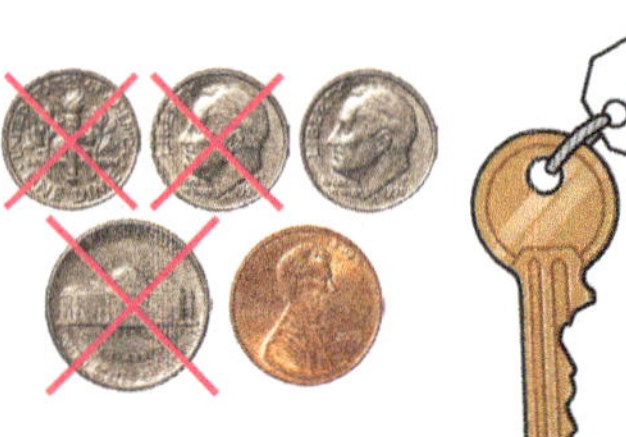

Provide items with price tags for the students to purchase at the general store. The prices needed for this chapter are 25¢, 30¢, 35¢, 45¢, and 60¢.

Practice & Review

Memorize Facts for the 6-7-13 Family

Introduce the following facts.

6 + 7 7 + 6 13 – 7 13 – 6

Display each addition flashcard slowly. Invite students to give the answers. Point out that 6 + 7 and 7 + 6 are near doubles.

Display each subtraction flashcard slowly and invite students to give the answers. Point out that 13 – 7 and 13 – 6 are related facts. Remind the students that these facts are part of the 6-7-13 fact family.

Distribute Number Cards 0–20. Display each flashcard again. Direct the students to hold up the correct Number Card to indicate each answer.

Practice these and recently memorized facts.

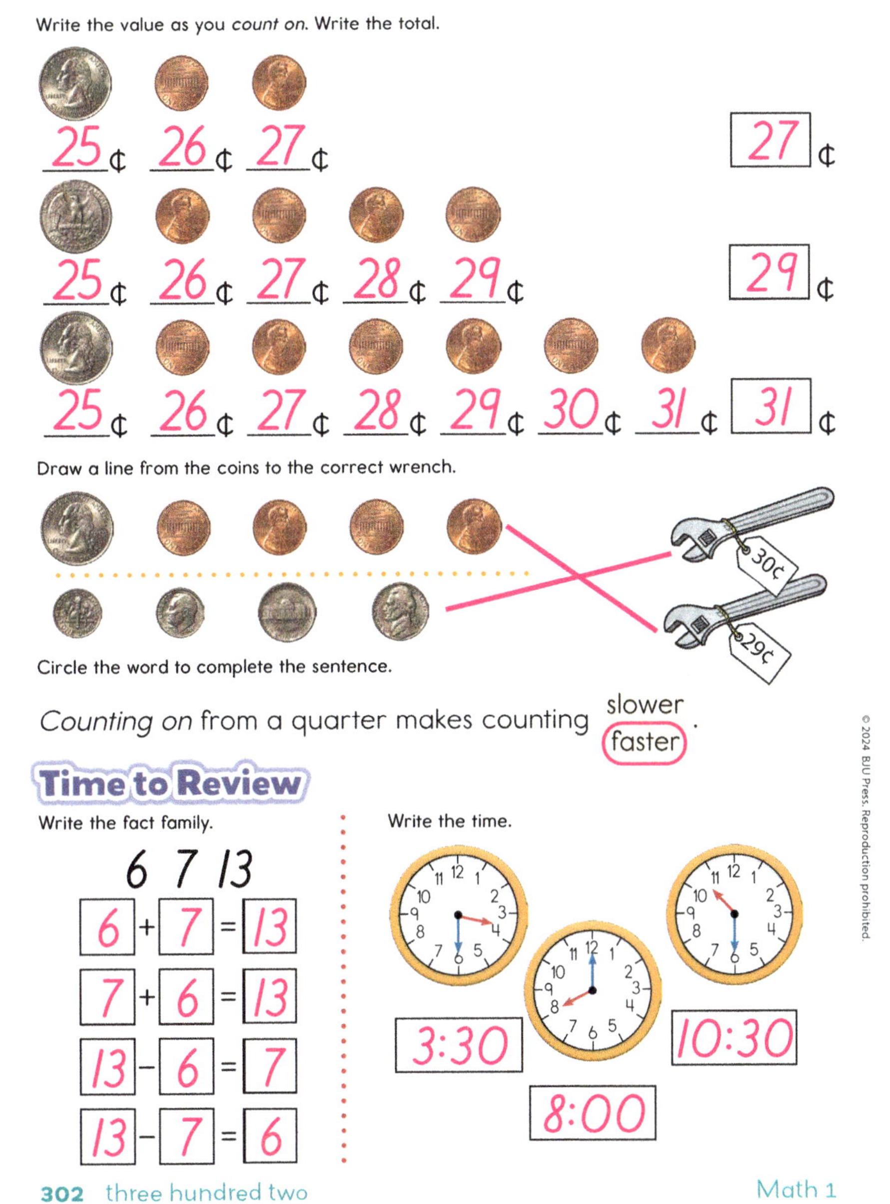

Write the value as you count on. Write the total.
25¢ 26¢ 27¢
27¢
25¢ 26¢ 27¢ 28¢ 29¢
29¢
25¢ 26¢ 27¢ 28¢ 29¢ 30¢ 31¢
31¢
Draw a line from the coins to the correct wrench.
30¢
29¢
Circle the word to complete the sentence.
Counting on from a quarter makes counting slower faster.
Time to Review
Write the fact family.
6 7 13
6 + 7 = 13
7 + 6 = 13
13 − 6 = 7
13 − 7 = 6
Write the time.
3:30
8:00
10:30
© 2024 BJU Press. Reproduction prohibited.
302 three hundred two
Math 1

DQ was happy to go along to the auto parts store. There were lots of interesting car parts to see.

Once they arrived at the store, it took a few minutes for Liam's dad to find the door hinge repair kit he was looking for. Liam and DQ walked the aisles that displayed headlights, fuel pumps, and radiators. Then DQ saw something he didn't recognize—a brake drum. He wondered if it was a new kind of musical instrument. After browsing for a while, they followed Liam's dad to the checkout counter.

Mr. Sawyer, the manager, was waiting. "Did you find the part you were looking for?" he asked.

"Yes, thank you," said Liam's dad. "I'm replacing one of my door hinges. It's old and cracked."

"Quack, quack," joked DQ playfully.

"That's *crack*, not *quack*, DQ!" said Liam.

"That will be $10.90," said Mr. Sawyer with a smile.

Liam's dad set a ten-dollar bill on the counter and gave DQ some coins to count.

"Can you find 90¢, DQ?" asked Liam.

DQ counted the dimes: "10, 20, 30, 40, 50, 60, 70, 80, 90¢. There you go. That's 90¢."

Liam's dad handed the money to Mr. Sawyer and thanked him for the parts.

As they turned to leave, Liam thought, *That's a lot of dimes to count. I wonder if there is a faster way to count 90¢?*

Invite a student to read aloud the essential question. Encourage the students to consider this question as you lead them to answer it by the end of the chapter.

Write Time to the Hour & Half-Hour

Explain that Dad is washing his truck at 9:30. Set the Judy Clock (or Visual 15) for 9:30 and write "9:30" for display.

What does the number to the left of the colon or two dots tell? the hour

What does the number to the right of the colon or two dots tell? the number of minutes

Where is the hour hand? between the 8 and the 9 or in the 8's "backyard"

Set the demonstration clock for 2:00, 10:00, and 3:30; ask volunteers to write the times for display for the students to check their answers.

Engage

Essential Question

Direct attention to the chapter opener on Worktext page 300 and **read aloud** the following scenario to introduce the chapter essential question.

"Dad, may I take DQ along to the store this afternoon?" asked Liam one day after school.

Liam's dad was going to buy a repair kit for his car door. One of the hinges needed to be replaced.

"Sure," he said. "We'll pick him up on the way. Let's go."

Instruct

Determine the Value of Coins by *Counting on*

Guide the students in a **discussion** to review the value of coins.

Display Visual 14. Sing together "Pennies, Nickels, Dimes" to the tune of "The Farmer in the Dell." Point to each coin on the visual as you sing the verse for it.

155

Pennies, nickels, dimes,
Pennies, nickels, dimes,
Pennies are each worth one cent,
Pennies, nickels, dimes.

Pennies, nickels, dimes,
Pennies, nickels, dimes,
Nickels are each worth five cents,
Pennies, nickels, dimes.

Pennies, nickels, dimes,
Pennies, nickels, dimes,
Dimes are each worth ten cents,
Pennies, nickels, dimes.

Display 3 dimes, 2 nickels, and 2 pennies in random order.

What do you do first to count the value of this set of coins? separate the dimes, nickels, and pennies

In what order should you count the coins? first the dimes, followed by nickels, and then pennies; because dimes have the greatest value

After you count the dimes, what should you do before you begin counting the nickels? I should take a deep breath to remind myself to stop counting by 10s and to begin *counting on* by 5s.

What other time should you take a deep breath? after I finish counting the nickels and before I begin *counting on* the pennies

Count together the total value: 10, 20, 30, (deep breath), 35, 40, (deep breath), 41, 42¢.

Display 5 dimes, 3 nickels, and 4 pennies in random order.

Ask a volunteer to separate the coins and determine the total value by *counting on*. 10, 20, 30, 40, 50, (deep breath), 55, 60, 65, (deep breath), 66, 67, 68, 69¢

Continue the activity with various combinations of dimes, nickels, and pennies.

Determine the Value of a Set of 1 Quarter & Pennies by *Counting on*

Use a **song** to help the students *count on* with a quarter.

Display a penny, a nickel, a dime, and a quarter in a row.

Sing a new verse for "Pennies, Nickels, Dimes."

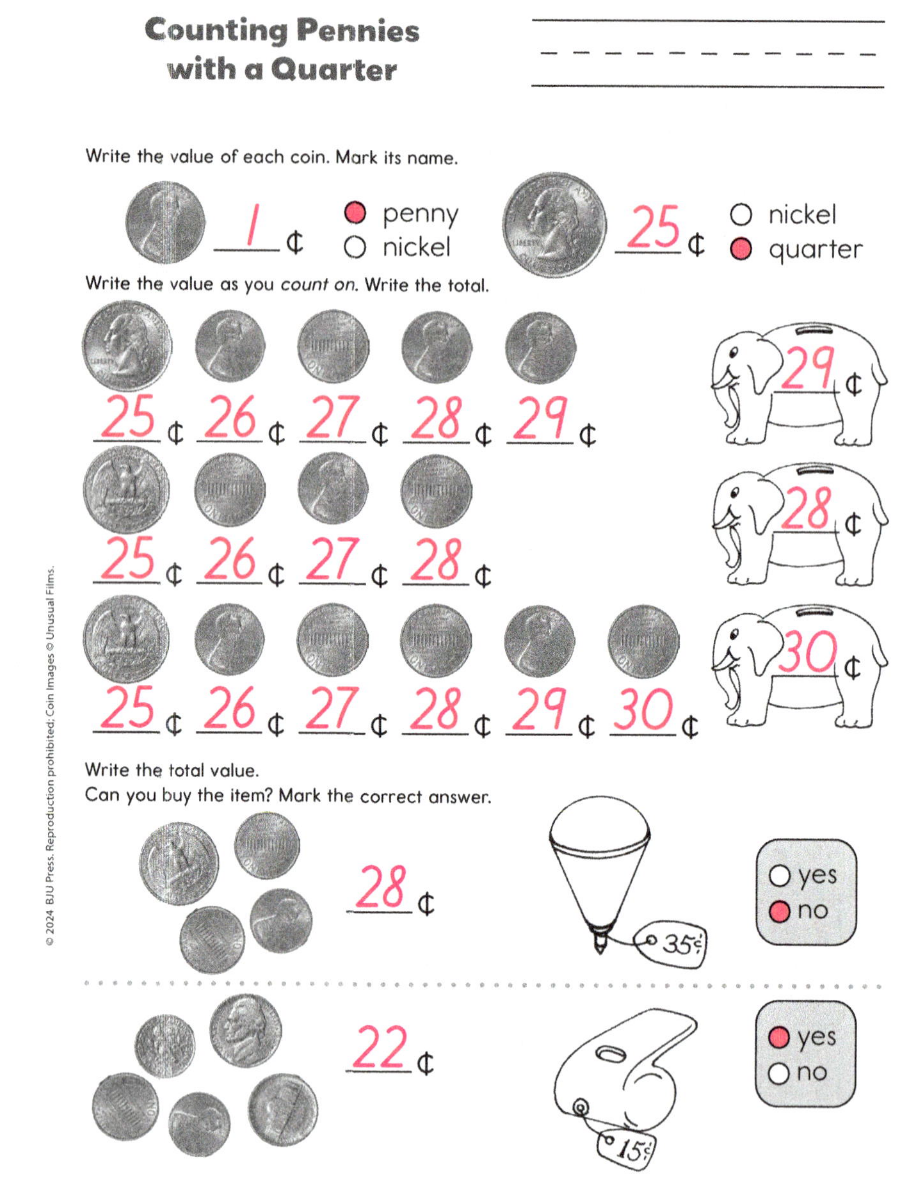

Pennies, nickels, dimes,
Now quarters join the line,
Their value is twenty-five cents,
Pennies, nickels, dimes.

Guide the students as they sing the new verse as you point to each coin.

What is the value of a quarter? 25¢

Display 1 quarter and 2 pennies.

When you count the value of a set of 1 quarter and 2 pennies, which coin do you count first? I count the quarter first because it has the greater value.

Count together the total value by *counting on*: 25, (deep breath), 26, 27¢.

Continue the activity with the following sets of coins.

1 quarter and 4 pennies 25, (deep breath), 26, 27, 28, 29¢

1 quarter and 5 pennies 25, (deep breath), 26, 27, 28, 29, 30¢

Determine Whether There Is Enough Money to Purchase an Item

Guide a **role-playing activity** to determine whether there is enough money.

Distribute the *Coin Purse Mat* page and a set of coins to each student. Direct the students to place 1 quarter and 6 pennies in the coin purse and count the total value.

What is the value? 31¢

Display a general store item with a price tag of 30¢.

What is the price of this item? 30¢

Chapter 2 Review

Read the word. Circle the correct picture.

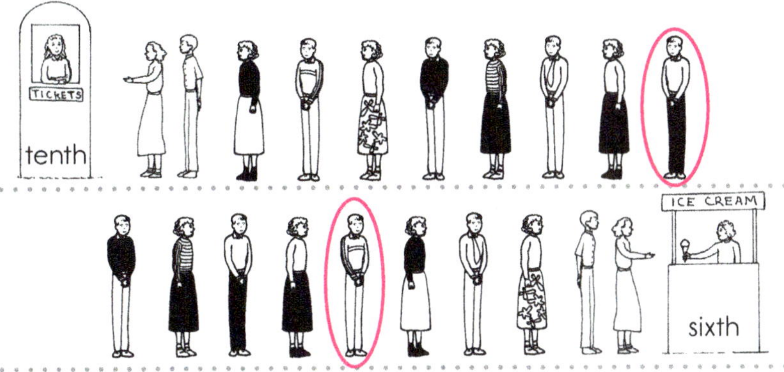

second

Addition Fact Review

Add.

5 + 4 = 9	9 + 3 = 12	6 + 6 = 12
9 + 2 = 11	4 + 6 = 10	7 + 4 = 11
4 + 4 = 8	7 + 5 = 12	2 + 8 = 10
6 + 3 = 9	0 + 9 = 9	3 + 5 = 8

290 two hundred ninety Math 1 Reviews

Do you have enough money to buy this item? yes

Continue the activity with the following sets of coins and price tags.

1 quarter, 4 pennies, 29¢; 25¢ price tag yes

1 quarter, 3 pennies 28¢; 35¢ price tag no

Identify the Coins Needed to Purchase an Item

Guide a **demonstration** to identify coins needed to purchase an item.

Write "28¢" above a coin purse drawn for display. Direct the students to place 3 dimes, 2 nickels, and 4 pennies in the coin purse on their mats. Demonstrate each step.

Explain that students are going to buy a ball for 28¢.

How many dimes do you need? 2

Direct the students to place 2 dimes on the hand on their mats.

Guide the students as they count the money placed on the hand: 10, 20¢.

How many nickels do you need? 1

Direct the students to place 1 nickel on the hand.

Count the money together: 10, 20, (deep breath), 25¢.

Do you need another nickel? No; that would be too much money.

How many pennies do you need to make 28¢? 3

Direct the students to place 3 pennies on the hand.

Count the money together: 10, 20, (deep breath), 25, (deep breath), 26, 27, 28¢.

Continue the activity with 1 quarter and 6 pennies and the ball for 28¢. 1 quarter and 3 pennies

Counting Helps Us Explore the World

Guide a **discussion** to help the students explain a biblical worldview shaping truth.

What did DQ need to do before he could help buy a part at the auto parts store? count the coins correctly

When you want to buy an item that costs 28¢, could you use 28 pennies? Yes; but it would take time to count them all.

What coins would be easier to use to buy the 28¢ item? a quarter and pennies

Which coin would you count first? the quarter, because it has the greater value

Explain that *counting on* from a quarter helps students understand the world. They can count more quickly, carry fewer coins, and know how much money they have.

How can counting pennies with a quarter help you explore the world? Understanding the value of coins helps me to buy something quickly by using the fewest number of coins.

Apply

Worktext pages 301–2

Read and guide completion of page 301.

Read and explain the directions for page 302. Assist the students as they complete the page independently.

Direct attention to the biblical worldview shaping statement and guide the students as they complete the sentence.

Assess

Reviews pages 289–90

Review identifying and locating ordinal positions first to tenth on page 290.

156

Worktext pages 303–4
Reviews pages 291–92

Objectives

- **156.1** Determine the value of a set of dimes, nickels, and pennies.
- **156.2** Determine the value of a set of 1 quarter and nickels.
- **156.3** Identify the coins needed to purchase an item.
- **156.4** Solve money word problems.

Printed Resources

- Instructional Aid 45: *Coin Purse Mat* (for each student)
- Visual 1: *Hundred Chart*
- Visual 14: *Money*
- Visuals: Money Kit (2 quarters, 8 dimes, 5 nickels, 6 pennies)
- Student Manipulatives: Number Cards (0–20)
- Student Manipulatives: Money Kit (2 quarters, 8 dimes, 5 nickels, 6 pennies)

Digital Resources

- Video: Nickels with Quarter
- Web Link: Learning Coins
- Web Link: Virtual Manipulatives: Money Pieces
- Games/Enrichment: Addition Flashcards
- Games/Enrichment: Subtraction Flashcards

Materials

- Addition flashcards
- Subtraction flashcards

Counting Nickels with a Quarter

Write the value of each coin.

5¢ nickel 25¢ quarter

Write the value as you count on. Write the total.

25¢ 30¢ 35¢ 40¢

40¢

25¢ 30¢ 35¢ 40¢ 45¢ 50¢

50¢

25¢ 30¢ 35¢ 40¢ 45¢

45¢

Mark an *X* on the coins needed to buy each item.

Practice & Review

Memorize Facts for the 4-9-13 Family

Introduce the following facts.

4 + 9 9 + 4 13 – 9 13 – 4

Display each addition and subtraction flashcard slowly. Invite students to give the answers.

Explain that making 10 may make it easier to add 9. Point out that 13 – 9 and 13 – 4 are related facts. Remind the students that these facts are part of the 4-9-13 fact family.

Distribute Number Cards 0–20. Display each flashcard again. Direct the students to hold up the correct Number Card to indicate each answer.

Practice these and recently memorized facts.

Engage

Count by 5s

Use a **visual** to review counting by 5s.

Display Visual 1. Count together by 5s from 15 to 60 as you point to the corresponding numbers on the chart. Continue the activity as you count from 25 to 80 and from 50 to 100.

Instruct

Determine the Value of a Set of Coins

Guide a **review** to help the students determine the value of a set of coins.

Display Visual 14. Display 3 dimes, 4 nickels, and 3 pennies in random order.

> **What must you do before you can count the value of this set of coins?**
> separate them into dimes, nickels, and pennies

156

Write the value as you *count on*.
Can you buy the item? Circle *yes* or *no*.

25¢ 30¢ 35¢

(yes) no

25¢ 30¢ 35¢ 40¢ 45¢ 50¢

(yes) no

25¢ 30¢ 35¢ 40¢

yes (no)

Write an equation for the word problem. Solve.

DQ found 27¢ in his truck.
Liam found 12¢ under the mat.
How much money did they find?

27 + 12 = 39¢

(work space)

$$\begin{array}{r} 27¢ \\ +\ 12¢ \\ \hline 39¢ \end{array}$$

Time to Review

Add.

6 + 7 = 13
7 + 6 = 13
4 + 9 = 13
9 + 4 = 13

Subtract.

13 − 7 = 6
13 − 6 = 7
13 − 9 = 4
13 − 4 = 9

304 three hundred four   Math 1

In what order do you count the coins? first the dimes, then the nickels, and then the pennies, because dimes have the greatest value

Count the coins together: 10, 20, 30, (deep breath), 35, 40, 45, 50, (deep breath), 51, 52, 53¢.

Continue the activity with 4 dimes, 3 nickels, and 4 pennies in random order. 10, 20, 30, 40, (deep breath), 45, 50, 55, (deep breath), 56, 57, 58, 59¢

Determine the Value of a Set of 1 Quarter & Nickels

Guide a **counting activity** to help the students determine the value of a set of coins.

Display 1 quarter and 1 nickel.

Which coin do you count first to find the value of this set? the quarter, because it has the greater value

What is the value of a quarter? 25¢

Explain that students should begin counting at 25 and then *count on* by 5s for each nickel. Demonstrate as you point to the coins: 25, (deep breath), 30¢. Guide the students as they count the coins.

Display 1 quarter and 4 nickels. Count together as you point to each coin: 25, (deep breath), 30, 35, 40, 45¢.

Continue the activity with 1 quarter and 3 nickels. 25, (deep breath), 30, 35, 40¢

Identify the Coins Needed to Purchase an Item

Guide a **demonstration** to identify coins needed to purchase an item.

Distribute the *Coin Purse Mat* page and a set of coins to each student. Write "47¢" above a coin purse drawn for display.

Direct the students to place 4 dimes, 4 nickels, and 4 pennies in the coin purse on their mats. Demonstrate each step.

Explain that students are going to buy a key chain for 47¢.

How many dimes do you need? 4

Direct the students to place 4 dimes on the hand on their mats.

Count the money together: 10, 20, 30, 40¢.

How many nickels do you need? 1

Direct the students to place 1 nickel on the hand.

Count the money together: 10, 20, 30, 40, (deep breath), 45¢.

Do you need another nickel? No; that would be too much money.

How many pennies do you need to make 47¢? 2

Direct the students to place 2 pennies on the hand on their mats.

Count the money together: 10, 20, 30, 40, (deep breath), 45, (deep breath), 46, 47¢.

Direct the students to move the money on the hand to their desks. Ask them to place 4 dimes, 4 nickels, and 4 pennies on the coin purse.

Encourage the students to think of another way to make 47¢ by using these same coins.

Guide the students as they place 3 dimes, 3 nickels, and 2 pennies onto the hand. Guide the students as they count the coins: 10, 20, 30, (deep breath), 35, 40, 45, (deep breath), 46, 47¢.

Explain that they can make 47¢ both ways, by using different combinations of dimes, nickels, and pennies.

Direct the students to move all the coins to their desks and to place 1 quarter, 3 dimes, and 1 nickel on their coin purses. Write "30¢" above your coin purse on display.

Explain that students are going to buy a cleaning cloth for 30¢.

156

How many quarters do you need? 1

Direct the students to place 1 quarter on the hand.

Guide the students as they count the money: 25¢.

How many nickels do you need? 1

Direct the students to place 1 nickel on the hand.

Count the money together: 25, (deep breath), 30¢.

Direct the students to move the money on the hand to their desks. Ask them to place 1 quarter, 3 dimes, and 1 nickel on the coin purse again.

Ask the students to think of another way to make 30¢.

Guide the students as they place 3 dimes on the hand and count the coins: 10, 20, 30¢.

Solve Word Problems

Use the **Problem-Solving Plan** to solve money word problems.

Read aloud the following word problem.

Sam had 33¢ in his pocket. While he was playing, he lost 12¢. How much money does Sam have left?

What is the question? How much money does Sam have left?

What information is given? He had 33¢ and lost 12¢.

Do you add or subtract to find the answer? subtract

What is the equation that you can use to solve this word problem?
33¢ – 12¢ = __¢

Write "33¢ – 12¢" in a Tens/Ones frame.

Differentiated Instruction

Count Coins for the Total Value

Give the students pennies, nickels, and dimes. Direct them to count out a set of coins with the value of 25¢.

Discuss how many pennies, nickels, and dimes they used. Instruct the students to use a different combination of coins with the value of 25¢. Continue the activity, directing the students to use at least 2 different coin combinations for each value.

Counting Nickels with a Quarter

Write the value as you *count on*.
Can you buy the item? Mark the correct answer.

Write an equation for each word problem. Solve.

Mark spent 49¢ on plates.
He spent 30¢ on cups.
How much money did he spend in all?

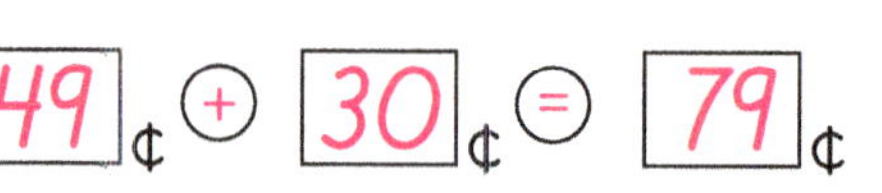

work space

49¢
+ 30¢
79¢

Bryce had 69¢.
He spent 33¢ on a gift.
How much money does he have left?

69¢ − 33¢ = 36¢

work space

69¢
− 33¢
36¢

Chapter 19 • Lesson 156 two hundred ninety-one **291**

Ask a volunteer to solve the equation. 21¢

Does the answer make sense? yes

Continue the activity with the following word problem.

Jen has saved 57¢ for a new book. Her grandmother gave her 22¢. How much money does Jen have in all?
57¢ + 22¢ = 79¢

Apply

Worktext pages 303–4

Read and guide completion of page 303.

Read and explain the directions for page 304. Assist the students as they complete the page independently.

Assess

Reviews pages 291–92

Review reading a calendar on page 292.

Chapter 8 Review

October						
Sunday	Monday	Tuesday	Wednesday	Thursday	Friday	Saturday
				1	2	3
4	5	6	7	8	9	10
11	12	13	14	15	16	17
18	19	20	21	22	23	24
25	26	27	28	29	30	31

Mark the correct answer.

How many days are in October? ○ 29 ○ 30 ● 31

How many Thursdays are in October? ○ 3 ○ 4 ● 5

What day is the 21st? ○ Tuesday ● Wednesday ○ Thursday

Subtraction Fact Review

Subtract.

12 − 6 = 6　　9 − 2 = 7　　10 − 2 = 8

11 − 4 = 7　　10 − 6 = 4　　12 − 4 = 8

8 − 5 = 3　　7 − 4 = 3　　9 − 5 = 4

10 − 3 = 7　　11 − 5 = 6　　11 − 9 = 2

157

Worktext pages 305–6
Reviews pages 293–94

Counting Dimes with a Quarter

Write the value of each coin.

25 ¢ quarter
10 ¢ dime

Write the value as you *count on*. Write the total.

25 ¢ 35 ¢ 45 ¢ 55 ¢ 65 ¢ — 65¢

25 ¢ 35 ¢ 45 ¢ 55 ¢ 65 ¢ 75 ¢ — 75¢

25 ¢ 35 ¢ 45 ¢ 55 ¢ 65 ¢ 75 ¢ 85 ¢ — 85¢

25 ¢ 35 ¢ 45 ¢ — 45¢

Mark an *X* on the coins needed to buy each item.

45¢ 65¢ 25¢

Objectives

- **157.1** Determine the value of a set of dimes, nickels, and pennies up to 99¢.
- **157.2** Determine the value of a set of 1 quarter and dimes.
- **157.3** Identify the coins needed to purchase an item.
- **157.4** Solve money word problems.

Printed Resources

- Instructional Aid 45: *Coin Purse Mat* (for each student)
- Visual 1: *Hundred Chart*
- Visuals: Money Kit (1 quarter, 6 dimes, 4 nickels, 7 pennies)
- Student Manipulatives: Number Cards (0–20)
- Student Manipulatives: Money Kit (1 quarter, 4 dimes, 4 nickels, 5 pennies)

Digital Resources

- Video: Dimes with Quarter
- Web Link: Learning Coins
- Web Link: Virtual Manipulatives: Money Pieces
- Games/Enrichment: Addition Flashcards
- Games/Enrichment: Subtraction Flashcards

Materials

- Addition flashcards
- Subtraction flashcards
- A general store item with a price tag of 45¢

Practice & Review

Memorize Facts for the 5-9-14 Family

Introduce the following facts.

5 + 9 9 + 5 14 – 9 14 – 5

Display each addition and subtraction flashcard slowly. Invite students to give the answers.

Explain that making 10 may make it easier to add 9. Point out that 14 – 9 and 14 – 5 are related facts. Remind the students that these facts are part of the 5-9-14 fact family.

Distribute Number Cards 0–20. Display each flashcard again. Direct the students to hold up the correct Number Card to indicate each answer.

Practice these and recently memorized facts.

Count 700–750

Arrange the students in 2 groups. Explain that students in Group 1 can pretend that they are DQ counting in a duck voice (nasal sound) and the students in Group 2 can pretend that they are Liam counting in a normal voice. Instruct Group 1 to begin counting with the number 700 and to continue counting until you clap your hands. Instruct Group 2 to *count on* from the last number counted by Group 1 until you clap your hands. Continue to take turns as you count to 750.

Engage

Giving to God

Read the following **retelling** of 2 Chronicles 24:1–14 to emphasize the importance of giving.

King Joash wanted to repair God's house. The people had allowed the temple to become dirty and rundown. King Joash asked the people to give money to help make God's house beautiful. The priests made a wooden box with a hole in the top. Every time the people came to worship the Lord, they placed their offerings in the box. Soon there was enough money to repair the temple. The people were happy to have a beautiful place to worship the Lord.

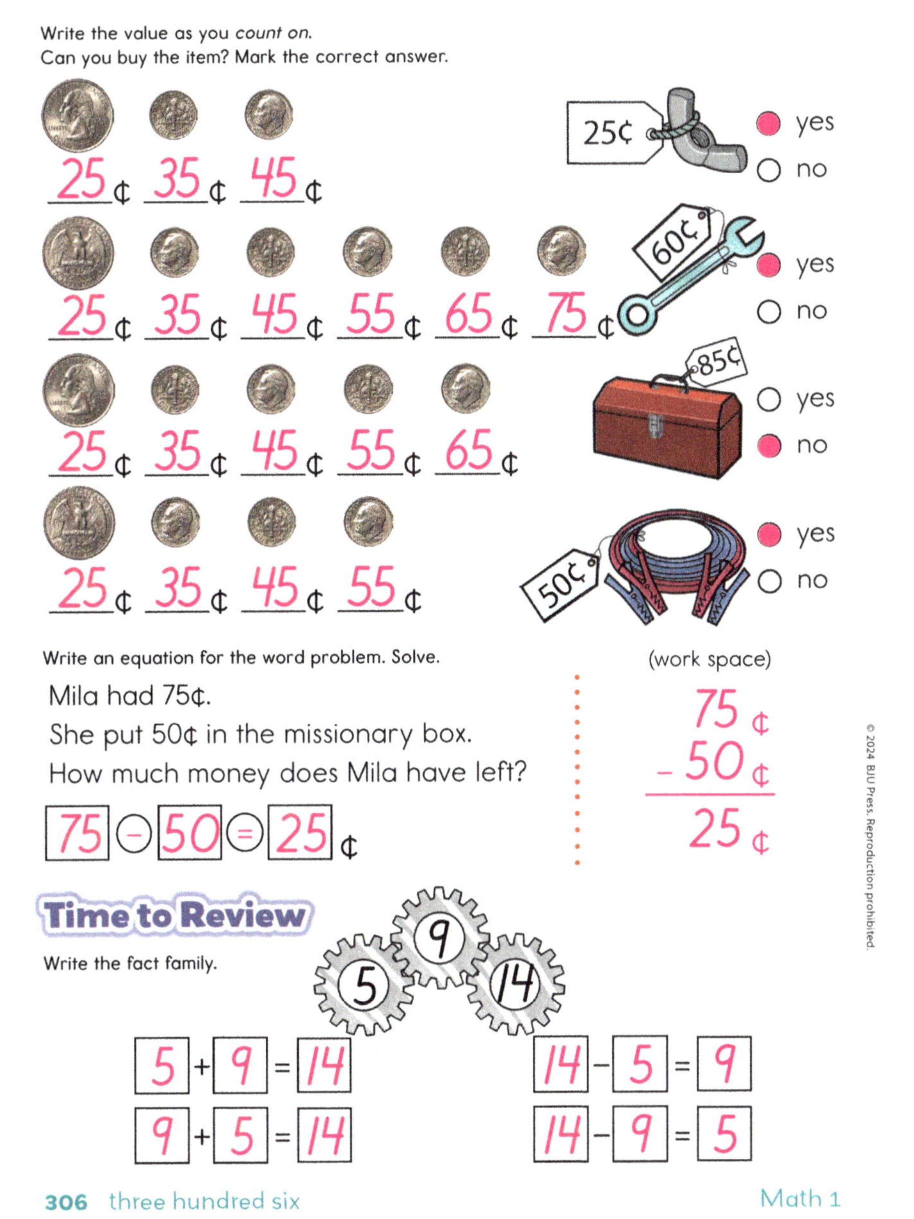

Write the value as you count on.
Can you buy the item? Mark the correct answer.
25¢ 35¢ 45¢
25¢
yes
no
25¢ 35¢ 45¢ 55¢ 65¢ 75¢
60¢
yes
no
25¢ 35¢ 45¢ 55¢ 65¢
85¢
yes
no
25¢ 35¢ 45¢ 55¢
50¢
yes
no
Write an equation for the word problem. Solve.
Mila had 75¢.
She put 50¢ in the missionary box.
How much money does Mila have left?
75 − 50 = 25 ¢
(work space)
75¢
− 50¢
25¢
© 2024 BJU Press. Reproduction prohibited.
Time to Review
Write the fact family.
5 9 14
5 + 9 = 14
9 + 5 = 14
14 − 5 = 9
14 − 9 = 5
306 three hundred six
Math 1

The Bible tells us in 2 Corinthians 9:7 that God loves a cheerful giver. What are some things that our offerings might be used for? paying the church's electric bills, repairing the roof when needed, paying someone to clean the building each week

Instruct

Determine the Value of Coins

Use a **song** to review counting a set of coins.

Take 7 pennies, 4 nickels, and 5 dimes and distribute each coin to a different student. Guide the students as they sing together "Pennies, Nickels, Dimes" to the tune of "The Farmer in the Dell," instructing each student to "pop up" every time the word for his or her coin is mentioned.

Pennies, nickels, dimes,
Pennies, nickels, dimes,
Pennies are each worth one cent,
Pennies, nickels, dimes.

Pennies, nickels, dimes,
Pennies, nickels, dimes,
Nickels are each worth five cents,
Pennies, nickels, dimes.

Pennies, nickels, dimes,
Pennies, nickels, dimes,
Dimes are each worth ten cents,
Pennies, nickels, dimes.

What should you do before you count the value of these coins? separate them into groups

Direct the students with coins to arrange themselves in groups. Guide the students as they count the coins: 10, 20, 30, 40, 50, (deep breath), 55, 60, 65, 70, (deep breath), 71, 72, 73, 74, 75, 76, 77¢.

Continue the activity, using the following sets of coins.

6 dimes, 3 nickels, 3 pennies 78¢

4 dimes, 8 nickels, 5 pennies 85¢

Guide the students as they count by 10s as you point to the corresponding numbers on Visual 1. Begin at 15 and count by 10s to 65. Repeat the activity beginning at 25 and counting to 95.

Determine the Value of a Set of 1 Quarter & Dimes

Guide a **counting activity** to determine the value of a set of coins.

Display a quarter as you sing together the quarter verse for "Pennies, Nickels, Dimes."

Pennies, nickels, dimes,
Now quarters join the line,
Their value is twenty-five cents,
Pennies, nickels, dimes.

Display 1 quarter and 1 dime.

Which coin do you count first to find the value of this set? the quarter, because the quarter has a greater value than a dime

What is the value of a quarter? 25¢

Explain that students begin counting at 25; then *count on* by 10s for each dime. Demonstrate as you point to the coins: 25, (deep breath), 35¢. Count the coins together.

Display 1 quarter and 3 dimes. Count the coins together as you point to each coin: 25, (deep breath), 35, 45, 55¢.

Continue the activity with 1 quarter and 5 dimes: 25, (deep breath), 35, 45, 55, 65, 75¢.

Identify the Coins Needed to Purchase an Item

Guide a **demonstration** to identify the coins needed to purchase an item.

Distribute the *Coin Purse Mat* page and a set of coins to each student. Direct the students to place all the coins in the purse on their mats.

Demonstrate each step with a coin purse drawn for display.

Display a general store item with the price of 45¢.

How much does this item cost? 45¢

Explain that students are going to buy this item by using only 1 quarter and dimes.

What is the value of 1 quarter? 25¢

Direct the students to place 1 quarter on the hand on their mats.

How many dimes do you need to make 45¢? 2

Direct the students to place 2 dimes on the hand.

Count the money together: 25, (deep breath), 35, 45¢.

Direct the students to place the money back in the coin purse.

Explain that students are going to buy this item again by using only dimes and nickels.

How many dimes do you need? 4

Direct the students to place 4 dimes on the hand.

Count the money together: 10, 20, 30, 40¢.

How many nickels do you need to make 45¢? 1

Direct the students to place 1 nickel on the hand.

Guide the students as they count the money again: 10, 20, 30, 40, (deep breath), 45¢.

What other dime and nickel combination can you make to equal 45¢? 3 dimes and 3 nickels

Solve Word Problems

Guide a **discussion** to solve money word problems.

Read aloud the following word problem.

Seth has saved 55¢. He takes 2 dimes to put in the church offering plate. How much money does Seth have left?

What is the question? How much money does Seth have left?

How much money did Seth have? 55¢

Write "55¢" in a Tens/Ones frame.

How much money did Seth put in the offering plate? 2 dimes

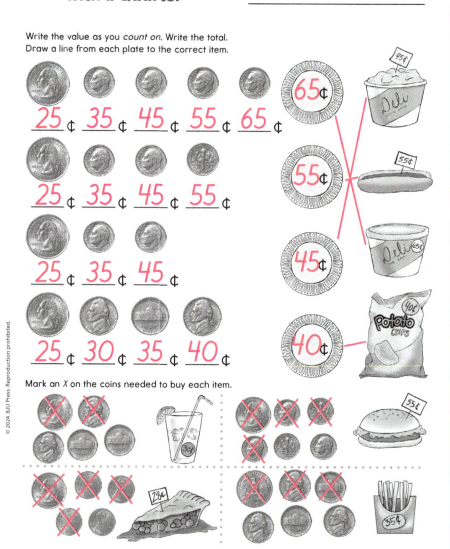

Counting Dimes with a Quarter

Write the value as you *count on*. Write the total.
Draw a line from each plate to the correct item.

25¢ 35¢ 45¢ 55¢ 65¢ — 65¢

25¢ 35¢ 45¢ 55¢ — 55¢

25¢ 35¢ 45¢ — 45¢

25¢ 30¢ 35¢ 40¢ — 40¢

Mark an *X* on the coins needed to buy each item.

What is the value of 2 dimes? 20¢

Write "20¢" in the Tens/Ones frame.

Do you add or subtract to find the answer? subtract

Ask a volunteer to solve the problem. 35¢

Does the answer make sense? Yes; he now has less money than he started with.

Continue the activity with the following word problem.

Mia found 60¢ in her desk drawer. Uncle Jim gave her 1 more quarter. How much money does Mia have in all?
60¢ + 25¢ = 85¢

Apply

Worktext pages 305–6

Read and guide completion of page 305.

Read and explain the directions for page 306. Assist the students as they complete the page independently.

Assess

Reviews pages 293–94

Review estimating and measuring the length of objects in inches on page 294.

Chapter 11 Review

Guess each length. Use an inch ruler to measure.

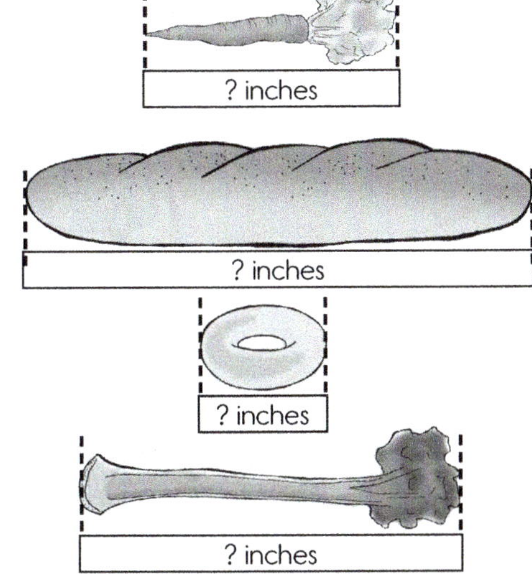

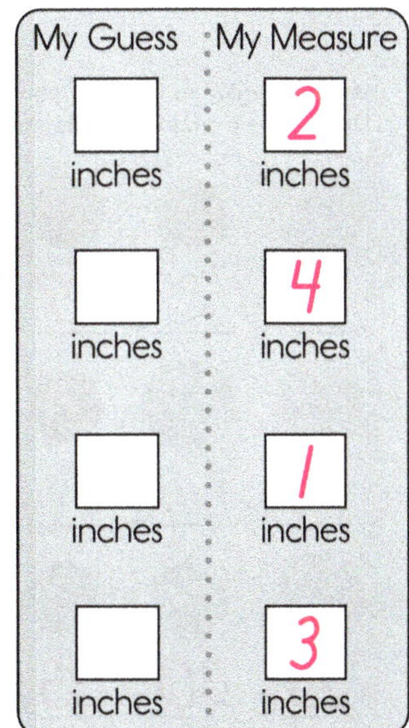

My Guess
My Measure
inches
2 inches
inches
4 inches
inches
1 inches
inches
3 inches

Addition Fact Review

Mark the correct answer.

Problem	Choices	Marked
$0 + 8 =$	7, 8, 9	8
$5 + 3 =$	7, 8, 9	8
$3 + 6 =$	7, 8, 9	9
$3 + 9 =$	10, 11, 12	12
$8 + 4 =$	10, 11, 12	12
$5 + 7 =$	10, 11, 12	12
$2 + 8 =$	10, 11, 12	10
$7 + 4 =$	10, 11, 12	11

158 Worktext pages 307–8 Reviews pages 295–96

Objectives

- **158.1** Determine the value of a set of 1 quarter and pennies and a set of 2 quarters and pennies.
- **158.2** Determine the value of a set of 1 quarter and nickels and a set of 2 quarters and nickels.
- **158.3** Determine the value of a set of 1 quarter and dimes and a set of 2 quarters and dimes.
- **158.4** Identify the coins needed to purchase an item.
- **158.5** Solve money word problems.
- **158.6** Express praise to God for a world where math works so well. BWS

Biblical Worldview Shaping

- **Exploring** (formulate): The more we explore God's world, the more we are amazed that math works so well. When we understand that this good gift comes from God, we should praise Him with thankful hearts (158.6).

Printed Resources

- Instructional Aid 45: *Coin Purse Mat* (for each student)
- Visuals: Money Kit (2 quarters, 4 dimes, 6 nickels, 10 pennies)
- Student Manipulatives: DQ Clock (for each student and 2 for the teacher)
- Student Manipulatives: Number Cards (0–20)
- Student Manipulatives: Money Kit (2 quarters, 4 dimes, 4 nickels, 10 pennies)

Digital Resources

- Web Link: Learning Coins
- Web Link: Virtual Manipulatives: Money Pieces
- Games/Enrichment: Addition Flashcards
- Games/Enrichment: Subtraction Flashcards

Materials

- Addition flashcards
- Subtraction flashcards
- A general store item with a price tag of 60¢

Counting with 2 Quarters

Write the value as you *count on*.
Draw a line from the coins to the correct car.

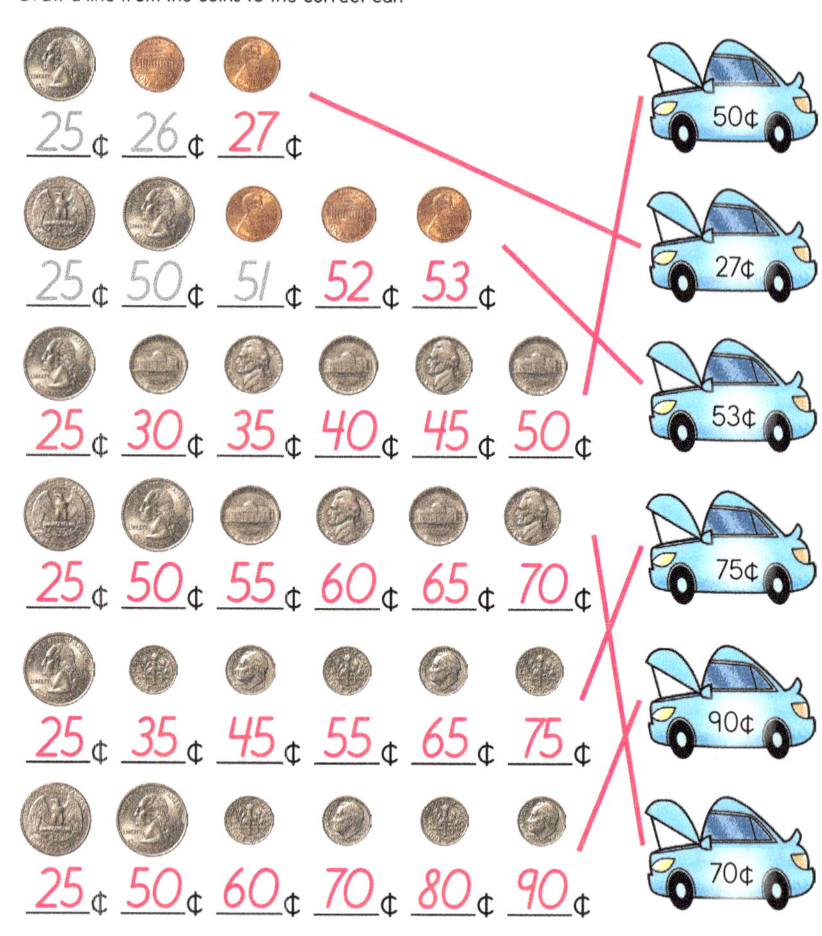

Practice & Review

Determine the Time Elapsed

Display 2 DQ Clocks, the first clock set for 2:00 and the second clock set for 3:00; then read the following word problem.

DQ began painting at 2:00. He finished painting at 3:00. How much time passed between the time DQ started painting and when he finished? 1 hour

Set the DQ Clocks for 5:00 and 5:30; then read the following word problem.

DQ began cleaning up his painting mess at 5:00. He finished cleaning up at 5:30. How much time passed between the time that DQ started cleaning and the time he finished? 30 minutes

Distribute the DQ Clocks and direct the students to set them for 1:00. Explain that Reagan began reading her library book at 1:00. She finished reading her book at 2:00.

Direct the students to move the minute hand on their clocks until it is at 2:00.

How much time passed while Reagan was reading her library book? 1 hour

Continue the activity for Reagan starting breakfast at 8:00 and finishing at 8:30. 30 minutes

Engage

Memorize Facts for the 6-9-15 Family

Introduce the following facts.

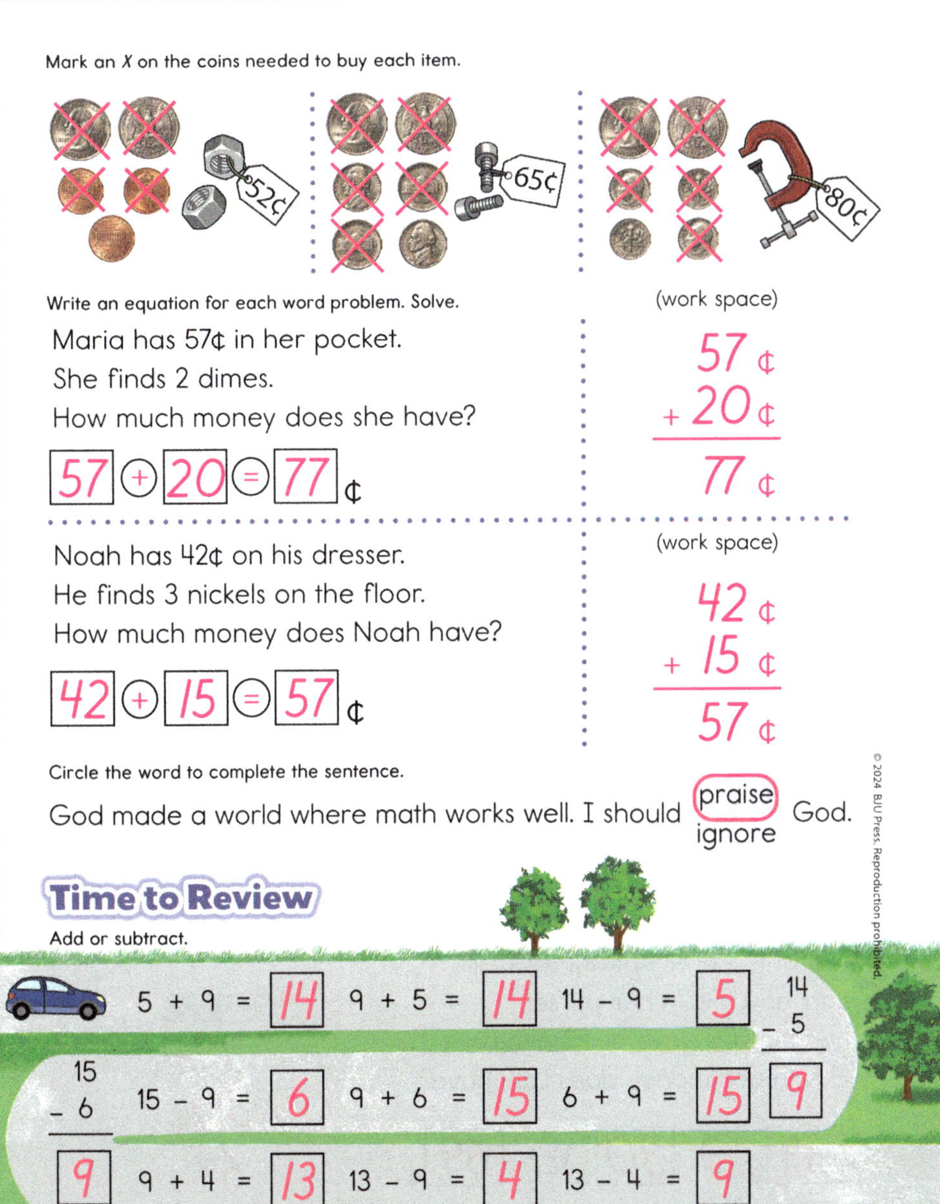

Mark an X on the coins needed to buy each item.

52¢ 65¢ 80¢

Write an equation for each word problem. Solve.

Maria has 57¢ in her pocket.
She finds 2 dimes.
How much money does she have?

57 + 20 = 77 ¢

(work space)

57 ¢
+ 20 ¢
77 ¢

Noah has 42¢ on his dresser.
He finds 3 nickels on the floor.
How much money does Noah have?

42 + 15 = 57 ¢

(work space)

42 ¢
+ 15 ¢
57 ¢

Circle the word to complete the sentence.

God made a world where math works well. I should (praise / ignore) God.

Time to Review

Add or subtract.

5 + 9 = 14 9 + 5 = 14 14 − 9 = 5 14 − 5 = 9

15 − 6 = 9 15 − 9 = 6 9 + 6 = 15 6 + 9 = 15

9 + 4 = 13 13 − 9 = 4 13 − 4 = 9

308 three hundred eight Math 1

6 + 9 9 + 6 15 − 9 15 − 6

Display the addition and subtraction **flashcards** slowly to help the students memorize the facts. Invite students to give the answers. Explain that making 10 may make it easier to add 9. Point out that 15 − 9 and 15 − 6 are related facts. Remind the students that these facts are part of the 6-9-15 fact family.

Distribute Number Cards 0–20. Display each flashcard again. Direct the students to hold up the correct Number Card to indicate each answer.

Practice these and recently memorized facts.

Instruct

Determine the Value of a Set of Quarters & Pennies

Sing a **song** to help the students *count on* with 1 or 2 quarters.

Display 1 quarter and 4 pennies.

What is the value of 1 quarter? 25¢

How do you count the pennies? by 1s

Count together the value of the set of coins: 25, (deep breath), 26, 27, 28, 29¢.

Display 2 quarters. Guide the students as they sing "A Quarter's Value" together to the tune of "London Bridge."

One quarter equals 25 cents, 25 cents, 25 cents.
Two quarters equal 50 cents, 50 cents.

Hold 1 quarter in one hand and 2 quarters in the other.

Raise one hand and then the other randomly, encouraging the students to call out 25¢ or 50¢.

Display 2 quarters and 1 penny in random order.

Which coin do you count first to find the value of this set? the quarter, because a quarter has a greater value than a penny

What is the value of 1 quarter? 25¢

What is the value of 2 quarters? 50¢

Separate the coins and demonstrate counting the set: 25, 50, (deep breath), 51¢.

Display 2 quarters and 6 pennies in random order.

Which coin do you count first in this set? the quarter, because a quarter has a greater value than a penny

Ask a volunteer to separate the coins. Count together the value of the set: 25, 50, (deep breath), 51, 52, 53, 54, 55, 56¢.

Continue the activity with 2 quarters and 4 pennies in random order. 25, 50, (deep breath), 51, 52, 53, 54¢

Determine the Value of a Set of Quarters & Nickels

Guide a **counting activity** to help the students determine the value of a set of coins.

Display 1 quarter and 2 nickels.

What is the value of 1 quarter? 25¢

How do you count nickels? by 5s

Count the coins together: 25, (deep breath), 30, 35¢.

Display 2 quarters and 3 nickels in random order.

Which coin do you count first in this set? the quarter, because a quarter has a greater value than a nickel

What is the value of 1 quarter? 25¢

What is the value of 2 quarters? 50¢

Separate the coins and demonstrate counting the set: 25, 50, (deep breath), 55, 60, 65¢.

158

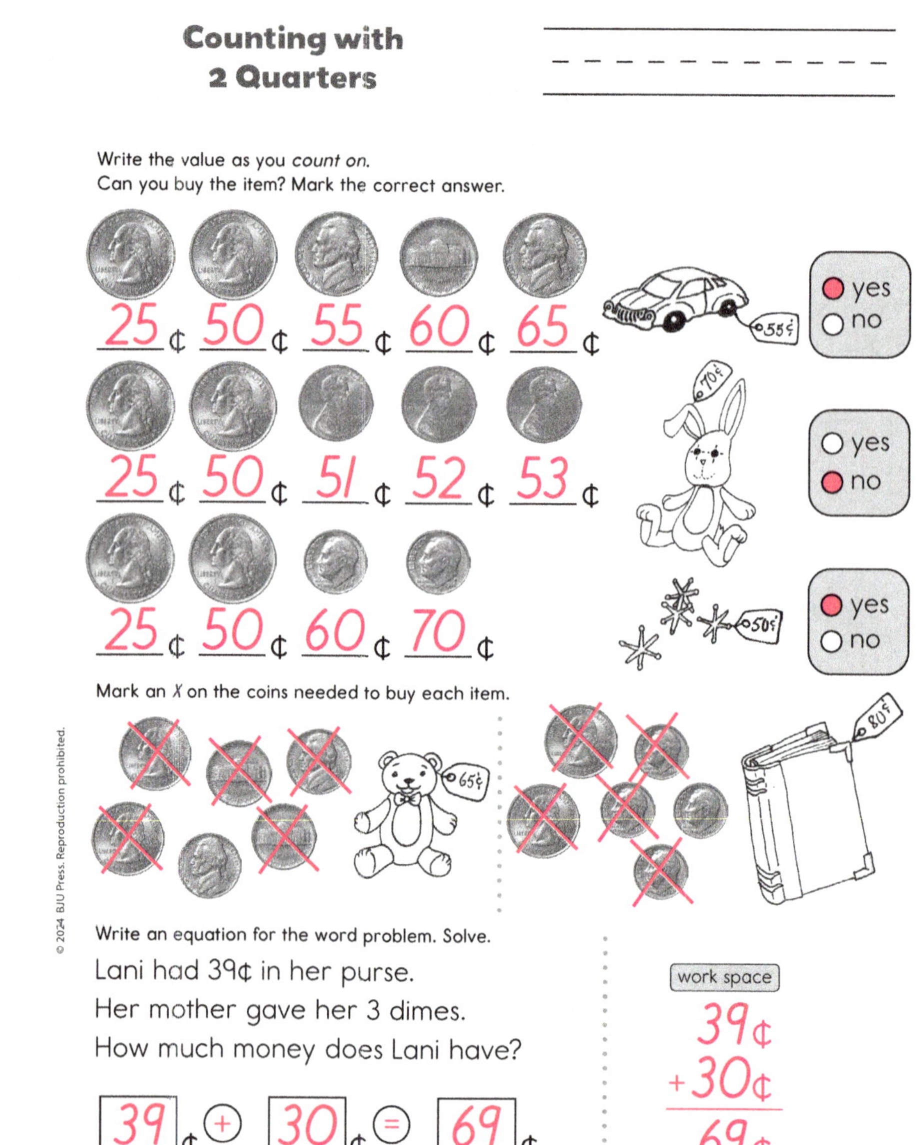

Counting with 2 Quarters

Write the value as you *count on.*
Can you buy the item? Mark the correct answer.

25¢ 50¢ 55¢ 60¢ 65¢ — 55¢ — ● yes ○ no

25¢ 50¢ 51¢ 52¢ 53¢ — 70¢ — ○ yes ● no

25¢ 50¢ 60¢ 70¢ — 50¢ — ● yes ○ no

Mark an *X* on the coins needed to buy each item.

65¢ — 80¢

Write an equation for the word problem. Solve.

Lani had 39¢ in her purse.
Her mother gave her 3 dimes.
How much money does Lani have?

39¢ + 30¢ = 69¢

work space

39¢
+30¢
69¢

Chapter 19 • Lesson 158 two hundred ninety-five 295

Display 2 quarters and 6 nickels in random order. Ask a volunteer to separate the coins. Count together the value of the set: 25, 50, (deep breath), 55, 60, 65, 70, 75, 80¢.

Continue the activity with 2 quarters and 4 nickels in random order. 25, 50, (deep breath), 55, 60, 65, 70¢

Determine the Value of a Set of Quarters & Dimes

Guide a **counting activity** to help the students determine the value of a set of coins.

Display 1 quarter and 2 dimes.

What is the value of 1 quarter? 25¢

How do you count dimes? by 10s

Count together the value of the set: 25, (deep breath), 35, 45¢.

Display 2 quarters and 3 dimes.

What is the value of 2 quarters? 50¢

Demonstrate counting the value of the set: 25, 50, (deep breath), 60, 70, 80¢.

Display 2 quarters and 2 dimes in random order. Ask a student to separate the coins. Guide the students as they count the total value of the set: 25, 50, (deep breath), 60, 70¢.

Continue the activity with 2 quarters and 4 dimes in random order. 25, 50, (deep breath), 60, 70, 80, 90¢

Identify the Coins Needed to Purchase an Item

Guide a **demonstration** to identify coins needed to purchase an item.

Distribute a *Coin Purse Mat* page and a set of coins to each student. Direct the students to place the money in the coin purse on their mats. Demonstrate each step. Display a general store item with a price of 60¢.

How much does this item cost? 60¢

Explain that students are going to buy this item by using only quarters and pennies.

How many quarters do you need? 2

Direct the students to place 2 quarters on the hand on their mats.

Count the money together: 25, 50¢.

How many pennies do you need to make 60¢? 10

Direct the students to place 10 pennies on the hand.

Count together the value of the money: 25, 50, (deep breath), 51, 52, 53, 54, 55, 56, 57, 58, 59, 60¢.

Direct the students to place the money back in the coin purse.

Explain that students are going to buy this item again. This time they will use only quarters and nickels.

How many quarters do you need? 2

Direct students to place 2 quarters on the hand.

Count the money together: 25, 50¢.

How many nickels do you need to make 60¢? 2

Direct the students to place 2 nickels on the hand.

Count together the value of the money: 25, 50, (deep breath), 55, 60¢.

Direct students to put the money back in the coin purse.

Explain that this time students will use only quarters and dimes.

How many quarters do you need? 2

Direct the students to place 2 quarters on the hand.

Count the money together: 25, 50¢.

How many dimes do you need to make 60¢? 1

Direct the students to place 1 dime on the hand.

Guide the students as they count the value of the money: 25, 50, (deep breath), 60¢.

Chapter 10 Review

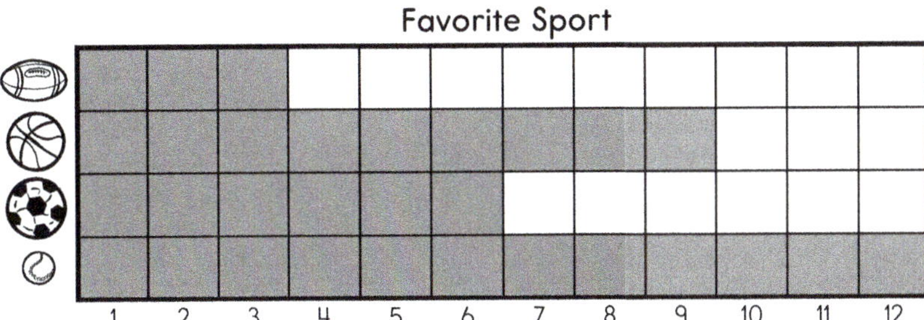

Use the graph to answer each question.

Some first graders were asked to pick the sport they liked the most.

How many children picked baseball? 12

How many children picked football?  3

How many more children picked baseball than football?

    12 − 3 = 9 children

How many more children picked basketball than soccer?

    9 − 6 = 3 children

Subtraction Fact Review

Mark the correct answer.

- $8 - 4 =$ 4 ● 5 ○ 6 ○
- $6 - 3 =$ 3 ● 4 ○ 5 ○
- $10 - 5 =$ 3 ○ 4 ○ 5 ●
- $12 - 6 =$ 4 ○ 5 ○ 6 ●
- $6 - 5 =$ 0 ○ 1 ● 2 ○
- $10 - 4 =$ 4 ○ 5 ○ 6 ●
- $4 - 2 =$ 2 ● 3 ○ 4 ○
- $11 - 5 =$ 4 ○ 5 ○ 6 ●

296 two hundred ninety-six Math 1 Reviews

DQ counted the dimes: "10, 20, 30, 40, 50, 60, 70, 80, 90¢. There you go. That's 90¢."

Liam's dad handed the money to Mr. Sawyer and thanked him for the parts.

As they turned to leave, Liam thought, *That's a lot of dimes to count. I wonder if there is a faster way to count 90¢?*

Just as they reached DQ's house, Liam had an idea.

"DQ, did you give 9 dimes to Mr. Sawyer?" Liam asked. "I just remembered a faster way to count."

Ask the students if they can tell a faster way to count to 90¢.

"We learned we can *count on* with two quarters. So we could have started with 2 quarters and then added some dimes: 25, 50, (deep breath) 60, 70, 80, 90. That's a lot faster," said Liam.

"Counting coins works the same way whether we are buying a snack, a gift, or a car part," said DQ. "It sure is helpful knowing how to use coins!"

"God made math a part of our world," Liam's dad explained. "The better we understand math, the better we can understand the world where we live. We should praise God that we can use math. We should thank Him for its usefulness."

What should be your response to God, who made a world where math works well? I should thank and praise God.

Solve Word Problems

Use the **Problem-Solving Plan** to solve money word problems.

Read aloud the following word problem.

Shiloh got 50¢ from his grandmother. His grandfather gave him 2 dimes. How much money does Shiloh have in all?

What is the question? How much money does Shiloh have in all?

What information is given? He has 50¢ from his grandmother and 2 dimes from his grandfather.

Do you add or subtract in this problem? add

What is the value of 2 dimes? 20¢

What equation do you write to solve this problem? 50¢ + 20¢ = ___

Write "50¢ + 20¢ = ___" for display.

Invite a student to write 50¢ + 20¢ vertically and solve the problem. Complete the equation. 70¢

Does the answer make sense? yes

Continue the activity with the following word problem.

Keith has 73¢ on his dresser. He also found 3 nickels. How much money does Keith have in all? 73¢ + 15¢ = 88¢

Express Praise to God

Share a **scenario** to help the students express a biblical worldview shaping truth.

Remind the students that DQ was able to count coins at the auto parts store. Read aloud part of the opening scenario before continuing the story.

Apply

Worktext pages 307–8

Read and guide completion of page 307.

Read and explain the directions for page 308. Assist the students as they complete the page independently.

Direct attention to the biblical worldview shaping statement and guide the students as they complete the sentence.

Assess

Reviews pages 295–96

Review interpreting a bar graph on page 296.

159

Worktext pages 309–10
Reviews pages 297–98

Objectives

- **159.1** Identify the number of sets and the number of objects in each set.
- **159.2** Count by 2s, 3s, or 5s to find the total.
- **159.3** Make equal sets.

Printed Resources

- Instructional Aid 46: *Repeated Addition Mat* (for each student)
- Visuals: Money Kit (20 pennies)
- Student Manipulatives: Money Kit (20 pennies)

Materials

- 6 books displayed in 3 stacks of 2 books

Lessons 159–62 are included as enrichment instruction. Present these lessons based on your schedule and the readiness of the students.

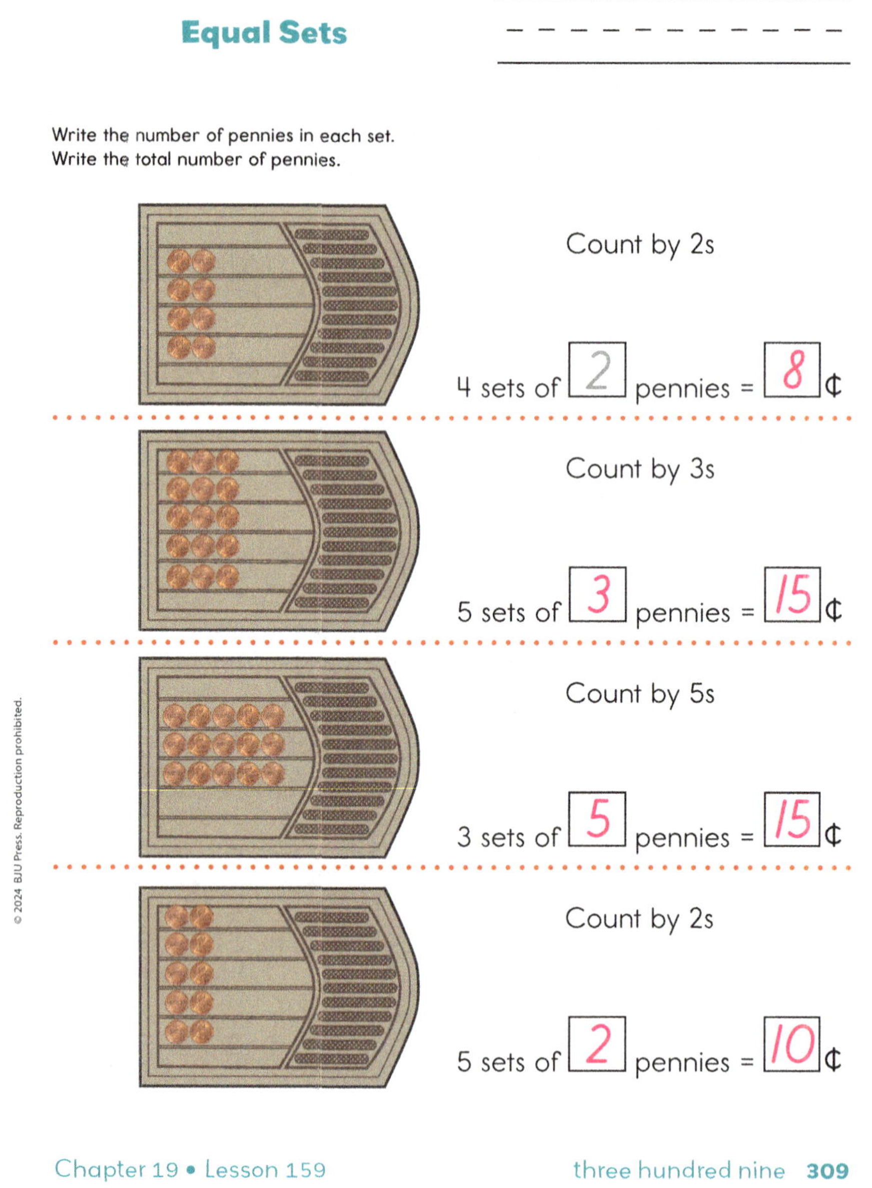

Equal Sets
Write the number of pennies in each set.
Write the total number of pennies.
Count by 2s
4 sets of 2 pennies = 8¢
Count by 3s
5 sets of 3 pennies = 15¢
Count by 5s
3 sets of 5 pennies = 15¢
Count by 2s
5 sets of 2 pennies = 10¢
© 2024 BJU Press. Reproduction prohibited.
Chapter 19 • Lesson 159
three hundred nine 309

Engage

Introduction to Equal Sets

Share a **Bible account** to introduce the concept of equal sets.

Explain that during the time when Jesus lived, people wore sandals everywhere, so their feet would be very dusty after walking all day with sandals and no socks.

Point out that servants would wash the feet of people who had walked a long way. Explain that Jesus showed His apostles the importance of helping others by kneeling down in front of each one of them and washing their feet; He then told them that they should be willing to wash each others' feet just as He had washed theirs (John 13:4–5, 14).

How many apostles did Jesus have? 12

Ask 12 students to stand beside their desks.

How many feet does each person have? 2

How can you find out the total number of feet that Jesus washed? count all the feet of the 12 standing students, count the feet by 2s

Guide the students as they count by 2s as you point to each of the 12 students: 2, 4, 6, 8, 10, 12, 14, 16, 18, 20, 22, 24. Conclude that Jesus washed 24 feet.

Instruct

Identify the Number of Sets & the Number of Objects in Each Set

Use **direct questioning** to identify the number of sets and the number of objects in each set.

Invite 6 students to come to the front of the room.

Arrange them so that they are standing in 2 groups with 3 students in each group.

How many groups or sets of students are there? 2

How many students are in each set? 3

How many students are there in all? 6

Write for display "2 sets of 3 = 6."

Direct attention to the 3 sets of 2 books on display.

How many sets of books are there? 3

How many books are in each set? 2

How many books are there in all? 6

Write "3 sets of 2 = 6."

Direct 3 students to each raise 1 hand.

How many sets of fingers are raised? 3

How many fingers are in each set (on each raised hand)? 5

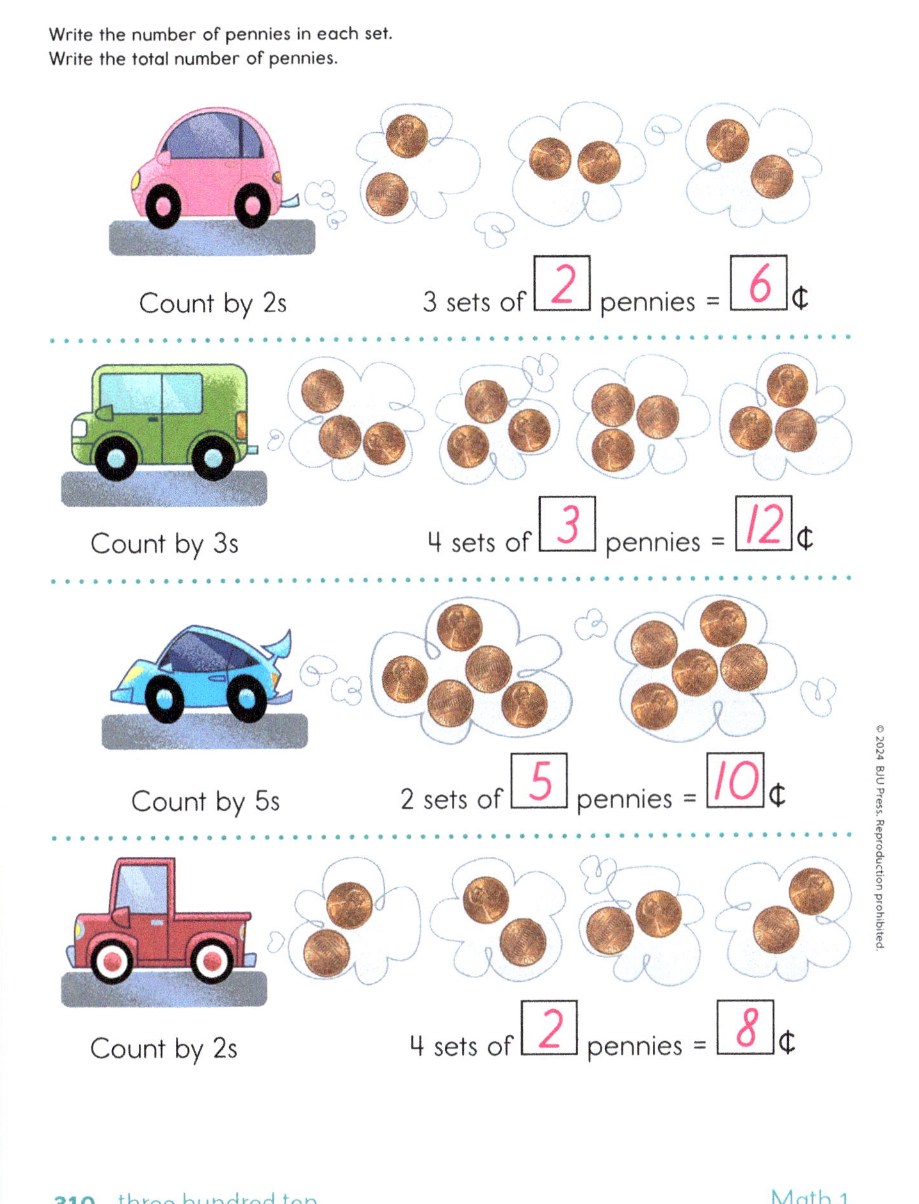

How many fingers are raised in all? 15

Write "3 sets of 5 = 15."

Count by 2s, 3s & 5s to Find the Total

Guide the students through a **discussion** to find the total in a repeated addition problem.

Display 5 sets of 2 pennies.

How many sets of pennies are there? 5

How many pennies are in each set? 2

Explain that students can find out the total number of pennies by counting by 2s. Point to each set as you guide the students as they count by 2s.

What is the total number of pennies in 5 sets of 2 pennies? 10

Write for display "5 sets of 2 = 10."

Display 3 sets of 3 pennies.

How many sets of pennies are there? 3

How many pennies are in each set? 3

Explain that students can find the total number of pennies by counting by 3s. Point to each set as you count by 3s together.

What is the total number of pennies in 3 sets of 3 pennies? 9

Write "3 sets of 3 = 9."

Display 4 sets of 5 pennies.

How many sets of pennies are there? 4

How many pennies are in each set? 5

How should you count to find the total number of pennies when there are 5 pennies in each set? by 5s

159

Point to each set as you count by 5s together.

What is the total number of pennies in 4 sets of 5 pennies? 20

Write "4 sets of 5 = 20."

Make Equal Sets

Model making sets with manipulatives.

Distribute a set of pennies and the *Repeated Addition Mat* page to each student.

Write for display "6 sets of 2 = ___." Direct the students to place 2 pennies in each of the 6 sections on their mats.

How should you count to find the total number of pennies on your mat? by 2s, because there are 2 pennies in each set

Guide the students as they count the sets by 2s.

What is the total number of pennies in 6 sets of 2? 12

Ask a student to write "12" in the blank.

Write "3 sets of 5 = ___." Direct students to place 5 pennies in each of 3 sections on their mats.

How should you count to find the total number of pennies on your mat? by 5s, because there are 5 pennies in each set

Guide the students as they count the sets by 5s.

What is the total number of pennies in 3 sets of 5? 15

Invite a student to write "15" in the blank.

Continue the activity for the following sets.

4 sets of 3 = 12; count by 3s

2 sets of 5 = 10; count by 5s

5 sets of 2 = 10; count by 2s

Apply

Worktext pages 309–10

Read and guide completion of page 309.

Read and explain the directions for page 310. Assist the students as they complete the page independently.

Assess

Reviews pages 297–98

Read the directions and guide completion of the pages.

159

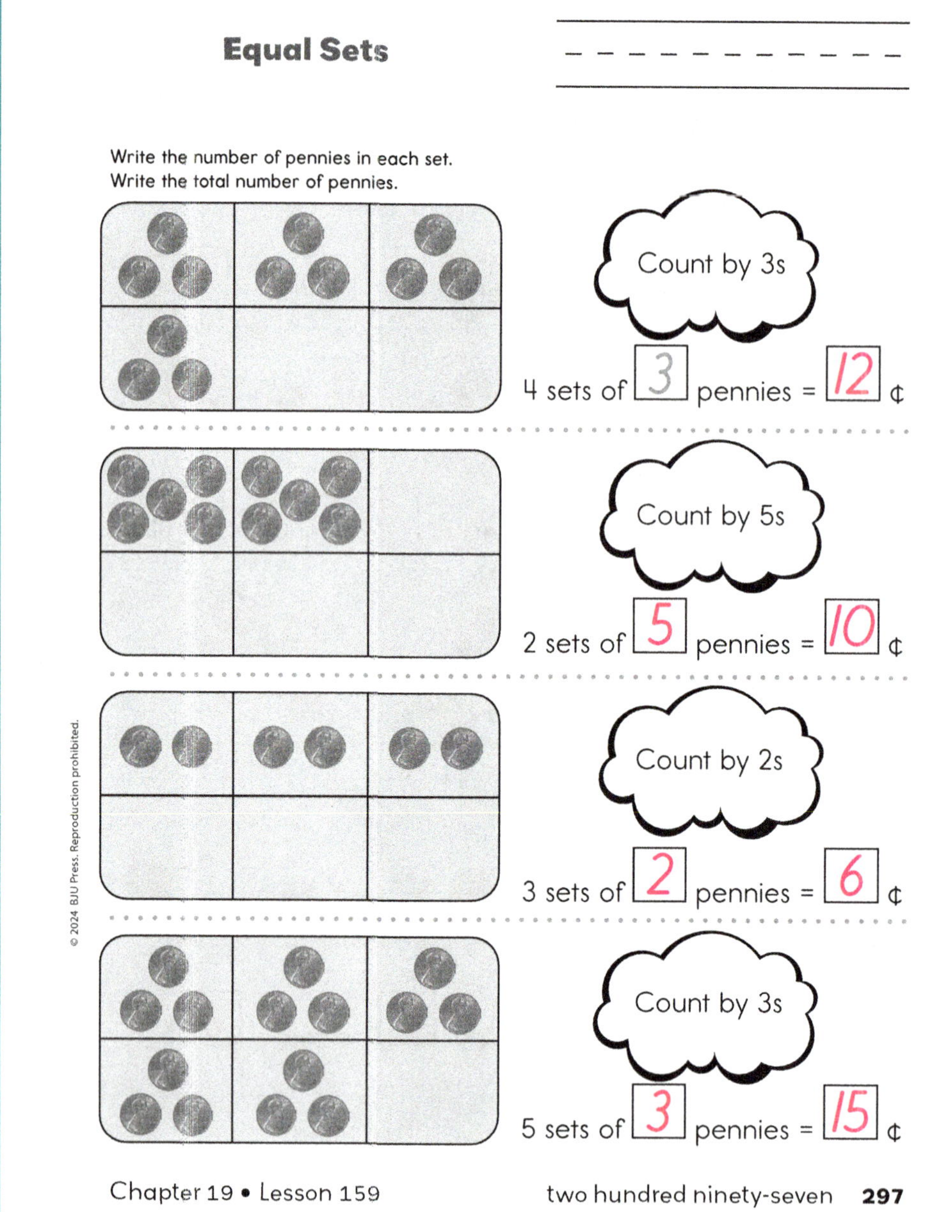
Equal Sets
Write the number of pennies in each set.
Write the total number of pennies.
Count by 3s
4 sets of 3 pennies = 12 ¢
Count by 5s
2 sets of 5 pennies = 10 ¢
Count by 2s
3 sets of 2 pennies = 6 ¢
Count by 3s
5 sets of 3 pennies = 15 ¢
© 2024 BJU Press. Reproduction prohibited.
Chapter 19 • Lesson 159
two hundred ninety-seven 297

Write the number of pennies in each set.
Write the total number of pennies.

Count by 2s

4 sets of 2 pennies = 8 ¢

Count by 2s

3 sets of 2 pennies = 6 ¢

Count by 5s

4 sets of 5 pennies = 20 ¢

Count by 3s

3 sets of 3 pennies = 9 ¢

160 Worktext pages 311–12 Reviews pages 299–300

Objectives

- **160.1** Identify the number of sets and the number of objects in each set.
- **160.2** Count by 2s, 3s, and 5s to find the total.
- **160.3** Make equal sets.
- **160.4** Complete a repeated addition problem.
- **160.5** Write a repeated addition equation for a word problem.

Printed Resources

- Instructional Aid 46: *Repeated Addition Mat* (for each student)
- Visuals: Counters (20 cars)
- Student Manipulatives: Counters (20 cars)

Engage

Introduction to Repeated Addition

Guide a **role-playing activity** to introduce repeated addition.

Explain that Jesus sent His 12 apostles to preach salvation to the people living in different villages and that He did not want the apostles to go alone, so He sent them in groups of 2 (Mark 6:7).

How many men were in each group? 2

How many men went out to preach in the villages? 12

Write for display "__ groups of 2 = 12 men."

How can you find out the number of groups of men there were? draw pictures, use counters, count by 2s

Invite 12 students to stand in the front of the room.

Arrange the students into groups of 2. Guide the students as they count to find out the number of groups of 2 in 12. 6

Finding the number of equal groups is a division concept.

Repeated Addition

Write the number in each set. Write the total number.

How many nuts are there?

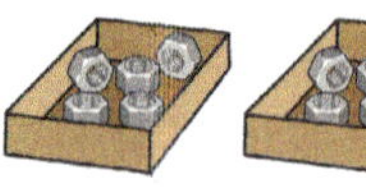 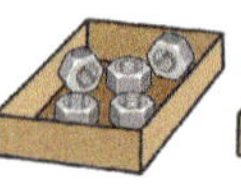  

3 sets of 5

5 + 5 + 5 = 15 nuts

How many bolts are there?

4 sets of 3

3 + 3 + 3 + 3 = 12 bolts

How many cars are there?

 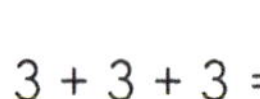  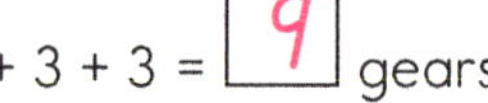

5 sets of 2

2 + 2 + 2 + 2 + 2 = 10 cars

How many gears are there?

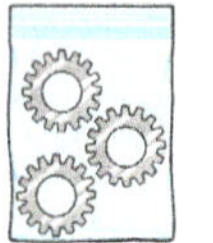  

3 sets of 3

3 + 3 + 3 = 9 gears

Picture the problem.
Write an equation for the word problem. Solve.

Mother served 3 plates of cookies.
She put 2 cookies on each plate.
How many cookies did she serve?

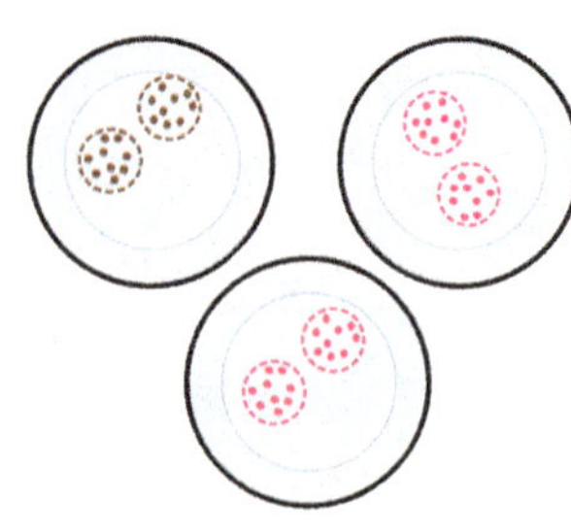

 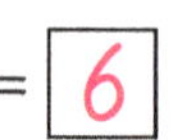

2 + 2 + 2 = 6 cookies

Instruct

Make Equal Sets; Count by 2s, 3s, or 5s to Find the Total

Use **manipulatives** to make equal sets and find the total.

Distribute a set of counters and the *Repeated Addition Mat* page to each student.

Write for display "4 sets of 2 = __." Direct the students to place 2 cars in each of 4 sections on their mats. Remind the students that to find the total number of cars they can count by 2s because there are 2 cars in each set. Guide the students as they count the sets by 2s.

What is the total number of cars in 4 sets of 2? 8

Ask a student to write "8" in the blank.

Write for display "4 sets of 5 = __." Direct the students to place 5 cars in each of 4 sections on their mats.

How should you count to find the total number of cars on your mat? by 5s, because there are 5 cars in each group

Guide the students as they count the sets by 5s.

What is the total number of cars in 4 sets of 5? 20

Invite a student to write "20" in the blank.

Continue the activity for the following sets.

5 sets of 3 = 15; count by 3s

2 sets of 3 = 6; count by 3s

3 sets of 5 = 15; count by 5s

Write the number in each set. Write the total number.

How many gloves are there?

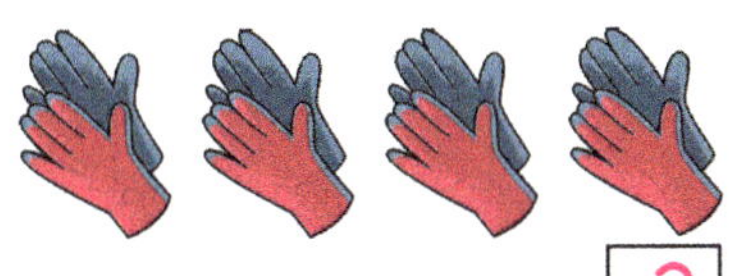

4 sets of 2

2 + 2 + 2 + 2 = 8 gloves

How many wax containers are there?

3 sets of 5

5 + 5 + 5 = 15 containers

How many spray bottles are there?

5 sets of 3

3 + 3 + 3 + 3 + 3 = 15 bottles

How many sponges are there?

3 sets of 3

3 + 3 + 3 = 9 sponges

Picture the problem.
Write an equation for the word problem. Solve.

Miguel has 3 toy boxes.
He has 5 balls in each toy box.
How many balls does Miguel have?

5 + 5 + 5 = 15 balls

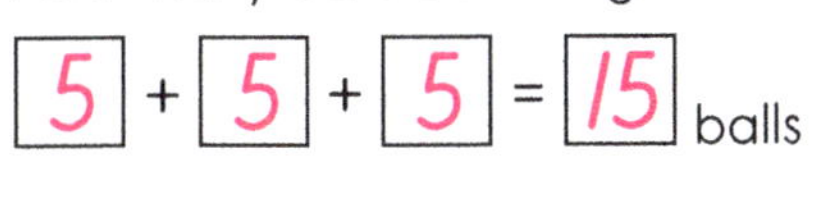

Write a Repeated Addition Equation

Use a **guided practice** to write a repeated addition equation.

Read aloud the following word problem several times.

Last week workers repaired cars on Monday, Tuesday, and Wednesday. They repaired 5 cars each day. How many cars did they repair during these 3 days?

How many cars were repaired on Monday? 5 cars

Display a set of 5 cars.

How many cars were repaired on Tuesday? 5 cars

Display a second set of 5 cars.

How many cars were repaired on Wednesday? 5 cars

Display a third set of 5 cars.

What addition equation could you write to find out how many cars were repaired during the 3 days? 5 + 5 + 5 = __

Write for display "5 + 5 + 5 = __ cars." Invite a student to write the answer and explain how he or she found the answer. 15; I counted the sets of cars by 5s.

Read aloud the following word problem several times.

This week workers will spend 5 days painting cars. They can paint 2 cars each day. How many cars will they be able to paint during this time?

How many cars will they paint each day? 2 cars

How many days will they paint cars? 5 days

Display 5 sets of 2 cars.

What addition equation could you write to find out how many cars were painted during the 5 days? 2 + 2 + 2 + 2 + 2 = __

Write for display "2 + 2 + 2 + 2 + 2 = __ cars." Ask a volunteer to write the answer and explain how he or she found the answer. 10; I counted the sets of cars by 2s.

Explain that this equation is sometimes called a *repeated addition equation* since the same addend is repeated.

Point out that in this equation, 2 was repeated 5 times because in the word problem 2 cars were painted each day for 5 days. Explain that students can find the answer by pointing to each addend as they count by 2s.

Continue the activity by using the following word problem.

During a big winter storm, there were 4 days in a row when the tow truck had to tow 3 cars each day. What was the total number of cars towed during these 4 days? 3 + 3 + 3 + 3 = 12 cars

Apply

Worktext pages 311–12

Read and guide completion of page 311.

Read and explain the directions for page 312. Assist the students as they complete the page independently.

Assess

Reviews pages 299–300

Read the directions and guide completion of the pages.

160

Extended Activity

Illustrate a Word Problem

Procedure

Distribute a sheet of drawing paper to each student. Instruct the students to solve the following word problem by drawing an illustration. Direct them to write and solve a repeated addition equation below the picture.

Lily bought 3 boxes of cupcakes. Each box contains 5 cupcakes. What is the total number of cupcakes Lily purchased?
5 + 5 + 5 = 15 cupcakes

Repeated Addition

Write the number in each set. Write the total number.

How many cupcakes are there?

3 sets of 2

2 + 2 + 2 = 6 cupcakes

How many bags of popcorn are there?

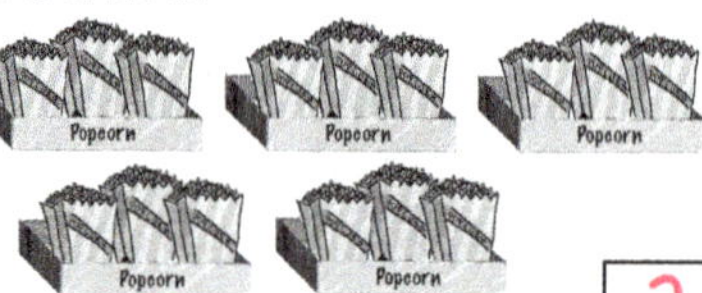

5 sets of 3

3 + 3 + 3 + 3 + 3 = 15 bags of popcorn

How many candy apples are there?

4 sets of 3

3 + 3 + 3 + 3 = 12 candy apples

How many pretzels are there?

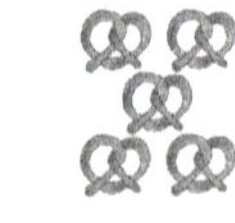

3 sets of 5

5 + 5 + 5 = 15 pretzels

Picture the problem. Write an equation for the word problem. Solve.

Dad bought 4 boxes of ice cream bars. There were 2 ice cream bars in each box. How many ice cream bars did he buy?

2 + 2 + 2 + 2 = 8 ice cream bars

Write the number in each set. Write the total number.

How many gumballs are there?

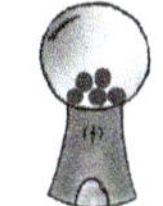

4 sets of [5]

5 + 5 + 5 + 5 = [20] gumballs

How many candles are there?

3 sets of [3]

3 + 3 + 3 = [9] candles

How many cookies are there?

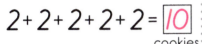

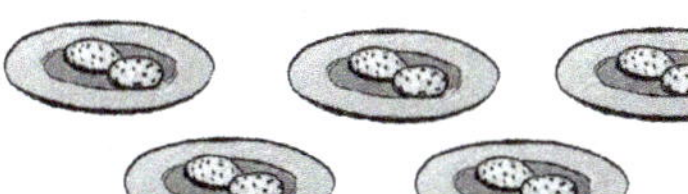

5 sets of [2]

2 + 2 + 2 + 2 + 2 = [10] cookies

How many pieces of candy are there?

4 sets of [3]

3 + 3 + 3 + 3 = [12] pieces of candy

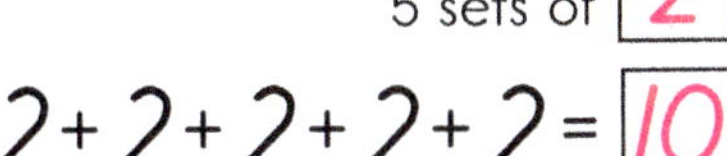

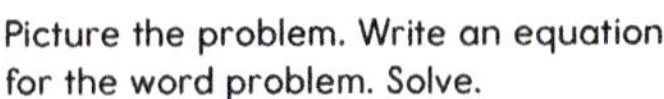

Picture the problem. Write an equation for the word problem. Solve.

Uncle Blake bought 3 ice cream cones. Each cone had 2 scoops of ice cream. How many scoops did Uncle Blake buy?

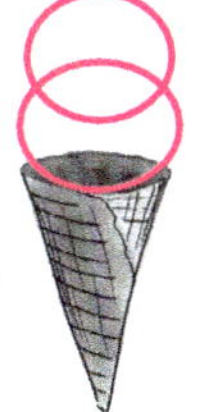
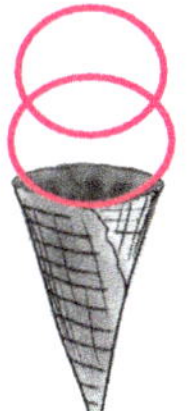
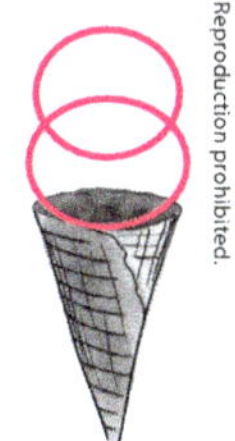

[2] + [2] + [2] = [6] scoops of ice cream

161–62

Worktext pages 313–14
Reviews pages 301–2

Objectives

- **161–62.1** Identify the number of sets and the number of objects in each set.
- **161–62.2** Count by 2s, 3s, and 5s to find the total.
- **161–62.3** Write a repeated addition equation for sets of objects.
- **161–62.4** Write a repeated addition equation for a word problem.

Printed Resources

- Visuals: Counters (20 candies)

Depending on your schedule and the students' abilities, you may take either 1 or 2 days to present Lessons 161–62.

Engage

Introduction to Repeated Addition Equations

Share an **example** to introduce the students to repeated addition equations.

What is the purpose of windshield wipers on a car? to make it easier to see when it is raining or snowing, to clean the windshield

What happens when the wipers get old? They start to squeak; they do not keep the windshield clear.

Explain that most cars have 2 windshield wipers for the front windshield, but many SUVs and minivans have 3 windshield wipers: 2 for the front and 1 for the back.

How many windshield wipers would be needed to replace all the wipers on 3 cars? 6

How many windshield wipers would be needed to replace all the wipers on 3 minivans? 9

You may ask volunteers to draw pictures to answer the questions.

Writing Repeated Addition Equations

Write the number in each set.
Complete the repeated addition equation.

How many wrenches are there?

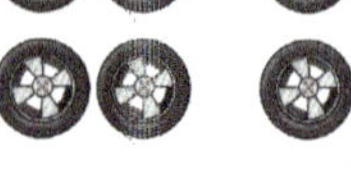

3 sets of 5

5 + 5 + 5 = 15 wrenches

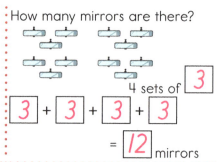

How many mirrors are there?

4 sets of 3

3 + 3 + 3 + 3 = 12 mirrors

How many tires are there?

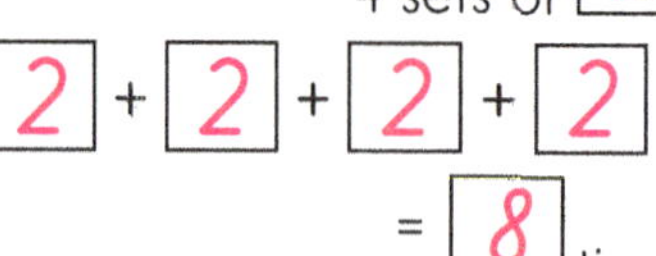

4 sets of 2

2 + 2 + 2 + 2 = 8 tires

How many license plates are there?

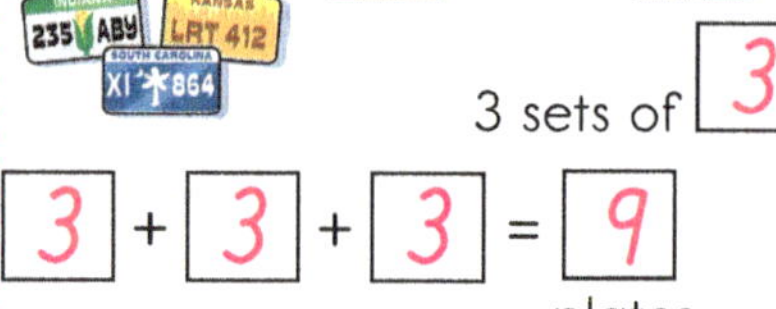

3 sets of 3

3 + 3 + 3 = 9 plates

Picture the problem.
Write an equation for the word problem. Solve.

I saw 5 birds' nests.
There were 2 eggs in each nest.
How many eggs were there?

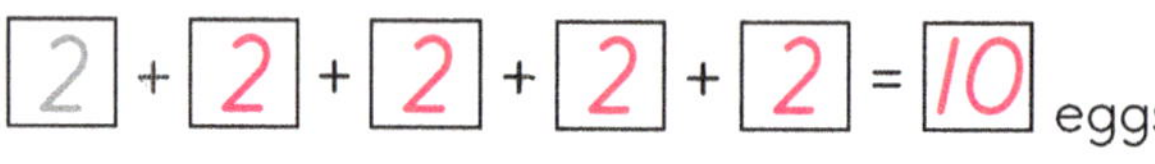

2 + 2 + 2 + 2 + 2 = 10 eggs

Instruct

Write a Repeated Addition Equation for Sets of Objects

Model a repeated addition equation by using manipulatives.

Display 4 sets of 3 candies.

How many sets of candies are there? 4

How many candies are in each set? 3

What repeated addition equation can you write to find the total number of candies? 3 + 3 + 3 + 3 = __

Write "3 + 3 + 3 + 3 = __ candies."

What addend is repeated? 3

How should you count to find the answer? by 3s

Guide the students as they count by 3s as you point to each addend.

Complete the equation, explaining that 4 groups of 3 is equal to 12.

Display 3 sets of 5 candies.

How many sets of candies are there? 3

How many candies are in each set? 5

What repeated addition equation can you write to find the total number of candies? 5 + 5 + 5 = __

Write "5 + 5 + 5 = __ candies."

What addend is repeated? 5

How should you count to find the answer? by 5s

Guide the students as they count by 5s as you point to each addend.

Write the number in each set.
Complete the repeated addition equation.

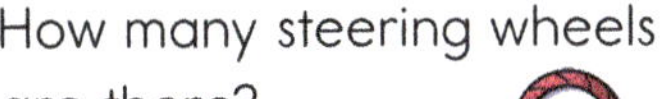

How many steering wheels are there?

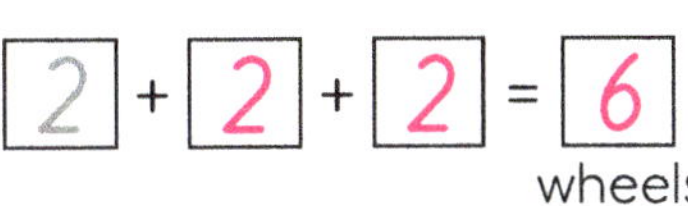

How many lug nuts are there?

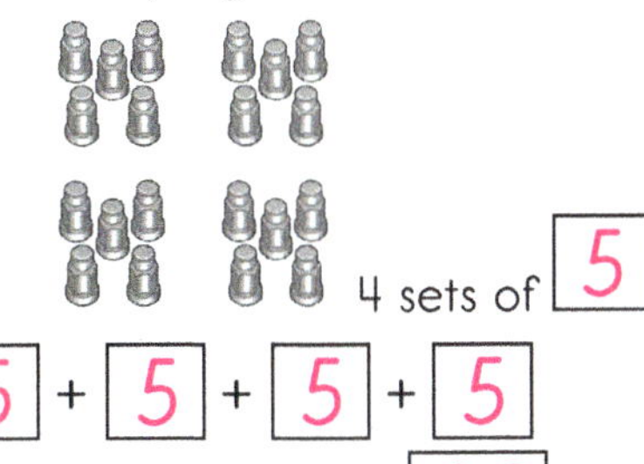

How many air fresheners are there?

How many jumper cables are there?

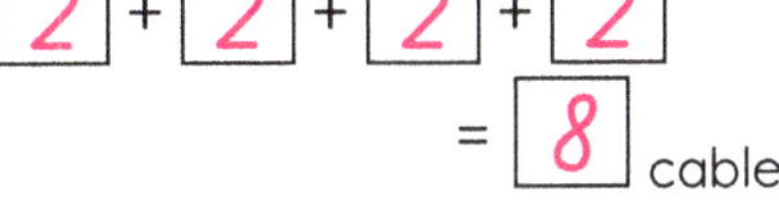

Picture the problem.
Write an equation for the word problem. Solve.

Aunt Marcie made 5 mugs of orange juice.
She used 3 oranges for each mug.
How many oranges did she use in all?

3 + 3 + 3 + 3 + 3
= 15 oranges

Complete the equation and point out that 3 groups of 5 is equal to 15.

Continue the activity for the following sets, asking volunteers to write the repeated addition equations and to lead in counting to find the answer.

4 sets of 2 = 2 + 2 + 2 + 2 = 8; count by 2s

2 sets of 5 = 5 + 5 = 10; count by 5s

5 sets of 3 = 3 + 3 + 3 + 3 + 3 = 15; count by 3s

Write a Repeated Addition Equation for a Word Problem

Work through the word problems as a **guided practice** for writing a repeated addition equation.

Read aloud the following word problem several times.

Mother gave her 4 children 5 candies each. How many pieces of candy did Mother give away?

How many children did Mother give candy to? 4 children

How many pieces of candy did each child receive? 5 pieces

Display 4 sets of 5 candies.

What repeated addition equation can you write to find the number of candies Mother gave away? 5 + 5 + 5 + 5 = __

Write "5 + 5 + 5 + 5 = __ candies."

What addend is repeated? 5

How should you count to find the answer? by 5s

Guide the students as they count by 5s as you point to each addend.

Complete the equation, explaining that 4 groups of 5 is equal to 20.

Read aloud the following word problem several times.

On Saturday, Luke's father told Luke that he would give him 2 pieces of candy for every row in the garden that Luke weeded. He weeded 5 rows. How many pieces of candy did Luke receive?

How many rows in the garden did Luke weed? 5 rows

How many pieces of candy did Luke receive for weeding each row? 2 pieces

Display 5 sets of 2 candies. Invite a student to write the repeated addition equation and to lead in counting to solve the equation.
2 + 2 + 2 + 2 + 2 = 10 pieces; count by 2s

Continue the activity by using the following word problem.

When Mother is out of town, she gives Gabby 3 pieces of candy for each meal that she remembers to set and clear the table. How many pieces of candy will Gabby receive if she does this job for 3 meals?
3 + 3 + 3 = 9 pieces of candy

Apply

Worktext pages 313–14

Read and guide completion of page 313.

Read and explain the directions for page 314. Assist the students as they complete the page independently.

Assess

Reviews pages 301–2

Read the directions and guide completion of the pages.

161–62

Extended Activity

Repeated Addition on a Number Line

Materials

- Number Line for the teacher and for each student

Procedure

Distribute a Number Line to each student and display your Number Line.

Write for display "2 + 2 + 2 = ___." Explain that students can count by 2s on the number line by making jumps that are 2 spaces each. Start at 0 and draw an arc to 2, an arc from 2 to 4, and another arc from 4 to 6 while you count by 2s.

What is the answer? 6

Direct the students to put their fingers on 0 on their number lines and make 3 jumps of 2 while they count by 2s.

Continue the activity for 2 + 2 + 2 + 2 + 2 + 2 = 12.

Write "3 + 3 + 3 = ___." Explain that students can count by 3s on the number line by making jumps that are 3 spaces each. Start at 0 and draw an arc to 3, an arc from 3 to 6, and another arc from 6 to 9 while you count by 3s.

What is the answer? 9

Direct the students to put their fingers on 0 on their number lines and make 5 jumps of 3 while they count by 3s.

Continue the activity for 3 + 3 + 3 + 3 + 3 = 15.

Write "5 + 5 + 5 = ___."

How should you count to find the answer to this repeated addition equation? by 5s

If you use the number line, how large will each jump be? 5 spaces

Start at 0 and draw an arc to 5, an arc from 5 to 10, and another arc from 10 to 15 while you count by 5s.

What is the answer? 15

Direct the students to put their fingers on 0 on their number lines and make 3 jumps of 5 while they count by 5s.

Continue the activity for 5 + 5 + 5 + 5 + 5 = 25.

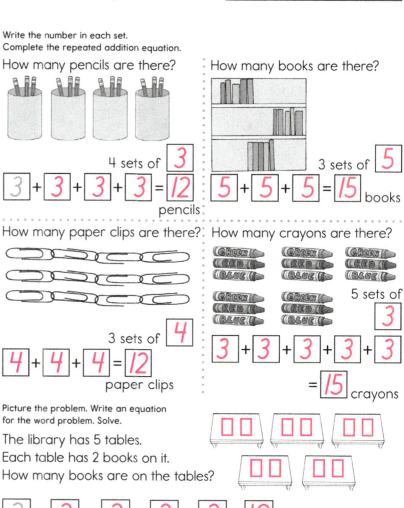

Writing Repeated Addition Equations

Write the number in each set.
Complete the repeated addition equation.

How many pencils are there?

4 sets of 3

3 + 3 + 3 + 3 = 12 pencils

How many books are there?

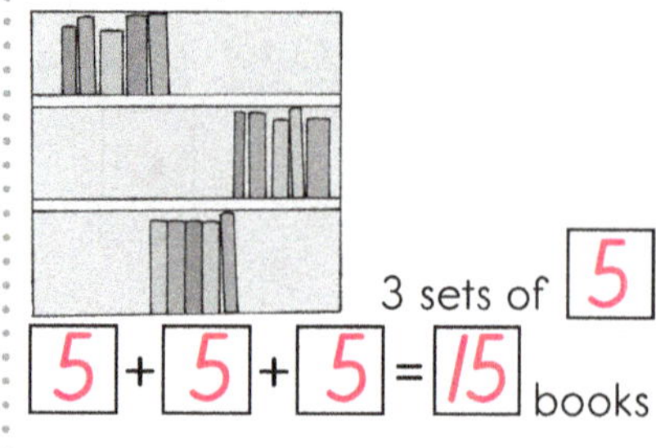

3 sets of 5

5 + 5 + 5 = 15 books

How many paper clips are there?

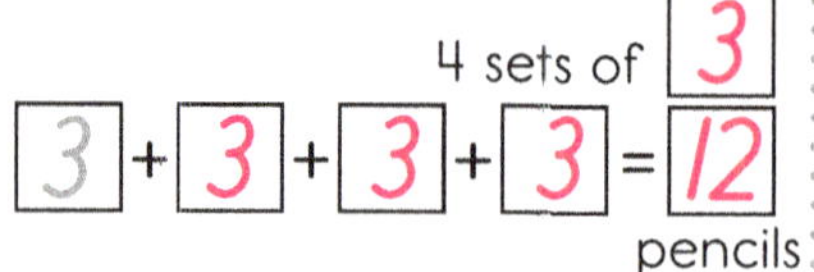

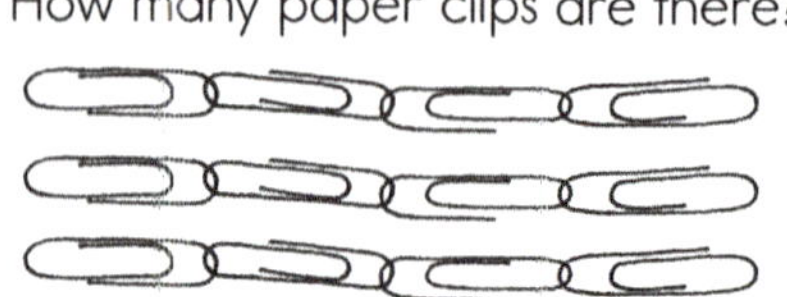

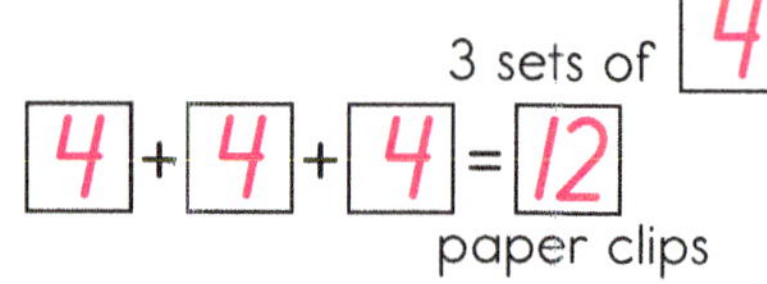

3 sets of 4

4 + 4 + 4 = 12 paper clips

How many crayons are there?

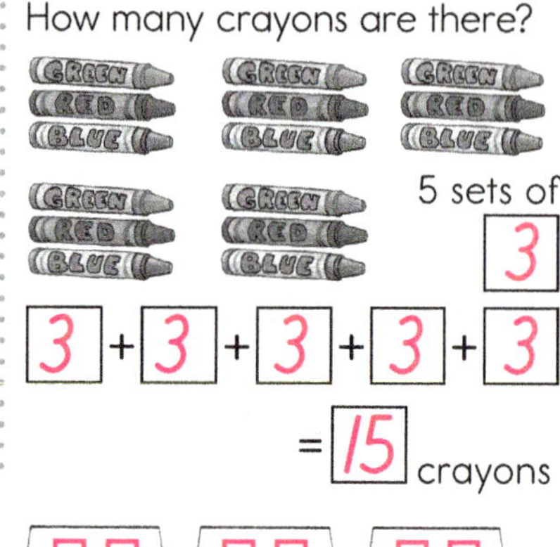

5 sets of 3

3 + 3 + 3 + 3 + 3 = 15 crayons

Picture the problem. Write an equation for the word problem. Solve.

The library has 5 tables.
Each table has 2 books on it.
How many books are on the tables?

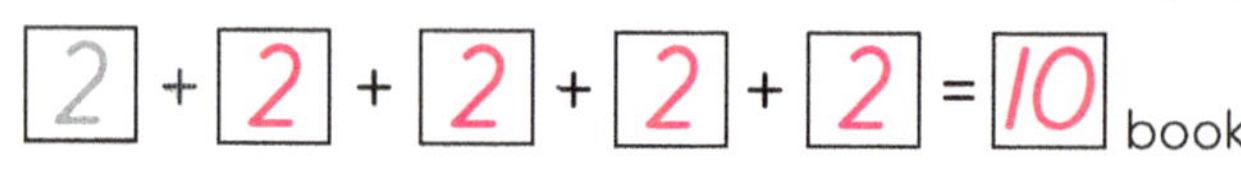

2 + 2 + 2 + 2 + 2 = 10 books

Write the number in each set.
Complete the repeated addition equation.

How many crayons are there?

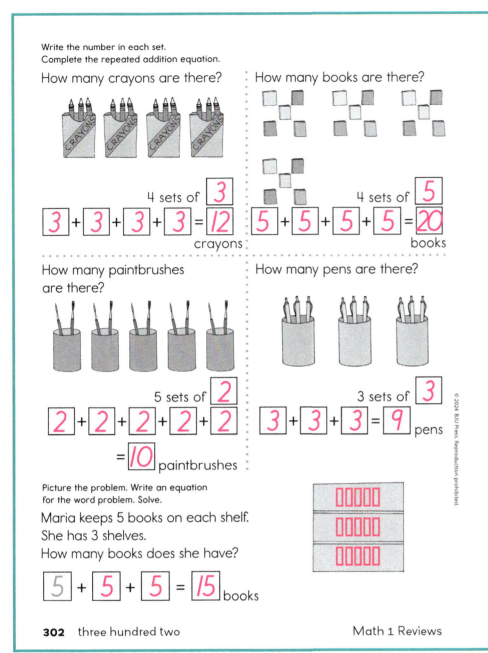

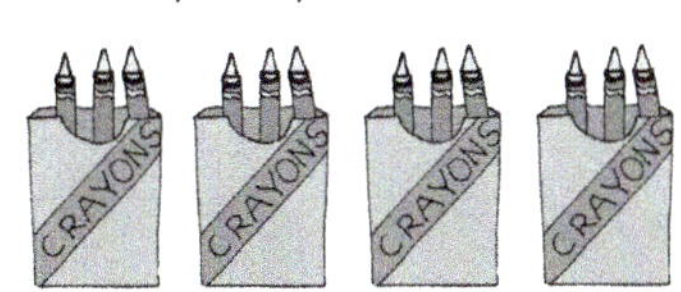

4 sets of 3

3 + 3 + 3 + 3 = 12 crayons

How many books are there?

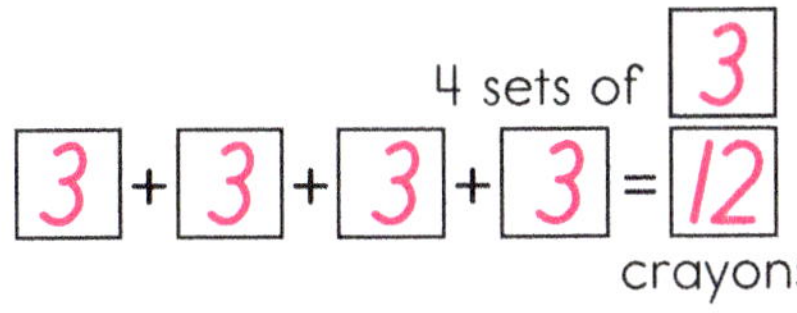

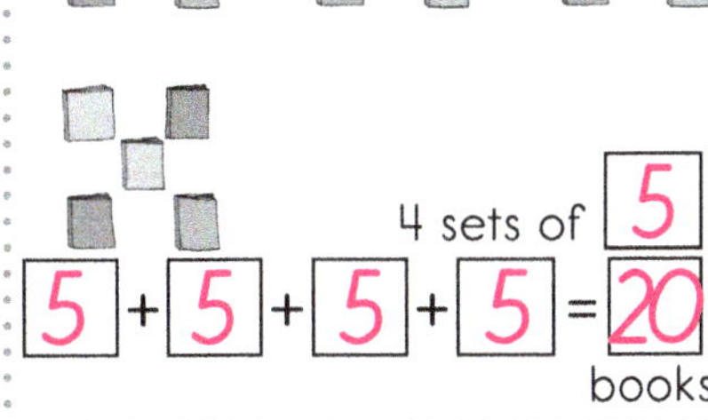

4 sets of 5

5 + 5 + 5 + 5 = 20 books

How many paintbrushes are there?

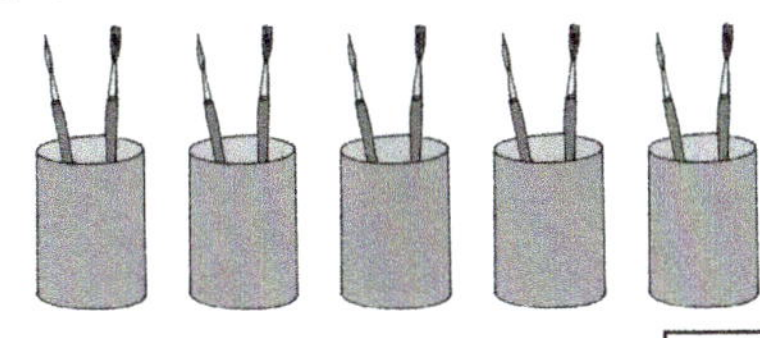

5 sets of 2

2 + 2 + 2 + 2 + 2 = 10 paintbrushes

How many pens are there?

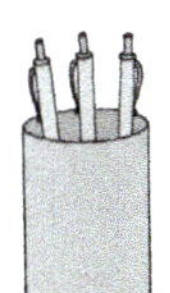

3 sets of 3

3 + 3 + 3 = 9 pens

Picture the problem. Write an equation for the word problem. Solve.

Maria keeps 5 books on each shelf.
She has 3 shelves.
How many books does she have?

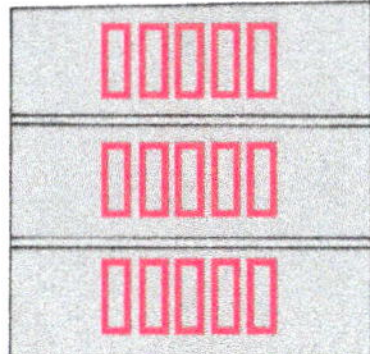

5 + 5 + 5 = 15 books

163

Worktext pages 315–16
Reviews pages 303–4

What should I tell God for giving me a world where math works well?

Chapter Concept Review

- Practice concepts from Chapter 19 to prepare for the test.

Printed Resources

- Instructional Aid 45: *Coin Purse Mat* (for each student)
- Visuals: Money Kit (2 quarters, 7 dimes, 7 nickels, 5 pennies)
- Student Manipulatives: Money Kit (2 quarters, 7 dimes, 7 nickels, 5 pennies)

Digital Resources

- Web Link: Virtual Manipulatives: Money Pieces
- Games/Enrichment: Addition Flashcards
- Games/Enrichment: Subtraction Flashcards

Materials

- Games/Enrichment: Addition and Subtraction flashcards

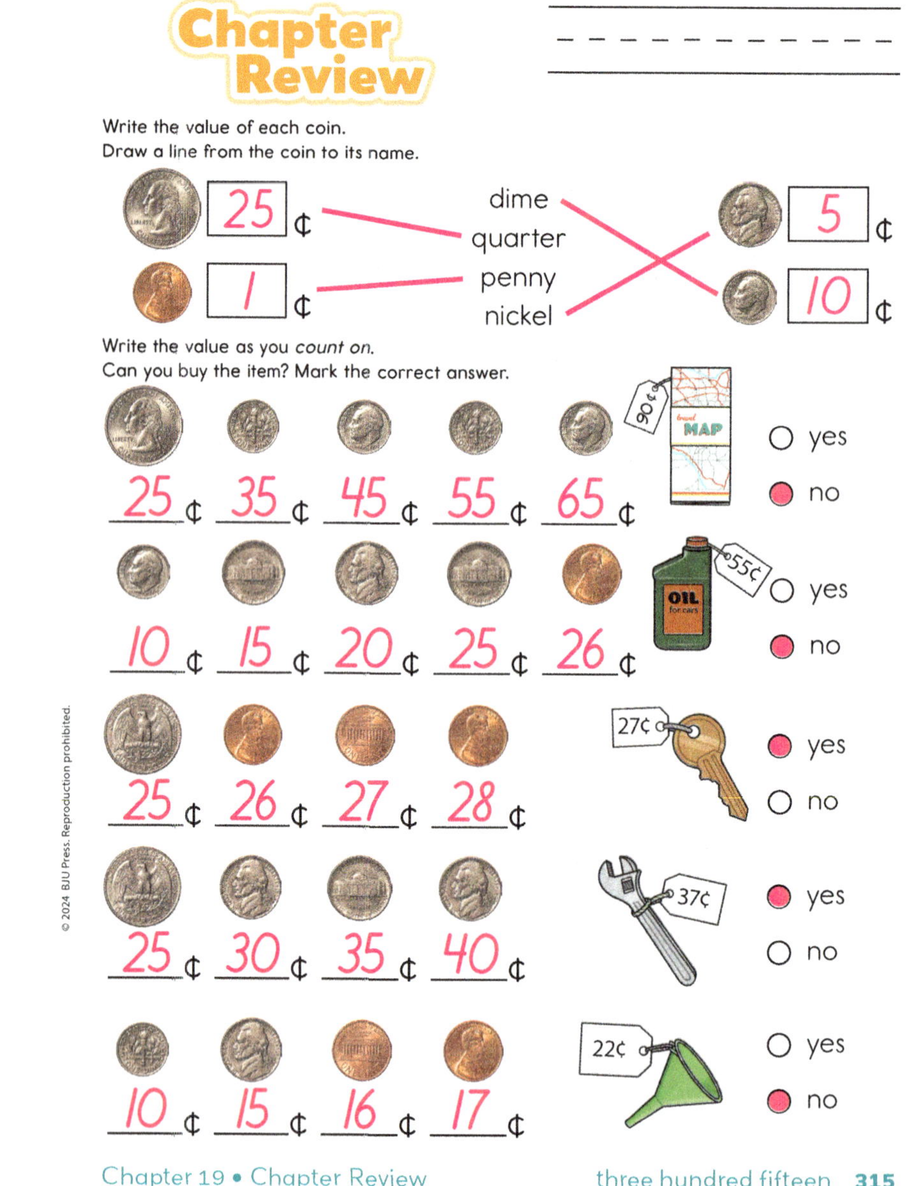

Practice & Review

Memorize Facts for the 7-9-16 Family

Introduce the following facts.

7 + 9   9 + 7   16 – 9   16 – 7

Display each addition and subtraction flashcard slowly. Invite students to give the answers. Explain that making 10 may make it easier to add 9. Point out that 16 – 9 and 16 – 7 are related facts. Remind the students that these facts are part of the 7-9-16 fact family.

Distribute Number Cards 0–20. Display each flashcard again. Direct the students to hold up the correct Number Card to indicate each answer.

Practice these and recently memorized facts.

Count 750–800

Guide the students as you take turns counting together. Explain that you will start by saying 750; then they will say the next number. Continue to take turns as you count to 800.

Instruct

Recognize & Identify the Value of a Penny, a Nickel, a Dime & a Quarter

Distribute a *Coin Purse Mat* page and a set of coins to each student. Direct the students to hold up the correct coins in response to the following questions.

Which coin's value is 25¢? quarter

Which coin is worth 1¢? penny

Which coin's value is 5¢? nickel

Which coin's value is 10¢? dime

Determine the Value of a Set of Coins; Determine Whether There Is Enough Money to Purchase an Item

Write for display as a list: "gum 30¢, crackers 35¢, water 45¢, nuts 55¢." Explain that these are the prices for the refreshments at Mr. Sawyer's auto parts shop.

Direct the students to put 7 dimes, 5 nickels, and 4 pennies in the coin purse on their mats. Demonstrate each step.

What should you do before you count the value of this set? separate the dimes, nickels, and pennies into groups

In what order do you count the groups? dimes, nickels, pennies; because I count the coins in order of their value—greatest to least

Mark an *X* on the coins needed to buy each item.

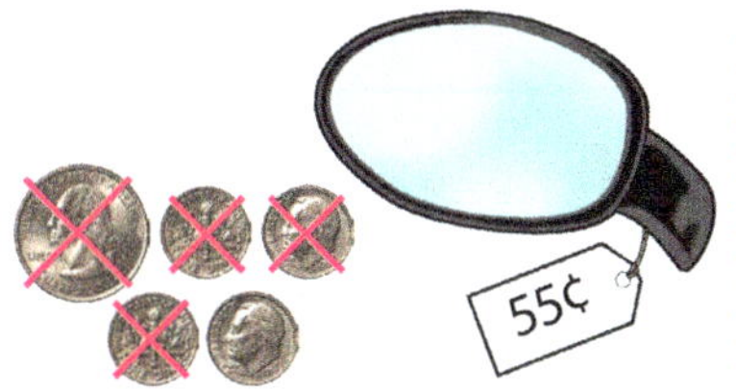

Write an equation for each word problem. Solve.

Henry has 65¢ in his bank.
He puts in 4 nickels.
How much money is in his bank?

65 + 20 = 85 ¢

(work space)

$$\begin{array}{r} 65¢ \\ +\ 20¢ \\ \hline 85¢ \end{array}$$

Elena had 79¢ in her pocket.
She gave the clerk 3 nickels for gum.
How much money does she have left?

79 − 15 = 64 ¢

(work space)

$$\begin{array}{r} 79¢ \\ -\ 15¢ \\ \hline 64¢ \end{array}$$

Trace the word to complete the sentence.

I should say "thank you" to God for making a world where math works well.

Separate your coins.

Guide the students as they count the total value: 10, 20, 30, 40, 50, 60, 70, (deep breath), 75, 80, 85, 90, 95, (deep breath), 96, 97, 98, 99¢.

Look at the snack items listed. Do you have enough money to buy any of the items? yes, any one of them

Direct the students to remove the money from the coin purse and then to put 1 quarter and 5 pennies in it.

What is the value of 1 quarter? 25¢

How do you count pennies? by 1s

Guide the students as they count the value of the set: 25, (deep breath), 26, 27, 28, 29, 30¢.

Do you have enough money to buy any of the snack items? Which ones? yes, gum

Direct the students to remove the money from the coin purse and then to put 1 quarter and 7 nickels in it.

How do you count nickels? by 5s

Invite a student to count the value of the money. 25, (deep breath), 30, 35, 40, 45, 50, 55, 60¢

Do you have enough money to buy the nuts? yes

Direct the students to remove the money from the coin purse and then to put 1 quarter and 2 dimes in it.

How do you count dimes? by 10s

Ask a volunteer to count the total value of the set. 25, (deep breath), 35, 45¢

Do you have enough money to buy the nuts? no

Do you have enough money to buy the water? yes

Direct the students to remove the money from the coin purse and then to put 2 quarters and 2 dimes in it.

What is the value of 2 quarters? 50¢

Guide the students as they count the total value of the coins: 25, 50, (deep breath), 60, 70¢.

Are there any snack items that you cannot buy with this amount? no

Identify the Coins Needed to Purchase an Item

Direct the students to place all of the money in the coin purse on their mats. Demonstrate each step.

How much does the gum cost? 30¢

Explain that students are going to buy this item by using only quarters and pennies.

How many quarters do you need? 1

Direct the students to place 1 quarter on the hand on their mats.

How many pennies do you need to make 30¢? 5

Direct the students to place 5 pennies on the hand.

Guide the students as they count the money: 25, (deep breath), 26, 27, 28, 29, 30¢.

Continue the activity with the following snack items and coins.

buy gum, using quarters and nickels
1 quarter and 1 nickel

buy water, using quarters and dimes
1 quarter and 2 dimes

buy nuts, using quarters and nickels
2 quarters and 1 nickel

Solve Word Problems

Read aloud the following word problem.

DQ had 55¢ in his pocket. Rosa gave him 3 dimes. How much money does DQ have in all?

What is the question? How much money does DQ have in all?

What information is given? There was 55¢ in DQ's pocket; then Rosa gave him 3 dimes.

Do you add or subtract in this problem? add

163

What is the value of 3 dimes? 30¢

What equation do you write to solve this problem? 55¢ + 30¢ = __

Write "55¢ + 30¢ = __" for display.

Invite a student to write 55¢ + 30¢ vertically and solve the problem. Complete the equation.

What is the answer to the problem? 85¢

Does the answer make sense? yes

Continue the activity with the following word problem.

Grandpa has 75¢ in his pocket. He used 4 dimes to buy a treat for his grandson. How much money does Grandpa have left? 75¢ – 40¢ = 35¢

Apply

Worktext pages 315–16

Read the directions and guide completion of the pages.

Direct attention to the biblical worldview shaping statement on page 316. Display the picture from the chapter opener and remind the students that DQ counted coins at the auto parts store. Point out that we should praise and thank God for making a world where math works well.

Provide prompts as necessary to guide the students to the answer on the Worktext page.

What should I tell God for giving me a world where math works well? I should say "thank you" to God for making a world where math works well.

Assess

Reviews pages 303–4

Use the Reviews pages to provide additional preparation for the chapter test.

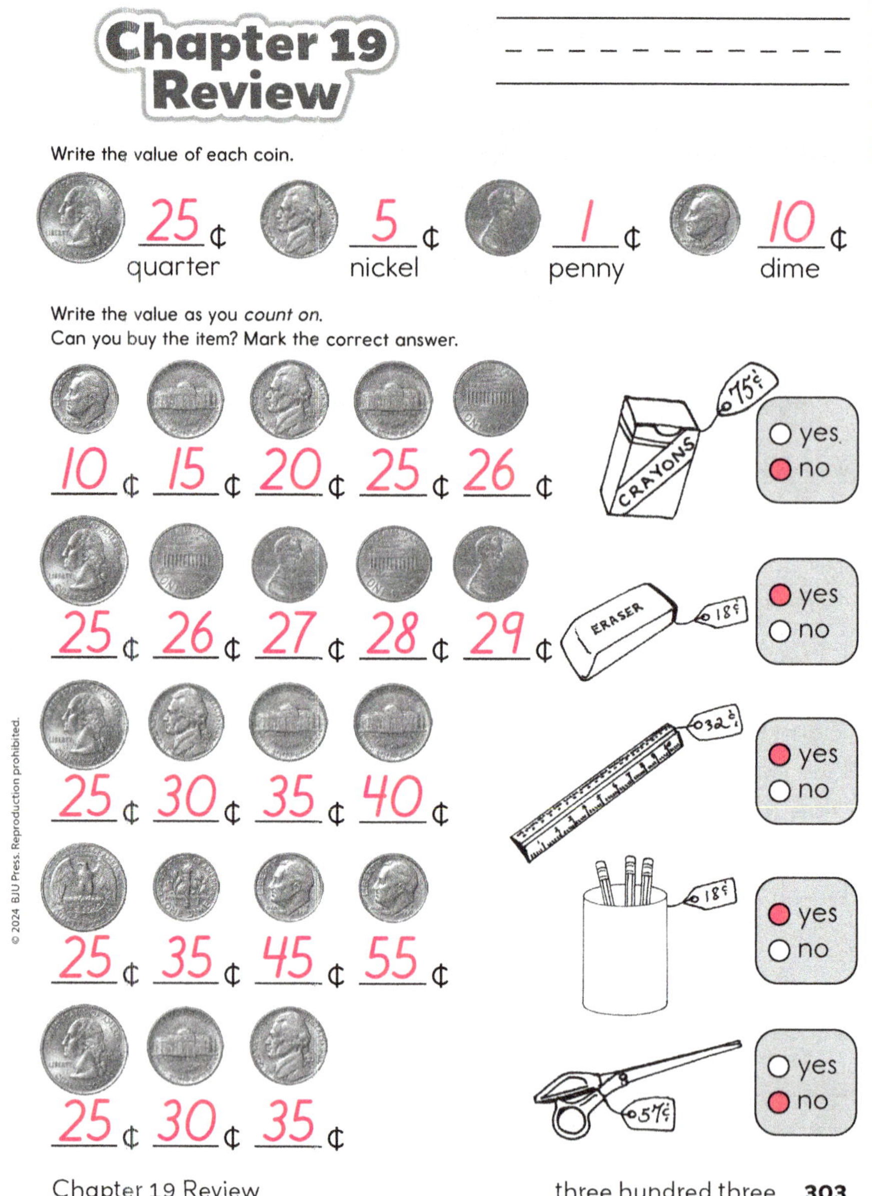

Mark an *X* on the coins needed to buy each item.

Write an equation for each word problem. Solve.

Shelby had 65¢ for a hot dog.
Her mother gave her 3 dimes.
How much money does Shelby have?

65 ¢ + 30 ¢ = 95 ¢

work space

65¢
+ 30¢
95¢

Daniel spent 15¢ for a glue stick.
He spent 54¢ for stickers.
How much money did he spend in all?

   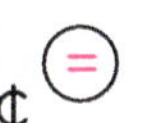 

15 ¢ + 54 ¢ = 69 ¢

work space

15¢
+ 54¢
69¢

Tamara had 90¢ in her purse.
She gave 4 dimes to her brother.
How much money does she have left?

  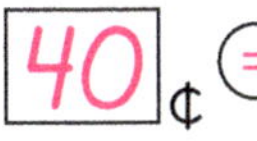  

90 ¢ − 40 ¢ = 50 ¢

work space

90¢
− 40¢
50¢

164 Worktext page 317 Reviews pages 305–6

Chapter 19 Test

Administer the Chapter 19 Test.

Cumulative Concept Review

Worktext page 317

Review the following concepts. Adapt instructions and activities and provide reteaching as needed to meet the specific needs of your students.

- Determining whether a container has a capacity of more than or less than 1 liter (Lesson 138)
- Determining whether an object has a mass of more than or less than 1 kilogram (Lesson 139)
- Identifying cubes, spheres, cones, cylinders, and rectangular prisms (Lesson 115)
- Completing a 2-digit addition problem in vertical form (Lesson 93)

Retain a copy of Worktext page 318 for the discussion of the Chapter 20 essential question during Lessons 165 and 171.

Reviews pages 305–6

Use the Reviews pages to help students retain previously learned skills.

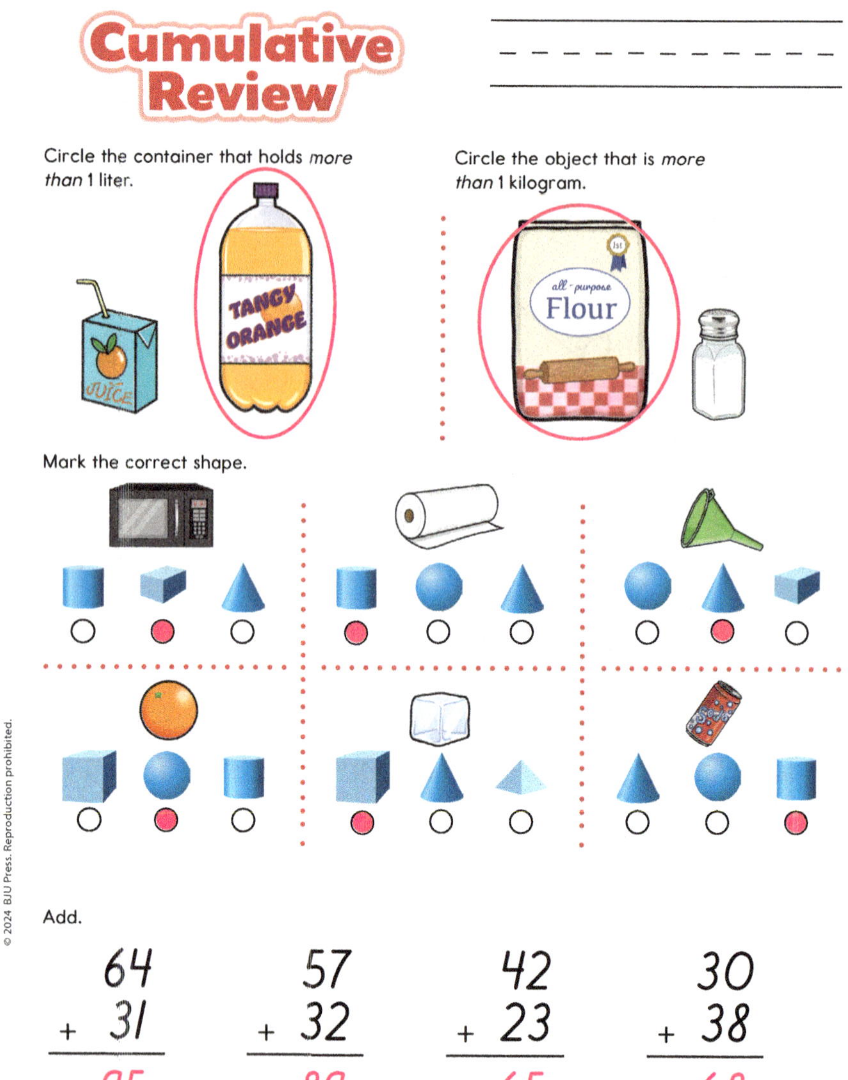

Add.

$$\begin{array}{r} 64 \\ +\ 31 \\ \hline 95 \end{array} \quad \begin{array}{r} 57 \\ +\ 32 \\ \hline 89 \end{array} \quad \begin{array}{r} 42 \\ +\ 23 \\ \hline 65 \end{array} \quad \begin{array}{r} 30 \\ +\ 38 \\ \hline 68 \end{array}$$

Write the number that comes just *before*.

347 348 | 618 619

Write the number that comes just *after*.

876 877 | 293 294

Write the number that comes *between*.

489 490 491 | 724 725 726 | 560 561 562

Apples Sold	
Monday	🍎🍎🍎🍎🍎🍎
Tuesday	🍎🍎
Wednesday	🍎🍎🍎🍎🍎
Thursday	🍎🍎🍎🍎
Friday	🍎🍎🍎

🍎 = 5 apples

Use the pictograph to answer each question. Mark the correct answer.

How many apples did the farmer sell on Monday?

- ○ 20
- ● 30
- ○ 40

How many more apples did the farmer sell on Wednesday than on Tuesday?

- ● 25 − 10 = 15 apples
- ○ 25 + 10 = 35 apples

164

Subtraction Fact Review

Subtract.

9 − 7 = 2	5 − 2 = 3	10 − 3 = 7	9 − 6 = 3	11 − 8 = 3
7 − 3 = 4	8 − 6 = 2	7 − 5 = 2	10 − 9 = 1	8 − 5 = 3
8 − 8 = 0	9 − 1 = 8	6 − 4 = 2	7 − 0 = 7	10 − 6 = 4
10 − 2 = 8	7 − 4 = 3	9 − 8 = 1	11 − 3 = 8	7 − 2 = 5
9 − 4 = 5	8 − 2 = 6	12 − 8 = 4	6 − 5 = 1	9 − 9 = 0
8 − 4 = 4	12 − 6 = 6	9 − 2 = 7	8 − 7 = 1	12 − 5 = 7

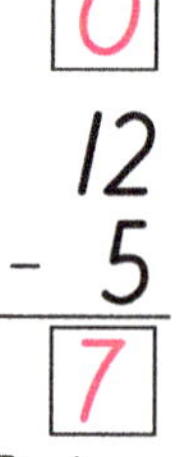

306 three hundred six   Math 1 Reviews

CHAPTER 20: TIME & CALENDAR

PAGES	OBJECTIVES	RESOURCES & MATERIALS	ASSESSMENTS
Lesson 165 Time to the Hour & Half-Hour			
Teacher Edition 703–7 **Worktext** 318–20	**165.1** Distinguish the parts of an analog clock. **165.2** Identify 1 hour as 60 minutes and a half-hour as 30 minutes. **165.3** Write time to the hour and the half-hour. **165.4** Demonstrate time to the hour and the half-hour on a clock. **165.5** Draw hands on a clock to show time to the hour and the half-hour. **165.6** Explain how telling time to the half-hour can be used to care for others. BWS Caring (explain)	**Visuals** • Visual 1: *Hundred Chart* **Student Manipulatives Packet** • DQ Clock • Number Cards: 0–20 **BJU Press Trove*** • Video: Ch 20 Intro • Web Link: Virtual Manipulatives: Analog Clock • Web Link: Learn to Tell Time • Games/Enrichment: Addition Flashcards • Games/Enrichment: Subtraction Flashcards • PowerPoint® presentation	**Reviews** • Pages 307–8
Lessons 166–67 Time to 5 Minutes			
Teacher Edition 708–11 **Worktext** 321–22	**166–67.1** Determine the minutes in 1 hour or a half-hour by counting by 5s. **166–67.2** Tell time to the nearest 5-minute interval. **166–67.3** Demonstrate time to the nearest 5-minute interval.	**Teacher Edition** • Instructional Aid 1: *Bar Graph* **Student Manipulatives Packet** • DQ Clock • Number Cards: 0–20 **BJU Press Trove** • Web Link: Virtual Manipulatives: Analog Clock • Games/Enrichment: Addition Flashcards • Games/Enrichment: Subtraction Flashcards • PowerPoint® presentation	**Reviews** • Pages 309–10
Lessons 168–69 Using a Schedule			
Teacher Edition 712–15 **Worktext** 323–24	**168–69.1** Write time to the nearest 5-minute interval. **168–69.2** Determine the elapsed time. **168–69.3** Read a table to use a schedule. **168–69.4** Use a schedule to care for others. BWS Caring (apply)	**Teacher Edition** • Instructional Aid 47: *School Schedule* **Visuals** • Visual 1: *Hundred Chart* **Student Manipulatives Packet** • Number Cards: 0–20 • DQ Clock **BJU Press Trove** • Video: Time Passed • Web Link: Virtual Manipulatives: Analog Clock • Games/Enrichment: Addition Flashcards • Games/Enrichment: Subtraction Flashcards • PowerPoint® presentation	**Reviews** • Pages 311–12

*Digital resources for homeschool users are available on Homeschool Hub.

PAGES	OBJECTIVES	RESOURCES & MATERIALS	ASSESSMENTS
Lesson 170	**Using Calendars**		
Teacher Edition 716–19 **Worktext** 325–26	**170.1** Name the days of the week and the months of the year in order. **170.2** Name the day of the week or the month of the year that comes next. **170.3** Read a calendar.	**Teacher Edition** • Song: "Days of the Week" • Song: "There Are Twelve Months" • Instructional Aid 2: *Hundred Chart* • Instructional Aid 12: *Days of the Week Cards* • Instructional Aid 13: *Months of the Year Cards* **Visuals** • Visual 1: *Hundred Chart* • Visual 16: *12-Month Calendar* **Student Manipulatives Packet** • Number Cards: 0–20 **BJU Press Trove** • Web Link: Calendar Quiz • Games/Enrichment: Addition Flashcards • Games/Enrichment: Subtraction Flashcards • PowerPoint® presentation	**Reviews** • Pages 313–14
Lesson 171	**Chapter 20 Review**		
Teacher Edition 720–23 **Worktext** 327–28	**171.1** Recall concepts and terms from Chapter 20.	**Teacher Edition** • Instructional Aid 2: *Hundred Chart* • Instructional Aid 47: *School Schedule* • Instructional Aid 12: *Days of the Week Cards* • Instructional Aid 13: *Months of the Year Cards* **Visuals** • Visual 1: *Hundred Chart* **Student Manipulatives Packet** • Number Cards: 0–20 **BJU Press Trove** • Web Link: Virtual Manipulatives: Analog Clock • Web Link: Learn to Tell Time • Web Link: Calendar Quiz • Games/Enrichment: Addition Flashcards • Games/Enrichment: Subtraction Flashcards • PowerPoint® presentation	**Worktext** • Chapter 20 Review **Reviews** • Pages 315–16
Lesson 172	**Test, Cumulative Review**		
Teacher Edition 724–26 **Worktext** 329	**172.1** Demonstrate knowledge of concepts from Chapter 20 by taking the test.		**Assessments** • Chapter 20 Test **Worktext** • Cumulative Review **Reviews** • Pages 317–18

Chapter 20

How can I help others by telling time?

Chapter Objectives

- Demonstrate time to the nearest 5 minutes.
- Tell time to the nearest 5 minutes from analog and digital clocks.
- Write time to the nearest 5 minutes from analog and digital clocks.
- Determine elapsed time to the half-hour.
- Use calendar terms fluently.
- Use knowledge of time to care for others.

20 Time & Calendar

How can I help others by telling time?

Spot 10 differences in the pictures below.

165 Worktext pages 318–20 Reviews pages 307–8

Objectives

- **165.1** Distinguish the parts of an analog clock.
- **165.2** Identify 1 hour as 60 minutes and a half-hour as 30 minutes.
- **165.3** Write time to the hour and the half-hour.
- **165.4** Demonstrate time to the hour and the half-hour on a clock.
- **165.5** Draw hands on a clock to show time to the hour and the half-hour.
- **165.6** Explain how telling time to the half-hour can be used to care for others. BWS

Biblical Worldview Shaping

- **Caring** (explain): Sometimes caring for others can be as simple as being on time so that things can run smoothly. Since many appointments start or end on the half-hour, telling time to the half-hour is an important skill (165.6).

Printed Resources

- Visual 1: *Hundred Chart*
- Student Manipulatives: Number Cards (0–20)
- Student Manipulatives: DQ Clock

Digital Resources

- Video: Ch 20 Intro
- Web Link: Virtual Manipulatives: Analog Clock
- Web Link: Learn to Tell Time
- Games/Enrichment: Addition Flashcards
- Games/Enrichment: Subtraction Flashcards

Materials

- Addition flashcards
- Subtraction flashcards
- Judy® Clock (or Visual 15: *Clock*)

Time to the Hour & Half-Hour

Color the hour hand red.
Color the minute hand blue.
Count the minutes by 5s.

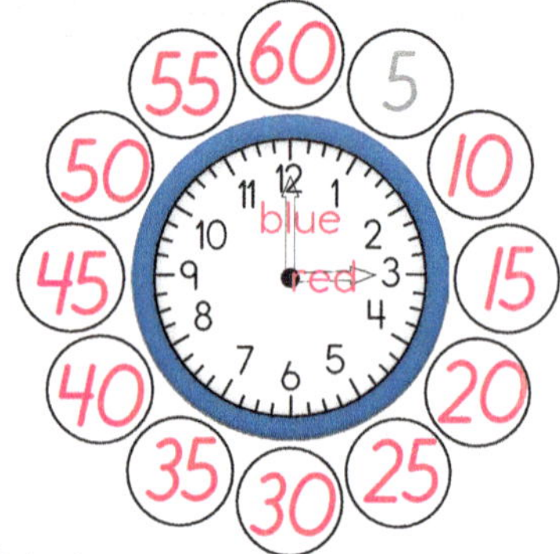

There are 60 minutes in 1 hour.

There are 30 minutes in a half-hour.

Write the time.

 12:00

 5:30

 6:00

 11:30

 2:30

Draw hands on each clock.

3:00

3:30

Practice & Review

Count by 5s to 150

Display Visual 1. Count together by 5s as you point out the corresponding numbers on the visual.

Count by 10s from a 3-Digit Number

Write “537” for display.

Which digit is in the Tens place? 3

Explain that the students will be counting by 10s starting with 537. Write “547” below 537 and point out that 547 is 10 more than 537. Guide the students as they count to 637: 537, 547, 557, 567, 577, 587, 597, 607, 617, 627, 637.

You may write all the numbers for display so the students can see them as they count.

Continue the activity, starting with a different number.

Memorize Facts for the 7-8-15 Family

Introduce the following facts.

7 + 8   8 + 7   15 – 7   15 – 8

Display the addition flashcards slowly. Invite students to give the answers. Point out that 7 + 8 and 8 + 7 are near doubles.

Display the subtraction flashcards slowly and invite students to give the answers. Point out that 15 – 7 and 15 – 8 are related facts. Remind the students that these facts are part of the 7-8-15 fact family.

Distribute Number Cards 0–20. Display each flashcard again. Direct the students to hold up the correct Number Card to indicate each answer.

Practice these and recently memorized facts.

165

Mark the correct time.

- ○ 4:30
- ● 4:00
- ○ 5:00

- ● 10:30
- ○ 9:30
- ○ 10:00

- ○ 4:30
- ○ 3:00
- ● 3:30

- ○ 8:00
- ● 7:00
- ○ 9:00

Write the time.

11:00

4:30

7:30

5:00

1:30

Write the word to complete the sentence.

I can care for others by telling time.

Time to Review

Write each missing number.

437 438 439 440 441 442 443

685 686 687 688 689 690 691

Write the fact family.

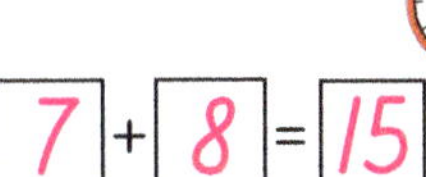

7 8 15

7 + 8 = 15

8 + 7 = 15

15 − 7 = 8

15 − 8 = 7

Engage

Essential Question

Direct attention to the chapter opener on Worktext page 318 and **read aloud** the following scenario to introduce the chapter essential question.

DQ looked out the classroom window at the raindrops splashing against the glass. Recess was in the classroom again today because of the rain. DQ wouldn't have minded splashing in the rain and watching the water slide off his well-oiled feathers. The other students would've enjoyed splashing in the puddles too, but they didn't have feathers or a change of clothes to keep them dry.

"It's your turn, DQ," Liam said as he slid a red checker from one square to another on the checkerboard.

DQ jumped Liam's red checker and placed his black checker into position in Liam's back row. "King me!" he quacked triumphantly.

"How did I miss that one?" Liam groaned, placing another checker on top of DQ's. Liam studied the board intently to figure out his next move. DQ's head feathers fluttered as he planned his strategy for winning the game.

"Class," Miss Markle interrupted the boys' thoughts, "game time ends at 12:30. Please have everything put away and your desks in order by that time, so you'll be ready for the next chapter in our story."

DQ checked the clock on the wall. The hour hand was on the 12, and the minute hand was on the 3. He and Liam still had a few minutes left to play. DQ hoped they could finish their game, but he couldn't wait to hear what would happen next in the book Miss Markle was reading to the class each day. He would be sure to clean up at the right time so Miss Markle would have plenty of time to read!

Invite a student to read aloud the essential question. Encourage the students to consider this question as you guide them to answer it by the end of the chapter.

Instruct

Distinguish the Parts of an Analog Clock

Guide a **discussion** to help the students review the parts of an analog clock.

Display a demonstration clock or Visual 15.

Which hand of the clock shows the hour of the day? the short hour hand

What does the long hand show? the minute of the hour

How are the hour hand and minute hand different? The hour hand is short, and the minute hand is long.

Guide the students as they count together 1–12 as you point out the numbers on the clock.

Identify 1 Hour as 60 Minutes & a Half-Hour as 30 Minutes

Guide a **visual analysis** of an analog clock to help the students identify 1 hour as 60 minutes and a half-hour as 30 minutes.

Display the demonstration clock set for 2:00. Move the minute hand around the clock until it reaches the 12 again and the clock is set for 3:00.

Explain that 1 hour or 60 minutes have passed. Direct attention to the minute marks on the demonstration clock. Ask the students to tell how many minute marks they think there are.

165

Count the minute marks from 12 to 1 to remind the students that 5 minutes pass as the minute hand moves from one number to the next one.

What is a quicker way to count all the minutes on the clock? count by 5s

Move the minute hand around the clock, counting by 5s with the students until the minute hand reaches the 12 again and the clock is set for 4:00.

What time is on the clock now? 4:00

How many minutes are in 1 hour? 60 minutes

Move the hands around the clock again, counting by 5s with the students until the minute hand reaches the number 6 and the clock is set for 4:30.

What time is it now? 4:30

Explain that when the minute hand has moved halfway around the clock from the 12, the hour hand moves halfway between the 4 and the 5, and the time is 4:30 or *half past* 4. Point out that the minute hand is on the 6 when 30 minutes have passed.

Set the demonstration clock for 9:00. Count by 5s again, moving the minute hand to the 6.

What time is it now? 9:30 or half past 9

How many minutes are in a half-hour? 30 minutes

Write Time to the Hour & Half-Hour

Guide a **discussion** to help the students write time to the hour and half-hour.

Display the demonstration clock set to 9:30 and write "9:30" for display.

Which number is to the left of the colon? 9

What does this number represent? the hour; 9 o'clock

Which number is to the right of the colon? 30

What does this number represent? the minutes; 30 minutes past 9 o'clock

Set the demonstration clock for the following times: 2:00, 12:00, 10:30, and 6:30. Direct the students to write the times and invite volunteers to write the times for display.

Time to the Hour & Half-Hour

Color the hour hand red.
Color the minute hand blue.
Count the minutes by 5s.

There are 60 minutes in 1 hour.

There are 30 minutes in a half-hour.

Write each time.

2:00	2:30	12:00	8:00
11:30	4:30	10:00	3:30

Demonstrate Time to the Hour & Half-Hour

Use **manipulatives** to help the students demonstrate time.

Distribute the DQ Clocks.

Which number is the long minute hand placed on to set the clock for an "o'clock" time? 12

Instruct the students to set their clocks for 8:00. Set the demonstration clock for 8:00 for the students to check their clocks.

Ask the students to change their clocks to show 11:00.

Which number should the long minute hand be on when the hand is halfway around the clock or has moved 30 minutes past 11:00? 6

Instruct the students to set their clocks for 11:30 as you set the demonstration clock for 11:30. Remind the students that a half-hour is 30 minutes.

Where does the hour hand move when the minute hand moves halfway around the clock to the number 6? The hour hand moves toward the next number; it is halfway between the number 11 and the number 12; it is in the "backyard" of the number 11.

Continue the activity, instructing the students to set their clocks as you set the demonstration clock for 7:00, 7:30, 12:00, and 12:30.

Chapter 17 Review

Circle each container that holds *less than* 1 liter.

Mark an *X* on each object that is *more than* 1 kilogram.

Circle the correct temperature.

30°C 40°C 50°C

80°C 90°C 100°C

20°C 30°C 40°C

Fact Review

0 1 2 3 4 5 6 7 8 9 10 11 12 13 14 15 16 17 18 19 20

Add or subtract. Use the number line if needed.

$6 + 7 = 13$ $4 + 9 = 13$ $14 - 9 = 5$ $15 - 9 = 6$ $7 + 9 = 16$

$7 + 8 = 15$ $13 - 6 = 7$ $5 + 9 = 14$ $13 - 4 = 9$ $9 + 6 = 15$

How did DQ show care for Miss Markle and the other students? He used his skill in telling time to the half-hour so he could obey Miss Markle and clean up in time for the story. He showed his care for the other students by setting a good example and helping the class have their full amount of story time by starting on time.

Explain that being on time is one way of showing respect and care for others' time. Invite the students to tell ways they could care for others by telling time. getting ready for school on time, finishing my chores or other activities on time, returning home at the time I am instructed to return, being on time for events

Apply

Worktext pages 319–20

Read and guide completion of page 319.

Read and explain the directions for page 320. Assist the students as they complete the page independently.

Direct attention to the biblical worldview shaping statement and guide the students as they complete the sentence.

Assess

Reviews pages 307–8

Review metric measurements on page 308.

Draw Hands on the Clock

Guide a **drawing activity** to help the students draw the hands on the clock to show time to the hour and the half-hour.

Draw for display a clock without the hands.

> **What is missing on this clock?** the clock hands

Explain that when they draw hands on the clock, they should draw the short hand to point to the hour but not touch the number. Point out that they should draw the long hand to show the number of minutes and that it should touch the number.

Ask a student to draw the hands on the clock so that it shows 10:00. Continue the activity for 1:00, 6:30, and 8:30.

Caring for Others by Telling Time

Guide a **discussion** to help the students explain a biblical worldview shaping truth.

Remind the students that DQ and Liam were playing checkers during recess in the opener scenario.

> **What time did Miss Markle say the students needed to have their games put away and their desks put in order so they would be ready for story time?** 12:30
>
> **Why did DQ check the clock on the wall?** to check the time so he could know how much longer he and Liam could play checkers

Extended Activity

Match Time on an Analog Clock to Digital Form

Materials

- Index cards
- Clothespins

Procedure

Make several cards with clockfaces set to various hour or half-hour times. Write on clothespins the digital times represented by the clockfaces. Direct the students to clip the clothespin showing the correct time onto the matching clockface card. A self-check can be made by writing the digital time on the back of each card.

166–67

Worktext pages 321–22
Reviews pages 309–10

Objectives

- **166–67.1** Determine the minutes in 1 hour or a half-hour by counting by 5s.
- **166–67.2** Tell time to the nearest 5-minute interval.
- **166–67.3** Demonstrate time to the nearest 5-minute interval.

Printed Resources

- Instructional Aid 1: *Bar Graph*
- Student Manipulatives: Number Cards (0–20)
- Student Manipulatives: DQ Clock

Digital Resources

- Web Link: Virtual Manipulatives: Analog Clock
- Games/Enrichment: Addition Flashcards
- Games/Enrichment: Subtraction Flashcards

Materials

- Addition flashcards
- Subtraction flashcards
- Judy® Clock (or Visual 15: *Clock*)

You may take either 1 or 2 days to present Lessons 166–67.

Time to 5 Minutes

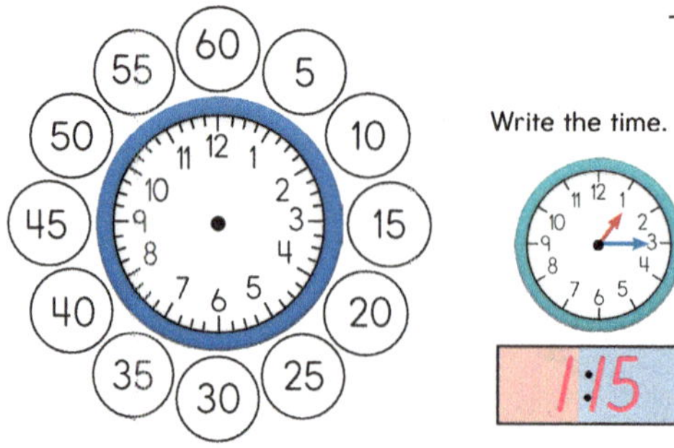

Write the time.

1:15

8:45

3:25

Circle the correct time.

Practice & Review

Bar Graphs

Write *red, blue, green, yellow,* and *purple* in a column for display.

Read the name of each color. Ask the students to raise their hands when you read the name of their favorite colors. Make tally marks next to the colors to represent the students who chose the colors as their favorites.

Display the *Bar Graph* page and label headings for the colors. Ask students to count the number of tallies for each color. Fill in the spaces, making bars of the corresponding numbers on the graph.

Invite volunteers to give equations to use to find the answers to the following questions. Modify the questions to match your results.

How many more (or fewer) students chose green than yellow?

How many more (or fewer) students chose red than blue?

How many more students chose (the most favorite color) than (the least favorite color)?

Memorize Facts for the 8-9-17 Family

Introduce the following facts.

8 + 9 9 + 8 17 – 8 17 – 9

Display the addition flashcards slowly. Invite students to give the answers. Point out that 8 + 9 and 9 + 8 are near doubles.

Display each subtraction flashcard slowly and invite students to give the answers. Point out that 17 – 8 and 17 – 9 are related facts. Remind the students that these facts are part of the 8-9-17 fact family.

Distribute Number Cards 0–20. Display each flashcard again. Direct the students to hold up the correct Number Card to indicate each answer.

Practice these and recently memorized facts.

Engage

A Time to Keep Silent or a Time to Speak

Guide a **discussion** to help the students identify appropriate times to speak or to keep silent.

Read aloud Ecclesiastes 3:7*b*.

Ask the students to share times when someone should be silent. during church, when

Write the time.

6:45

4:10

8:15

1:50

Mark the correct time.

● 1:05
○ 1:10

● 7:35
○ 7:30

○ 6:25
● 4:25

● 8:40
○ 8:45

Time to Review

Mark the correct answer.

How many more worms did they use on Friday than on Monday?
- ● 25 – 10 = 15 worms
- ○ 25 + 10 = 35 worms
- ○ 20 – 10 = 10 worms

Fishing Worms

Day	Worms
Monday	2 worms
Tuesday	3 worms
Wednesday	4 worms
Thursday	2 worms
Friday	5 worms

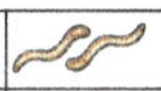

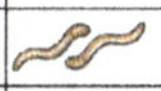

(one worm = 5 worms)

How many worms did they use in all on Tuesday and Thursday?
- ○ 20 – 10 = 10 worms
- ○ 15 – 10 = 5 worms
- ● 15 + 10 = 25 worms

Which day did they use the most worms?
- ○ Monday
- ○ Wednesday
- ● Friday

Set the demonstration clock for 2:00. Count by 5s with the students as you move the minute hand around the clock until it reaches the 6 and the clock is set for 2:30.

What time is it? 2:30 or half past 2

How many minutes are in a half-hour? 30 minutes

Continue the activity by setting the demonstration clock for 4:00 and counting by 5s as you move the minute hand until the clock is set for 4:30.

Tell Time to the Nearest 5-Minute Interval

Demonstrate telling time to the nearest 5-minute interval.

Set the demonstration clock for 1:05. Explain that the time on the clock is 5 minutes after 1 o'clock.

Write "1:05" for display. Remind the students that the numbers to the left of the colon tell the hour and the numbers to the right of the colon tell the minutes past the hour. Explain that there are always 2 digits to the right of the colon for the minutes past the hour. Point out that since it is 5 minutes past 1, there is a 0 in front of the 5 and we say the time as "one O (oh) five." Say the time together.

Set the demonstration clock for 1:10. Guide the students as they count the minutes together by 5s.

What is the time? 10 minutes after 1 or 1:10

Write "1:10" for display.

Set the demonstration clock for 1:15.

What is the time? 15 minutes after 1 or 1:15

Instruct the students to write the time at their desks as you invite a volunteer to write the time for display. Direct the students to check their answers.

Continue the activity for 2:35, 4:20, and 6:45.

Demonstrate Time to the Nearest 5-Minute Interval

Guide the students as they use **clock manipulatives** to demonstrate time to the nearest 5-minute interval.

someone else is talking, when the teacher is teaching the class, when I am supposed to go to sleep

Ask the students to share times when they may speak in school. during lunch and recess, to answer a question, to ask a question after raising my hand, when I have permission to speak

Remind the students that it is important to speak and to keep silent at the right times.

Instruct

Determine the Minutes in 1 Hour or a Half-Hour

Guide a **discussion** to help the students determine the minutes in an hour or half-hour.

Display the demonstration clock set for 10:00.

What time is it? 10:00

Move the minute hand around the clock until it reaches the 12 again and the clock is set for 11:00.

What time is it now? 11:00

Direct attention to the minute marks on the demonstration clock.

What is the quickest way to count all the minutes that pass between 10:00 and 11:00? count them by 5s

Count by 5s with the students as you move the minute hand completely around the clock until it reaches the 12 again and the clock is set for 12:00.

What time is on the clock now? 12:00

How many minutes are in 1 hour? 60 minutes

Distribute the DQ Clocks. Instruct the students to set their clocks for 7:00. Set the demonstration clock for 7:00 for the students to check their clocks.

Direct the students to change their clocks to show 7:15. Guide them as they count the minutes together by 5s as you set the demonstration clock for 7:15.

Ask the students to change their clocks to show 2:45. Guide them as they count the minutes together by 5s, beginning with the minute hand at the 12, while you set the demonstration clock for 2:45.

How do you say the time? 2:45

Invite a volunteer to write "2:45" for display.

Remind the students that as the minute hand moves around the clock, the hour hand moves into the "backyard" of the number of the hour. Point out that for the time 2:45, the hour hand is in the "backyard" that belongs to the 2; the time is 2:45, not 3:45. Conclude together that the hour hand will reach the 3 when the minute hand reaches the 12.

Continue the activity, directing the students to set their clocks as you set the demonstration clock for 12:25, 2:55, 5:10, and 8:40.

Differentiated Instruction

Tell Time to the Nearest 5-Minute Interval

Prepare circles about an inch larger than a DQ Clock with 5-minute intervals written on them. Attach the circles to DQ Clocks and distribute the prepared clocks to the students. Show 6:15 on a demonstration clock. Help the students count by 5s to find the minutes after the hour. Continue the activity with different times. Remove the circles behind the clocks and direct the students to count by 5s to find the minutes after the hour.

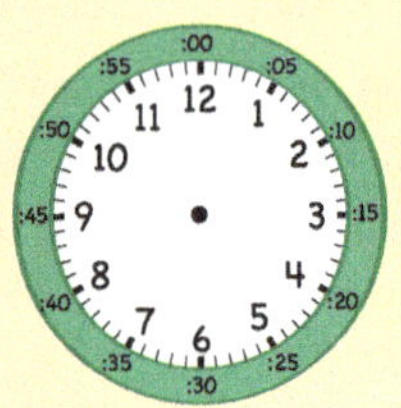

Time to 5 Minutes

Write each time.

Color the correct clock yellow.

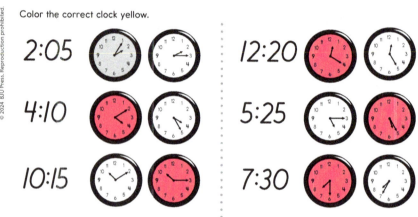

Chapter 20 • Lessons 166–67 three hundred nine **309**

Apply

Worktext pages 321–22

Read and guide completion of page 321.

Read and explain the directions for page 322. Assist the students as they complete the page independently.

Assess

Reviews pages 309–10

Review the total value of coins on page 310.

Extended Activity

Demonstrate Time to the Nearest 5-Minute Interval

Materials

- Judy® Clock (or Visual 15: *Clock*)

Procedure

Ask the students to tell about the events of a typical school day as you write the events in story form for display.

Include the time of day for each event. After the story is complete, invite students to read it aloud. Whenever a time of day is mentioned, ask a student to show the time on the demonstration clock. Duplicate the story for display in the room or make a copy for the students to take home and read to their families.

Chapter 17 Review

Write the total value.

35 ¢

55 ¢

30 ¢

41 ¢

27 ¢

Fact Review

0 1 2 3 4 5 6 7 8 9 10 11 12 13 14 15 16 17 18 19 20

Add or subtract. Use the number line if needed.

$7 + 8 = \boxed{15}$ $17 - 9 = \boxed{8}$ $18 - 9 = \boxed{9}$

$8 + 7 = \boxed{15}$ $17 - 8 = \boxed{9}$ $15 - 7 = \boxed{8}$

168–69

Worktext pages 323–24
Reviews pages 311–12

Objectives

- **168–69.1** Write time to the nearest 5-minute interval.
- **168–69.2** Determine the elapsed time.
- **168–69.3** Read a table to use a schedule.
- **168–69.4** Use a schedule to care for others. BWS

Biblical Worldview Shaping

- **Caring** (apply): Using a schedule helps us plan ahead and be on time so that things can run smoothly. Planning ahead is one way we can care for others by respecting their time (168-69.4).

Printed Resources

- Instructional Aid 47: *School Schedule*
- Visual 1: *Hundred Chart*
- Student Manipulatives: Number Cards (0–20)
- Student Manipulatives: DQ Clock (2 for the teacher)

Digital Resources

- Video: Time Passed
- Web Link: Virtual Manipulatives: Analog Clock
- Games/Enrichment: Addition Flashcards
- Games/Enrichment: Subtraction Flashcards

Materials

- Addition flashcards
- Subtraction flashcards
- Judy® Clock (or Visual 15: *Clock*)

You may take either 1 or 2 days to present Lessons 168–69.

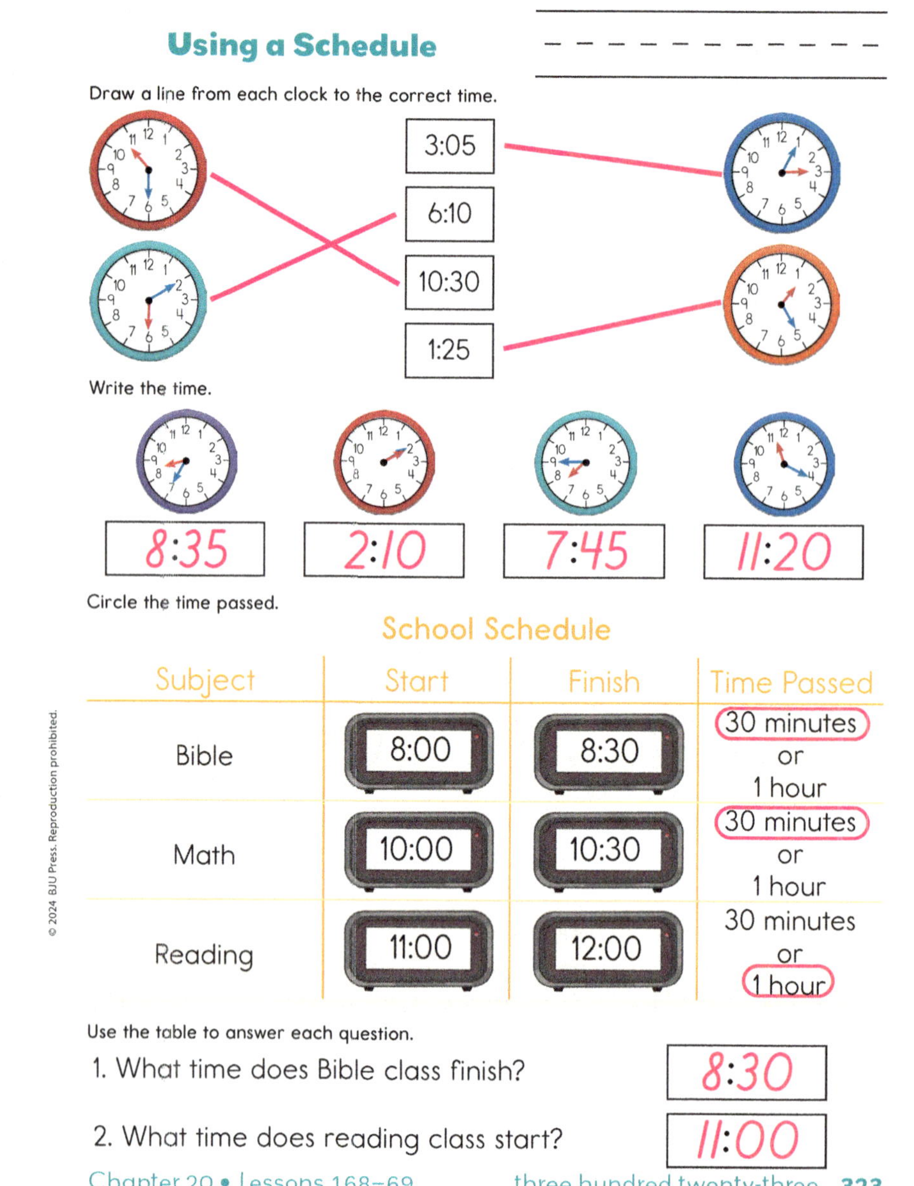

Using a Schedule

Draw a line from each clock to the correct time.

3:05
6:10
10:30
1:25

Write the time.

8:35 2:10 7:45 11:20

Circle the time passed.

School Schedule

Subject	Start	Finish	Time Passed
Bible	8:00	8:30	30 minutes or 1 hour
Math	10:00	10:30	30 minutes or 1 hour
Reading	11:00	12:00	30 minutes or 1 hour

Use the table to answer each question.

1. What time does Bible class finish? 8:30
2. What time does reading class start? 11:00

Practice & Review

Count by 2s to 50

Display Visual 1. Explain that the students will be counting by 2s. Point out that when they count by 1s, they count every number; but when they count by 2s, they skip a number and count every other number.

Count together by 2s as you slowly point to every other number.

Count 800–900

Divide the students into 3 groups. Instruct Group 1 to begin counting with the number 800 and to continue counting until you clap your hands. Instruct the next group to *count on* from the last number counted by the previous group until you clap your hands and say the group to count next. Alternate among groups until you reach 900.

Memorize Facts for the 7-7-14, 8-8-16 & 9-9-18 Families

Introduce the following facts.

7 + 7 14 − 7 8 + 8 16 − 8 9 + 9 18 − 9

Display the addition flashcards slowly. Invite students to say the answers.

What do you notice about these addition facts? They are doubles.

Display the subtraction flashcards slowly and invite students to say the answers.

How many facts are in each of these fact families? 2

Remind the students that there are only 2 facts in each of these fact families since each fact is a double.

Distribute Number Cards 0–20. Display each flashcard again. Direct the students to hold up the correct Number Card to indicate each answer.

Practice these and recently memorized facts.

Write the time.

3:30 | 2:20 | 10:15 | 8:40

Write the time. Circle the time passed.

School Schedule

Subject	Start	Finish	Time Passed
Art	10:00	11:00	30 minutes or (1 hour)
Spelling	1:00	1:30	(30 minutes) or 1 hour

Use the table to answer each question.

1. What time does spelling start? 1:00
2. What time does art finish? 11:00

Circle the word to complete the sentence.

I can follow a (schedule) / game to help me arrive on time.

Time to Review

Add.

$8 + 9 = 17$ $9 + 8 = 17$ $9 + 9 = 18$

Subtract.

$17 - 8 = 9$ $17 - 9 = 8$ $18 - 9 = 9$

324 three hundred twenty-four — Math 1

Engage

Use a Schedule

Use a **real-world setting** to introduce the lesson topic of using a schedule. Read aloud the following scenario.

Lily smiled as she looked at the calendar on her wall. She had drawn a heart around today's date. Tonight was the school spring program! It was the last program of the year. Lily had practiced her part and was ready. She even had a line that she would say alone. She looked over her schedule one more time.

6:45 p.m. Arrive at the school and meet with Miss Markle; line up in order.

6:55 p.m. Enter the auditorium with the class and be seated.

7:00 p.m. Program begins.

It was important that everyone be there on time and prepared to do his or her part in the program so it would go smoothly!

Lily glanced at the clock. *6:30.* Only 30 minutes until the program would begin. It was time to leave for the school. Lily straightened her hair bow and ran to the car. She was ready to go!

Explain that in this lesson the students will read a table to use a schedule.

168–69

Instruct

Time to the Nearest 5-Minute Interval

Guide an **interactive exercise** to help the students write time to the nearest 5-minute interval.

Display the demonstration clock.

How many minutes are between each number? 5

What is an easier way to count the minutes in an hour? count by 5s

Count together by 5s the minute marks on the clock.

Set the demonstration clock for 9:35.

What is the time? 9:35

Write "9:35" for display.

Set the demonstration clock for 7:05.

What is the time? 7:05

Instruct the students to write the time at their desks as you invite a volunteer to write "7:05" for display. Direct the students to check their answers.

Continue the activity for 10:15, 4:25, 3:40, and 11:05.

Determine the Elapsed Time

Use a **visual display** to help the students determine the elapsed time.

Display 2 DQ Clocks; set the first clock for 10:00 and the second clock for 11:00.

Explain that in DQ's school, art class starts at 10:00 and finishes at 11:00. Ask the students to think about how much time passes while the students are in art class.

Demonstrate the time passed by moving the minute hand around the first clock until 1 hour has passed and both clocks are set at 11:00.

How much time passes while the students are in art class? 1 hour

Set the clocks for 12:00 and 12:30 and explain that this is DQ's lunchtime. Discuss the amount of time that passes from when DQ starts to when he finishes eating lunch. 30 minutes or a half-hour

Continue the discussion for similar activities in which the amount of time that passes is 1 hour or a half-hour.

168–69

Read a Table to Use a Schedule

Guide a **visual analysis** of a table to help the students use a schedule.

Display the *School Schedule* page. Direct attention to the title of the schedule and read it aloud.

> **What are the headings above the columns?** Subject, Start, Finish, Time Passed

Remind the students of how to read a schedule, modeling each step. Put your finger on *Bible*. Demonstrate how to find the time that Bible class starts by sliding your finger to the right until it is under the word *Start*.

> **What time does the teacher at this school start her Bible lesson?** 8:30
>
> **What time does she finish the Bible lesson?** 9:00
>
> **How much time has passed during the Bible lesson?** 30 minutes

Circle "30 minutes" in the *Time Passed* column.

Continue the activity for the other subjects on the schedule. Invite students to circle the time passed for each subject.

Care for Others

Guide a **discussion** to help the students apply a biblical worldview shaping truth.

Remind the students that Lily checked a schedule as she prepared to leave for the school spring program.

> **Why did Lily check the schedule?** so she would know what time to be at school for the program

Read aloud Philippians 2:4. Explain that God's Word teaches us to think about how our actions affect others.

> **How does being on time help you show others that you care about them?** It shows I care about their time and plans.
>
> **How did Lily show she cared about others by using a schedule?** She used the schedule to help her know when to leave her house so she could arrive at the program on time.
>
> **How can using a schedule help you show that you care for others?** I can follow a schedule to help me arrive on time.

Using a Schedule

Color the ball below the correct clock orange.

Write each time.

6:15

11:50

3:40

Circle the time passed.

Basketball Practice Schedule

Activity	Start	Finish	Time Passed
Dribbling	7:00	7:30	(30 minutes) or 1 hour
Shooting	9:00	10:00	30 minutes or (1 hour)
Passing	8:00	9:00	30 minutes or (1 hour)

Use the table to answer each question.

What time do they finish shooting baskets? 10:00

What time do they start passing drills? 8:00

Apply

Worktext pages 323–24

Read and guide completion of page 323.

Read and explain the directions for page 324. Assist the students as they complete the page independently.

Direct attention to the biblical worldview shaping statement and guide the students as they complete the sentence.

Assess

Reviews pages 311–12

Review determining the coins needed to buy an item on page 312.

Extended Activity

Determine the Elapsed Time

Materials

- Judy® Clock (or Visual 15: *Clock*)

Procedure

Display the demonstration clock set for 3:00. Ask the students what time it will be 30 minutes from now. 3:30 Invite a student to check the response by moving the hands on the clock to show the time 30 minutes later.

Continue the activity with different start times and elapsed times of 30 minutes or 1 hour.

For a challenging activity, ask about an elapsed time of 2 hours.

Chapter 19 Review

Mark an *X* on the coins needed to buy each item.

Fact Review

Add or subtract. Use the number line if needed.

$9 + 9 = 18$ $7 + 8 = 15$ $8 + 9 = 17$ $8 + 7 = 15$ $9 + 8 = 17$

$15 - 8 = 7$ $17 - 8 = 9$ $18 - 9 = 9$ $15 - 7 = 8$ $17 - 9 = 8$

170

Worktext pages 325–26
Reviews pages 313–14

Objectives

- **170.1** Name the days of the week and the months of the year in order.
- **170.2** Name the day of the week or the month of the year that comes next.
- **170.3** Read a calendar.

Printed Resources

- Song: "Days of the Week"
- Song: "There Are Twelve Months"
- Instructional Aid 2: *Hundred Chart* (for each student)
- Instructional Aids 12a–b: *Days of the Week Cards* (for the teacher and for each student)
- Instructional Aids 13a–c: *Months of the Year Cards* (for the teacher and for each student)
- Visual 1: *Hundred Chart*
- Visual 16: *12-Month Calendar*
- Student Manipulatives: Number Cards (0–20)

Digital Resources

- Web Link: Calendar Quiz
- Games/Enrichment: Addition Flashcards
- Games/Enrichment: Subtraction Flashcards

Materials

- Addition flashcards
- Subtraction flashcards
- A large classroom calendar for the current month

Using Calendars

Mark the day that comes next.

Monday
- ○ Sunday
- ● Tuesday

Thursday
- ● Friday
- ○ Wednesday

Friday
- ● Saturday
- ○ Sunday

Mark the month that comes next.

February
- ● March
- ○ April

April
- ○ March
- ● May

August
- ○ July
- ● September

May						
Sunday	Monday	Tuesday	Wednesday	Thursday	Friday	Saturday
				1	2	3
4	5	6	7	8	9	10
11	12	13	14	15	16	17
18	19	20	21	22	23	24
25	26	27	28	29	30	31

Mark the correct answer.

1. How many Wednesdays are there? ○ 3 ● 4 ○ 5
2. What day is May 15? ○ Tuesday ○ Wednesday ● Thursday
3. On what day does May begin? ○ Wednesday ● Thursday ○ Friday

Practice & Review

Count by 2s to 50

Display Visual 1 and distribute the *Hundred Chart* pages. Remind the students that when they count by 1s, they count every number, but when they count by 2s, they skip a number and count every other number. Direct the students to point to the number 2 on their pages while you point to it on the visual. Count together by 2s as the students slowly point to every other number. Repeat the activity.

Count by 3s to 30

Point out that they can also count by 3s. Explain that when they count by 3s, they skip 2 numbers and count every third number. Direct the students to point to the number 3 on their *Hundred Chart* pages while you point to it on Visual 1. Instruct the students to skip 2 numbers and follow along as you count by 3s: 3, 6, 9, 12, 15, 18, 21, 24, 27, 30.

Guide the students as they count together by 3s as you point to every third number. Repeat the activity.

Memorize Facts for the 5-8-13 Family

Introduce the following facts.

$5 + 8$ $8 + 5$ $13 - 5$ $13 - 8$

Display the addition and subtraction flashcards slowly. Invite students to give the answers. Explain that they may want to use the strategy of making 10. Point out that $13 - 5$ and $13 - 8$ are related facts. Remind the students that these facts are part of the 5-8-13 fact family.

Distribute Number Cards 0–20. Display each flashcard again. Direct the students to hold up the correct Number Card to indicate each answer.

Practice these and recently memorized facts.

Mark an *X* on each Tuesday.

March						
Sunday	Monday	Tuesday	Wednesday	Thursday	Friday	Saturday
	1	2 X	3	4	5	6
7	8	9 X	10	11	12	13
14	15	16 X	17	18	19	20
21	22	23 X	24	25	26	27
28	29	30 X	31			

Mark the correct answer.

1. What day is March 26? Wednesday ○ Thursday ○ Friday ●
2. What is the date of the first Sunday? 1 ○ 7 ● 14 ○
3. On what day does March begin? Sunday ○ Monday ● Tuesday ○

Write each fact family.

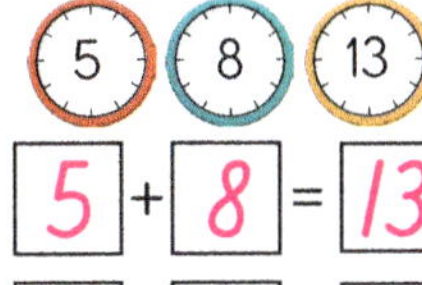

5 + 8 = 13
8 + 5 = 13
13 − 5 = 8
13 − 8 = 5

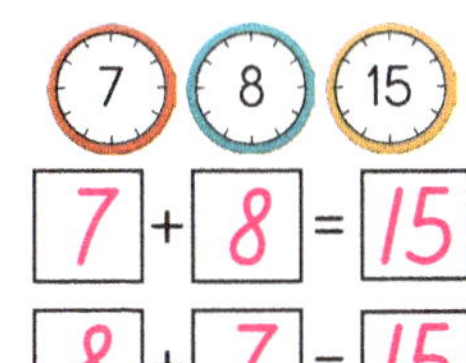

7 + 8 = 15
8 + 7 = 15
15 − 7 = 8
15 − 8 = 7

326 three hundred twenty-six Math 1

Engage

Song

Guide the students as they **sing** "Days of the Week," found in the back of this book, for the current day.

How can we spend our days for the Lord? tell others of Christ; listen and obey at home and school; be a help and encouragement to others

Instruct

Days of the Week

Guide a **discussion** to help the students name the days of the week in order.

Direct attention to the large classroom calendar. Say the days of the week in order together as you point to each day.

What do you do on Sunday? go to Sunday school; go to church; go to a restaurant for dinner; go to Grandma's house

What do you do on Monday? go to school; go to baseball practice; go to art class

Continue the activity by discussing what happens on each remaining day of the week.

Distribute sets of *Days of the Week Cards* to the students. Instruct the students to say the days of the week as they put their cards in order.

Invite a student to arrange your *Days of the Week Cards* in order for display so the students can check the order of their cards.

Instruct the students to name the days of the week in order again as you point to the names on the classroom calendar.

The Day That Comes Next

Guide a **sequencing activity** to help the students name the day of the week that comes next.

Display the card for *Tuesday*. Direct the students to look at their cards that they put in order.

Which day comes after Tuesday? Wednesday

Continue the activity with different days of the week.

Read a Calendar

Guide a **discussion** to help the students read a calendar.

Direct attention to the classroom calendar. Ask a student to locate the *7* on the calendar.

Remind the students that to find out the name for the 7th day of this month, they locate the *7* and look up to the top of the calendar and read the name of the day.

Which day of the week is the 7th of this month?

What day of the week is (name of the month) 21?

What is the date of the fourth Sunday?

How many Fridays are in (name of the month)?

What are the dates of all the Fridays in (name of the month)?

Continue the activity, using different days and dates.

Months of the Year

Guide a **discussion** to help the students name the months of the year in order.

Display Visual 16 and explain that this calendar shows a whole year. Say the months of the year in order together as you point out each one.

Sing together "There Are Twelve Months" found in the back of this book.

Explain that each month is known for something special. Direct attention to each month again as you ask a question about that month.

170

Which month is first in a new year? January

Which month includes Valentine's Day? February

Which month is windy and good for flying kites? March

Which month is known for many rain showers? April

Which month includes Mother's Day? May

Which month includes Father's Day? June

In which month do you celebrate the United States' independence? July

In which month could you take a vacation or start a new school year? August

Which month includes Labor Day? September

Which month includes Columbus Day? October

Which month includes Thanksgiving? November

Which month includes Christmas? December

Distribute sets of *Months of the Year Cards* to the students. Instruct the students to say the months of the year as they put their cards in order.

Invite students to arrange your *Months of the Year Cards* in order for display so the students can check the order of their cards.

The Month That Comes Next

Guide a **sequencing activity** to help the students name the month of the year that comes next.

Display the card for *May*. Direct the students to look at the cards that they put in order.

Which month comes next after May? June

Continue the activity by using different months of the year.

Apply

Worktext pages 325–26

Read and guide completion of page 325.

Read and explain the directions for page 326. Assist the students as they complete the page independently.

Using Calendars

Color the sun beside the day that comes next yellow.

Monday	Sunday	Tuesday (colored)
Wednesday	Tuesday	Thursday (colored)
Friday	Saturday (colored)	Sunday

Circle the month that comes next.

April	March	May (circled)
July	June	August (circled)
September	October (circled)	November
November	January	December (circled)

Mark the correct answer.

What month is this?
- ○ April
- ● July

How many days are in this month?
- ○ 29
- ● 31

What day is the 19th?
- ● Tuesday
- ○ Wednesday

What is the date of the second Wednesday?
- ○ 20
- ● 13

July						
Sunday	Monday	Tuesday	Wednesday	Thursday	Friday	Saturday
					1	2
3	4	5	6	7	8	9
10	11	12	13	14	15	16
17	18	19	20	21	22	23
24	25	26	27	28	29	30
31						

Chapter 20 • Lesson 170 three hundred thirteen **313**

Assess

Reviews pages 313–14

Review the number that is 1 more, 1 less, 10 more, or 10 less, and the value of a particular digit on page 314.

Extended Activity

Read a Calendar

Materials

- A small blank calendar labeled with the current month and the days of the week for each student
- Round stickers
- Assorted special stickers (optional)

Procedure

Provide a convenient place for the students to keep their calendars, such as the top corner of their desks. On the first day of the month, discuss on which day the month began. Instruct the students to put a round sticker on the first box under that day; then write the number 1 on the sticker. Continue the activity on each day of the month. Use special stickers for holidays, birthdays, or special school events.

Chapter 16 Review

Write the number that is 1 *more*, 1 *less*, 10 *more*, or 10 *less*.

1 more		1 less		10 more		10 less	
764	765	114	115	570	580	717	727
800	801	432	433	243	253	458	468
598	599	656	657	924	934	165	175

Write the value of each underlined digit.

326	20

475	400

628	8

Addition Fact Review

Circle the doubles.
Add.

$7 + 8 = 15$ $9 + 9 = 18$ $9 + 8 = 17$

$8 + 8 = 16$ $17 + 9 = 16$ $7 + 7 = 14$

$6 + 9 = 15$ $6 + 6 = 12$ $6 + 7 = 13$

171

Worktext pages 327–28
Reviews pages 315–16

How can I help others by telling time?

Chapter Concept Review

- Practice concepts from Chapter 20 to prepare for the test.

Printed Resources

- Instructional Aid 2: *Hundred Chart* (for each student)
- Instructional Aid 47: *School Schedule*
- Instructional Aids 12a–b: *Days of the Week Cards* (for the teacher and for each student)
- Instructional Aids 13a–c: *Months of the Year Cards* (for the teacher and for each student)
- Visual 1: *Hundred Chart*
- Student Manipulatives: Number Cards (0–20)

Digital Resources

- Web Link: Virtual Manipulatives: Analog Clock
- Web Link: Learn to Tell Time
- Web Link: Calendar Quiz
- Games/Enrichment: Addition Flashcards
- Games/Enrichment: Subtraction Flashcards

Materials

- Addition flashcards
- Subtraction flashcards
- Judy® Clock (or Visual 15: *Clock*)
- A large classroom calendar for the current month

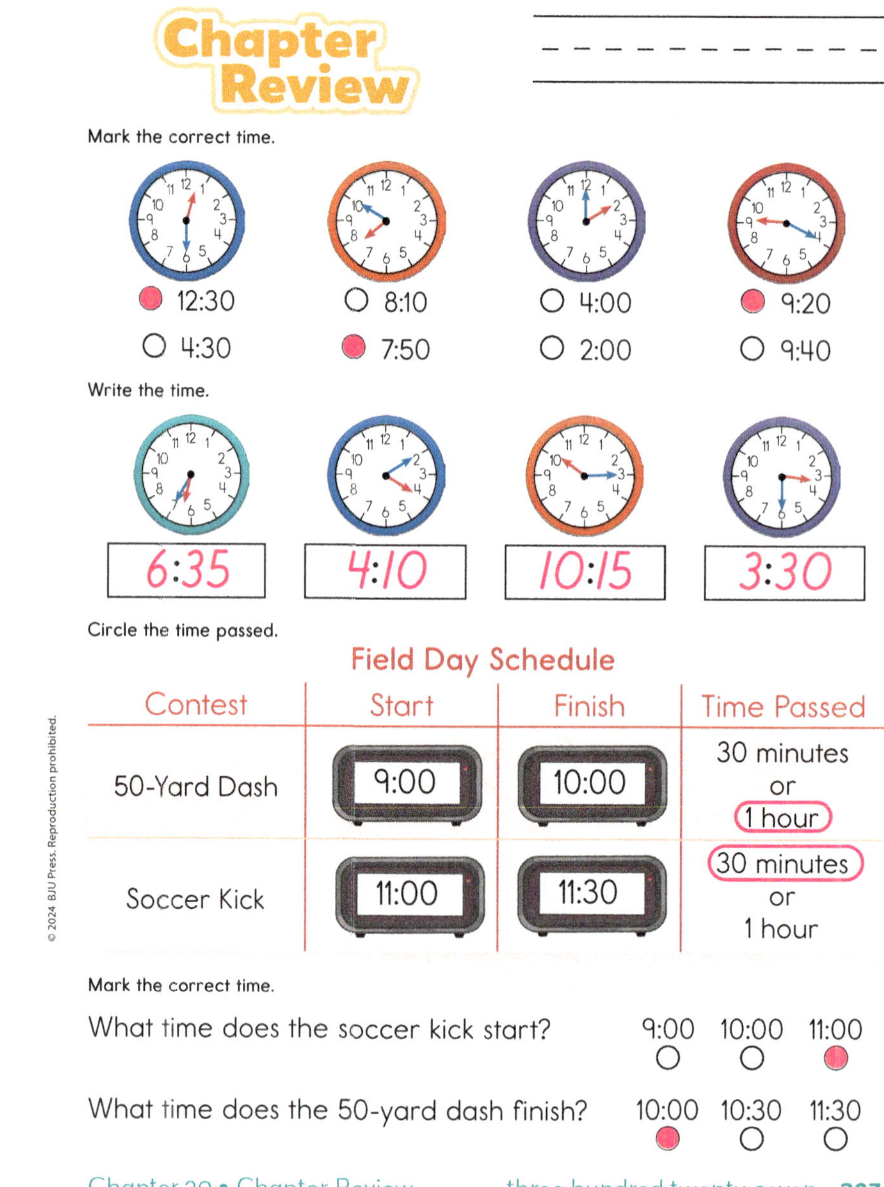

Chapter Review

Mark the correct time.

12:30 / 4:30

8:10 / 7:50

4:00 / 2:00

9:20 / 9:40

Write the time.

6:35 | 4:10 | 10:15 | 3:30

Circle the time passed.

Field Day Schedule

Contest	Start	Finish	Time Passed
50-Yard Dash	9:00	10:00	30 minutes or 1 hour
Soccer Kick	11:00	11:30	30 minutes or 1 hour

Mark the correct time.

What time does the soccer kick start? 9:00 10:00 11:00

What time does the 50-yard dash finish? 10:00 10:30 11:30

Practice & Review

Count by 2s to 50

Display Visual 1 and distribute the *Hundred Chart* pages. Remind the students that when they count by 1s, they count every number, but when they count by 2s, they skip a number and count every other number. Direct the students to point to the number 2 on their pages while you point to it on the visual. Count together by 2s as the students slowly point to every other number on their pages. Repeat the activity.

Count by 3s to 30

Point out that they can also count by 3s. Explain that when they count by 3s, they skip 2 numbers and count every third number. Direct the students to point to the number 3 on their *Hundred Chart* pages while you point to it on Visual 1. Instruct the students to skip 2 numbers and follow along as you count by 3s: 3, 6, 9, 12, 15, 18, 21, 24, 27, 30.

Guide the students as they count together by 3s as you point to every third number. Repeat the activity.

Memorize Facts for the 6-8-14 Family

Introduce the following facts.

6 + 8 8 + 6 14 – 6 14 – 8

Distribute Number Cards 0–20. Practice the family facts and previously memorized facts as in earlier lessons.

Instruct

Duration of One Hour & a Half-Hour in Minutes

Display the demonstration clock set for 9:00. Move the minute hand completely around the clock until it reaches the 12 again and the clock is set for 10:00.

What is the quickest way to count all the minutes that pass between 9:00 and 10:00? count them by 5s

Mark the day that comes next.

Sunday	Friday
● Monday	● Saturday
○ Tuesday	○ Sunday

Mark the month that comes next.

May	October
● June	○ September
○ July	● November

Mark an *X* on each Monday.

February						
Sunday	Monday	Tuesday	Wednesday	Thursday	Friday	Saturday
			1	2	3	4
5	6 X	7	8	9	10	11
12	13 X	14	15	16	17	18
19	20 X	21	22	23	24	25
26	27 X	28				

Mark the correct answer.

1. How many days are in February? ○ 26 ● 28 ○ 30
2. What day is February 14? ○ Sunday ○ Monday ● Tuesday
3. What day is February 9? ○ Tuesday ● Thursday ○ Saturday

Circle the phrase that completes the sentence.

DQ and Lily read the clock so they could obey and be ready on time. Telling time helps me be on time. I show that I care for others when I am not on time / (on time).

How many minutes are in 1 hour? 60 minutes

Set the demonstration clock for 4:00. Count together by 5s as you move the minute hand around the clock until it reaches the 6 and the clock is set for 4:30.

What time is it? 4:30

How many minutes are in a half-hour? 30 minutes

Time to the Hour, Half-Hour & Nearest 5-Minute Interval

Divide the class into 2 groups. Set the demonstration clock for 8:00.

Direct the first student in Group 1 to tell the time on the clock. Instruct the first student in Group 2 to write the time for display. Set the demonstration clock for 5:30. Direct the second student in Group 2 to tell the time; then direct the second student in Group 1 to write the time for display.

Continue the activity, using times to the hour, half hour, and nearest 5-minute interval. Alternate writing and telling time between the 2 groups.

A Schedule & Elapsed Time

Display the *School Schedule* page.

What information is in this table? school schedule; what time subjects are taught; how long the subjects are studied

Which subjects are listed in the schedule? Bible, Phonics, Reading, Math, Spelling

What does the schedule tell you about each subject? the time it starts, the time it finishes, and the time passed

What time does reading class start? 10:00

What time does reading class finish? 11:00

How much time passed from the start of reading class to when it finished? 1 hour

Continue the discussion with the other subjects.

Days of the Week

Guide the students as they name the days of the week in order.

Distribute sets of *Days of the Week Cards* to the students. Instruct the students to put their cards in order.

Ask a student to arrange your *Days of the Week Cards* in order for display.

Direct the students to hold up the correct cards to answer the following questions.

Which day of the week comes next after Tuesday? Wednesday

Which day comes next after Monday? Tuesday

Which day comes next after Friday? Saturday

Calendar

Direct attention to the classroom calendar.

Which day of the week is the 10th of this month?

How many days are in (name of the month)?

On which day does the month begin?

Which day of the week is (name of the month) 13?

How many Thursdays are in (name of the month)?

What are the dates of all the Thursdays in (name of the month)?

Months of the Year

Guide the students as they name the months of the year in order.

Distribute your *Months of the Year Cards* among 12 students. Ask these students to stand in front of the room in random order, displaying the cards. Invite volunteers to direct the movement of the students so that the *Months of the Year Cards* are in the correct order.

171

Distribute sets of *Months of the Year Cards* to the students. Direct the students to put their cards in the correct order.

Ask the students to hold up the correct cards to answer the following questions.

Which month of the year comes next after June? July

Which month comes next after October? November

Which month comes next after March? April

Which month comes next after July? August

Apply

Worktext pages 327–28

Read and explain the directions for the pages. Assist the students as they complete the pages independently.

Direct attention to the biblical worldview shaping section on page 328. Display the picture from the chapter opener.

Remind the students that DQ wanted to obey Miss Markle by cleaning up at the right time so he could be ready for story time. Also remind them that Lily checked her schedule so she could arrive at the school on time for the program.

What math skills did DQ and Lily use to help them be on time? telling the time, using a schedule

How did DQ and Lily show that they cared for others? They obeyed and they were on time.

Read aloud the chapter essential question on the opener page and guide the students as they complete the response on page 328.

Assess

Reviews pages 315–16

Use the Reviews pages to provide additional preparation for the chapter test.

Mark the correct time.

● 6:30	○ 8:50	○ 1:00	● 10:35
○ 5:30	● 7:55	● 12:00	○ 10:25

Write each time.

4:45	9:05	3:15	1:20

Circle the time passed.

Playground Schedule

Equipment	Start	Finish	Time Passed
Jungle Gym	2:00	3:00	30 minutes or (1 hour)
Swings	1:00	1:30	(30 minutes) or 1 hour
Seesaw	11:00	11:30	(30 minutes) or 1 hour

Mark the correct time.

What time does playing on the seesaw start?	11:00 ●	1:00 ○
What time does playing on the jungle gym finish?	11:30 ○	3:00 ●
What time does playing on the swings finish?	1:00 ○	1:30 ●

Mark the day that comes next.

Monday	○ Sunday	● Tuesday
Wednesday	○ Tuesday	● Thursday
Thursday	● Friday	○ Saturday

Mark the month that comes next.

February	● March	○ May
March	● April	○ August
July	○ June	● August
November	● December	○ January

Color all the Thursdays.

June						
Sunday	Monday	Tuesday	Wednesday	Thursday	Friday	Saturday
	1	2	3	4	5	6
7	8	9	10	11	12	13
14 Flag Day	15	16	17	18	19	20
21 Father's Day	22	23	24	25	26	27
28	29	30				

Mark the correct answer.

How many days are in June?	30 ●	31 ○
What is the date of the third Sunday?	14 ○	21 ●
What day is the 11th?	Thursday ●	Friday ○
What day is the 1st?	Sunday ○	Monday ●

172 Worktext page 329
Reviews pages 317–18

Chapter 20 Test

Administer the Chapter 20 Test.

Cumulative Concept Review

Worktext page 329

Review the following concepts. Adapt instructions and activities and provide reteaching as needed to meet the specific needs of your students.

- Writing addition and subtraction equations for a fact family (Lesson 155)
- Determining the value of a set of coins (Lessons 155–57)
- Measuring to find the perimeter of a figure, using centimeters (Lesson 141)

Retain a copy of Worktext page 329 for the discussion of the Chapter 21 essential question during Lessons 173 and 179.

Reviews pages 317–18

Use the Reviews pages to help students retain previously learned skills.

Extended Activity

Everyday Problem Solving

Materials

- A calendar for the current month
- Small cards or adhesive notes (one for each school day)

Procedure

Display a calendar for the current month. Write a word problem on each small card. Use word problems for addition, subtraction, money, fractions, and time. Write the answers on the back of each card for the students to check their answers. Each day of the month, choose one word problem and fasten it to that day on the calendar, or fasten all the word problems to the calendar at the beginning of the month. Guide the students as they read and solve the problem for the day.

Write each fact family.

(6) (8) (14)

6 + 8 = 14
8 + 6 = 14
14 − 6 = 8
14 − 8 = 6

(8) (9) (17)

8 + 9 = 17
9 + 8 = 17
17 − 8 = 9
17 − 9 = 8

Write the total value.

28¢

45¢

40¢

Use a centimeter ruler to find the perimeter.

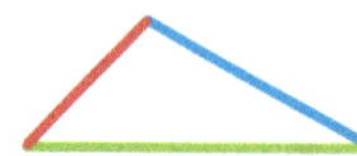

2 + 3 + 4 = 9 cm

Draw a dot next to the smaller number.
Draw the correct sign.

less than < greater than >

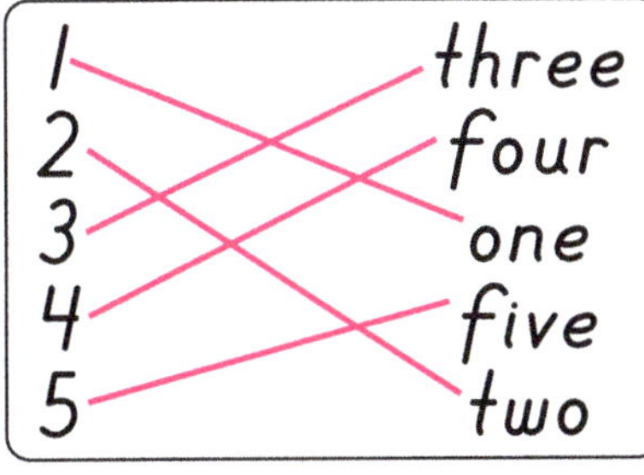

453	>	161
800	<	900
594	>	504
399	<	769

Match each number to the correct number word.

1	three
2	four
3	one
4	five
5	two

6	eight
7	nine
8	seven
9	ten
10	six

11	fifteen
12	eleven
13	twelve
14	thirteen
15	fourteen

16	eighteen
17	sixteen
18	twenty
19	seventeen
20	nineteen

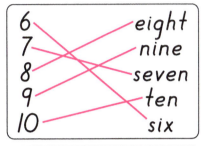

Subtract.

$93 - 41 = 52$ $57 - 36 = 21$ $86 - 43 = 43$ $29 - 15 = 14$ $61 - 20 = 41$

172

Addition Fact Review

Add.

6 + 7 = 13	5 + 2 = 7	6 + 6 = 12
3 + 9 = 12	4 + 6 = 10	6 + 5 = 11
7 + 0 = 7	5 + 9 = 14	4 + 8 = 12
9 + 2 = 11	7 + 5 = 12	3 + 3 = 6
6 + 3 = 9	9 + 4 = 13	3 + 7 = 10
2 + 8 = 10	9 + 3 = 12	9 + 7 = 16
5 + 7 = 12	6 + 2 = 8	6 + 9 = 15
8 + 4 = 12	3 + 5 = 8	9 + 5 = 14
4 + 2 = 6	7 + 9 = 16	5 + 6 = 11
5 + 4 = 9	9 + 6 = 15	7 + 6 = 13

CHAPTER 21: ADDITION WITH THREE-DIGIT NUMBERS

PAGES	OBJECTIVES	RESOURCES	ASSESSMENTS
Lesson 173	**Adding Hundreds**		
Teacher Edition 731–35 **Worktext** 330–32	**173.1** Add 3-digit numbers by using manipulatives. **173.2** Write an addition or subtraction equation for a word problem. **173.3** Explain how writing an addition or subtraction equation helps people do work. BWS Working (explain)	**Visuals** • Visual 21: *Problem-Solving Plan* • Place Value Kit: hundreds, tens, ones **Student Manipulatives Packet** • Hundreds/Tens/Ones Mat • Place Value Kit: hundreds, tens, ones **BJU Press Trove*** • Video: Ch 21 Intro • Web Link: Virtual Manipulatives: Base Ten Number Pieces • Games/Enrichment: Fact Reviews (Addition Facts to 18) • Games/Enrichment: Addition Flashcards • Games/Enrichment: Subtraction Flashcards • Games/Enrichment: Fact Fun Activities • PowerPoint® presentation	**Reviews** • Pages 319–20
Lesson 174	**Practicing Adding Hundreds**		
Teacher Edition 736–39 **Worktext** 333–34	**174.1** Add 3-digit numbers with and without manipulatives. **174.2** Write an addition or subtraction equation for a word problem.	**Visuals** • Visual 21: *Problem-Solving Plan* • Place Value Kit: hundreds, tens, ones **Student Manipulatives Packet** • Hundreds/Tens/Ones Mat • Place Value Kit: hundreds, tens, ones **BJU Press Trove** • Web Link: Virtual Manipulatives: Base Ten Number Pieces • Games/Enrichment: Fact Reviews (Addition Facts to 18) • Games/Enrichment: Addition Flashcards • Games/Enrichment: Subtraction Flashcards • Games/Enrichment: Fact Fun Activities • PowerPoint® presentation	**Reviews** • Pages 321–22

*Digital resources for homeschool users are available on Homeschool Hub.

PAGES	OBJECTIVES	RESOURCES	ASSESSMENTS
Lesson 175 **Problem Solving**			
Teacher Edition 740–43 **Worktext** 335–36	**175.1** Add 3-digit numbers. **175.2** Solve addition and subtraction problems by using the Problem-Solving Plan. **175.3** Use an addition or subtraction equation to do work. BWS Working (apply)	**Visuals** • Visual 21: *Problem-Solving Plan* • Place Value Kit: hundreds, tens, ones **Student Manipulatives Packet** • Hundreds/Tens/Ones Mat • Place Value Kit: hundreds, tens, ones **BJU Press Trove** • Web Link: Virtual Manipulatives: Base Ten Number Pieces • Games/Enrichment: Fact Reviews (Subtraction Facts to 18) • Games/Enrichment: Addition Flashcards • Games/Enrichment: Subtraction Flashcards • Games/Enrichment: Fact Fun Activities • PowerPoint® presentation	**Reviews** • Pages 323–24
Lesson 176 **Renaming 10 Ones**			
Teacher Edition 744–47 **Worktext** 337–38	**176.1** Rename 10 ones as 1 ten by using manipulatives. **176.2** Add 3-digit numbers with and without renaming.	**Visuals** • Place Value Kit: hundreds, tens, ones **Student Manipulatives Packet** • Hundreds/Tens/Ones Mat • Place Value Kit: hundreds, tens, ones **BJU Press Trove** • Web Link: Virtual Manipulatives: Base Ten Number Pieces • PowerPoint® presentation	**Reviews** • Pages 325–26
Lessons 177–78 **Adding Hundreds with Renaming**			
Teacher Edition 748–51 **Worktext** 339–40	**177–78.1** Add 3-digit numbers with and without renaming. **177–78.2** Solve a word problem by writing an addition equation.	**Visuals** • Visual 21: *Problem-Solving Plan* • Place Value Kit: hundreds, tens, ones **Student Manipulatives Packet** • Hundreds/Tens/Ones Mat • Place Value Kit: hundreds, tens, ones **BJU Press Trove** • Web Link: Virtual Manipulatives: Base Ten Number Pieces • PowerPoint® presentation	**Reviews** • Pages 327–28

PAGES	OBJECTIVES	RESOURCES	ASSESSMENTS
Lesson 179 **Chapter 21 Review**			
Teacher Edition 752–55 **Worktext** 341–42	**179.1** Recall concepts and terms from Chapter 21.	**Teacher Edition** • Instructional Aid 2: *Hundred Chart* **Visuals** • Visual 1: *Hundred Chart* **Student Manipulatives Packet** • Hundreds/Tens/Ones Mat • Place Value Kit: hundreds, tens, ones **BJU Press Trove** • Games/Enrichment: Fact Reviews (Subtraction Facts to 18) • Games/Enrichment: Addition Flashcards • Games/Enrichment: Subtraction Flashcards • Games/Enrichment: Fact Fun Activities • PowerPoint® presentation	**Worktext** • Chapter 21 Review **Reviews** • Pages 329–30
Lesson 180 **Test, Cumulative Review**			
Teacher Edition 756–59 **Worktext** 343–44	**180.1** Demonstrate knowledge of concepts from Chapter 21 by taking the test.		**Assessments** • Chapter 21 Test **Worktext** • Cumulative Review **Reviews** • Pages 331–32

21 Addition with Three-Digit Numbers

How can 3-digit addition help me complete my work?

There are 10 cupcakes hidden throughout the page. See if you can find them all.

Chapter 21

How can 3-digit addition help me complete my work?

Chapter Objectives

- Join sets to model 3-digit addition.
- Solve 3-digit addition problems in vertical form.
- Solve word problems by using the Problem-Solving Plan.
- Use equations to do work.

173

Worktext pages 330–32
Reviews pages 319–20

Objectives

- **173.1** Add 3-digit numbers by using manipulatives.
- **173.2** Write an addition or subtraction equation for a word problem.
- **173.3** Explain how writing an addition or subtraction equation helps people do work. BWS

Biblical Worldview Shaping

- **Working** (explain): Working involves solving problems, and many everyday problems require adding or subtracting large numbers. Knowing how to write equations helps us solve problems to complete our work (173.3).

Printed Resources

- Visual 21: *Problem-Solving Plan*
- Visuals: Place Value Kit (hundreds, tens, ones)
- Student Manipulatives: Hundreds/Tens/Ones Mat
- Student Manipulatives: Place Value Kit (hundreds, tens, ones)

Digital Resources

- Video: Ch 21 Intro
- Web Link: Virtual Manipulatives: Base Ten Number Pieces
- Games/Enrichment: Fact Reviews (Addition Facts to 18, page 29)
- Games/Enrichment: Addition Flashcards
- Games/Enrichment: Subtraction Flashcards
- Games/Enrichment: Fact Fun Activities

All problems in this chapter should be written vertically. All word-problem equations should be written horizontally.

Practice & Review

Count 900–1,000

Instruct the students to stand in a circle. Say a number between 900 and 1,000; then direct the student to your right to say the next number. Continue counting around the circle to 1,000.

Adding Hundreds

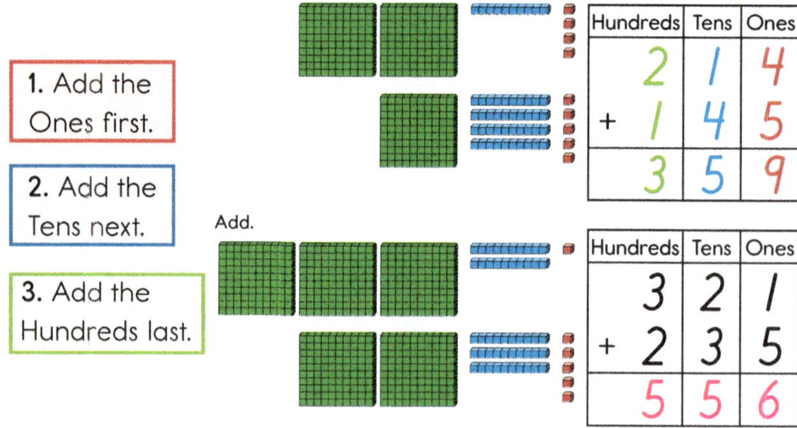
1. Add the Ones first.
2. Add the Tens next.
3. Add the Hundreds last.
Add.

	Hundreds	Tens	Ones
	2	1	4
+	1	4	5
	3	5	9

	Hundreds	Tens	Ones
	3	2	1
+	2	3	5
	5	5	6

Add.

	Hundreds	Tens	Ones
	5	6	2
+	4	2	4
	9	8	6

	Hundreds	Tens	Ones
	6	2	4
+	3	5	4
	9	7	8

	Hundreds	Tens	Ones
	7	8	3
+	2	0	2
	9	8	5

Write an equation for the word problem. Solve.

DQ gave out 152 balloons at the party.
Lily gave out 134 balloons.
How many balloons did they give out in all?

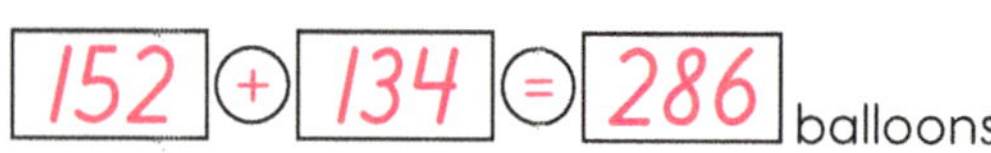

152 + 134 = 286 balloons

	Hundreds	Tens	Ones
	1	5	2
+	1	3	4
	2	8	6

Review Addition & Subtraction Facts (13–18)

Choose a game or activity from the "Fact Fun Activities" available in Trove.

Engage

Essential Question

Direct attention to the chapter opener on Worktext page 330 and **read aloud** the following scenario to introduce the chapter essential question.

"Come on, Liam, let's hurry," said Lily as the two friends followed the winding path to the duck pond.

Each year the community planned a party for all the children who were patients in Madison Children's Hospital, and Lily and Liam were helping.

They spotted DQ near the shore of the pond, finishing his breakfast of fish and algae.

"We didn't know you'd be up so soon. It's pretty early," said Lily.

"I'm a duck. I'm always up at the 'quack' of dawn, right?" joked DQ.

"We're helping with the party at the Children's Hospital today," Lily continued. "Do you want to come along?"

"Sure!" DQ waddled excitedly out of the pond, leaving a trail of webbed footprints behind him.

Add.

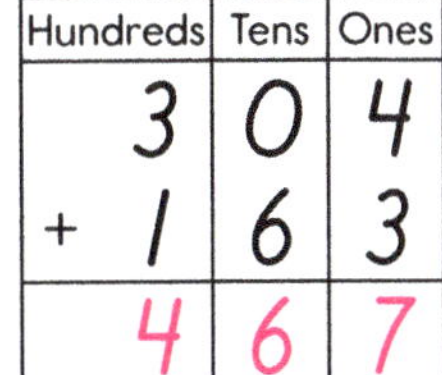

Hundreds	Tens	Ones
3	0	4
+ 1	6	3
4	6	7

Hundreds	Tens	Ones
2	8	1
+ 3	1	5
5	9	6

Hundreds	Tens	Ones
7	6	2
+ 2	1	4
9	7	6

Write an equation for the word problem. Solve.

There are 253 beds on the first floor of the hospital.
There are 312 beds on the second floor.
How many beds are there on these 2 floors?

253 + 312 = 565 beds

Hundreds	Tens	Ones
2	5	3
+ 3	1	2
5	6	5

Trace the word to complete the sentence.

I can write an equation that helps me complete my work.

Time to Review

Add.

6	4	7	9	8
3	4	1	4	0
+ 7	+ 8	+ 6	+ 1	+ 7
16	16	14	14	15

"What can we do to really cheer up the children besides having games and cake?" wondered Lily aloud. "Some children have been in the hospital for many weeks."

"Why don't we take bunches of balloons?" suggested DQ.

"What a great idea!" said Liam. "We could decorate the party room and save some balloons to give to the children."

Lily and Liam told their parents about their idea, and they agreed it was wonderful. They took the friends to the store to pick out the balloons.

"How many should we buy?" asked Liam. "The room where the party is held is very large, and I heard there are over 100 children."

"Wow! This is going to be a lot of work taking so many balloons," said DQ to Lily. "We may have to ask your dad to make several trips."

"We can use our math skills to figure this out," said Lily. "We know how to add large numbers now."

"That's right," said Liam. "Solving these problems will help us get this work done."

The friends arrived at the store, excited to choose a great number of colorful balloons to cheer the children.

"Now let's see," said DQ. "If we take 120 balloons to decorate and 114 balloons to hand out, how many is that?"

Invite a student to read aloud the essential question. Encourage the students to consider this question as you guide them to answer it by the end of the chapter.

Instruct

Add 3-Digit Numbers

Use **manipulatives** to add 3-digit numbers.

Distribute the Hundreds/Tens/Ones Mats and the Place Value Kits. Write "135 + 324" for display.

Display 1 hundred, 3 tens, and 5 ones in a Hundreds/Tens/Ones frame. Direct the students to put 1 hundred, 3 tens, and 5 ones near the top of their mats.

Display 3 hundreds, 2 tens, and 4 ones below the first addend in the frame. Direct the students to put 3 hundreds, 2 tens, and 4 ones near the bottom of their mats.

Remind the students that addition means joining groups.

How can these numbers be added? join the sets of ones first, the sets of tens next, and the sets of hundreds last

Hundreds	Tens	Ones

Explain that when joining sets with hundreds, tens, and ones, the squares in the Ones place are always joined first; then the strips in the Tens place are joined; finally, the squares in the Hundreds place are joined.

Direct the students to move all the ones in the Ones place together while you demonstrate.

How many ones are there in all? 9

Write "9" in the Ones place of the problem displayed.

Direct the students to move all the tens in the Tens place together while you demonstrate.

173

How many tens are there in all? 5

Write "5" in the Tens place.

Direct the students to move all the hundreds in the Hundreds place together while you demonstrate.

How many hundreds are there in all? 4

Write "4" in the Hundreds place.

What is the answer? 459

Model the following addition problems while the students do them on their mats. Emphasize adding the Ones place first, the Tens place next, and the Hundreds place last.

214	232	526
+ 362	+ 205	+ 163
576	437	689

Write an Equation to Solve a Word Problem

Use the **Problem-Solving Plan** to solve a word problem.

Display Visual 21 and read aloud the questions. Remind the students that solving word problems is made easier by following a plan. Read the following word problem several times. Emphasize that the students need to think about the information given.

The hospital has 56 nurses and 23 doctors. How many more nurses are there than doctors at this hospital?

What is the question in this word problem? How many more nurses are there than doctors at this hospital?

What information is given? The hospital has 56 nurses and 23 doctors.

Do you add or subtract to find the answer? I subtract because I am finding the difference between the groups.

What equation is needed to solve this problem? 56 – 23 = __

Write "56 – 23 = __" for display.

Should this equation be written in vertical form? Yes, it is easier to subtract 2-digit numbers in vertical form.

Invite a student to write the problem in vertical form.

What do you subtract first when there are 2-digit numbers? the ones

Ask a volunteer to solve the problem. 33 Write "33 nurses" to complete the equation.

Does the answer make sense? yes

Explain that if students had added 56 and 23, the answer would have been 79 and would not make sense.

Continue the activity for the following word problem.

A hospital in town made 332 meals on Tuesday and 265 meals on Wednesday. What is the total number of meals made on those 2 days?

What is the question in this word problem? What is the total number of meals made on those 2 days?

What information is given in the word problem? The hospital made 332 meals on Tuesday and 265 meals on Wednesday.

Do you add or subtract to find the answer? I add because I am joining the numbers to find the total.

What equation can be used to solve this problem? 332 + 265 = __

Write "332 + 265 = __" for display.

Should the equation be written in vertical form? Yes, it will be easier to add 3-digit numbers in vertical form.

Invite a student to write the problem in vertical form.

What do you add first when there are 3-digit numbers? the ones

Ask a volunteer to solve the problem. 597 Write "597 meals" to complete the equation.

Adding Hundreds

Add.

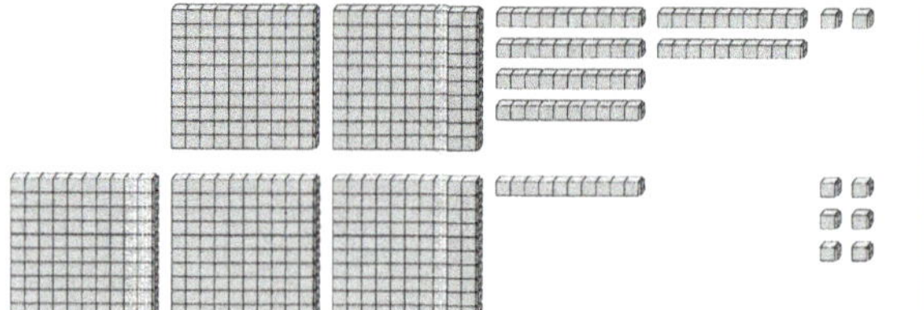

	Hundreds	Tens	Ones
	1	7	4
+	3	1	2
	4	8	6

	Hundreds	Tens	Ones
	2	6	2
+	3	1	6
	5	7	8

Add.

	Hundreds	Tens	Ones
	4	5	3
+	3	4	6
	7	9	9

	Hundreds	Tens	Ones
	7	8	2
+	2	0	5
	9	8	7

	Hundreds	Tens	Ones
	4	2	6
+	2	3	1
	6	5	7

Write an equation for the word problem.
Solve.

The men used 125 bricks to make some steps. They used 750 bricks to make a wall. How many bricks did they use in all?

   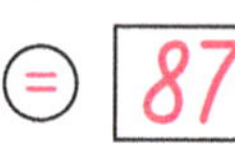 bricks

	Hundreds	Tens	Ones
	1	2	5
+	7	5	0
	8	7	5

Chapter 13 Review

Subtract.

97 − 64 = 33	38 − 17 = 21	59 − 42 = 17	68 − 21 = 47
87 − 46 = 41	95 − 13 = 82	86 − 60 = 26	49 − 13 = 36

Subtraction Fact Review

Subtract.

14 − 8 = 6	17 − 9 = 8	16 − 8 = 8	13 − 5 = 8	13 − 8 = 5
15 − 8 = 7	17 − 8 = 9	18 − 9 = 9	14 − 6 = 8	15 − 7 = 8

Does the answer make sense? yes

Explain that if students had subtracted 265 from 332, the answer would have been less than 332 and would not make sense.

Writing an Equation Helps People Work

Guide a **discussion** to help the students explain a biblical worldview shaping truth.

What were DQ and his friends planning in the opener story? to have a party for the children staying at the hospital

Read aloud the following word problem.

DQ wanted to take 120 balloons to the party to use for decorations. He wanted to take 114 balloons to give to the children. How many balloons in all does DQ need to take?

What is the question? How many balloons in all does DQ need to take?

How can DQ use math to find the answer? write and solve an addition equation

Read the word problem again slowly and ask a student to state the equation. Invite another student to write and solve the addition problem. 120 + 114 = 234 balloons

Remind the students that gathering and taking balloons to the party is a type of work. Point out that writing and solving equations can help us do our work.

How did writing an addition equation help DQ complete his work? He was able to find out how many balloons to take to the party.

Apply

Worktext pages 331–32

Read and guide completion of page 331.

Read and explain the directions for page 332. Assist the students as they complete the page independently.

Direct attention to the biblical worldview shaping statement and guide the students as they complete the sentence.

Assess

Reviews pages 319–20

Review subtracting a 2-digit number on page 320.

Fact Reviews

Use "Addition Facts to 18" page 29 in Trove.

Differentiated Instruction

Add 3-Digit Numbers

Direct the students to write "234 + 123" vertically. Provide the students with an index card to use as a "cover sheet." Instruct them to cover all the numbers except those in the Ones place. Instruct them to add the numbers in the Ones place and write the answer. Direct them to move the "cover sheet" over to show the Tens place. Direct them to add the numbers in the Tens place and write the answer. Instruct them to take the "cover sheet" away and add the numbers in the Hundreds place. Continue the activity with the other 3-digit addition problems.

174

Worktext pages 333–34
Reviews pages 321–22

Objectives

- **174.1** Add 3-digit numbers with and without manipulatives.
- **174.2** Write an addition or subtraction equation for a word problem.

Printed Resources

- Visual 21: *Problem-Solving Plan*
- Visuals: Place Value Kit (hundreds, tens, ones)
- Student Manipulatives: Hundreds/Tens/Ones Mat
- Student Manipulatives: Place Value Kit (hundreds, tens, ones)

Digital Resources

- Web Link: Virtual Manipulatives: Base Ten Number Pieces
- Games/Enrichment: Fact Reviews (Addition Facts to 18, page 30)
- Games/Enrichment: Addition Flashcards
- Games/Enrichment: Subtraction Flashcards
- Games/Enrichment: Fact Fun Activities

Practicing Adding Hundreds

1. Add the Ones first.
2. Add the Tens next.
3. Add the Hundreds last.

Add.

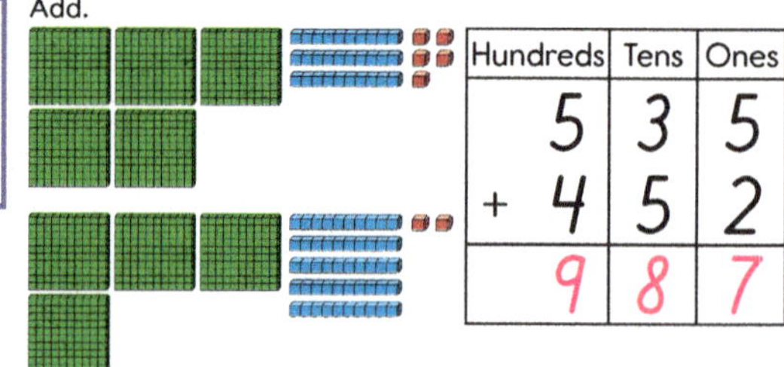

Hundreds	Tens	Ones
5	3	5
+ 4	5	2
9	8	7

Add.

Hundreds	Tens	Ones
1	6	4
+ 2	3	1
3	9	5

Hundreds	Tens	Ones
8	7	3
+ 1	0	6
9	7	9

Hundreds	Tens	Ones
7	1	4
+ 2	7	3
9	8	7

Write an addition or subtraction equation for each word problem. Solve.

The hospital received 384 phone calls on Monday. They received 412 phone calls on Tuesday. How many calls did they receive on these 2 days?

384 + 412 = 796 phone calls

(work space)

$$\begin{array}{r} 384 \\ +\ 412 \\ \hline 796 \end{array}$$

The nurses had 89 boxes of supplies. Today they used 17 boxes of supplies. How many boxes of supplies do the nurses have left?

89 − 17 = 72 boxes

(work space)

$$\begin{array}{r} 89 \\ -\ 17 \\ \hline 72 \end{array}$$

Practice & Review

Identify the Ordinal Positions of the Days of the Week

Guide the students as they say the days of the week in order. Then ask questions such as the following.

What is the first day of the week? Sunday

What is the third day of the week? Tuesday

What is the sixth day of the week? Friday

Which day of the week is Monday? second

Which day of the week is Saturday? seventh

Name the Months of the Year in Order

Guide the students as they say the months of the year in order.

Review Addition & Subtraction Facts (13–18)

Choose a game or activity from the "Fact Fun Activities" available in Trove.

Engage

Showing Kindness

Use **direct instruction** to reinforce the importance of showing kindness.

Have you ever visited someone in a hospital?

Have you ever been a patient in a hospital?

Explain that the hospital might seem like a scary place, but it really is not; it is a place for people to get the help they need to get well. Explain that Jesus taught that He views our going to visit sick people as a kindness shown to Him; He wants Christians to show kindness to all people. Explain that every person is important to Jesus, so every person should be important to Christians. Read aloud Matthew 25:34–40.

Instruct

Add 3-Digit Numbers

Use **manipulatives** to add 3-digit numbers.

Distribute the Hundreds/Tens/Ones Mats and the Place Value Kits. Write "203 + 124" for display.

Display 2 hundreds and 3 ones in a Hundreds/Tens/Ones frame. Direct the students to use their hundreds, tens, and ones to make the first addend near the top of their mats.

How many hundreds, tens, and ones are in 203? 2 hundreds, 0 tens, 3 ones

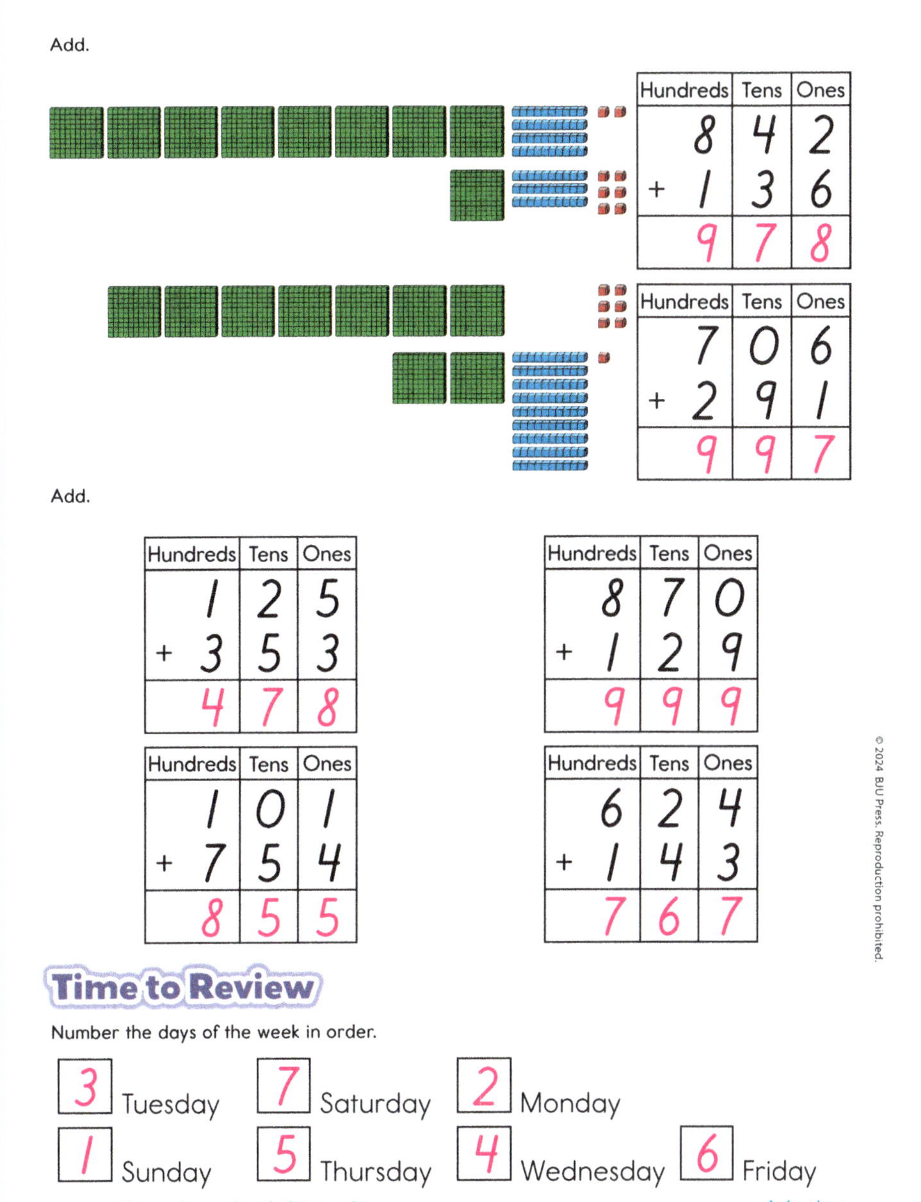

Add.

	Hundreds	Tens	Ones
	8	4	2
+	1	3	6
	9	7	8

	Hundreds	Tens	Ones
	7	0	6
+	2	9	1
	9	9	7

Add.

	Hundreds	Tens	Ones
	1	2	5
+	3	5	3
	4	7	8

	Hundreds	Tens	Ones
	8	7	0
+	1	2	9
	9	9	9

	Hundreds	Tens	Ones
	1	0	1
+	7	5	4
	8	5	5

	Hundreds	Tens	Ones
	6	2	4
+	1	4	3
	7	6	7

Time to Review

Number the days of the week in order.

3 Tuesday 7 Saturday 2 Monday

1 Sunday 5 Thursday 4 Wednesday 6 Friday

Display 1 hundred, 2 tens, and 4 ones below the first addend.

Direct the students to make the second addend near the bottom of their mats.

> **How many hundreds, tens, and ones are in 124?** 1 hundred, 2 tens, 4 ones
>
> **What does addition mean?** joining or putting sets together
>
> **Which digits should you add first?** Add the 3 and the 4 together because they are in the Ones place.

Direct the students to move all the ones together while you demonstrate.

> **How many ones are there in all?** 7

Write "7" in the Ones place.

Direct the students to move all the tens together on their mats while you demonstrate.

> **How many tens are there in all?** 2

Write "2" in the Tens place.

Direct the students to move all the hundreds together while you demonstrate.

> **How many hundreds are there in all?** 3

Write "3" in the Hundreds place.

> **What is the answer?** 327

Model the following addition problems while the students do them on their mats. Emphasize adding the Ones place first, the Tens place next, and the Hundreds place last.

157	240	435
+ 321	+ 316	+ 163
478	556	598

174

Write an Equation to Solve a Word Problem

Use the **Problem-Solving Plan** to solve a word problem.

Display Visual 21 and read aloud the questions. Read aloud the following word problem several times.

Supplies are delivered to the Madison Children's Hospital 2 times each week. On Monday 365 boxes were delivered, and on Thursday 232 boxes were delivered. What is the total number of boxes delivered during the week?

> **What is the question in this word problem?** What is the total number of boxes delivered during the week?
>
> **What information is given?** On Monday 365 boxes were delivered, and on Thursday 232 boxes were delivered.
>
> **Do you add or subtract to find the answer?** Add; I am finding the total number of boxes delivered on the 2 days.
>
> **What equation is needed to solve this problem?** 365 + 232 = __

Write "365 + 232 = __" for display.

> **Should you rewrite the equation in vertical form?** Yes, it is easier to add 3-digit numbers in vertical form.

Invite a student to write the problem in vertical form.

> **What do you add first when there are 3-digit addends?** the ones

Ask a volunteer to solve the problem. 597

Write "597 boxes" to complete the equation.

> **Does the answer make sense?** yes

Explain that if the students had subtracted 232 from 365, the answer would have been 133 and would not make sense.

Continue the activity for the following word problem.

When the boxes were delivered to the hospital, 65 boxes were placed in the supply closet on the second floor. By the end of the week, 35 boxes had been opened and the supplies were used. How many boxes remain in the closet?

> **What is the question in this word problem?** How many boxes remain in the closet?

174

What information is given in the word problem? They placed 65 boxes in the closet, and the supplies from 35 boxes were used.

Do you add or subtract to find the answer? Subtract; I need to find out how many are left.

What equation can be used to solve this problem? 65 – 35 = ___

Write "65 – 35 = ___" for display.

Should you rewrite the equation in vertical form? Yes, it is easier to subtract 2-digit numbers in vertical form.

Invite a student to write it in vertical form.

What do you subtract first when there are 2-digit numbers? the ones

Ask a volunteer to solve the problem. 30

Write "30 boxes" to complete the equation.

Does the answer make sense? yes

Explain that if the students had added 65 and 35, the answer would have been 100 and would not make sense.

Practicing Adding Hundreds

Add.

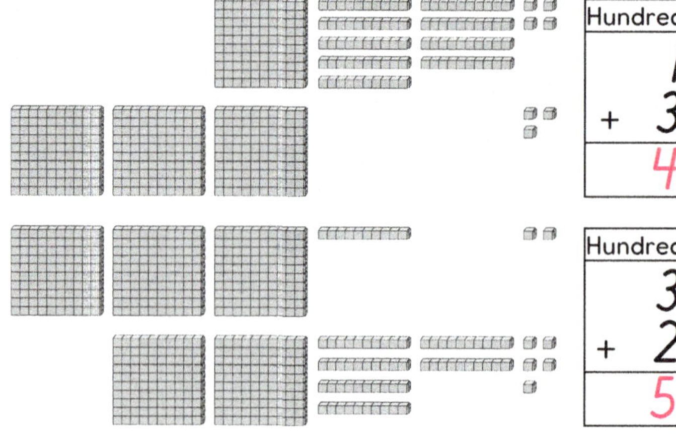

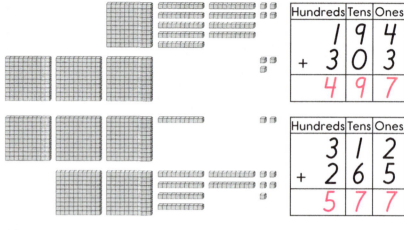

Hundreds	Tens	Ones
	1	9	4
+ 3	0	3
4	9	7

Hundreds	Tens	Ones
3	1	2
+ 2	6	5
5	7	7

Add.

Hundreds	Tens	Ones
7	2	5
+ 1	4	2
8	6	7

Hundreds	Tens	Ones
5	0	3
+ 2	9	2
7	9	5

Hundreds	Tens	Ones
3	4	5
+ 6	0	3
9	4	8

Write an equation for the word problem.
Solve.

Mrs. Smith's class read 453 books.
Mr. Crow's class read 426 books.
How many books did the classes read in all?

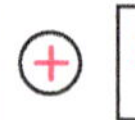 +  = 

453 + 426 = 879 books

Hundreds	Tens	Ones
4	5	3
+ 4	2	6
8	7	9

Apply

Worktext pages 333–34

Read and guide completion of page 333.

Read and explain the directions for page 334. Assist the students as they complete the page independently.

Assess

Reviews pages 321–22

Review writing the number that is 1 more, 1 less, 10 more, or 10 less and the value of a specified digit on page 322.

Fact Reviews

Use "Addition Facts to 18" page 30 in Trove.

Extended Activity

Add 3-Digit Numbers by Using a Spinner

Materials

- Instructional Aid 33: *Spinner* for each pair of students with one of the following numbers written in each section: 244, 113, 334, 143, 444, 322
- Instructional Aid 48: *Spin & Add* for each student

Procedure

Pair the students. Distribute a copy of the *Spin & Add* page to each student and the *Spinner* page to each pair of students.

Direct a student in each pair to spin for his or her 2 addends. Direct both students in the pair to write the addends on the Hundreds/Tens/Ones frame under the name of the student that spun the numbers. Instruct them to solve the problem and compare their answers. If their answers differ, they should solve the problem again or ask the teacher to identify whose answer is correct.

Allow the other student to spin for his or her 2 addends. Repeat the procedure.

Continue the activity until the page is completed or until both students have had the same number of opportunities to spin for addends.

You may make a game out of this activity. The winner could be the player whose problem resulted in the largest sum or the player who had the most correct answers.

Chapter 16 Review

Write the number that is 1 *more*, 1 *less*, 10 *more*, or 10 *less*.

1 more		1 less		10 more		10 less	
694	695	741	742	153	163	970	980
365	366	229	230	612	622	447	457
813	814	474	475	386	396	224	234

Mark the value of each underlined digit.

637 ○ 300 ● 30 ○ 3

415 ● 400 ○ 40 ○ 4

Subtraction Fact Review

Subtract.

8 − 6 = 2 10 − 2 = 8 9 − 4 = 5

12 − 6 = 6 14 − 8 = 6 10 − 5 = 5

11 − 6 = 5 13 − 5 = 8 11 − 8 = 3

9 − 6 = 3 15 − 7 = 8 10 − 3 = 7

175 Worktext pages 335–36 Reviews pages 323–24

Objectives

- **175.1** Add 3-digit numbers.
- **175.2** Solve addition and subtraction problems by using the Problem-Solving Plan.
- **175.3** Use an addition or subtraction equation to do work. BWS

Biblical Worldview Shaping

- **Working** (apply): Writing and solving an addition or subtraction equation is often an important part of completing our work efficiently and accurately. When we use equations to solve problems, we are using math to help us work (175.3).

Printed Resources

- Visual 21: *Problem-Solving Plan*
- Visuals: Place Value Kit (hundreds, tens, ones)
- Student Manipulatives: Hundreds/Tens/Ones Mat
- Student Manipulatives: Place Value Kit (hundreds, tens, ones)

Digital Resources

- Web Link Virtual Manipulatives: Base Ten Number Pieces
- Games/Enrichment: Fact Reviews (Subtraction Facts to 18, page 59)
- Games/Enrichment: Addition Flashcards
- Games/Enrichment: Subtraction Flashcards
- Games/Enrichment: Fact Fun Activities

Materials

- Judy® Clock (or Visual 15: *Clock*)

Problem Solving

Add.

Hundreds	Tens	Ones
2	5	0
+ 1	2	5
3	7	5

Hundreds	Tens	Ones
6	2	5
+ 2	3	4
8	5	9

Hundreds	Tens	Ones
3	1	5
+ 3	6	3
6	7	8

Add.

$743 + 214 = 957$ $653 + 345 = 998$ $724 + 123 = 847$ $536 + 232 = 768$

Write an equation for the word problem. Solve.

There are 58 people with wheelchairs and 43 people with crutches. How many more people are using wheelchairs than are using crutches?

58  43  15 people

(work space)

$58 - 43 = 15$

Practice & Review

Count by 10s from a 3-Digit Number

Write "456" for display.

Which digit is in the Tens place? 5

Explain that students will be counting by 10s starting with 456. Write "466" below 456 and point out that 466 is 10 more than 456. Guide the students as they count to 556: 456, 466, 476, 486, 496, 506, 516, 526, 536, 546, 556.

Continue the activity, starting with a different number.

Tell & Write Time to the Nearest 5-Minute Interval

Display the Judy Clock or Visual 15. Remind the students that there are 5 minutes between each number.

Set the demonstration clock for 8:05.

What is the time? 8:05

Ask a volunteer to write "8:05" for display.

Continue the activity for 12:15, 3:25, 5:40, and 11:05.

Review Addition & Subtraction Facts (13–18)

Choose a game or activity from the "Fact Fun Activities" available in Trove.

Engage

God Knows & Cares

Use **direct instruction** to explain a biblical truth.

Remind the students that this year they have learned how to add 1-digit numbers, 2-digit numbers, and 3-digit numbers. Point out that although students have learned how to add larger numbers, they should remember that what they know about numbers and can do with them is very small compared to what God knows and can do. Explain that the Bible teaches that God knows the number of all the hairs on a person's head (Luke 12:7). Because God knows and loves all people, He can forgive every single one

Add.

Hundreds	Tens	Ones
2	0	1
+ 4	8	3
6	8	4

Hundreds	Tens	Ones
1	2	7
+ 5	6	2
6	8	9

Hundreds	Tens	Ones
3	7	4
+ 2	1	5
5	8	9

Add.

$$\begin{array}{r} 765 \\ +\ 132 \\ \hline 897 \end{array} \qquad \begin{array}{r} 165 \\ +\ 314 \\ \hline 479 \end{array} \qquad \begin{array}{r} 894 \\ +\ 103 \\ \hline 997 \end{array}$$

Write an equation for the word problem. Solve.

Sid is helping at the hospital.
On Tuesday he counted 134 bandages.
On Friday he counted 242 bandages.
How many bandages did Sid count?

134 + 242 = 376 bandages

(work space)

$$\begin{array}{r} 134 \\ +\ 242 \\ \hline 376 \end{array}$$

Circle the word to complete the sentence.

I wrote an equation to 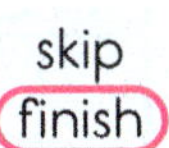 skip / finish my work.

Time to Review

Write the time.

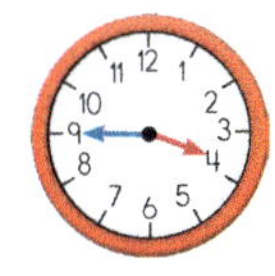

5:20 3:45 8:10 2:55

of their sins. When people go through a troubling time, such as having to stay in the hospital, God sees and cares.

Instruct

Add 3-Digit Numbers

Use **manipulatives** to add 3-digit numbers.

Distribute the Hundreds/Tens/Ones Mats and the Place Value Kits. Write "326 + 123" vertically for display.

Display 3 hundreds, 2 tens, and 6 ones in a Hundreds/Tens/Ones frame. Direct the students to use their hundreds, tens, and ones to make the first addend near the top of their mats.

How many hundreds, tens, and ones are in 326? 3 hundreds, 2 tens, 6 ones

Display 1 hundred, 2 tens, and 3 ones below the first addend. Direct the students to make the second addend near the bottom of their mats.

How many hundreds, tens, and ones are in 123? 1 hundred, 2 tens, 3 ones

What does addition mean? joining or putting sets together

Which digits should you add first? Add the 6 and the 3 together because they are in the Ones place.

Direct the students to move all the ones together while you demonstrate.

How many ones are there in all? 9

Write "9" in the Ones place.

Direct the students to move all the tens together on their mats while you demonstrate.

How many tens are there in all? 4

Write "4" in the Tens place.

Direct the students to move all the hundreds together while you demonstrate.

How many hundreds are there in all? 4

Write "4" in the Hundreds place.

What is the answer? 449

Model the following addition problems while the students do them on their mats. Emphasize adding the Ones place first, the Tens place next, and the Hundreds place last.

$$\begin{array}{r} 324 \\ +\ 462 \\ \hline 786 \end{array} \qquad \begin{array}{r} 230 \\ +\ 448 \\ \hline 678 \end{array} \qquad \begin{array}{r} 325 \\ +\ 532 \\ \hline 857 \end{array}$$

Solve a Word Problem

Use the **Problem-Solving Plan** to solve a word problem.

Display Visual 21 and read aloud the following word problem several times.

> The hospital cafeteria made 84 lunches for Tuesday, but only 61 lunches were sold. How many lunches did not sell?

What is the question in this word problem? How many lunches did not sell?

What information is given? 84 lunches were made and 61 were sold.

Do you add or subtract to find the answer? I subtract to find out how many lunches were left.

What equation can be used to solve this problem? 84 – 61 = __

Write "84 – 61 = __" for display.

Should the equation be written in vertical form? Yes, it is easier to subtract 2-digit numbers in vertical form.

Invite a student to write the problem in vertical form.

Ask a volunteer to solve the problem. 23 Write "23 lunches" to complete the equation.

Does the answer make sense? yes

Repeat the procedure for the following word problem.

> Last week the hospital clinic saw 295 patients. This week the clinic saw 302 patients. How many patients visited the hospital clinic in the past 2 weeks?

What is the question in this word problem? How many patients visited the hospital clinic in the past 2 weeks?

175

What information is given? 295 patients came to the clinic last week, and 302 patients came this week.

Do you add or subtract to find the answer? I add to find out the total number of patients for the past 2 weeks.

What equation can be used so solve this problem? 295 + 302 = __

Write "295 + 302 = __" for display.

Should the equation be written in vertical form? Yes, it is easier to add 3-digit numbers in vertical form.

Invite a student to write the problem in vertical form.

Ask a volunteer to solve the problem. 597 Write "597 patients" to complete the equation.

Does the answer make sense? yes

Use an Equation to Do Work

Guide a **discussion** to help the students apply a biblical worldview shaping truth.

Read aloud the following word problem.

Pretend you are going on a trip. There are 36 bottles of water on the table, and you must be sure to leave 15 for the next group. How many water bottles can you pack into your cooler?

What is the question? How many water bottles can you pack into your cooler?

What information is given? There are 36 bottles of water on the table, and I must leave 15.

Do you add or subtract to find the answer? Subtract; I am finding how many bottles I can pack after I set aside 15.

What equation can be used? 36 − 15 = __

Write the problem for display and invite a student to write the problem in vertical form and solve it. 21

Write "21 bottles of water" to complete the equation.

Remind the students that work such as finding out how many bottles can be packed into a cooler is accomplishing a task. Math makes this kind of work possible.

Point out that the more students understand math, the more opportunities they have to finish different kinds of tasks that God gives them to do.

Problem Solving

Add.

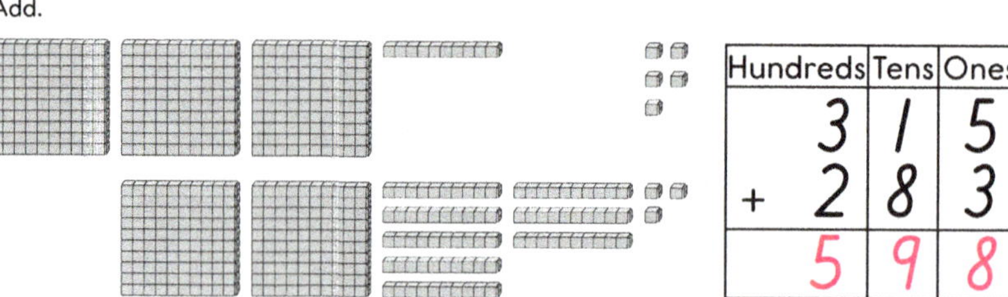

Hundreds	Tens	Ones
3	1	5
+ 2	8	3
5	9	8

Hundreds	Tens	Ones
6	3	4
+ 2	0	5
8	3	9

Hundreds	Tens	Ones
1	4	4
+ 7	2	3
8	6	7

Add.

503 + 472 = 975

464 + 234 = 698

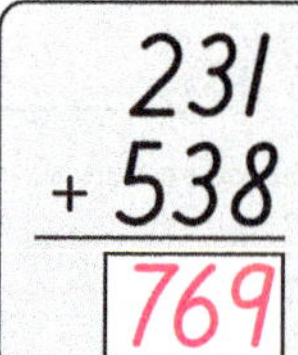

231 + 538 = 769

491 + 305 = 796

Write an equation for the word problem. Solve.

The hiker walked 362 miles last year.
This year he walked 435 miles.
How many miles did he walk in both years?

362 + 435 = 797 miles

work space

362
+ 435
797

Apply

Worktext pages 335–36

Read and guide completion of page 335.

Read and explain the directions for page 336. Assist the students as they complete the page independently.

Direct attention to the biblical worldview shaping statement and guide the students as they complete the sentence.

Assess

Reviews pages 323–24

Review writing a less-than or greater-than sign to compare 2 numbers on page 324.

Fact Reviews

Use "Subtraction Facts to 18" page 59 in Trove.

Extended Activity

Add 3-Digit Numbers

Materials

- A list of the students' addresses

Procedure

Use the students' addresses to write addition problems on the board. Refer to the students whose addresses you use (for example, Mia lives at 201 Main Street, and James lives at 125 Court Street. 201 + 125). Ask a volunteer to solve the problem (or allow Mia and James to solve it).

Chapter 16 Review

Draw a dot next to the smaller number.
Draw the correct sign.

< less than   > greater than

363 < 413	563 > 365
900 > 500	275 < 375
121 < 122	834 > 814

Subtraction Fact Review

Subtract.

15 − 8 = 7	14 − 6 = 8	17 − 8 = 9
17 − 9 = 8	15 − 7 = 8	14 − 8 = 6
13 − 5 = 8	9 − 9 = 0	13 − 8 = 5

176 Worktext pages 337–38 Reviews pages 325–26

Objectives

- **176.1** **Rename 10 ones as 1 ten by using manipulatives.**
- **176.2** **Add 3-digit numbers with and without renaming.**

Printed Resources

- Visuals: Place Value Kit (hundreds, tens, ones)
- Student Manipulatives: Hundreds/Tens/Ones Mat
- Student Manipulatives: Place Value Kit (hundreds, tens, ones)

Digital Resources

- Web Link: Virtual Manipulatives: Base Ten Number Pieces

Lessons 176–78 are included as enrichment instruction. Present these lessons based on your schedule and the readiness of the students.

Renaming 10 Ones

Add. Rename if needed.
Circle *yes* if you renamed.
Circle *no* if you did not rename.

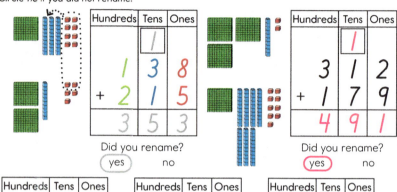

	Hundreds	Tens	Ones
		1	
	1	3	8
+	2	1	5
	3	5	3

Did you rename? (yes) no

	Hundreds	Tens	Ones
		1	
	3	1	2
+	1	7	9
	4	9	1

Did you rename? (yes) no

	Hundreds	Tens	Ones
		1	
	4	2	7
+	2	5	4
	6	8	1

Did you rename? (yes) no

	Hundreds	Tens	Ones
	6	4	5
+	2	4	2
	8	8	7

Did you rename? yes (no)

	Hundreds	Tens	Ones
		1	
	3	1	6
+	6	2	4
	9	4	0

Did you rename? (yes) no

	Hundreds	Tens	Ones
	1	5	6
+	7	4	3
	8	9	9

Did you rename? yes (no)

	Hundreds	Tens	Ones
		1	
	1	2	5
+	1	4	6
	2	7	1

Did you rename? (yes) no

	Hundreds	Tens	Ones
		1	
	2	7	3
+	7	1	9
	9	9	2

Did you rename? (yes) no

Engage

God Made People Special

Use **direct instruction** to explain a biblical truth.

Explain that God has made everybody special; in fact, God made the world in such a way that if all the people were the same, things would not work right at all. Remind the students that God showed His infinite love and wisdom in the way that He made them. Some children must spend many days in a hospital, and other children do not; some are better at playing soccer, and others are better at reading or doing math problems. Point out that God has a special plan for each person, and He wants each one to become a child of God. Read aloud 2 Peter 3:9.

Instruct

Add 3-Digit Numbers with & without Renaming

Model adding 3-digit numbers.

Distribute the Hundreds/Tens/Ones Mats and the Place Value Kits. Write "136 + 257" for display as shown.

	Hundreds	Tens	Ones
		☐	
	1	3	6
+	2	5	7

Display 1 hundred, 3 tens, and 6 ones in a Hundreds/Tens/Ones frame. Direct the students to use their hundreds, tens, and ones to make the first addend near the top of their mats.

How many hundreds, tens, and ones are in 136? 1 hundred, 3 tens, 6 ones

Display 2 hundreds, 5 tens, and 7 ones below the first addend. Direct the students to make the second addend near the bottom of their mats.

How many hundreds, tens, and ones are in 257? 2 hundreds, 5 tens, 7 ones

Direct the students to move all the ones together while you demonstrate.

How many ones are there in all? 13

Explain that when there are 10 or more ones, students can rename or trade 10 ones for 1 ten. Remove 10 ones from the Ones place and put 1 ten at the top of the Tens place. Point out that the 10 ones have received a new name and have been moved to a new home in the Tens place.

Direct the students to remove 10 ones, trade them for 1 ten, and place the 1 ten at the top of the Tens place.

What do you add after the ones have been added? the tens

Direct the students to move all the tens together while you demonstrate.

How many tens are there in all? 9

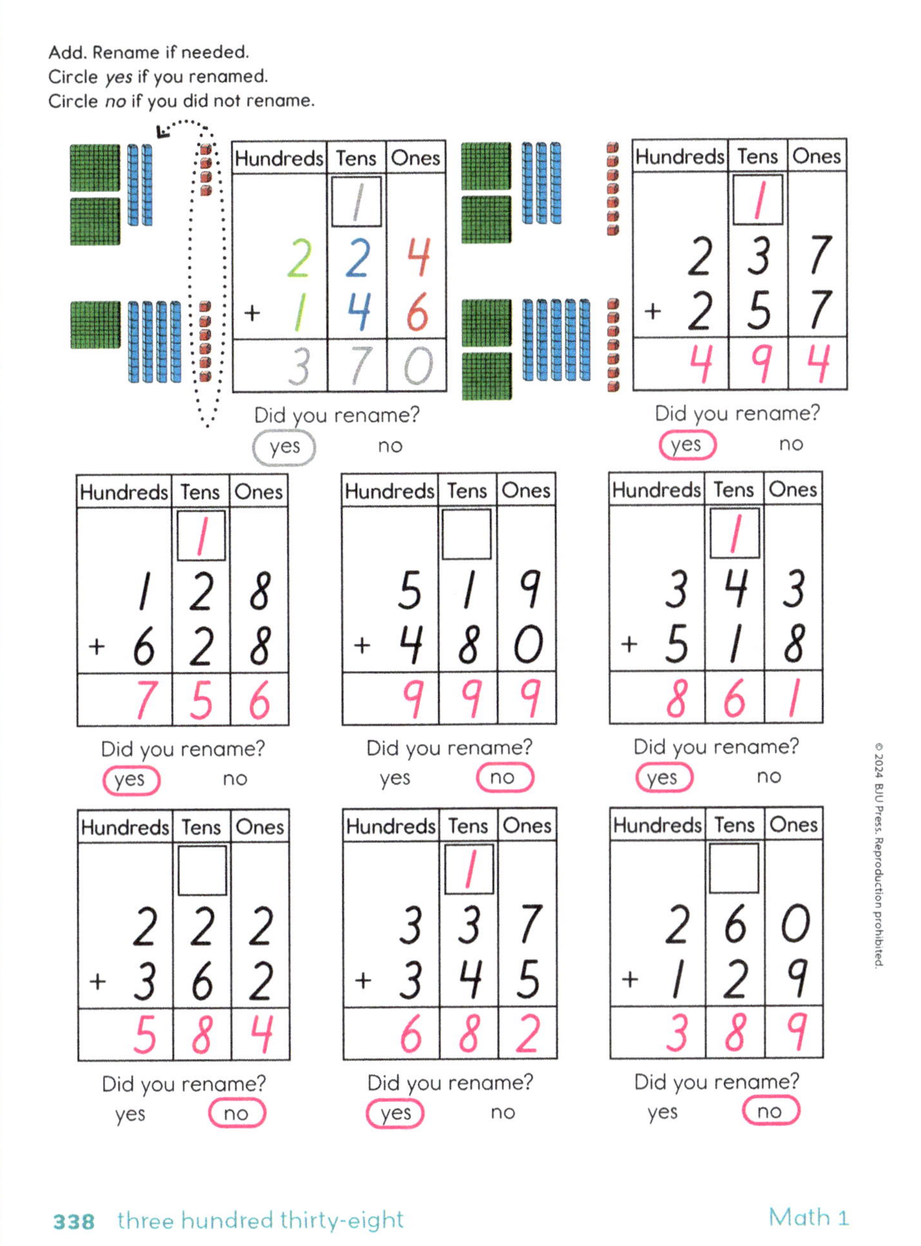
Add. Rename if needed.
Circle *yes* if you renamed.
Circle *no* if you did not rename.

Hundreds	Tens	Ones
	1	
2	2	4
+ 1	4	6
3	7	0

Did you rename? (yes) no

Hundreds	Tens	Ones
	1	
2	3	7
+ 2	5	7
4	9	4

Did you rename? (yes) no

Hundreds	Tens	Ones
	1	
1	2	8
+ 6	2	8
7	5	6

Did you rename? (yes) no

Hundreds	Tens	Ones
5	1	9
+ 4	8	0
9	9	9

Did you rename? yes (no)

Hundreds	Tens	Ones
	1	
3	4	3
+ 5	1	8
8	6	1

Did you rename? (yes) no

Hundreds	Tens	Ones
2	2	2
+ 3	6	2
5	8	4

Did you rename? yes (no)

Hundreds	Tens	Ones
	1	
3	3	7
+ 3	4	5
6	8	2

Did you rename? (yes) no

Hundreds	Tens	Ones
2	6	0
+ 1	2	9
3	8	9

Did you rename? yes (no)

338 three hundred thirty-eight Math 1

Direct the students to move all the hundreds together while you demonstrate.

How many hundreds are there in all? 3

Direct attention to the problem written for display.

Explain that when 6 ones are added to 7 ones, the answer is 13 ones, but you cannot write 13 in the Ones place. Point out that when 10 ones were renamed as 1 ten, there were 3 ones left in the Ones place. Write "3" in the Ones place of the answer.

What happened to the 1 ten that was traded for 10 ones? It was put in the Tens place.

Write "1" in the box in the Tens place to represent the 10 ones that became 1 ten.

What is 1 ten + 3 tens + 5 tens? 9 tens

Write "9" in the Tens place.

What is 1 hundred + 2 hundreds? 3 hundreds

Write "3" in the Hundreds place and read aloud the completed problem: "136 + 257 = 393."

Model the following addition problems while the students do them on their mats. Emphasize that they need to rename only when there are 10 or more ones in the Ones place.

124	236	325
+ 236	+ 413	+ 147
360	649	472

176

Apply

Worktext pages 337–38

Read and guide completion of page 337.

Read and explain the directions for page 338. Assist the students as they complete the page independently.

Assess

Reviews pages 325–26

Read the directions and guide completion of the pages.

Extended Activity

Complete a "Magic" Square by Adding 3-Digit Numbers

Preparation

Draw a grid for display; write the 4 black numbers as shown. Do not write the answers in the grid.

200	35	235
150	50	200
350	85	435

Procedure

Instruct the students to copy the grid on their papers. Explain that this is a special square that contains 6 different addition problems.

Direct attention to the first 2 numbers on the top row and direct the students to write and solve the problem "200 + 35" at the bottom of their papers and then to write the answer in the empty space on the top row. 235

Direct the students to write and solve the problem "150 + 50" and then to write the answer in the empty space on the second row. 200

Repeat the procedure for the 3 vertical problems. Explain that they will know that all their answers are correct if the answer to the problem in the third row is the same as the answer to the problem in the third column.

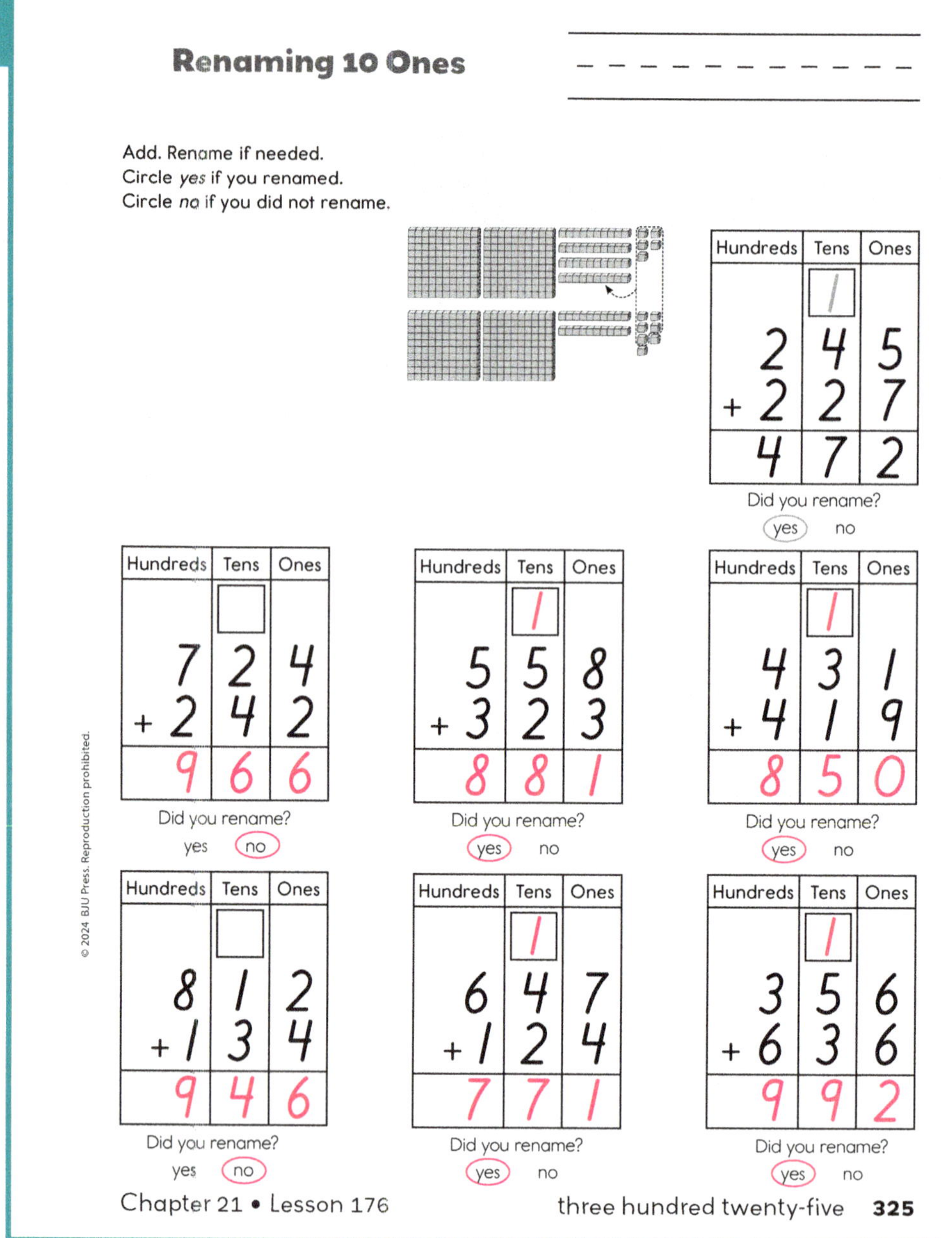

Renaming 10 Ones

Add. Rename if needed.
Circle *yes* if you renamed.
Circle *no* if you did not rename.

Hundreds	Tens	Ones
	1	
2	4	5
+ 2	2	7
4	7	2

Did you rename? (yes) no

Hundreds	Tens	Ones
7	2	4
+ 2	4	2
9	6	6

Did you rename? yes (no)

Hundreds	Tens	Ones
	1	
5	5	8
+ 3	2	3
8	8	1

Did you rename? (yes) no

Hundreds	Tens	Ones
	1	
4	3	1
+ 4	1	9
8	5	0

Did you rename? (yes) no

Hundreds	Tens	Ones
8	1	2
+ 1	3	4
9	4	6

Did you rename? yes (no)

Hundreds	Tens	Ones
	1	
6	4	7
+ 1	2	4
7	7	1

Did you rename? (yes) no

Hundreds	Tens	Ones
	1	
3	5	6
+ 6	3	6
9	9	2

Did you rename? (yes) no

Add. Rename if needed.
Circle *yes* if you renamed.
Circle *no* if you did not rename.

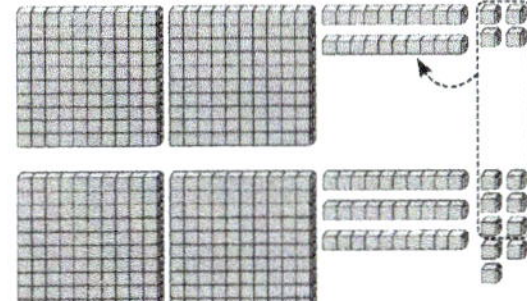

Hundreds	Tens	Ones
	1	
2	2	4
+ 2	3	9
4	6	3

Did you rename?
yes no

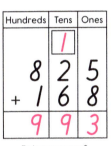

Hundreds	Tens	Ones
	1	
6	3	6
+ 2	4	4
8	8	0

Did you rename?
yes no

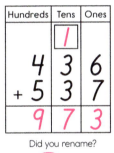

Hundreds	Tens	Ones
5	2	3
+ 3	1	1
8	3	4

Did you rename?
yes no

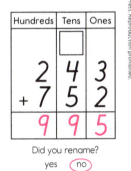

Hundreds	Tens	Ones
	1	
4	5	7
+ 4	3	8
8	9	5

Did you rename?
yes no

Hundreds	Tens	Ones
	1	
8	2	5
+ 1	6	8
9	9	3

Did you rename?
yes no

Hundreds	Tens	Ones
	1	
4	3	6
+ 5	3	7
9	7	3

Did you rename?
yes no

Hundreds	Tens	Ones
2	4	3
+ 7	5	2
9	9	5

Did you rename?
yes no

Worktext pages 339–40
Reviews pages 327–28

Objectives

- **177–78.1** Add 3-digit numbers with and without renaming.
- **177–78.2** Solve a word problem by writing an addition equation.

Printed Resources

- Visual 21: *Problem-Solving Plan*
- Visuals: Place Value Kit (hundreds, tens, ones)
- Student Manipulatives: Hundreds/Tens/Ones Mat
- Student Manipulatives: Place Value Kit (hundreds, tens, ones)

Digital Resources

- Web Link: Virtual Manipulatives: Base Ten Number Pieces

Depending on your schedule and the students' abilities, you may take either 1 or 2 days to present Lessons 177–78.

Adding Hundreds with Renaming

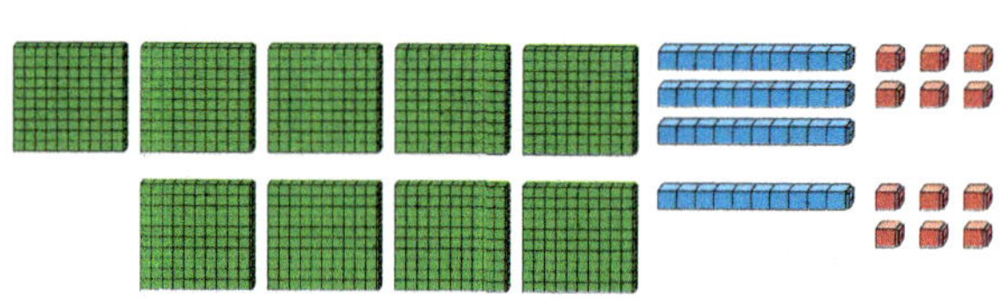

	Hundreds	Tens	Ones
		1	
	5	3	6
+	4	1	6
	9	5	2

Add. Rename if needed.
Circle *yes* if you renamed.
Circle *no* if you did not rename.

	Hundreds	Tens	Ones
		1	
	2	2	7
+	4	3	4
	6	6	1

Did you rename? (yes) no

Did you rename? yes (no)

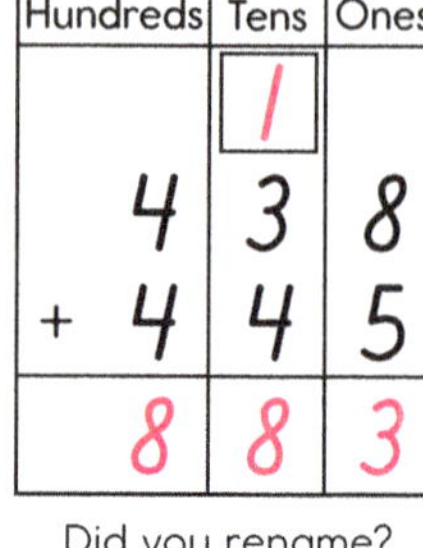

Hundreds Tens Ones
1
4 3 8
+ 4 4 5
8 8 3

Did you rename? (yes) no

	Hundreds	Tens	Ones
	6	2	6
+	3	1	2
	9	3	8

Did you rename? yes (no)

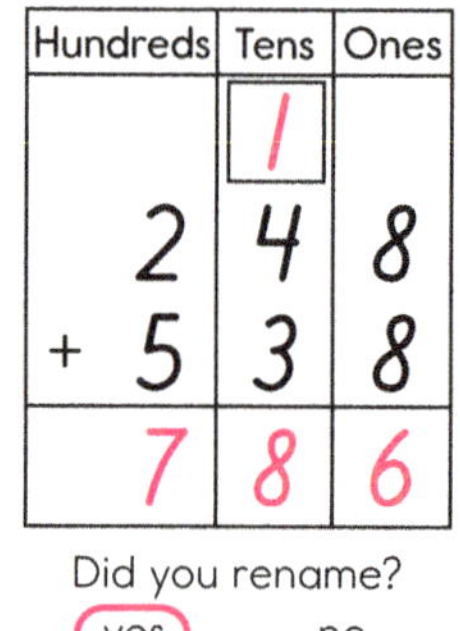

Hundreds Tens Ones
1
2 4 8
+ 5 3 8
7 8 6

Did you rename? (yes) no

	Hundreds	Tens	Ones
		1	
	6	1	7
+	1	3	5
	7	5	2

Did you rename? (yes) no

Engage

Numbers in the Bible

Guide a **discussion** to help the students determine Adam's age at death.

Explain that the Bible has examples of 3-digit numbers. Read aloud Genesis 5:3.

How old was Adam when his son Seth was born? 130 years old

Write "130" for display. Read aloud Genesis 5:4.

How long did Adam live after Seth was born? 800 years

Write "800" below 130. Explain that God told Adam that he would die, and his body would return to dust because of his sin in the Garden of Eden.

If you want to find out how old Adam was when he died, would you add or subtract these numbers? add

Complete the problem by writing a plus sign in front of the 800 and a line below the 800. Ask a volunteer to solve the problem. 930

How old was Adam when he died? 930 years old

Read aloud Genesis 5:5.

Read aloud Genesis 5:6–8 and continue the activity to write an addition problem to find out how old Seth was when he died. 105 + 807 = 912

Instruct

Add 3-Digit Numbers with & without Renaming

Model adding 3-digit numbers.

Distribute the Hundreds/Tens/Ones Mats and the Place Value Kits. Write "245 + 126" for display as shown.

Draw a Hundreds/Tens/Ones frame. Direct the students to use their hundreds, tens, and ones to make both addends on their mats as you demonstrate.

How many hundreds, tens, and ones are in 245? 2 hundreds, 4 tens, 5 ones

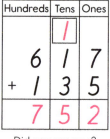

	Hundreds	Tens	Ones
		□	
	2	4	5
+	1	2	6

How many hundreds, tens, and ones are in 126? 1 hundred, 2 tens, 6 ones

What do you add first? the ones

Direct the students to move all the ones together while you demonstrate.

Add. Rename if needed.

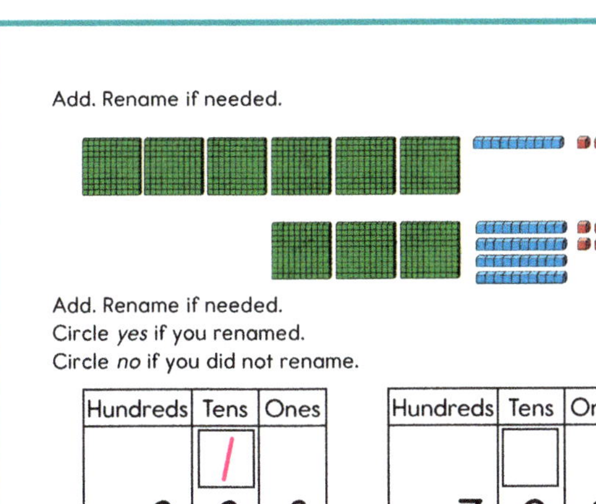

Hundreds	Tens	Ones
	1	
6	1	5
+ 3	4	7
9	6	2

Add. Rename if needed.
Circle *yes* if you renamed.
Circle *no* if you did not rename.

Hundreds	Tens	Ones
	1	
3	3	8
+ 4	1	4
7	5	2

Did you rename?
(yes) no

Hundreds	Tens	Ones
7	2	3
+ 1	5	6
8	7	9

Did you rename?
yes (no)

Hundreds	Tens	Ones
	1	
2	4	7
+ 4	3	5
6	8	2

Did you rename?
(yes) no

Write an equation for the word problem. Solve.

The hospital workers served 519 meals in a week. The next week, they served 356 meals. How many meals did the workers serve?

519 + 356 = 875 meals

Hundreds	Tens	Ones
	1	
5	1	9
+ 3	5	6
8	7	5

How many ones are there in all? 11

What do you do when you have 10 or more ones in the Ones place? rename or trade 10 ones for 1 ten

Direct the students to remove 10 ones, trade them for 1 ten, and then place the 1 ten at the top of the Tens place as you demonstrate.

What do you add after the ones have been added? the tens

Direct the students to move all the tens together while you demonstrate.

How many tens are there in all? 7

Direct the students to move all the hundreds together while you demonstrate.

How many hundreds are there in all? 3

Direct attention to the problem written for display. Ask a volunteer to solve the problem.

Why was a 1 written in the box above the Tens place? It represents the 1 ten that was renamed from the 10 ones.

Read aloud the completed problem: "245 + 126 = 371."

Model the following addition problems while the students do them on their mats. Emphasize that they need to rename only when there are 10 or more ones in the Ones place.

$$\begin{array}{r}342\\+\,136\\\hline 478\end{array}\qquad\begin{array}{r}257\\+\,308\\\hline 565\end{array}\qquad\begin{array}{r}235\\+\,248\\\hline 483\end{array}$$

Solve a Word Problem

Use the **Problem-Solving Plan** to solve a word problem.

What are the steps for solving word problems?

Write the students' answers for display.

Display Visual 21 and compare it to the students' answers; then read aloud the following word problem.

On Monday, Floor 1 of the hospital had 145 patients. Floor 2 had 238 patients. How many patients were on these floors on Monday?

What is the question in this word problem? How many patients were on these floors on Monday?

What information is given? Floor 1 had 145 patients, and Floor 2 had 238 patients.

Do you add or subtract to find the answer? I add because I am finding the total number of patients on the 2 floors.

What equation is needed to solve this problem? 145 + 238 = __

Write "145 + 238 = __" for display.

Should the equation be written in vertical form? Yes, it is easier to add 3-digit numbers in vertical form.

Invite a student to write the equation in vertical form.

What do you add first when there are 3-digit addends? the ones

Ask a volunteer to solve the problem. 383 Write "383 patients" to complete the equation.

Does the answer make sense? yes

Continue the activity with the following word problem.

The patients on Floor 2 had 87 visitors. The patients on Floor 1 had 45 visitors. How many more people visited the patients on Floor 2 than on Floor 1? 87 – 45 = 42 visitors

177–78

Apply

Worktext pages 339–40

Read and guide completion of page 339.

Read and explain the directions for page 340. Assist the students as they complete the page independently.

Assess

Reviews pages 327–28

Read the directions and guide completion of the pages.

Adding Hundreds with Renaming

Add. Rename if needed.

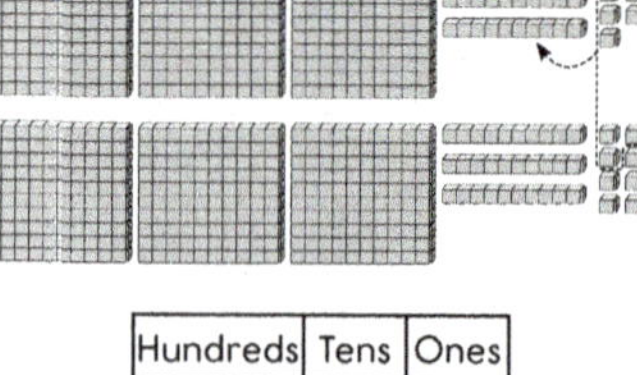

Hundreds	Tens	Ones
	1	
3	3	7
+ 3	3	8
6	7	5

Hundreds	Tens	Ones
	1	
5	3	7
+ 2	4	7
7	8	4

Did you rename? (yes) no

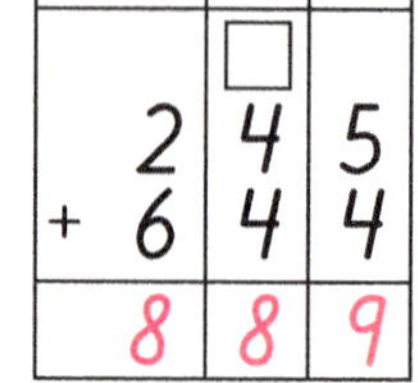

Hundreds	Tens	Ones
2	4	5
+ 6	4	4
8	8	9

Did you rename? yes (no)

Hundreds	Tens	Ones
	1	
4	2	7
+ 5	2	5
9	5	2

Did you rename? (yes) no

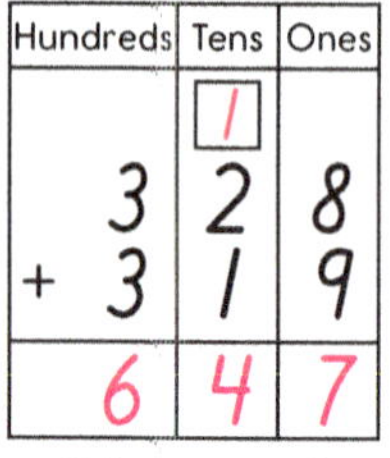

Hundreds	Tens	Ones
	1	
3	2	8
+ 3	1	9
6	4	7

Did you rename? (yes) no

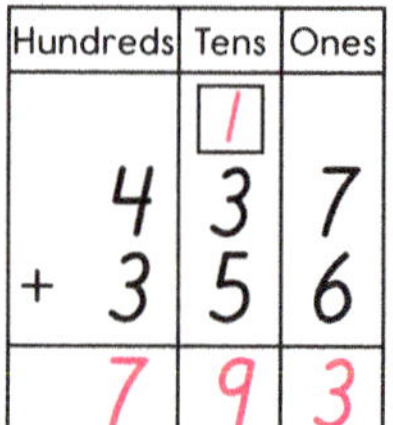

Hundreds	Tens	Ones
	1	
4	3	7
+ 3	5	6
7	9	3

Did you rename? (yes) no

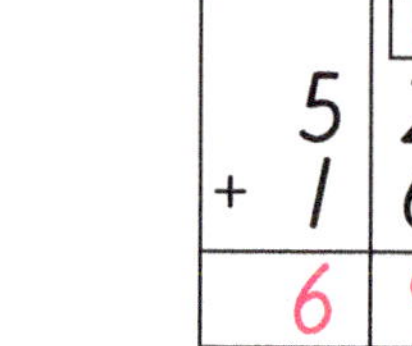

Hundreds	Tens	Ones
	1	
5	2	9
+ 1	6	4
6	9	3

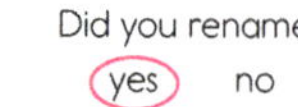

Did you rename? (yes) no

Write an equation for the word problem. Solve.

Dan sold 347 tickets.

Ann sold 134 tickets.

How many tickets did they sell in all?

347 + 134 = 481 tickets

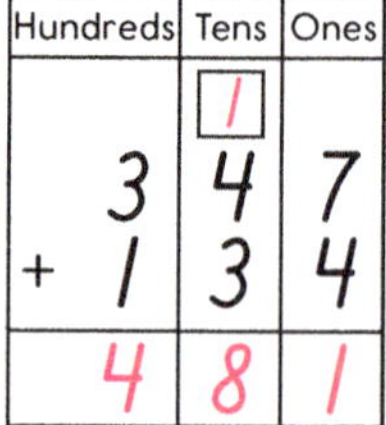

Hundreds	Tens	Ones
	1	
3	4	7
+ 1	3	4
4	8	1

Add. Rename if needed.

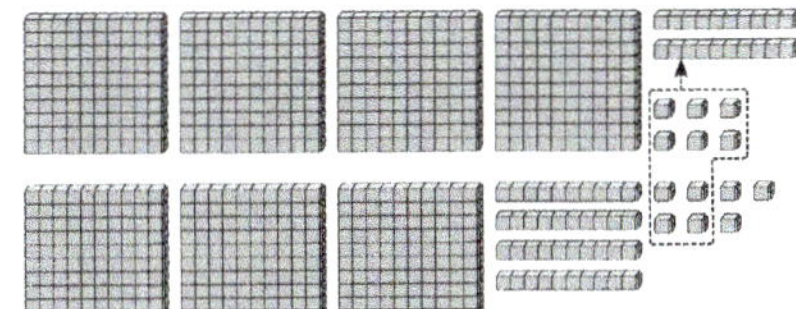

Hundreds	Tens	Ones
	1	
4	2	6
+ 3	4	7
7	7	3

Add. Rename if needed.
Circle *yes* if you renamed.
Circle *no* if you did not rename.

Hundreds	Tens	Ones
	1	
3	4	5
+ 6	2	7
9	7	2

Did you rename? (yes) no

Hundreds	Tens	Ones
4	6	4
+ 2	1	5
6	7	9

Did you rename? yes (no)

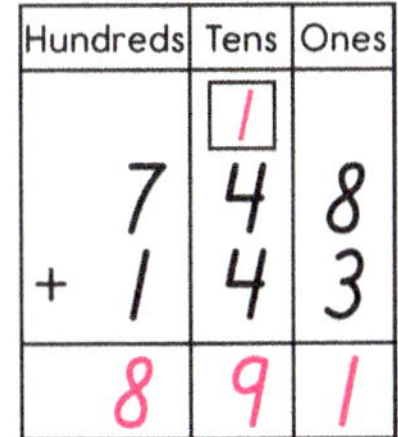

Hundreds	Tens	Ones
	1	
7	4	8
+ 1	4	3
8	9	1

Did you rename? (yes) no

Hundreds	Tens	Ones
4	2	8
+ 3	6	1
7	8	9

Did you rename? yes (no)

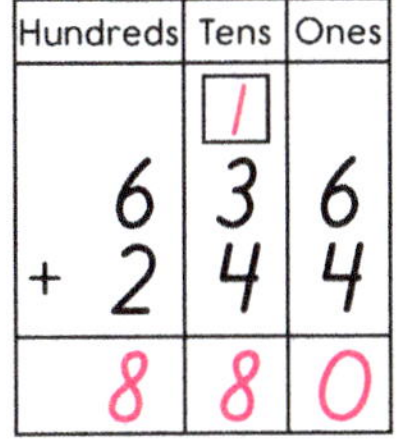

Hundreds	Tens	Ones
	1	
6	3	6
+ 2	4	4
8	8	0

Did you rename? (yes) no

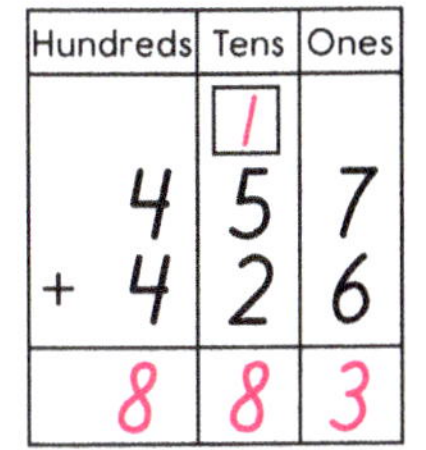

Hundreds	Tens	Ones
	1	
4	5	7
+ 4	2	6
8	8	3

Did you rename? (yes) no

Write an equation for the word problem. Solve.

On Friday 645 people came to the school play.
On Saturday 347 people came to the school play.
How many people came in all?

645  347 = 992 people

Hundreds	Tens	Ones
	1	
6	4	5
+ 3	4	7
9	9	2

179 Worktext pages 341–42 Reviews pages 329–30

How can 3-digit addition help me complete my work?

Chapter Concept Review

- Practice concepts from Chapter 21 to prepare for the test.

Printed Resources

- Instructional Aid 2: *Hundred Chart* (for each student)
- Visual 1: *Hundred Chart*
- Student Manipulatives: Hundreds/Tens/Ones Mat
- Student Manipulatives: Place Value Kit (hundreds, tens, ones)

Digital Resources

- Games/Enrichment: Fact Reviews (Subtraction Facts to 18, page 60)
- Games/Enrichment: Addition Flashcards
- Games/Enrichment: Subtraction Flashcards
- Games/Enrichment: Fact Fun Activities

Chapter Review

Add.

Hundreds	Tens	Ones
2	6	1
+ 1	3	7
3	9	8

Hundreds	Tens	Ones
4	2	5
+ 5	3	2
9	5	7

Add.

Hundreds	Tens	Ones
2	5	0
+ 3	4	5
5	9	5

Hundreds	Tens	Ones
3	7	4
+ 5	1	2
8	8	6

Hundreds	Tens	Ones
5	4	2
+ 4	3	2
9	7	4

Hundreds	Tens	Ones
6	3	2
+ 3	4	2
9	7	4

Hundreds	Tens	Ones
1	6	3
+ 8	2	3
9	8	6

Hundreds	Tens	Ones
3	5	9
+ 2	3	0
5	8	9

Practice & Review

Count by 2s to 50

Display Visual 1 and distribute the *Hundred Chart* pages. Remind the students that when they count by 2s, they skip a number and count every other number. Direct the students to point to the number 2 while you point on the visual. Guide the students as they count together by 2s as they slowly point to every other number. Continue the activity as you count together again.

Count by 3s to 30

Remind the students to count every third number when they count by 3s. Direct the students to point to the number 3 while you point on the visual. Direct the students to skip 2 numbers and follow along as you count by 3s: 3, 6, 9, 12, 15, 18, 21, 24, 27, 30.

Slowly count together by 3s as they skip 2 numbers and point to every third number. Continue the activity as you count together again.

Review Addition & Subtraction Facts (13–18)

Choose a game or activity from the "Fact Fun Activities" available in Trove.

Instruct

Add 3-Digit Numbers

Divide the class into teams and guide them in a review game. Explain that each team will receive a point for each correct answer on the team members' mats.

Distribute the Hundreds/Tens/Ones Mats and the Place Value Kits. Write "256 + 123" for display. Direct the students to solve the problem on their mats, using their hundreds, tens, and ones. 379

Continue the game with the following problems.

145	203	317	425	326
+ 432	+ 246	+ 542	+ 204	+ 172
577	449	859	629	498

Write an Equation to Solve a Word Problem

Read aloud the following word problem several times. Ask the students to think about the information so that they can develop a plan to find the answer.

Liam's family was printing flyers to advertise the children's party at the hospital. They printed 262 flyers in the morning and 127 flyers in the afternoon. How many flyers were printed during the day?

Add.

$$\begin{array}{r} 674 \\ +\,205 \\ \hline 879 \end{array} \qquad \begin{array}{r} 546 \\ +\,143 \\ \hline 689 \end{array} \qquad \begin{array}{r} 351 \\ +\,438 \\ \hline 789 \end{array}$$

$$\begin{array}{r} 432 \\ +\,224 \\ \hline 656 \end{array} \qquad \begin{array}{r} 304 \\ +\,185 \\ \hline 489 \end{array} \qquad \begin{array}{r} 546 \\ +\,403 \\ \hline 949 \end{array}$$

Write an equation for each word problem. Solve.

In March 436 X-rays were taken at the hospital. In April 353 X-rays were taken. How many X-rays were taken in March and April?

436 + 353 = 789 X-rays

(work space)

$$\begin{array}{r} 436 \\ +\,353 \\ \hline 789 \end{array}$$

The nurse took 87 X-rays of people's legs. She took 43 X-rays of people's arms. How many more X-rays of legs than arms were taken?

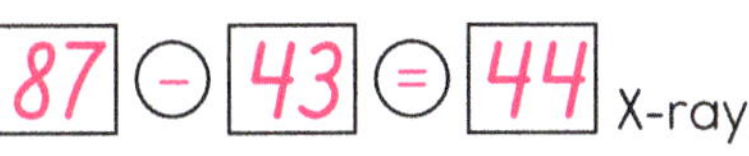

X-rays

(work space)

$$\begin{array}{r} 87 \\ -\,43 \\ \hline 44 \end{array}$$

Circle the phrase that completes the sentence.

DQ added to find how many balloons to take to the party.

Three-digit addition can tell me 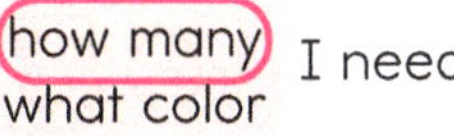 I need.

What is the question? How many flyers were printed during the day?

What information is given? Liam's family printed 262 flyers in the morning and 127 flyers in the afternoon.

Do you add or subtract to find the answer? I add because I am finding the total number of flyers printed.

What equation is needed to solve this problem? 262 + 127 = __

Write "262 + 127 = __" for display.

Should the equation be written in vertical form? Yes, it is easier to add 3-digit numbers in vertical form.

Invite a student to write the problem in vertical form.

What do you add first when there are 3-digit numbers? the ones

Ask a volunteer to solve the problem. 389 Write "389 flyers" to complete the equation.

Does this answer make sense? yes

Repeat the procedure with the following word problems.

DQ placed 55 flyers on the front desk of the hospital lobby. Later in the day, people took 45 of the flyers to read. How many flyers remained on the desk? 55 – 45 = 10 flyers

When the party started, 253 people entered. After lunch, 125 more people came. How many people attended the party in all? 253 + 125 = 378 people

Apply

Worktext pages 341–42

Read the directions and guide completion of the pages.

Direct attention to the biblical worldview shaping statement on page 342. Display the picture from the chapter opener and remind the students that DQ and his friends added large numbers. In doing so, they were able to finish the work of taking balloons to the party.

Read aloud the essential question and provide prompts as necessary to guide the students to the answer on page 342.

How can 3-digit addition help me complete my work? Three-digit addition can show me how many I need.

Assess

Reviews pages 329–30

Use the Reviews pages to provide additional preparation for the chapter test.

Fact Reviews

Use "Subtraction Facts to 18" page 60 in Trove.

Extended Activity

Illustrate 3-Digit Addition Problems

Materials

- A sheet of drawing paper for each student

If you have Place Value rubber stamps and a stamp pad, students can take turns illustrating problems by stamping the ones, tens, and hundreds.

Procedure

Distribute the drawing paper. Guide the students as they fold the paper in half lengthwise and then into thirds. Unfold the paper and write "224 + 165" in the bottom half of the paper.

Instruct the students to illustrate the addends in the top half of the paper. They should draw squares to represent the hundreds, draw lines to represent the tens, and draw dots for the ones.

179

Direct the students to solve the problem. 389

Encourage each student to make up his or her own problem and illustrate it on the back of the paper.

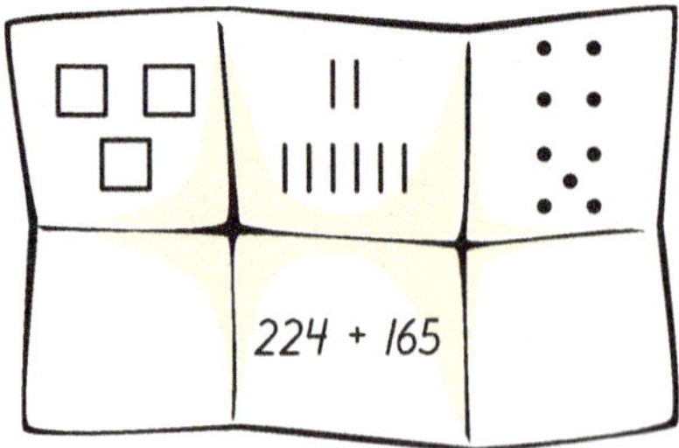

Add.

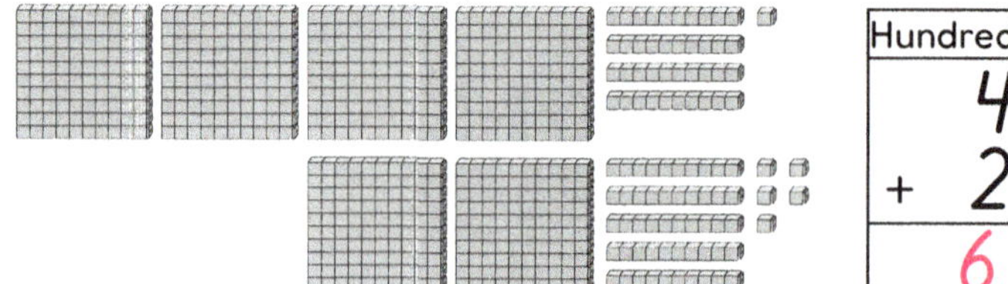

Hundreds	Tens	Ones
4	4	1
+ 2	5	5
6	9	6

Add.

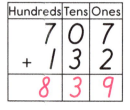

Hundreds	Tens	Ones
7	0	7
+ 1	3	2
8	3	9

Hundreds	Tens	Ones
6	1	4
+ 2	5	3
8	6	7

Hundreds	Tens	Ones
4	3	6
+ 2	0	1
6	3	7

$236 + 432 = 668$

$713 + 164 = 877$

$531 + 262 = 793$

Write an equation for the word problem. Solve.

The boys set up 250 chairs this morning.
They need to set up 125 more chairs this afternoon.
How many chairs will they set up today?

250 + 125 = 375 chairs

work space

$250 + 125 = 375$

Add.

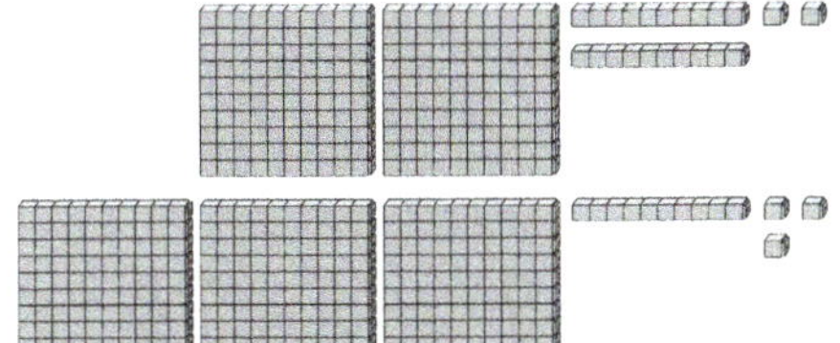

Hundreds	Tens	Ones
2	2	2
+ 3	1	3
5	3	5

Hundreds	Tens	Ones
8	1	4
+ 1	6	1
9	7	5

Hundreds	Tens	Ones
6	4	2
+ 2	2	5
8	6	7

Hundreds	Tens	Ones
8	3	5
+ 1	6	2
9	9	7

$$\begin{array}{r} 534 \\ +\,442 \\ \hline 976 \end{array} \qquad \begin{array}{r} 261 \\ +\,128 \\ \hline 389 \end{array} \qquad \begin{array}{r} 724 \\ +\,264 \\ \hline 988 \end{array}$$

Write an equation for the word problem. Solve.

The church had 345 song books.
They ordered 150 more.
How many song books do they have in all?

345   150  495 song books

work space

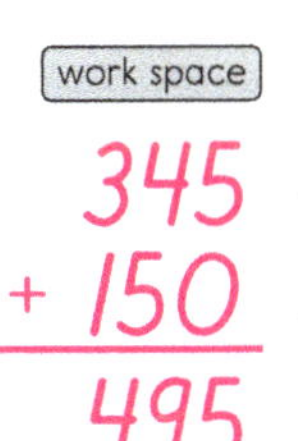

$$\begin{array}{r} 345 \\ +\,150 \\ \hline 495 \end{array}$$

180 Worktext pages 343–44 Reviews pages 331–32

Chapter 21 Test

Administer the Chapter 21 Test.

Cumulative Concept Review

Worktext pages 343–44

Review the following concepts. Adapt instructions and activities and provide reteaching as needed to meet the specific needs of your students.

- Subtracting a 2-digit number (Lessons 102–3)
- Identifying even and odd numbers (Lesson 23)
- Identifying and making one-fourth, one-half, two-thirds, and three-fourths of an object (Lessons 67–69)
- Identify one-fourth, one-half, two-thirds, and three-fourths of a set (Lesson 70)

Reviews pages 331–32

Use the Reviews pages to help students retain previously learned skills.

Subtract.

54 − 13 = 41	78 − 24 = 54	69 − 42 = 27	38 − 26 = 12
85 − 14 = 71	29 − 15 = 14	48 − 16 = 32	96 − 43 = 53

Write the number. Circle each pair of cubes.
Circle *even* if the number is even.
Circle *odd* if the number is odd.

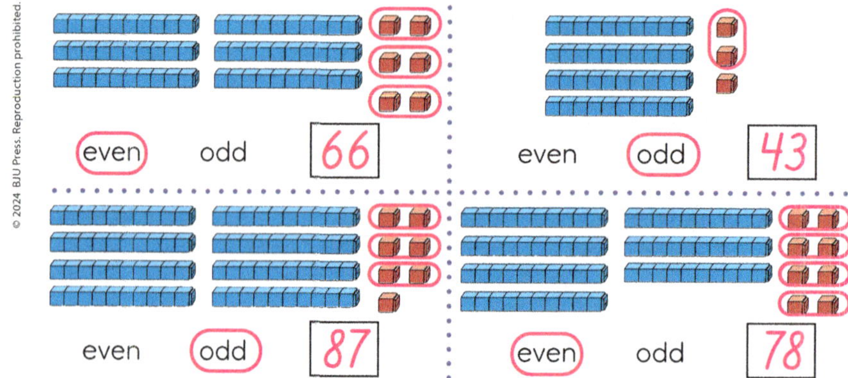

even (circled) odd 66

even odd (circled) 43

even odd (circled) 87

even (circled) odd 78

Circle the fraction that names the blue part.

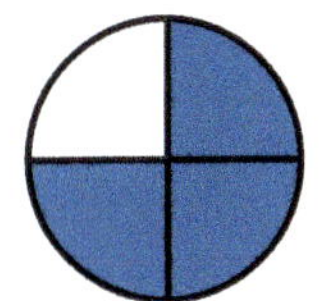 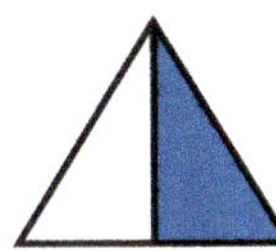 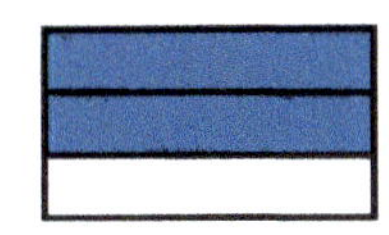 

$\frac{2}{3}$ $\frac{3}{4}$ | 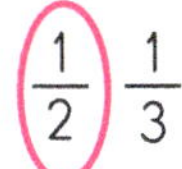 $\frac{1}{2}$ $\frac{1}{3}$ | $\frac{1}{3}$ $\frac{2}{3}$ | $\frac{1}{4}$ $\frac{1}{2}$

Color the shape to match the fraction.

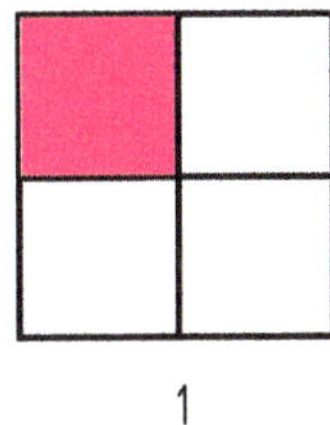   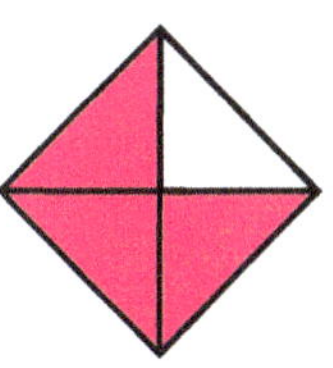

$\frac{1}{4}$ | $\frac{1}{2}$ | $\frac{2}{3}$ | $\frac{3}{4}$

Circle the fraction to identify the blue part of each set.

$\frac{1}{4}$ $\frac{1}{2}$

$\frac{1}{3}$ $\frac{1}{2}$

 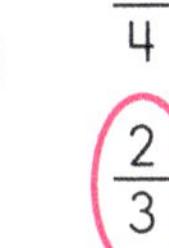 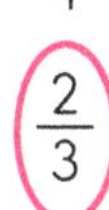

$\frac{2}{4}$ $\frac{2}{3}$

$\frac{2}{4}$ $\frac{3}{4}$

Mark an *X* in the box if the solid figure has curves, faces, or corners.

	Curves (curved sides)	Faces (flat sides)	Corners
Sphere	X		
Cone	X	X	
Cylinder	X	X	
Rectangular Prism		X	X
Pyramid		X	X

Write the number of sides and corners.

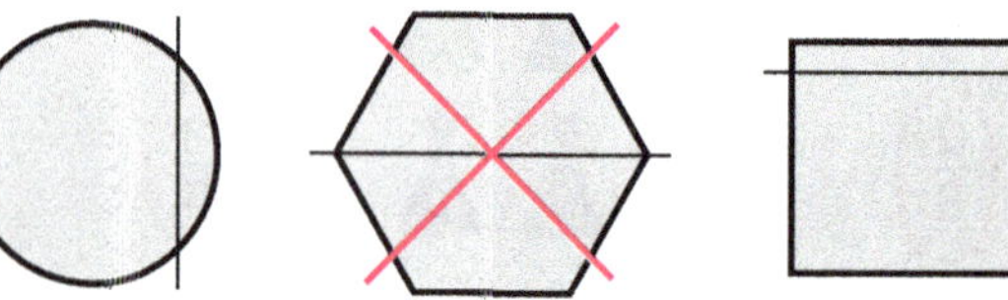

4 sides
4 corners

3 sides
3 corners

4 sides
4 corners

Look at the line on each shape.
Mark an *X* on each shape that has matching equal parts.

Subtraction Fact Review

Subtract.

12 − 9 = 3	7 − 2 = 5	6 − 4 = 2	10 − 2 = 8	12 − 5 = 7
11 − 6 = 5	10 − 8 = 2	9 − 9 = 0	15 − 8 = 7	12 − 6 = 6
12 − 3 = 9	7 − 6 = 1	12 − 8 = 4	7 − 0 = 7	4 − 3 = 1
9 − 3 = 6	13 − 5 = 8	10 − 5 = 5	5 − 3 = 2	14 − 6 = 8
10 − 1 = 9	17 − 9 = 8	5 − 5 = 0	10 − 3 = 7	7 − 5 = 2
8 − 3 = 5	11 − 5 = 6	8 − 7 = 1	9 − 2 = 7	10 − 4 = 6

Pictograph

Stamp Collections

Ken	
Meg	
Liam	
Sam	
Jill	
Lily	
Rosa	

= 5 stamps

Compass Rose

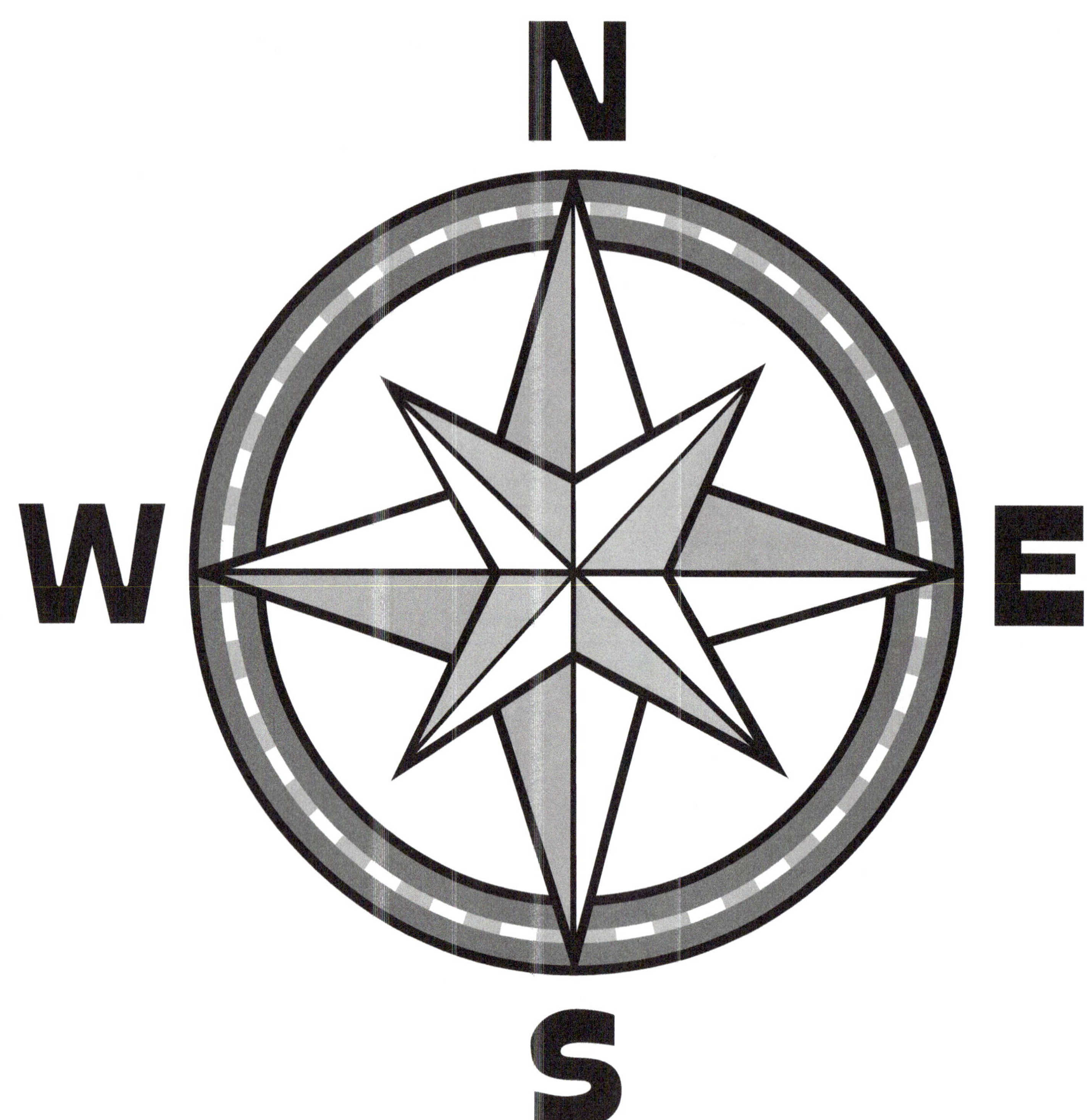

Trampoline Park Destination

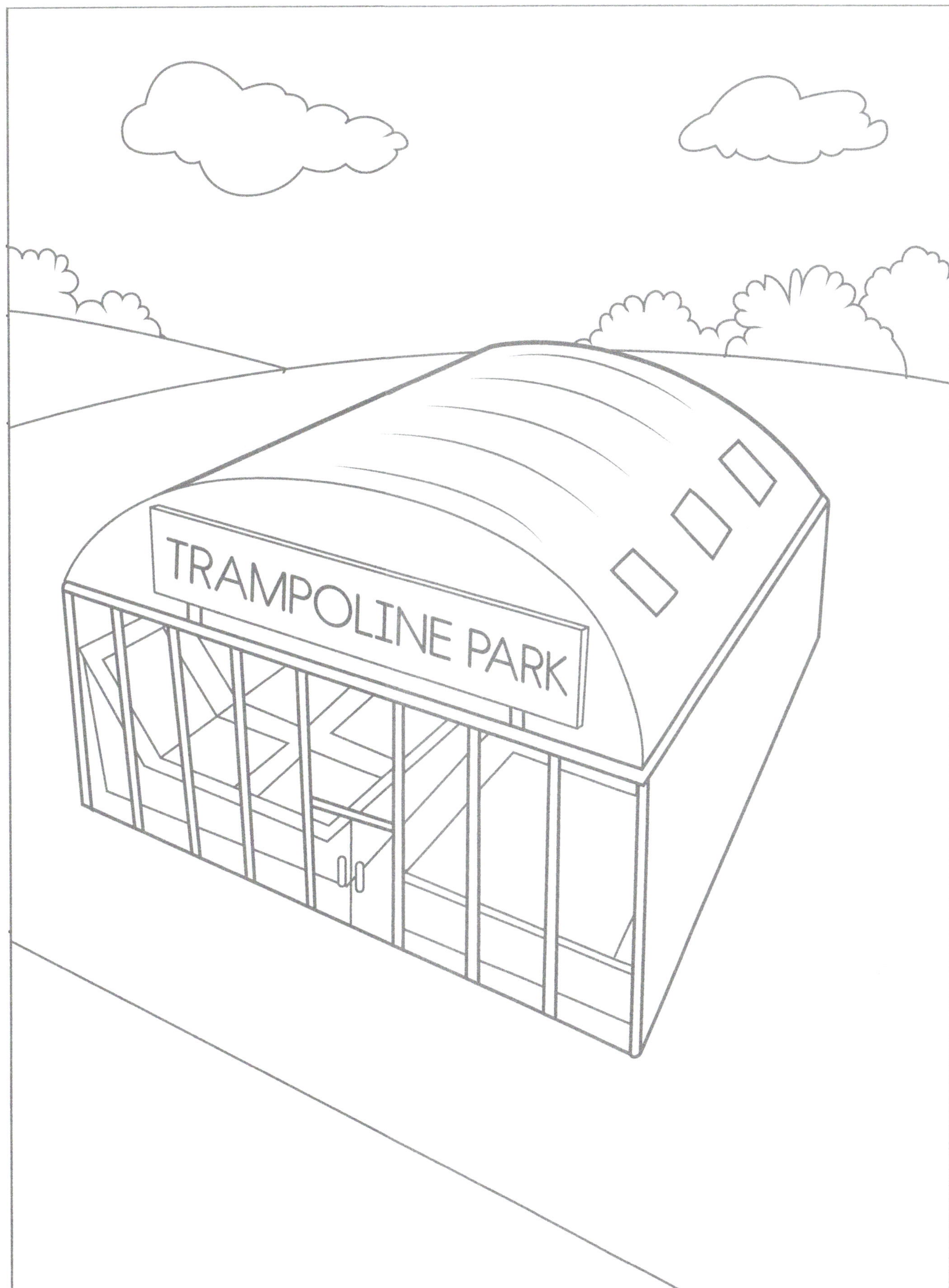

Coding Rubric: Guiding a Friend

Name: ____________________

	Excellent (3)	Good (2)	Needs Improvement (1–0)	Points Earned
The Engineering Design Process	Uses all the steps of the Engineering Design Process to solve the problem of guiding a friend to find a destination.	Uses some of the steps of the Engineering Design Process to solve the problem of guiding a friend to find a destination.	Uses few or none of the steps of the Engineering Design Process to solve the problem of guiding a friend to find a destination.	
Model	Design shows full understanding of concepts.	Design shows some understanding of concepts.	Design shows lack of understanding of concepts, or model is missing.	
Teamwork	Participates fully and shares the workload fairly.	Participates but sometimes does not share the workload fairly.	Relies on others to carry the workload most of the time or does not participate.	
Comments			**Total**	

Grocery Store Items

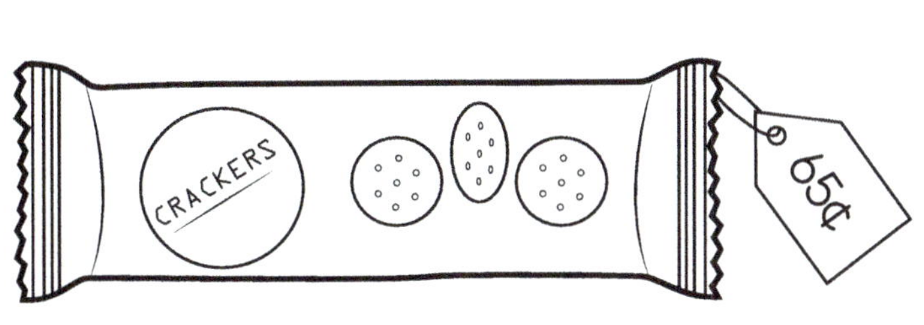

Pyramid Pattern

Copy the pattern onto regular paper, construction paper, or cardstock.

Cut along solid lines and fold on dotted lines.

Glue or tape the sides together to form a paper model.

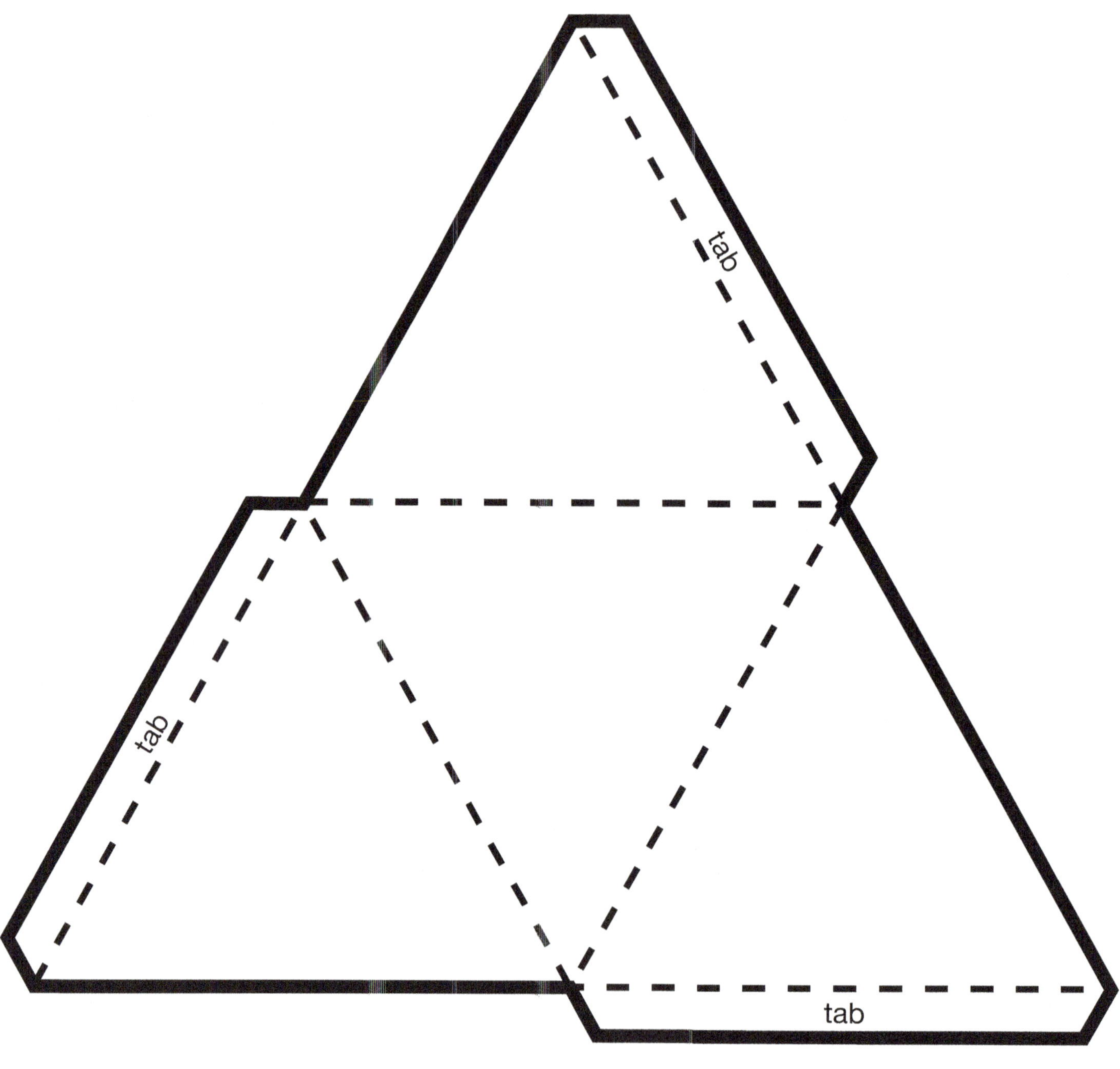

Spinner

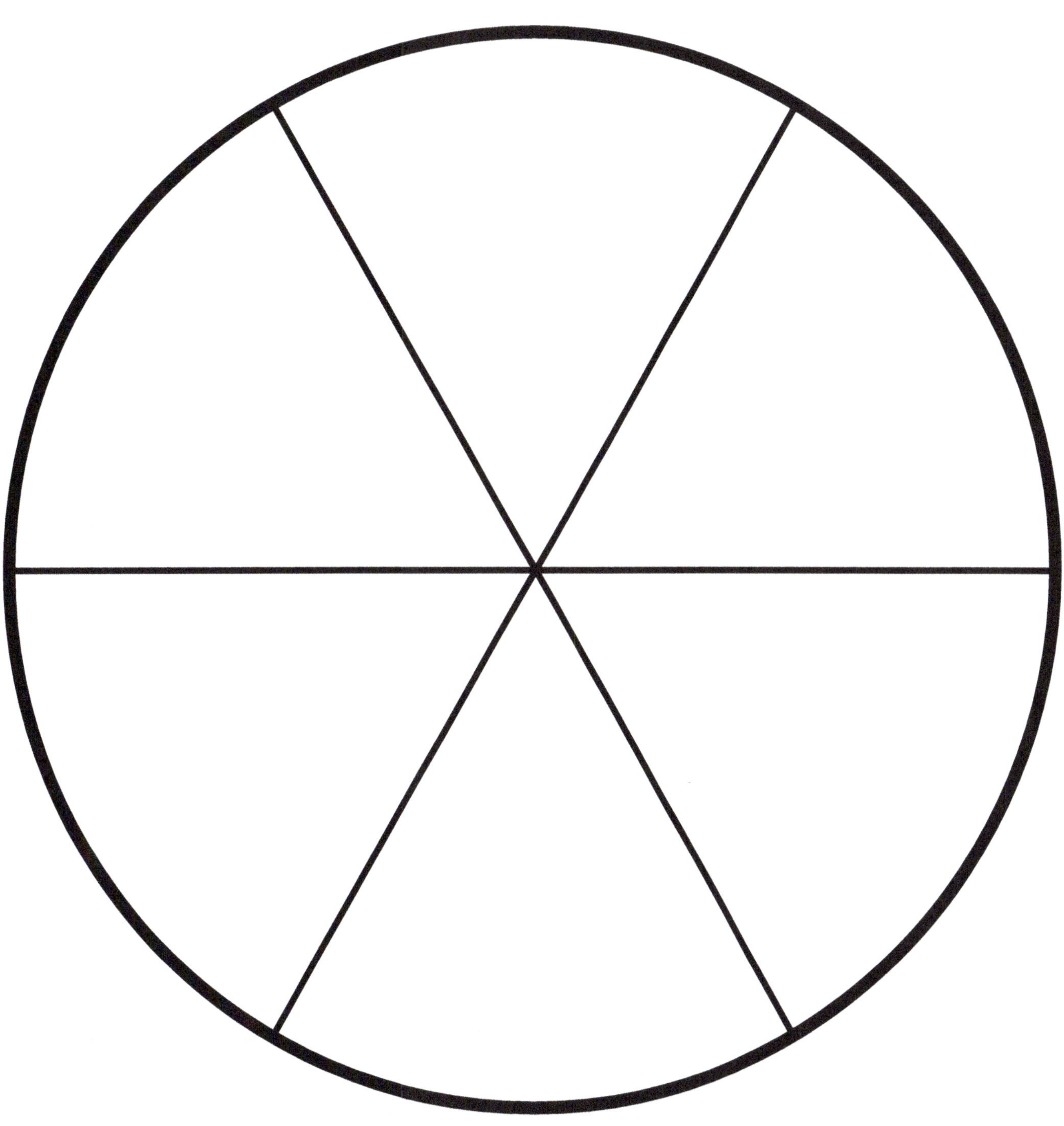

Dot Grid

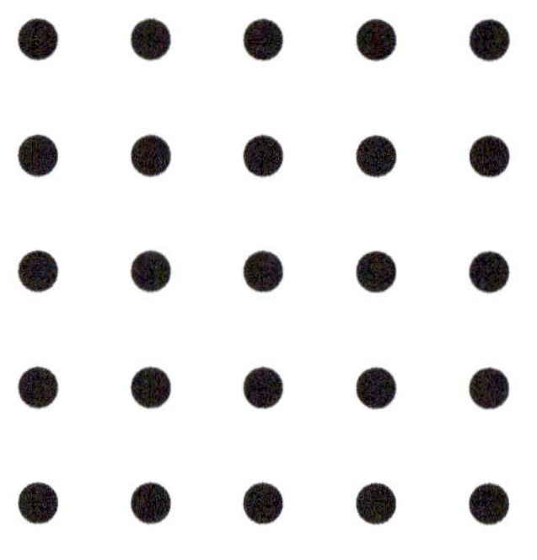

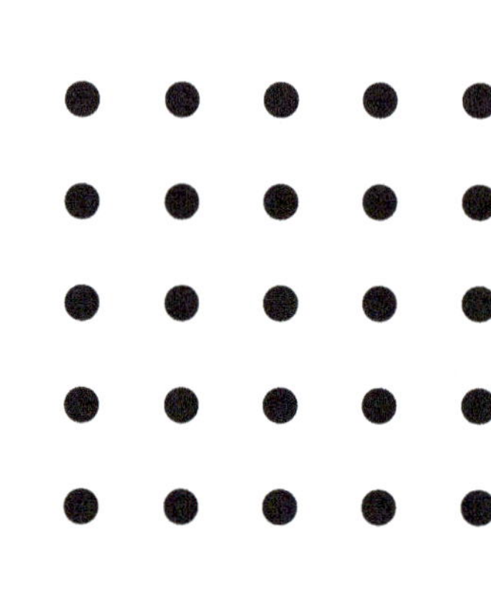

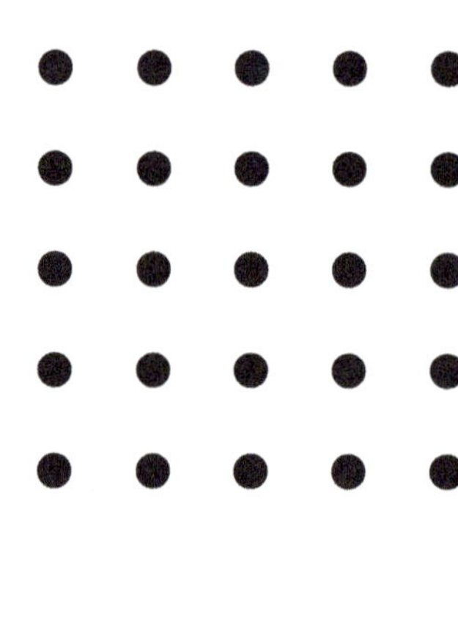

Symmetrical Shapes

Cut apart.

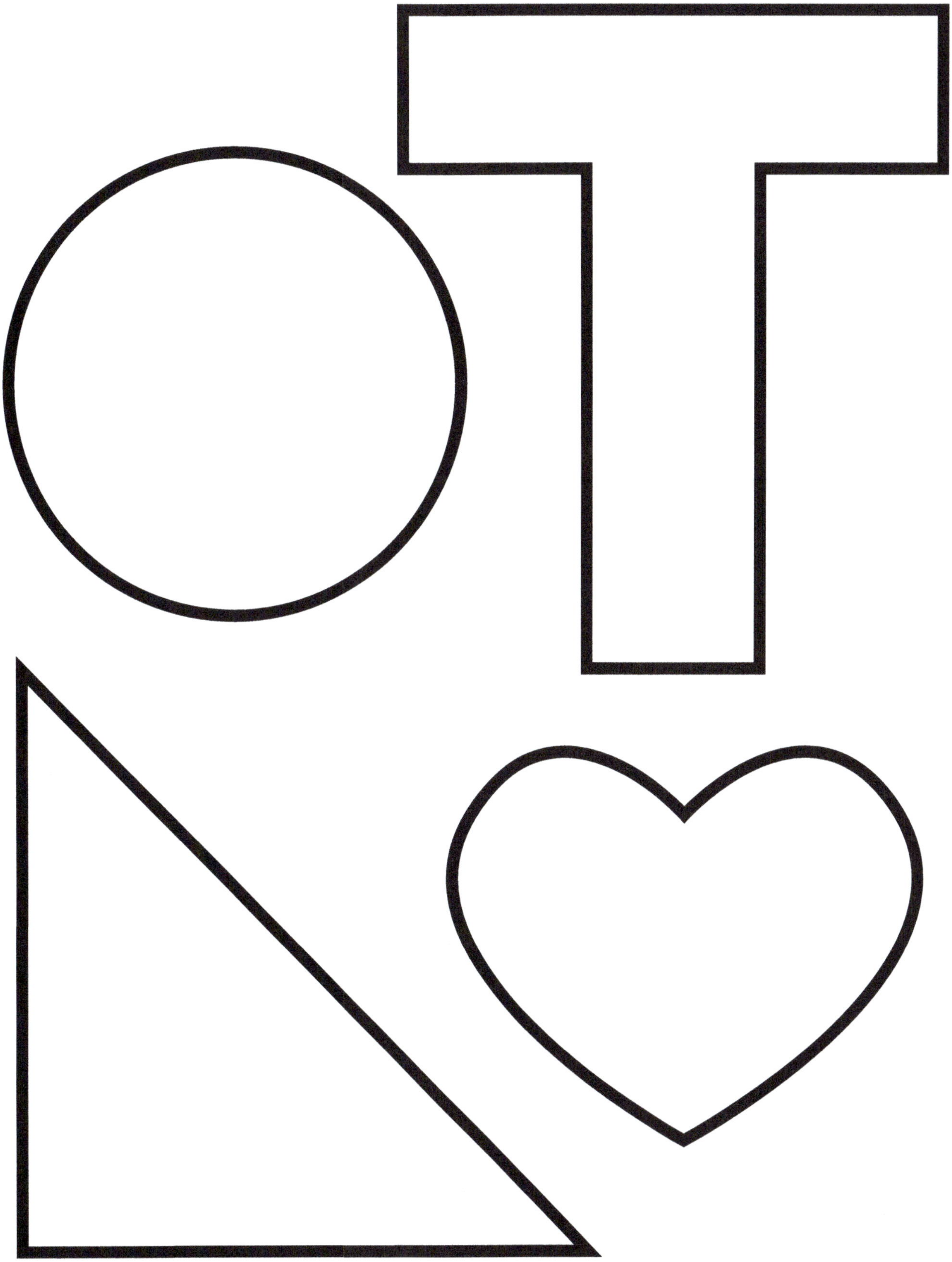

Asymmetrical Shapes

Cut apart.

Line of Symmetry

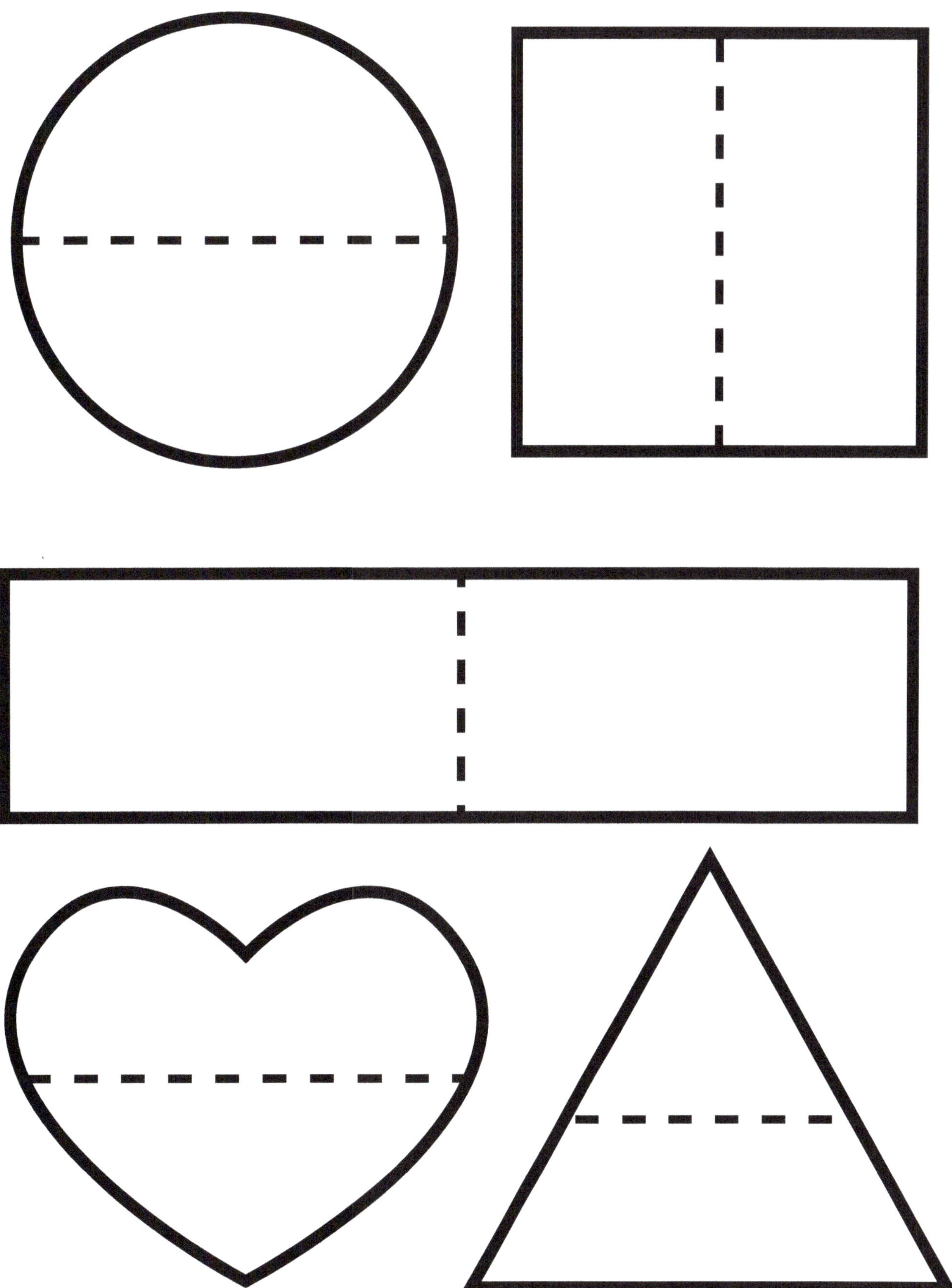

Clapping Games

Claps

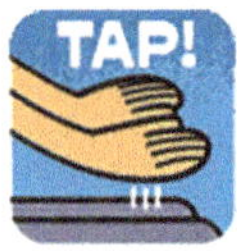

Bingo

Clapping Games

More Claps

A Sailor Went to Sea, Sea, Sea

A sailor went to sea, sea, sea
To see what he could see, see, see.
But all that he could see, see, see
Was the bottom of the deep blue sea, sea, sea.

Coding Rubric: Playing a Game

Name: ______________________

	Excellent (3)	Good (2)	Needs Improvement (1–0)	Points Earned
The Engineering Design Process	Uses all the steps of the Engineering Design Process to solve the problem of writing a program for a clapping game by using loops.	Uses some of the steps of the Engineering Design Process to solve the problem of writing a program for a clapping game by using loops.	Uses few or none of the steps of the Engineering Design Process to solve the problem of writing a program for a clapping game by using loops.	
Model	Design shows full understanding of concepts.	Design shows some understanding of concepts.	Design shows lack of understanding of concepts, or model is missing.	
Teamwork	Participates fully and shares the workload fairly.	Participates but sometimes does not share the workload fairly.	Relies on others to carry the workload most of the time or does not participate.	

Comments

Total ______

Centimeter Measurement Worksheet

★ 1.

centimeters

★ 2.

centimeters

♥ 3.

centimeters

♥ 4.

centimeters

▲

centimeters

▲

centimeters

Celsius Thermometer

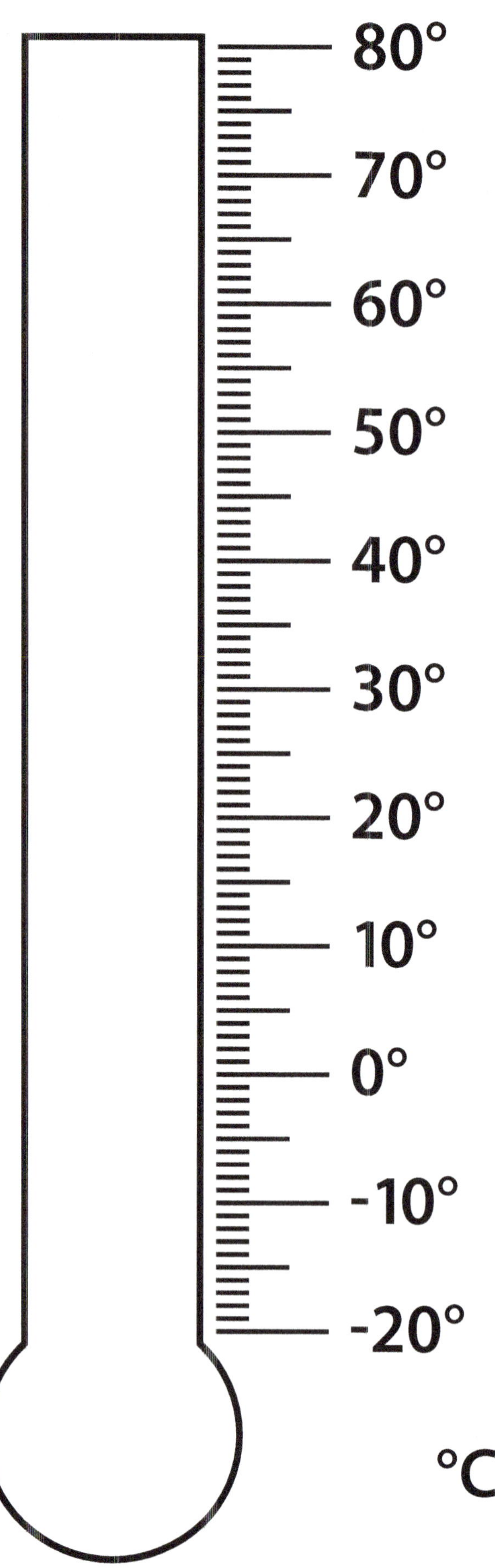

Metric Worksheet

Square Grid

Ten Bar

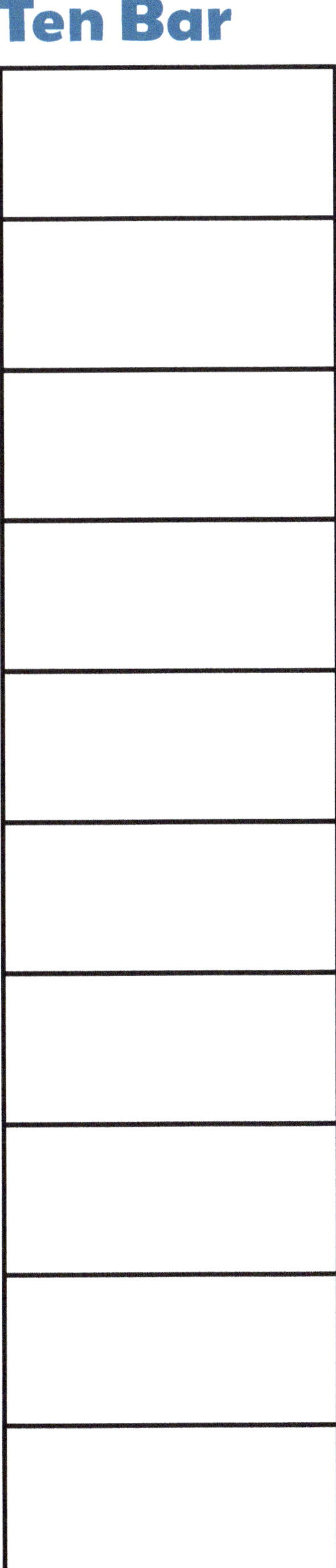

Coin Purse Mat

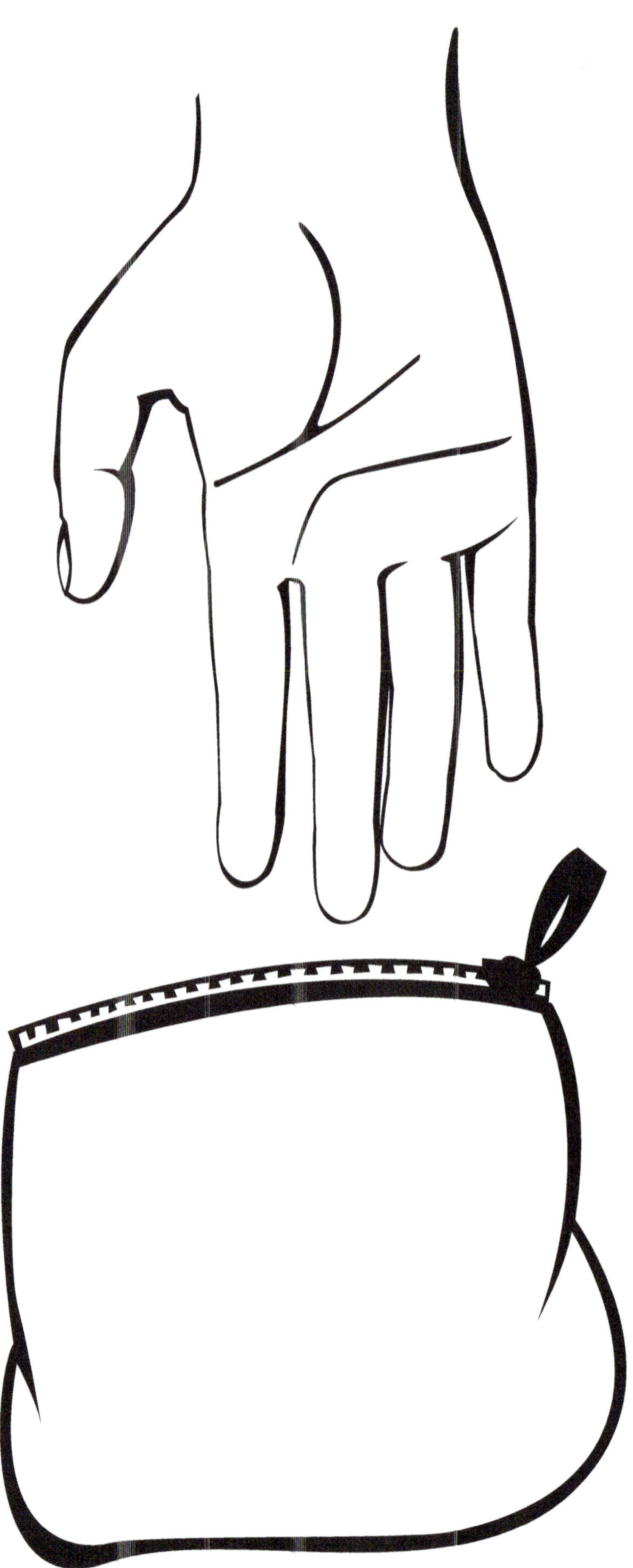

Repeated Addition Mat

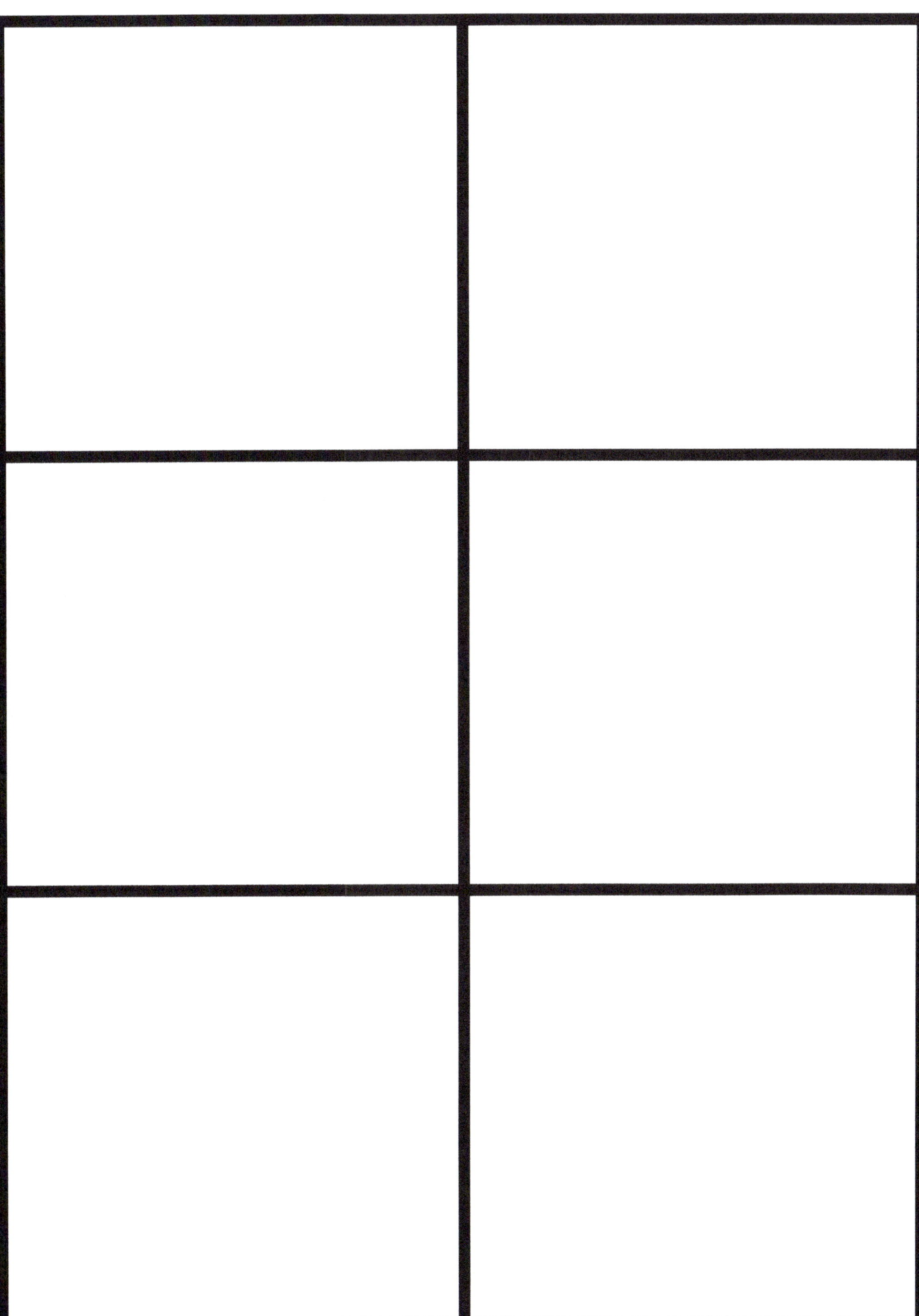

School Schedule

Subject	Start	Finish	Time Passed
Bible	8:30	9:00	30 minutes or 1 hour
Phonics	9:00	10:00	30 minutes or 1 hour
Reading	10:00	11:00	30 minutes or 1 hour
Math	11:00	11:30	30 minutes or 1 hour
Spelling	1:00	1:30	30 minutes or 1 hour

Spin & Add

your name ________________ your friend ________________

Hundreds	Tens	Ones
+		

Hundreds	Tens	Ones
+		

Hundreds	Tens	Ones
+		

Hundreds	Tens	Ones
+		

Hundreds	Tens	Ones
+		

Hundreds	Tens	Ones
+		

Hundreds	Tens	Ones
+		

Hundreds	Tens	Ones
+		

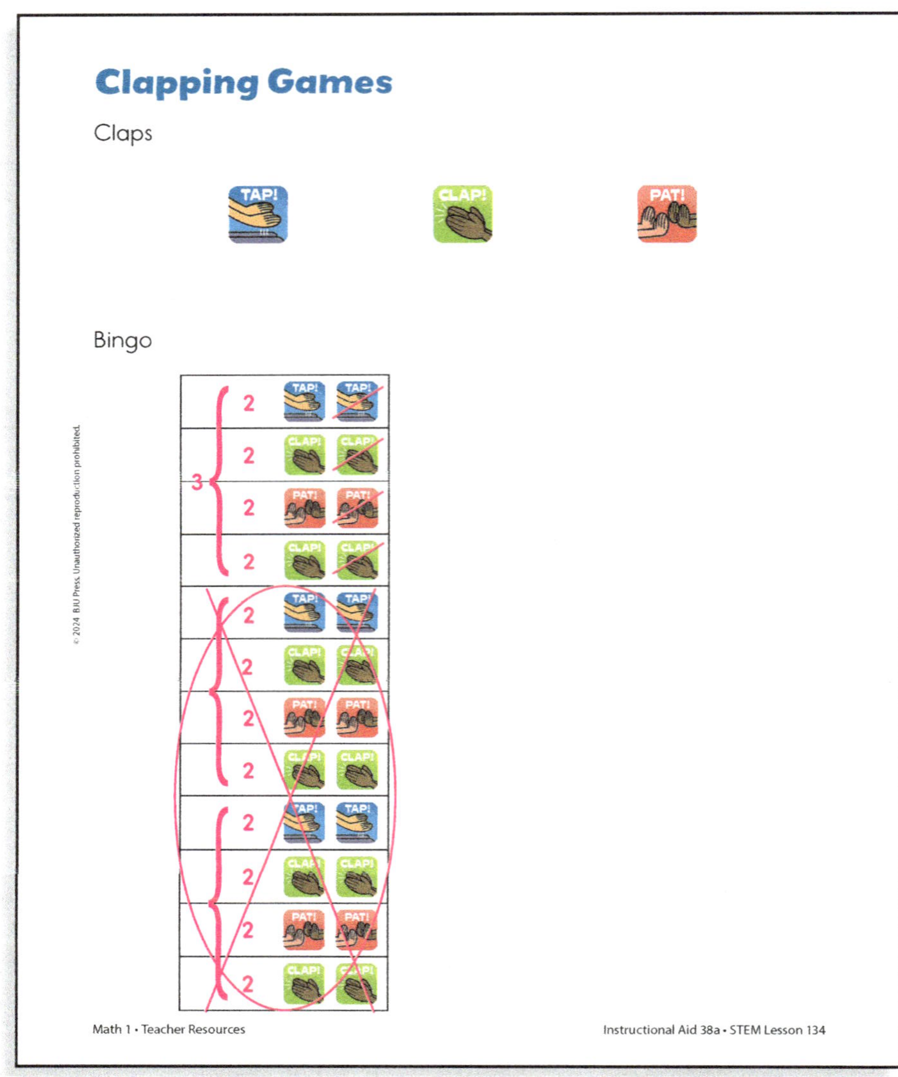

Clapping Games

Claps

Bingo

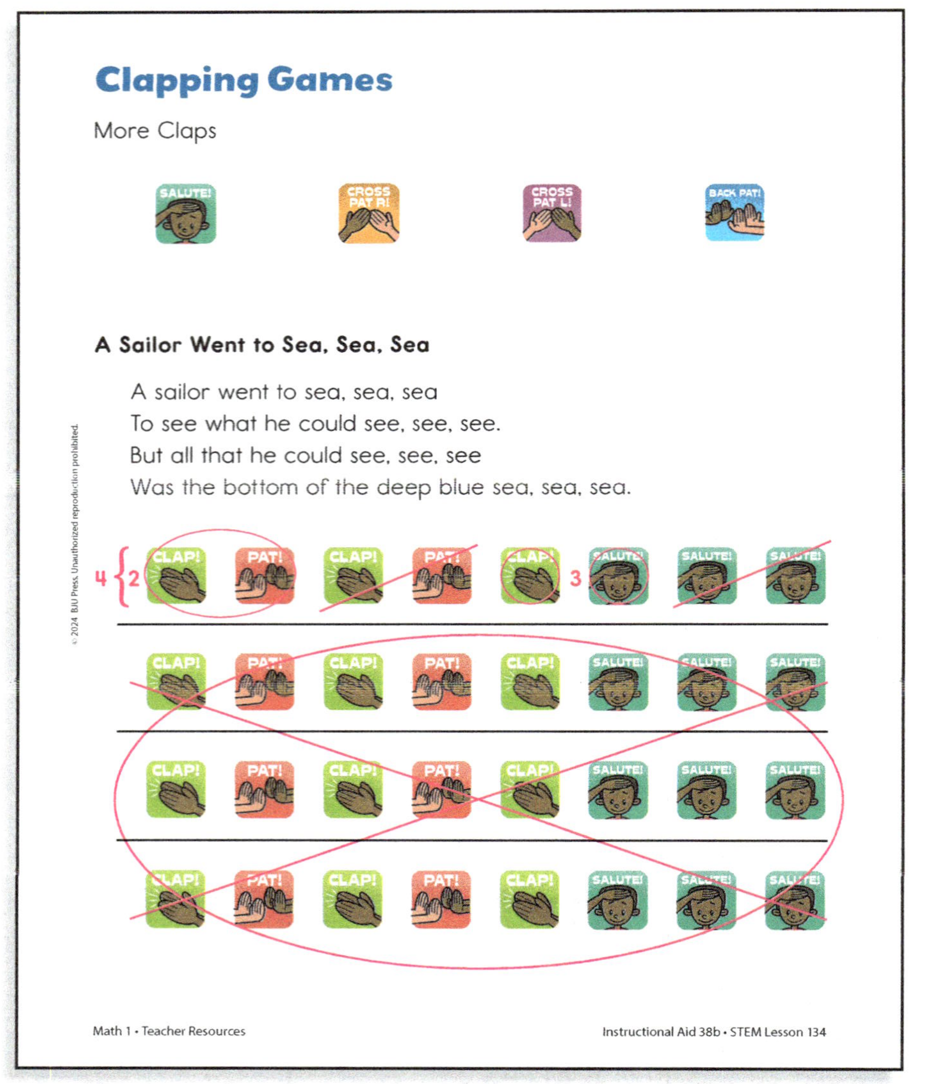

Clapping Games

More Claps

A Sailor Went to Sea, Sea, Sea

A sailor went to sea, sea, sea
To see what he could see, see, see.
But all that he could see, see, see
Was the bottom of the deep blue sea, sea, sea.

PreCursive Number Strokes

1. Begin near the top; swing around to touch.

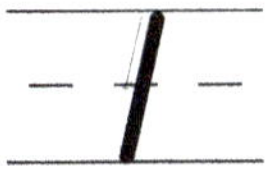

1

1. Drop.

2

1. Begin near the top; swing right and down to the left; glide right.

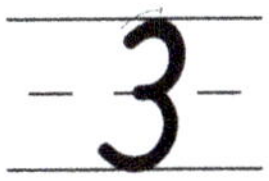

3

1. Begin near the top; swing around to the middle; swing around again.

4

1. Drop to the middle and glide right.
2. Drop.

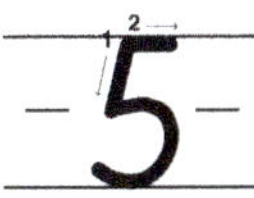

1. Drop to the middle and swing around.
2. Glide right.

6

1. Begin near the top; swing down and around to touch.

7

1. Glide right; drop left.

1. Begin near the top; swerve around and back; then up and around to touch.

9

1. Begin near the top; swing around to touch; drop.

Handwriting Pencil Hold—In the accepted position for pencil hold, the thumb and the index finger grasp the pencil, letting it rest on the middle finger. The last two fingers arch under the middle finger to support it. The hand rests on its side. Students should hold the pencil about one inch from the writing point. The pencil will point toward the shoulder. A student should hold his pencil lightly enough so that you can pull it out of his hand with little resistance. In general, low or medium pressure produces better writing. Teaching correct pencil hold is an important responsibility of the teacher.

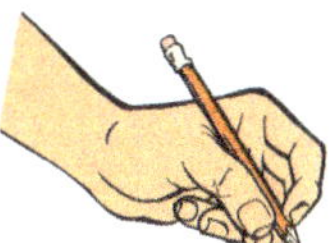

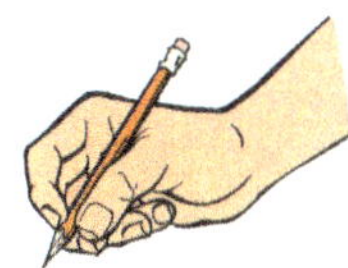

Paper Hold—A right-handed student's paper should be parallel to his right arm; a left-handed student's paper should be at a forty-five-degree angle to the edge of the desk.

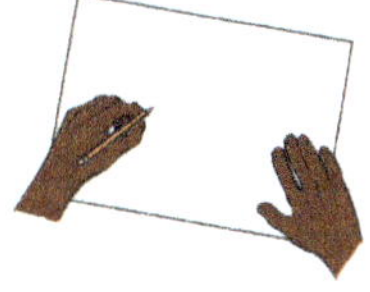

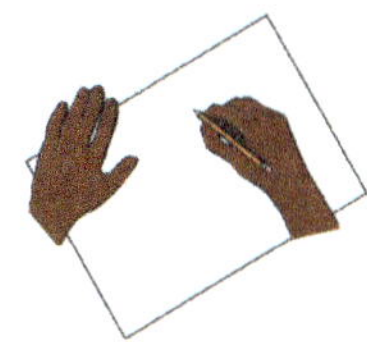

Posture—Good posture affects handwriting. Each child should sit comfortably in his chair with his feet touching the floor. The desk should be slightly higher than the student's waist. The student should sit, not leaning to the left or to the right, but bending slightly forward. His forearms should rest on his desk.

Handedness—By the time a child enters to kindergarten, right- or left-hand dominance is usually evident. If you observe a child who switches back and forth from his right to his left or shows mixed dominance, use the following procedures to help determine which hand the child should be using. These activities should be carried out as a game, without the child's knowing the purpose.

1. Eye preference: Direct the child to pick up a paper-towel tube and use it as a telescope to look at an object. Notice which hand and which eye the child uses.
2. Hand preference: Give the child a toy broom and ask him to sweep the floor. Observe which hand he uses to power the motion. Handedness may also be determined by observing with which hand a child eats, draws, or throws a ball.
3. Foot preference: Ask the child to walk across the room toward you; direct him to walk across a balance beam or walk on a line taped on the floor. Notice which foot he starts with. Foot preference may also be observed by asking a child to stamp his foot or kick a ball.

If a child continues to use both hands, use the following suggestions to help him establish a dominant hand.

1. Place a rubber band, a colored piece of tape, or a sticker on the dominant hand as a reminder to the child.
2. Direct the child to use his dominant hand to finger-paint, write in sand, clip clothespins around the top edge of a box or on a clothesline, or use clip clothespins or tongs to move cotton balls from one container to another.

If you have observed that a child is definitely left-handed, give him special help in placement of his paper, arm, wrist, and hand for handwriting activities. Do not try to force him into writing with his right hand.

Days of the Week

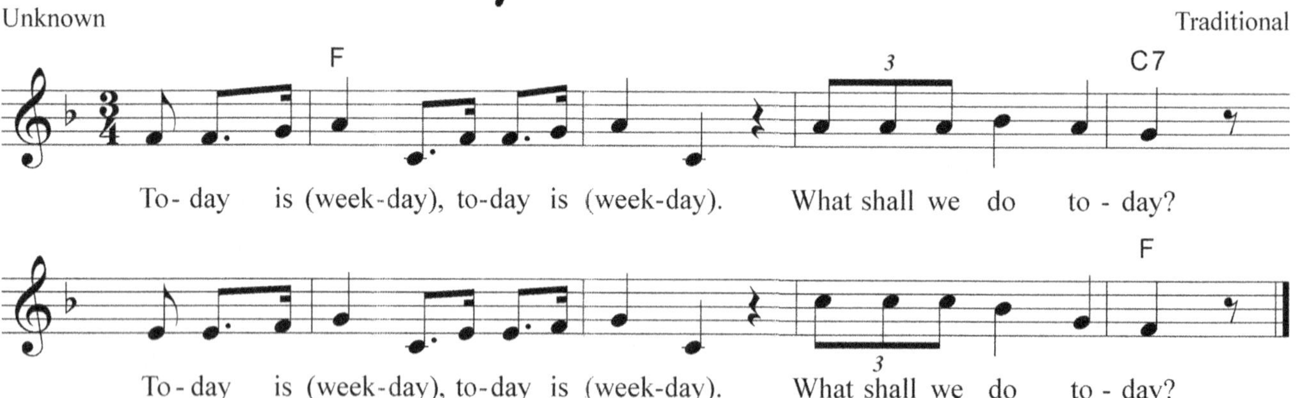

Unknown
Traditional
F
C7
3
To- day is (week-day), to-day is (week-day). What shall we do to - day?
F
3
To - day is (week-day), to-day is (week-day). What shall we do to - day?

God Made Seven Days in a Week

Muriel Murr

Adapted

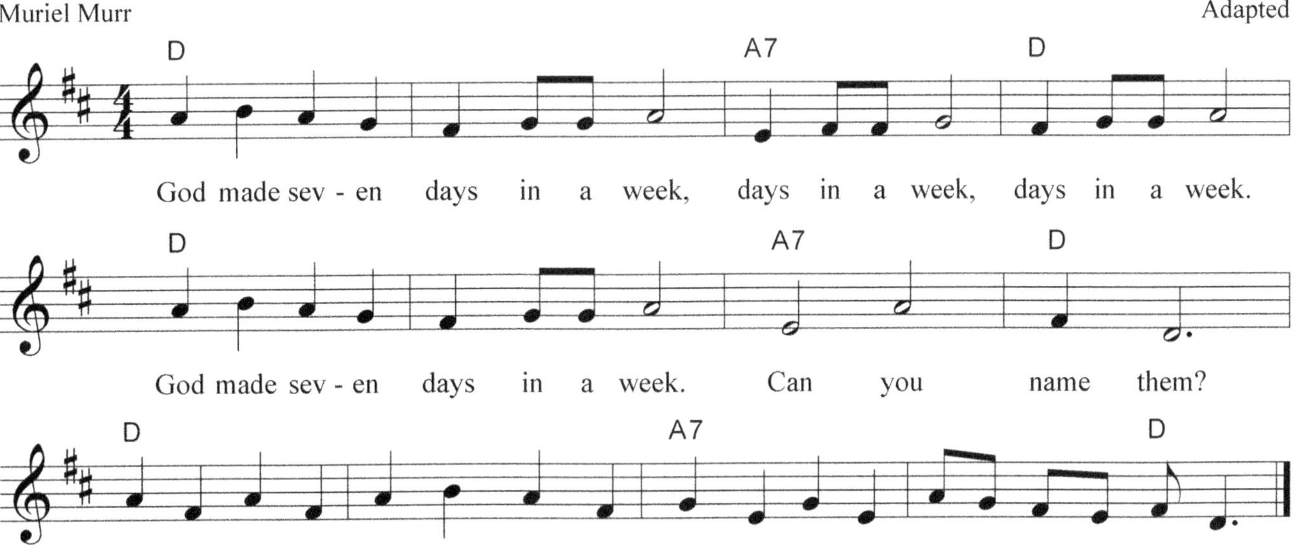
D
A7
D
God made sev - en days in a week, days in a week, days in a week.
D
A7
D
God made sev - en days in a week. Can you name them?
D
A7
D
Sun-day, Mon-day, Tues - day, Wednes - day, Thurs-day, Fri - day, Sat - ur - day make sev - en.

There Are Twelve Months

BJU Press Author

Adapted

There are twelve months in a year, Months in a year, months in a year.
There are twelve months in a year. Can you name them?
Jan - u - ar - y, Feb - ru - ar - y, March, and A - pril, May, June, and Ju - ly,
Au - gust, Sep - tem - ber, Oc - to - ber, No - vem - ber, De - cem - ber when Christ was born.

Wide, Wide as the Ocean

C. Austin Miles
C. Austin Miles
C F
Wide, wide as the o - cean, High as the hea-ven a - bove;
G7 C G
Deep, deep as the deep - est sea Is my Sa - vior's love;
C F
I, though so un - worth - y, Still am a child of His care;
C Dm G7 C
For His Word teach-es me that His love reach-es me eve - ry - where.

Teacher Edition Index

The numbers listed after each entry indicate the lesson(s) in which the entry appears.

Worktext Photo Credits

Cover: FatCamera/E+ via Getty Images

All coin images in this textbook: Fat Jackey/Shutterstock.com

Chapter 15
221, 228 (soccer ball) Smileus/Shutterstock.com; **221–22** (megaphone) Creatas/Thinkstock; **221–22** (glass jar) AlenKadr/Shutterstock.com; **221** (box of crackers) Walter Cicchetti/Shutterstock.com; **221, 228, 237** (soft drink can) scanrail/iStock/Getty Images Plus via Getty Images; **221–22, 228, 237** (basketball) FocusStocker/Shutterstock.com; **221, 237** (parking cone) AleksMaks/Shutterstock.com; **221, 223, 225, 228, 237** (tissue box) ©phanasitti/123RF; **222** (peach) PixaHub/Shutterstock.com; **222** (log) JIANG HONGYAN/Shutterstock.com; **222** (wigwam) DNY59/E+ via Getty Images; **222–23, 228, 237** (wrapped present) GCapture/Shutterstock.com; **222, 237** (water bottle) alexmisu/Shutterstock.com; **222, 237** (cereal box) george tsartsianidis/iStock/Getty Images Plus via Getty Images; **223, 225** (book) Anton Starikov/Shutterstock.com; **223** (cereal box) DenisMArt/Shutterstock.com; **223, 237** (pyramid-shaped puzzle) DENIS ESAULOV 1987/Shutterstock.com; **223, 228** (letter block) Blueprint Characters/Shutterstock.com; **223** (wooden pyramid block) Fishman64/Shutterstock.com; **225** (party hat) Dawn Poland/E+ via Getty Images; **225, 237** (spool of thread) ChaiKetsiam/Shutterstock.com; **228, 237** (brick) Lev Kropotov/Shutterstock.com; **228, 237** (globe) Elnur/Shutterstock.com; **237** (ball of yarn) Nataliia K/Shutterstock.com

Explaining the Gospel

One of the greatest desires of Christian teachers is to lead students to faith in the Savior. God has called you to present the gospel to your students so that they may repent and trust Christ, thereby being acceptable to God through Christ.

Relying on the Holy Spirit, you should take advantage of the opportunities that arise for presenting the good news of Jesus Christ. Ask questions to personally apply the Ten Commandments to your students (e.g., What is sin? Have you ever told a lie or taken something that wasn't yours? Are you a sinner?). You may also ask questions to discern the student's sincerity or to reveal any misunderstanding he or she might have (e.g., What is the gospel? What does it mean to repent? Can you do anything to save yourself?). Read verses from your Bible. You may find the following outline helpful, especially when dealing individually with a student.

1. **I have sinned (Romans 3:23).**
    - Sin is disobeying God's Word (1 John 3:4). I break the Ten Commandments (Exodus 20:2–17) by loving other people or things more than I love God, worshiping other things or people, using God's name lightly, disobeying and dishonoring my parents, lying, stealing, cheating, thinking harmful and sinful thoughts, or wanting something that belongs to somebody else.
    - Therefore, I am a sinner (Psalm 51:5; 58:3; Jeremiah 17:9).
    - God is holy and must punish me for my sin (Isaiah 6:3; Romans 6:23).
    - God hates sin, and there is nothing that I can do to get rid of my sin by myself (Titus 3:5; Romans 3:20, 28). I cannot make myself become a good person.
2. **Jesus died for me (Romans 5:8).**
    - God loves me even though I am a sinner.
    - He sent His Son, Jesus Christ, to die on the cross for me. Christ is sinless and did not deserve death. Because of His love for me, Christ took my sin on Himself and was punished in my place (1 Peter 2:24a; 1 Corinthians 15:3; John 1:29).
    - God accepted Christ's death as the perfect substitution for the punishment of my sin (2 Corinthians 5:21).
    - Three days later, God raised Jesus from the dead. Jesus is alive today and offers salvation to all. This is the gospel of Jesus Christ: He died on the cross for our sins according to the Scriptures, and He rose again the third day according to the Scriptures (1 Corinthians 15:1–4; 2 Peter 3:9; 1 Timothy 2:4).
3. **I need to put my trust in Jesus (Romans 10:9–10, 13–14a).**
    - I must repent (turn away from my sin) and trust only Jesus Christ for salvation (Mark 1:15).
    - If I repent and believe in what Jesus has done, I am putting my trust in Jesus.
    - Everyone who trusts in Jesus is forgiven of sin (Acts 2:21) and will live forever with God (John 3:16). I am given Christ's righteousness and become a new creation, with Christ living in me (2 Corinthians 5:21; Colossians 1:27).

If a student shows genuine interest and readiness, ask, "Are you ready to put your trust in Jesus and depend on only Christ for salvation?" If the student says yes, then ask him or her to talk to God about this. Perhaps he or she will pray something like the following:

> God, I know that I've sinned against You and that You hate sin, but that You also love me. I believe that Jesus died to pay for my sin and that He rose from the dead, so I put my trust in Jesus to forgive me and give me a home with You forever. In Jesus' name I pray. Amen.

Show the student how to know from God's Word whether he or she is forgiven and in God's family (1 John 5:12–13; John 3:18). Encourage the student to follow Jesus by obeying Him each day. Tell the student that whenever he or she sins, God will grant forgiveness as he or she confesses those sins to God (1 John 1:9).